ADAPTIVE AND LEARNING SYSTEMS

Theory and Applications

ADAPTIVE AND LEARNING SYSTEMS

Theory and Applications

Edited by

KUMPATI S. NARENDRA

Center for Systems Science
Yale University
New Haven, Connecticut

SPRINGER SCIENCE+BUSINESS MEDIA, LLC

Library of Congress Cataloging in Publication Data

Workshop on Adaptive Systems Control Theory (4th: 1985: Yale University)
Adaptive and learning systems.

"Proceedings of the Fourth Workshop on Adaptive Systems Control Theory, held May 29–31, 1985, at Yale University, New Haven, Connecticut"—T.p. verso.
Includes bibliographies and index.
1. Adaptive control systems—Congresses. I. Narendra, Kumpati S. II. Title.
TJ217.W68 1985 629.8′36 86-4962

ISBN 978-1-4757-1897-3 ISBN 978-1-4757-1895-9 (eBook)
DOI 10.1007/978-1-4757-1895-9

Proceedings of the Fourth Workshop on Adaptive Systems Control Theory, held May 29–31, 1985, at Yale University, New Haven, Connecticut

Preface

This volume offers a glimpse of the status of research in adaptive and learning systems in 1985. In recent years these areas have spawned a multiplicity of ideas so rapidly that the average research worker or practicing engineer is overwhelmed by the flood of information. The Yale Workshop on Applications of Adaptive Systems Theory was organized in 1979 to provide a brief respite from this deluge, wherein critical issues may be examined in a calm and collegial environment. The fourth of the series having been held in May 1985, it has now become well established as a biennial forum for the lively exchange of ideas in the ever changing domain of adaptive systems.

The scope of this book is broad and ranges from theoretical investigations to practical applications. It includes twenty eight papers by leaders in the field, selected from the Proceedings of the Fourth Yale Workshop and divided into five sections. I have provided a brief introduction to each section so that it can be read as a self-contained unit. The first section, devoted to adaptive control theory, suggests the intensity of activity in the field and reveals signs of convergence towards some common themes by workers with rather different motivation. Preliminary results concerning the reduced order model problem are dramatically changing the way we view the field and bringing it closer to other areas such as robust linear control where major advances have been recently reported. The importance of this problem is immediately evident from a cursory acquaintance with the assumptions made in many of the applications and in particular the control of flexible space structures.

Practical applications considered in the second, fourth, and fifth sections include areas such as process control, where adaptive methods are quite mature, and areas such as flexible space structures and robotics which appear to be promising for applications of developing but closely related theory. The gap between adaptive theory and practice is brought to focus in the second section where the practical steps required to assure satisfactory performance in the presence of real world constraints are discussed. The recent advent of commercially developed industrial adaptive regulators also serves to underscore the fact that the time from discovery to application in the engineering disciplines is constantly decreasing and that industry is poised to exploit new ideas.

The emphasis of the Workshop has shifted adaptively from year to year to include new areas of research. This is reflected in part in the third section (learning systems) and the fifth section (robotics) where the viewpoints of control engineering and artificial intelligence are brought together. While the problem definition and methods in these fields do not easily fit into the traditional paradigm of adaptive systems theory, these papers will serve to remind control theorists that adaptation appears in a variety of guises and the fundamental problems encountered are quite similar regardless of the approach.

Several new results are reported in this book for the first time. Since they have not gone through a formal peer review process, the opinions and conclusions expressed are the sole responsibility of the authors and do not necessarily reflect my opinion or that of the Center for Systems Science.

It is a privilege to acknowledge my debt of gratitude to the many friends and colleagues who have graciously helped both in the organization of the Workshop and in the preparation of this book. I cannot list them all but I record with pleasure my chief creditors. Richard Wheeler, a true believer in learning automata theory and Dan Koditschek, an applied mathematician turned roboticist, gave much help in the organization of the Workshop. The numerous discussions I enjoyed with Richard were to a large extent responsible for the expanded scope of the session on learning systems. Apart from organizing a successful session on robotics, Dan has helped me on numerous occasions with valuable comments and suggestions during the preparation of this book. My secretary, Jean Gemmell, who typed the entire book under fairly adverse conditions and went through successive revisions and corrections of hundreds of equations unflinchingly with her usual thoroughness, has my admiration and my vote of thanks. Above all, I am profoundly obligated to Anu Annaswamy, who worked closely with me on every aspect of the Workshop and played a major role in the preparation of this book – from its initial planning to its final shepherding to print. Without her help, I can safely say that this book would not exist.

K.S. Narendra
Editor

Contents

ADAPTIVE CONTROL THEORY

ADAPTIVE CONTROL APPLICATIONS

LEARNING SYSTEMS

CONTROL OF FLEXIBLE SPACE STRUCTURES

ROBOTICS

ADAPTIVE CONTROL THEORY

We dance around in a ring and suppose,
But the secret sits in the middle and knows.

Robert Frost

The current state of ferment in adaptive control theory is reflected by the numerous publications in leading control journals of new methods, perspectives, and simulation studies. In 1980 it was shown for the first time that a linear time-invariant plant with unknown parameters could be stabilized by an adaptive controller, provided no external disturbances are present and some prior information regarding the plant transfer function is available. Since the former condition is not satisfied by most practical systems and the latter is generally not available, much of the research activity in the recent past has been directed towards extending the class of systems to which adaptive control can be successfully applied. Six papers which provide a fairly up-to-date coverage of the more salient lines of recent research are included in this section.

In the first paper, Narendra and Annaswamy review some of the developments with an emphasis on global results where the signals in the system remain bounded for arbitrary initial conditions. They show that such global boundedness can be achieved either by modifying the adaptive laws or by requiring that the reference input be sufficiently persistently exciting. The paper by Kosut presents a summary of averaging methods used in the local stability analysis of adaptive systems. Based on these, he derives frequency domain conditions which explain the behavior of the system in the vicinity of tuned solutions and argues that these results may be necessary for good performance.

The problem of adaptively controlling a plant using a partial representation is perhaps the one that has attracted the most attention in recent years. It is also referred to in the literature as the reduced order model problem, the problem of unmodeled dynamics or the problem of state-dependent disturbances. The papers by Kreisselmeier, Praly, and Ioannou and Tsakalis, as well as the last part of the paper by Narendra and Annaswamy are devoted to this problem.

Kreisselmeier uses an indirect approach and assumes that the unknown parameters lie in a convex set where unstable pole-zero cancellations are not possible. Using a time-varying dead-zone in the adaptive law, he shows that the adaptive control system is globally stable in the presence of small unmodeled uncertainties. Praly, using a graph topology which defines an open neighborhood of the nominal plant transfer function, establishes the robustness of the adaptive system for all plants in this neighborhood. Ioannou and Tsakalis consider the same problem and establish global stability in the presence of sufficiently small unmodeled dynamics using the σ-modification in the adaptive law and assuming a bound on the magnitude of the control parameter vector. The paper by Narendra and Annaswamy demonstrates that, under suitable assumptions on the output of the unmodeled part, all the methods suggested in the literature to assure boundedness of signals when bounded external disturbances are present can be suitably modified to apply to the reduced order model problem as well. As in the three earlier papers, the proof of boundedness in these extensions requires that the output of the unmodeled part be small in comparison with the state of the system.

In contrast to the first five papers which deal with deterministic systems, the last paper by Kumar is concerned with identification and adaptive control of stochastic systems. In the first part of the paper, he discusses recursive identification schemes with regard to their consistency and efficiency. In the last part of the paper some of the recent advances in stochastic adaptive control theory are outlined and it is shown that a unified treatment of the regulation and tracking problems is possible when the stochastic gradient algorithm is used for parameter estimation.

Robust Adaptive Control

Kumpati S. Narendra and Anuradha M. Annaswamy
Center for Systems Science, Yale University

Abstract

Recent developments in robust adaptive control in the presence of bounded as well as state-dependent disturbances are discussed in the paper. The various approaches that have been suggested in the literature using modifications of a standard adaptive law, as well as arguments involving persistent excitation of the reference input, are presented. While the paper is partly tutorial, it also contains several new results due to the authors. These include a new adaptive law, the proof of global boundedness of all signals in the presence of bounded disturbances using persistent excitation, and the extension of these results to the adaptive control problem with state-dependent disturbances. Both local results and global results reported in the literature are discussed, but a greater emphasis is placed on those that are global in character. The implications of these results for future research on the reduced order model problem, are discussed towards the end of the paper.

1. Introduction

The global stability problem of both discrete and continuous adaptive control systems was resolved in 1980 [1]-[4] and represents a landmark in the development of the field of adaptive control. In this problem, which is now termed the "ideal case", it is assumed that some prior information regarding the plant transfer function $W_p(s)$ is available and that no external disturbances are present. The prior information includes knowledge of

(i) the sign of the high frequency gain k_p,

(ii) relative degree n^* (difference between the number of poles and zeros of $W_p(s)$),

(iii) an upper bound n on the order of the system (I)

as well as the fact that

(iv) all zeros of $W_p(s)$ lie in the open left half plane.

However, in practice these assumptions are seldom valid; a physical plant is rarely linear, time-invariant, finite dimensional, or disturbance free in the entire range of its operation. Further, since its behavior can only be approximated by a differential equation, its order as well as relative degree are never known exactly. Also, in most practical systems, adaptive control is attractive precisely in those cases where the plant parameters vary widely in an unknown fashion. In all the above cases, the adaptive controller has to perform satisfactorily in the presence of both operating and structural uncertainties. The algorithms suggested in [1]-[4] for the ideal system no longer assure the boundedness of the signals in the adaptive loop. It is therefore not surprising that in the early eighties most of the attention of research workers in the field shifted to questions of robustness, or the satisfactory performance of the system in the presence of different types of perturbations. This period witnessed considerable activity in the area and a better theoretical understanding of the basic questions involved was also gained.

An adaptive control system is one which is made intentionally nonlinear to obtain improved performance. The control parameters are adjusted as functions of the plant outputs as

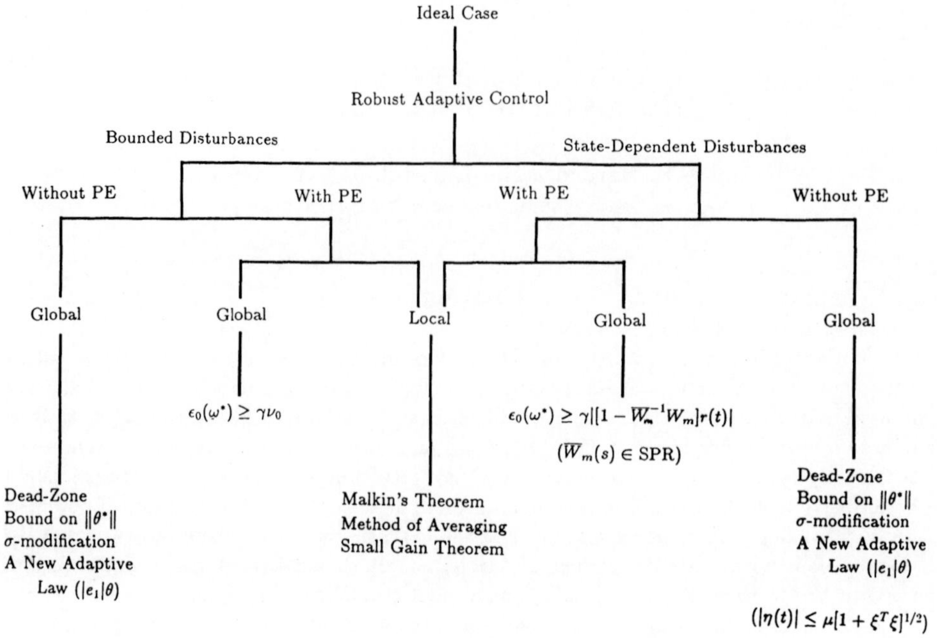

Figure 1: Recent Developments in Robust Adaptive Control.

well as the output errors, and hence the overall system is invariably nonlinear. The presence of the reference input as well as external disturbances also make the system nonautonomous. As in any control system, an adaptive system is required to be stable and respond rapidly and accurately to command inputs. In the model reference adaptive system, this implies that all the parameters as well as the signals in the the overall system should be bounded and the output error between plant and model should reach its final value(which satisfies specifications), as rapidly as possible. Most of the difficulties that arise in the analysis of adaptive systems in the presence of perturbations can be attributed to the fact that the differential equations needed to model them are both nonlinear and time-varying.

One of the aims of this paper is to provide a comprehensive review of the principal developments in robust adaptive control theory during the past five years. Such a review can be based on the type of problems that are of interest (e.g. bounded disturbances, time-varying parameters, reduced order model of the plant), the analytic methods used (Lyapunov's method, stability under persistent disturbances [5], averaging methods [6], growth rates of unbounded signals [7] and extensions of Lyapunov's methods based on persistent excitation [8]), whether local or global stability is of interest and whether or not the arguments call for the persistent excitation of the reference inputs. In this paper we shall follow the tree structure indicated in Figure 1 to categorize the various developments since 1980. After presenting briefly the principal theorems used in the different analytic methods in section 2, the stability problem of the ideal system is considered in section 3 since it forms the starting point for the analysis of almost all the problems in robust adaptive control. The robustness of adaptive systems in the presence of bounded disturbances is the subject of section 4. Methods for assuring global boundedness of signals without using persistently exciting reference inputs, but by suitably modifying the adaptive laws, are discussed in section 4.1. Both local and global stability results are then derived in section 4.2 using arguments involving persistent excitation. Finally, in section 5, the problem of controlling an unknown plant adaptively using a reduced order model is formulated and methods for extending the results given in section 4 to this problem are discussed.

While the paper is to some extent tutorial in nature, its main intent is to present the authors' general approach to robust adaptive control. As a consequence, the results reported in the literature which are in congruence with the authors' philosophy are discussed in detail, while others are merely touched upon. The paper also contains several new results due to the authors including a new adaptive law in section 4.1, proof of global boundedness of signals using persistent excitation in section 4.2, and the application of these to adaptive control problems with state-dependent disturbances in section 5.

2. Mathematical Preliminaries

To address the different aspects of the robust adaptive control problem a variety of mathematical approaches are currently being explored. Even in the ideal case it is known that, while the methods of Lyapunov and Popov can be directly applied to assure the boundedness of the control parameter vector, more involved arguments based on order relationships between inputs and outputs of linear systems are needed to demonstrate that all the signals in the system are bounded. This also applies to problems where bounded disturbances are present and the adaptive laws are suitably modified. While discussing global stability based on persistent excitation, extensions of Lyapunov theory which use integral rather than pointwise conditions are found to be pertinent. The relevant definitions as well as the principal theorems related to Lyapunov's method and its extensions are given in section 2.1. The concepts used in arguments related to order relationships between unbounded signals in a system are briefly presented in section 2.2. Finally, in section 2.3, work related to total stability and ideas involved in the method of averaging are included. These are useful in deriving local stability results when the reference input is persistently exciting.

2.1 Lyapunov Stability

We start with the definitions of uniform asymptotic and exponential stability of the solution $x(t) \equiv 0$ of an equation $\dot{x} = f(x,t)$, $f(0,t) = 0$. We assume that f is continuous and satisfies conditions which guarantee the existence and uniqueness of solutions and continuity of their dependence on the initial conditions. The general solution of the differential equation with an initial condition $x(t_0) = x_0$ is denoted by $x(t, x_0, t_0)$ with $x(t_0, x_0, t_0) = x_0$.

Definition 1 (Uniform Asymptotic Stability): The equilibrium $x = 0$ of the differential equation

$$\dot{x} = f(x,t) \qquad f : \mathcal{R}^n \times \mathcal{R}^+ \to \mathcal{R}^n \tag{1}$$

is uniformly asymptotically stable (u.a.s.) if it is uniformly stable and for some $\epsilon_1 > 0$ and all $\epsilon_2 > 0$, there is a $T(\epsilon_1, \epsilon_2) > 0$ such that

$$||x_0|| < \epsilon_1 \text{ implies } ||x(t, x_0, t_0)|| < \epsilon_2 \text{ for all } t \geq t_0 + T.$$

Definition 2 (Exponential Stability): The equilibrium state of the equation $\dot{x} = f(x,t)$ is exponentially stable if two positive constants α and β which are independent of the initial values exist such that for sufficiently small initial values,

$$||x(t, x_0, t_0)|| < \beta ||x_0|| e^{-\alpha(t-t_0)}.$$

For linear systems, all stability properties hold in the large. If the system is autonomous, all stability properties are uniform. Exponential stability implies u.a.s. while the converse is true in general only for linear systems.

Using Lyapunov's Direct Method, it is well known [9] that the existence of a scalar function $V(x,t)$ which is (i) positive definite, (ii) decrescent, and (iii) radially unbounded, and whose time-derivative $\dot{V}(x,t)$ along a trajectory of the system (1) is (iv) negative definite, assures the uniform asymptotic stability in the large (u.a.s.l) of the dynamical system (1).

While (i) and (iv) result in asymptotic stability, (ii) makes the stability uniform and the condition (iii) assures that the properties hold in the large.

An adaptive control problem can be expressed as the stability problem associated with a nonlinear time-varying differential equation. Even in the ideal case (ref. section 3.1), when no external disturbances are present, an adaptive law can be determined which assures the existence of a function V satisfying (i)-(iii) whose time-derivative is only negative semi-definite. Hence Lyapunov's Direct Method guarantees only uniform stability. It is well-known that for autonomous and periodic nonlinear systems satisfying the above condition, LaSalle's theorem [10,11] can be used to determine whether or not the system is u.a.s.l. However, in adaptive systems, the time-varying part of the field is due to the external input which, in general, is not periodic. Hence extensions of the ideas contained in [11] are needed to conclude u.a.s.l. These are found to be related to the concept of persistent excitation and allied theorems given below.

The following theorem yields sufficient conditions under which a system with $V > 0$ and $\dot{V} \leq 0$ is u.a.s.l.

Theorem 1: The equilibrium state of (1) is uniformly asymptotically stable in the large if a function $V(x,t)$ defined for all x and t with $V(0,t) = 0$ is (i) positive definite, (ii) decrescent, (iii) radially unbounded, and (iv) $\dot{V}(x,t)$ is negative semi-definite, and (v) $\int_t^{t+T} \dot{V}(x(\tau),\tau)d\tau \leq -\gamma(\|x(t)\|) < 0$ for some T and all $t \geq t_0$ where $\gamma(.)$ is a positive monotonic function with $\gamma(0) = 0$.

Theorem 1 states that for a class of nonautonomous systems, even if the time-derivative $\dot{V}$ of the Lyapunov function is only negative semi-definite, using integral properties of $\dot{V}$, uniform asymptotic stability can be concluded. At first glance this may appear tautological since by condition (v), u.a.s. follows if $V(t) - V(t+T) \geq \gamma(\|x(t)\|)$. The weakness of condition (v) is that the integral of $\dot{V}$ has to be evaluated along a solution of the differential equation over a finite period. However (v) can be satisfied in many problems without explicit knowledge of the solutions as shown in section 4, by making use of the persistent excitation of the reference signals. The reader is referred to [8] for further details.

Persistent Excitation: The concept of persistent excitation(PE) arose in the context of adaptive identification and control, and in particular while determining the conditions for u.a.s. of the following two linear differential equations, which have played a major role in the stability analysis of continuous time adaptive systems:

$$\dot{x} = -u(t)u^T(t)x \tag{2}$$

where $x, u : \mathcal{R}^+ \to \mathcal{R}^n$, and $u(.)$ is piecewise continuous and bounded, and

$$\dot{x} = \begin{bmatrix} A & -bu^T(t) \\ u(t)b^T & 0 \end{bmatrix} x \tag{3}$$

where $x^T \triangleq [x_1^T, x_2^T], x_1 : \mathcal{R}^+ \to \mathcal{R}^m, x_2 : \mathcal{R}^+ \to \mathcal{R}^n, A$ is an $m \times m$ dimensional stable matrix with $A + A^T = -Q < 0, (A,b)$ is controllable, and $u : \mathcal{R}^+ \to \mathcal{R}^n$ is piecewise continuous and bounded. The necessary and sufficient conditions for the u.a.s. of the equilibrium state of (2) is that positive constants t_0, T_0 and α exist so that $\int_t^{t+T_0} u(\tau)u^T(\tau)d\tau \geq \alpha I \quad \forall t \geq t_0$[12]. Equivalently, for every unit vector w in $\mathcal{R}^n$,

$$\frac{1}{T_0}\int_t^{t+T_0} |u^T(\tau)\mathrm{w}|d\tau \geq \epsilon_0 \tag{4}$$

for some constant ϵ_0. Similarly, in equation (3), the equilibrium state $x = 0$ is uniformly asymptotically stable iff positive constants T_0, ϵ_0 and δ_0 exist so that $\exists t_2 \in [t, t+T_0]$ with

$$\left|\frac{1}{T_0}\int_{t_2}^{t_2+\delta_0} u^T(\tau)\mathrm{w}d\tau\right| \geq \epsilon_0 \;\; \forall t \geq t_0 \tag{5}$$

for every unit vector $w \in \mathcal{R}^n$ [13]. Since both systems (2) and (3) are linear, the stability is exponential and global.

The conditions (4) and (5) which assure the global exponential stability of systems (2) and (3) have been used in the literature to define PE of continuous time signals. A unified definition of PE can be given if we restrict ourselves to the class of piecewise continuous and piecewise differentiable functions $\mathcal{P}_{[0,\infty)}$ defined by Yuan and Wonham [14]. For the purposes of our discussions in this paper, we shall restrict our attention to signals in $\mathcal{P}_{[0,\infty)}$ and use (4) and (5) interchangeably, depending upon the context, to denote a persistently exciting signal. We shall refer to ϵ_0 in (5) as the degree of PE of $u(.)$.

2.2 Growth Rates of Unbounded Signals

For the most part adaptive control problems have been addressed and solved individually. However, it has become apparent that seemingly different proofs share many common features. For example, the same general approach suggested in [1] to prove global stability in the ideal case was later used in [15] and [16] to deal with situations where bounded disturbances are present. Recent work has revealed that the same approach is also applicable in other adaptive contexts. The main aim of the approach is to bring together two independent analyses: one concerning the rate of growth of signals in the adaptive loop, assuming certain conditions on the parameter vector and the second based on the effect of the adaptive scheme on the growth rates of the signals. As a result of this, it is shown that two unbounded signals in the system grow at different rates, resulting in a contradiction. This in turn assures that all the signals in the system are bounded. For a detailed discussion of this approach the reader is referred to [7]. In this section we present only the principal definitions and results that are used later in this paper. In the following definitions all the signals are assumed to belong to the set $\mathcal{PC}_{[0,\infty)} \stackrel{\Delta}{=} \{x+y|\ x,y:\mathcal{R}^+\rightarrow\mathcal{R},\ x\in\mathcal{P}_{[0,\infty)},\ y\in\mathcal{C}_{[0,\infty)}\}$ which includes unbounded continuous functions with bounded discontinuities.

<u>Definition 3:</u> Let $x(.), y(.) \in \mathcal{PC}_{[0,\infty)}$. We denote $y(t) = O[x(t)]$ if there exist positive constants M_1, M_2 and $t_0 \in \mathcal{R}^+$ such that $|y(t)| \leq M_1|x(t)| + M_2 \qquad \forall t \geq t_0$.

<u>Definition 4:</u> Let $x(.), y(.) \in \mathcal{PC}_{[0,\infty)}$. We denote $y(t) = o[x(t)]$ if there exists a function $\beta(t) \in \mathcal{PC}_{[0,\infty)}$ and $t_0 \in \mathcal{R}^+$ such that $|y(t)| = \beta(t)x(t) \quad \forall t \geq t_0$, and $\lim_{t\rightarrow\infty}\beta(t) = 0$.

<u>Definition 5:</u> Let $x(.), y(.) \in \mathcal{PC}_{[0,\infty)}$. If $y(t) = O[x(t)]$ and $x(t) = O[y(t)]$ then $x(.)$ and $y(.)$ are said to be equivalent and denoted by $x(t) \sim y(t)$.

The above definitions are found to have a wider application when the class is limited to monotonic functions. Defining $x_s(t) \stackrel{\Delta}{=} \sup_{\tau\leq t}|x(\tau)|,\ \tau, t \in \mathcal{R}^+$, we have the following definition of two signals growing at the same rate.

<u>Definition 6:</u> Let $x(.), y(.) \in \mathcal{PC}_{[0,\infty)}$. $x(.)$ and $y(.)$ are said to grow at the same rate if $x_s(t) \sim y_s(t)$.

From the above definitions, we see that $\sim$ is reflexive, symmetric and transitive and hence an equivalence relation while O is reflexive, transitive and anti-symmetric and therefore is a partial order. Using these definitions, the following can be shown:

(i) In the system described by

$$\dot{x}(t) = Ax(t) + bu(t) \qquad x(t_0) = x_0$$

where $u \in \mathcal{PC}_{[0,\infty)}$, A is asymptotically stable, then

$$\|x_s(t)\| = O[u_s(t)].$$

(ii) In a system described by the differential equation

$$\dot{x}(t) = Ax(t) + b[\phi^T(t)x(t) + r(t)], \qquad x, \phi : \mathcal{R}^+ \to \mathcal{R}^n$$

where A is asymptotically stable and $\|\phi(t)\|$ and $r(t)$ are uniformly bounded, $|\phi^T x|$ and $\|x\|$ grow at the same rate.

(iii) In a linear time-varying system described by

$$\begin{aligned} \dot{x}_1(t) &= A_1(t)x_1(t) + b_1(t)y_2(t) + r_1(t) \qquad & y_1(t) = c_1^T(t)x_1(t) \\ \dot{x}_2(t) &= A_2(t)x_2(t) + b_2(t)y_1(t) + r_2(t) \qquad & y_2(t) = c_2^T(t)x_2(t) \end{aligned}$$

where $A_1(t)$ and $A_2(t)$ are exponentially stable, b_1, b_2, c_1 and c_2 are uniformly bounded for all $t \in \mathcal{R}^+$ and r_1 and r_2 are piecewise continuous uniformly bounded inputs,

$$\|x_1(t)\|, \|x_2(t)\|, y_1(t) \text{ and } y_2(t) \text{ grow at the same rate.}$$

(iv) If $\mathcal{E}$ is defined as the class of signals $\mathcal{E} = \{x(.) | \|\dot{x}\| \leq M_1 \sup_{\tau \leq t} \|x(\tau)\| + M_2,\ M_1, M_2 \in \mathcal{R}^+\}$ then in the linear system described by $\dot{x} = Ax + bu$, where (A, b) is controllable and $u \in \mathcal{E}$, then

$$u_s(t) = O[\|x(t)\|_s].$$

(iv) If $\beta \in \mathcal{L}^1$ or $\mathcal{L}^2, u(.) \in \mathcal{PC}_{[0,\infty)}$, and $\beta(.)u(.)$ is the input to an asymptotically stable linear time-invariant system and $y(.)$ is the corresponding output,

$$|y(t)| = o[u_s(t)].$$

(v) In the system $\dot{x} = Ax + b[\phi^T x + r]$ where A is asymptotically stable and $r(.)$ is uniformly bounded, if $\phi(.) \in \mathcal{L}^1$ or $\mathcal{L}^2$ or $\lim_{t\to\infty} \phi(t) = 0$ or $|\phi^T(t)x(t)| = o[\|x_s(t)\|]$, then $x(t)$ is uniformly bounded.

2.3 Stability Under Persistent Disturbances

In robust adaptive control, we are often interested in deducing the properties of the solutions of a perturbed system (S_p) from the behavior of the solutions of an unperturbed system (S). These are described by the differential equations

$$\dot{x} = f(x,t) \qquad (S); \qquad \dot{x} = f(x,t) + g(x,t). \qquad (S_p) \tag{6}$$

Let the equilibrium state of (S) be exponentially stable. If $\|g(x,t)\| < b\|x\|$ for a sufficiently small b, and δ, where $\|x\| < \delta$, then the equilibrium state of (S_p) is also exponentially stable [17]. The condition on $g(x,t)$ implies that $g(0,t) = 0$ or $x = 0$ is an equilibrium state of the perturbed system as well. In physical situations this condition is not generally met. Instead, all that can be assured is that $g(x,t)$ is small in some neighborhood of the origin. This gives rise to the concept of total stability.

Definition 7 (Total Stability)[18]: The equilibrium state $x = 0$ of (S) is totally stable if for every $\epsilon > 0$ two positive numbers $\delta_1(\epsilon)$ and $\delta_2(\epsilon)$ exist such that every solution $x(t, x_0, t_0)$ of (S_p) satisfies $\|x(t, x_0, t_0)\| < \epsilon,\ t \geq t_0$ provided $\|x_0\| < \delta_1$ and $\|g(x,t)\| < \delta_2$.

In the Russian literature this is also referred to as stability under persistent disturbances. The fact that the u.a.s. of the unperturbed system implies total stability was shown by Malkin [5] and is frequently used to prove robustness of adaptive systems in the presence of a persistently exciting reference input and sufficiently small perturbations.

Theorem 2: If the equilibrium state of (S) is uniformly asymptotically stable, then it is totally stable.

In practical systems we are interested in the uniform boundedness of the solutions in the presence of perturbations as well as in the magnitude of this bound. This leads to the concept of practical stability defined below.

Definition 8 (Practical Stability)[17]: Let $Q_0 = \{x \mid \|x\| < \delta_1\}$ be an open set in $\mathcal{R}^n$ and $\delta_2 > 0$ a constant such that $\|g(x,t)\| < \delta_2$ for all x and $t \geq t_0$. If the solutions of (S_p) lie within a closed bounded set $Q \supset Q_0$ for $x_0 \in Q_0$ then the system (S) is said to be practically stable.

Total stability merely assures the existence of Q_0 and δ_2 but provides no way of estimating their magnitudes. In the design of robust adaptive systems one is more interested in practical stability which provides an estimate of Q from a knowledge of Q_0 and δ_2.

Method of Averaging: The method of averaging was used for the first time by van der Pol in the study of nonlinear oscillations. A full justification of this method was made by Krylov and Bogoliuboff [6] and subsequently improved by Bogoliuboff and Mitropolsky [19] and more recently by Hale [20] and Arnold [21]. The method involves the study of differential equations where the right hand side involves a small parameter ϵ:

$$\dot{x} = \epsilon f(t, x, \epsilon) \qquad\qquad x(0) = x_0 \tag{7}$$

Hence from (7), since $\dot{x}$ is small, it is logical to expect that the rapidly varying terms in x do not affect the slow variation of x in the long run. By a process of averaging, the nonautonomous differential equation can be approximated by an autonomous differential equation in the variable x_{av} which represents the average value of x. More precisely, if

$$f_{av}(x) \triangleq \lim_{T\to\infty} \frac{1}{T} \int_t^{t+T} f(\tau, x, 0)\,d\tau \qquad \text{exists uniformly in } t \text{ and } x$$

the averaged system

$$\dot{x}_{av} = \epsilon f_{av}(x_{av}) \qquad\qquad x_{av}(0) = x_0 \tag{8}$$

is the approximation of (7) and the problem of stability of (7) can be analyzed by comparing it with that of the averaged system (8). In a recent report [22] Fu et al. provide a very good introduction to the method of averaging and the following theorem in [22] is found to be relevant for some of the discussions in section 5.

Theorem 3: Let $x = 0$ be an equilibrium point of (7), $f(t, x, \epsilon)$ be Lipschitz in x, and ϵ, $f_{av}(0) = 0$ and $f_{av}(x)$ be Lipschitz in x and $d(t, x) \triangleq f(t, x, 0) - f_{av}(x)$ be piecewise continuous with respect to t, have bounded and continuous first partials with respect to x. Let x_0 be such that $\|x_{av}(t)\| \leq \delta \ \ \forall t \in [0, \frac{T}{\epsilon}]$ for some $\delta > 0$. Then there exists an $\epsilon^* > 0$ such that

$$\|x(t) - x_{av}(t)\| \leq k\epsilon$$

for all $t \in [0, \frac{T}{\epsilon}]$, for some constant $k > 0$ and for all $\epsilon \leq \epsilon^*$.

BIBO Stability: The well known concept of bounded input - bounded output (BIBO) stability is a convenient vehicle for discussing the effect of bounded perturbations in adaptive systems.

Definition 9: A system $\dot{x} = f(x, u, t)$ with $f(0, 0, t) = 0$ is BIBO stable if for every $\alpha \geq 0$ and every $a \geq 0$ there is a $\beta = \beta(\alpha, a)$ such that $\|x_u(\tau, x_0, t_0)\| \leq \beta$ for all $\tau \geq t_0$ for every initial condition (x_0, t_0) with $\|x_0\| \leq \alpha$ and $\sup_t \|u(t)\| \leq a$, where $x_u(t, x_0, t_0)$ is the solution of the system with input $u(.)$.

A linear system $\dot{x} = A(t)x + b(t)u$ is BIBO stable if the homogeneous part is u.a.s. This is a property which is frequently used in robust adaptive control when some of the relevant signals are persistently exciting and the overall system is linearized around a nominal solution. In

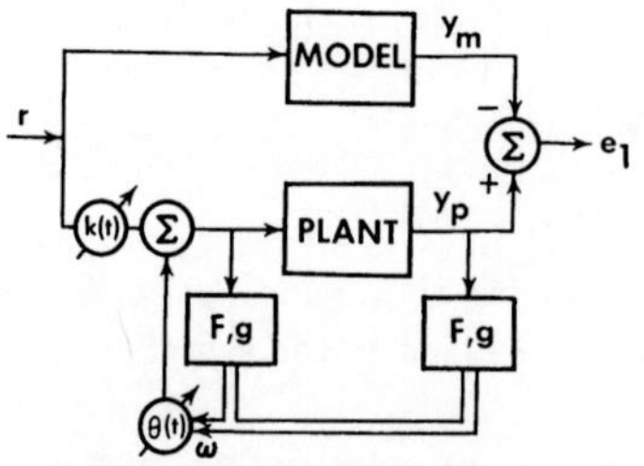

Figure 2: The Ideal System

contrast to the above, it is well known that u.a.s. of a nonlinear system does not imply BIBO stability. In fact several efforts have been made in recent years to determine conditions under which such a property holds [23,24]. In section 4.2 it is shown that if the ideal system is u.a.s.l., there exists a constant a_{max} such that the adaptive system is BIBO stable according to definition 9 for all $\sup_t |u(t)| \leq a_{max}$ where the external disturbance is considered as the input $u(.)$. Other results have also been reported in the literature [26,27] based on the small gain theorem [25] which restrict both the bound a on the input as well as that on the initial conditions. In such cases the boundedness of the output is established provided both a and α in definition 9 are sufficiently small (also ref. Total Stability).

3. The Ideal System

The starting point for most of the investigations in robust adaptive control is the ideal adaptive control problem. We briefly state the problem here along with its solution.

a. Statement of the Problem: The plant to be controlled is represented by a linear time-invariant differential equation

$$\begin{aligned} \dot{x}_p &= A_p x_p + b_p u \\ y_p &= h_p^T x_p \end{aligned} \tag{9}$$

with a transfer function $W_p(s) = h_p^T(sI - A_p)^{-1}b_p = k_p \frac{Z_p(s)}{R_p(s)}$, where $Z_p(s)$ and $R_p(s)$ are monic polynomials of degrees m and $n(> m)$ respectively, and $Z_p(s)$ is Hurwitz. The sign of k_p and the values of m and n are specified but the coefficients of $Z_p(s)$ and $R_p(s)$ are unknown. A reference model is set up with a transfer function $W_m(s) = \frac{k_m}{R_m(s)}$, where $R_m(s)$ is a known Hurwitz polynomial of degree $n^* = n - m$. A piecewise continuous uniformly bounded reference input $r(t)$ to the model yields the desired output $y_m(t)$. The adaptive control problem is to determine a bounded control input $u(.)$ to the plant so that the output error $e_1(t) \triangleq y_p(t) - y_m(t)$ tends to zero asymptotically with all the signals remaining bounded. For ease of exposition, we assume that $k_p = k_m = 1$.

b. Adaptive Controller: The adaptive controller is described by the equations [Fig. 2]

$$\begin{aligned} \dot{\omega}^{(1)} &= F\omega^{(1)} + gu \\ \dot{\omega}^{(2)} &= F\omega^{(2)} + gy_p \\ u(t) &= \theta^T(t)\omega(t) + r(t) \end{aligned} \tag{10}$$

where $\omega^T \triangleq [\omega^{(1)T}, \omega^{(2)T}], \theta^T = [\theta^{(1)T}, \theta^{(2)T}]$, $\omega^{(1)}$ and $\omega^{(2)}$ are n-dimensional vectors, F is an arbitrary asymptotically stable matrix, (F, g) is controllable, and $\theta(t)$ is the control parameter vector. When $\theta(t) \equiv \theta^*$, where θ^* is a constant vector, the transfer function of the plant together with the controller matches that of the model exactly [28]. The problem therefore can be restated as the determination of the adaptive law for adjusting $\theta(t)$ so that $\lim_{t\to\infty} e_1(t) = 0$ with all signals remaining bounded.

c. Error Models: It has been shown [29,30] that the stability of adaptive systems is most readily analyzed by considering error models relating an output error to the parameter error vector. Defining $\theta(t) = \theta^* + \phi(t)$, where $\phi(t)$ represents the parameter error vector, $\phi(t)$ and the output error $e_1(t)$ are related by the equation

$$W_m(s)\phi^T(t)\omega(t) = e_1(t). \tag{11a}$$

If $W_m(s)$ is strictly positive real, a stable adaptive control law can be realized using the signals in the adaptive loop as

$$\dot{\theta} = \dot{\phi} = -e_1\omega. \tag{11b}$$

For the general case when $W_m(s)$ has a relative degree ≥ 2, an augmented error ϵ_1 is generated as follows:

$$\begin{aligned} W_m(s)\omega &\triangleq \varsigma \\ \epsilon_1 &= e_1 + \theta^T\varsigma - W_m(s)\theta^T\omega \\ &= \phi^T\varsigma. \end{aligned} \tag{12a}$$

In this case the parameter vector is adjusted as

$$\dot{\theta} = \dot{\phi} = -\frac{\epsilon_1\varsigma}{1+\varsigma^T\varsigma}. \tag{12b}$$

The adaptive laws (11b) and (12b) are referred to as "the standard adaptive laws" in the following sections. In both cases, it can be shown that all the signals in the system are bounded [28,30] and that $\lim_{t\to\infty} e_1(t) = 0$. Since many of the results derived in the context of robust adaptive systems in this section use arguments similar to those used in the ideal case when $n^* \geq 2$, we briefly outline the various steps involved.

d. Outline of Proof: (i) If $V(\phi) = \frac{1}{2}\phi^T\phi$, the adaptive law (12b) yields

$$\dot{V} = -\frac{\epsilon_1^2}{1+\varsigma^T\varsigma} \leq 0.$$

From this it follows that $\|\phi\|$, $\|\dot{\phi}\|$ are bounded, and $\dot{\phi} \in \mathcal{L}^2$ which implies that $\epsilon_1(t) = \beta(t)[1+\varsigma^T(t)\varsigma(t)]^{1/2}$ where $\beta \in \mathcal{L}^2$.
(ii) Since the parameter vector θ is bounded, it can be shown using the arguments in [7] that if the signals y_p, $\|\omega\|$, $\|\omega^{(2)}\|$ and $\|\varsigma\|$ are unbounded, they grow at the same rate or

$$\sup_{\tau\leq t}|y_p(\tau)| \sim \sup_{\tau\leq t}\|\omega^{(2)}(\tau)\| \sim \sup_{\tau\leq t}\|\omega(\tau)\| \sim \sup_{\tau\leq t}\|\varsigma(\tau)\|. \tag{13}$$

(iii) Since $\dot{\phi} \in \mathcal{L}^2$, it can be shown that

$$e_1(t) = W_m(s)\phi^T(t)\omega(t) = \beta(t)[1+\varsigma^T(t)\varsigma(t)]^{1/2} + o[\sup_{\tau\leq t}\|\omega(\tau)\|].$$

This in turn results in $\|\omega^{(2)}(t)\| = o[\sup_{\tau\leq t}\|\omega(\tau)\|]$ which implies that $\|\omega^{(2)}(t)\|$ grows at a slower rate than $\|\omega(t)\|$. Since this contradicts (13), we conclude that all signals are bounded. This yields that $\lim_{t\to\infty} e_1(t) = \lim_{t\to\infty}\epsilon_1(t) = 0$.

It is seen from the above discussion that the boundedness of the parameter vector ϕ and the fact that in some sense it changes by small amounts (in this case assured by the condition $\dot{\phi} \in \mathcal{L}^2$) are the two principal features assured by the adaptive law which result in the stability of the adaptive system.

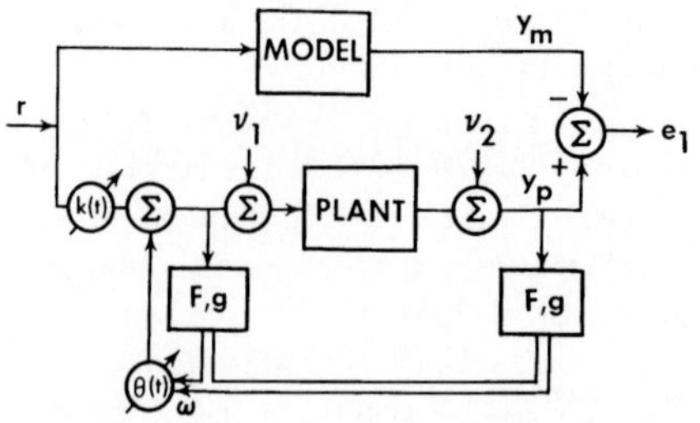

Figure 3: Robust Adaptive Control in the Presence of Bounded Disturbances.

e. Convergence of parameters: If in addition to the output error tending to zero, it is desired that the control parameters tend to their true values, (i.e., $\lim_{t\to\infty} \phi(t) = 0$), then $\omega(t)$ needs to be persistently exciting [12,13]. However, it is clear that the persistent excitation of $\omega(t)$ cannot be demanded directly since it is a dependent variable and the output of a nonlinear system. This fact was recognized by many authors [31]-[33] who later derived the convergence of $\theta(t)$ based on the persistent excitation of the reference input $r(t)$. Based on these results it can be stated that $\lim_{t\to\infty} \theta(t) = \theta^*$ if $r(t)$ contains n distinct frequencies (or n spectral lines according to [33]).

4. Robustness In The Presence Of Bounded Disturbances

When external disturbances or other perturbations such as time-variations in the parameters, or unmodeled dynamics of the plant are present, it is no longer possible to ensure that the error between plant and model outputs will tend to zero asymptotically if the standard adaptive law (12b) is used. In fact, even around 1980 it was realized that in the case of bounded external disturbances, the adaptive law (12b) could result in the parameter error ϕ growing in an unbounded fashion [34,35]. Similarly, other perturbations such as time-varying parameters or unmodeled dynamics can also result in all the signals of the system becoming unbounded. Hence new approaches were needed to assure the boundedness of the signals in the system.

The various approaches that have been suggested during the past few years to address the above problem can be divided into two classes. In the first are included methods in which the adaptive law is suitably modified, without using arguments related to persistent excitation of the reference input. The methods in the second class on the other hand use primarily the persistent excitation property of the external input to assure the boundedness of the signals in the adaptive system, without any modifications in the standard adaptive law.

When the external disturbances present are bounded, the plant to be adaptively controlled can be described by the differential equations

$$\begin{aligned} \dot{x}_p &= A_p x_p + b_p u + d_p \nu_1 \\ y_p &= h_p^T x_p + \nu_2 \end{aligned} \tag{14}$$

where ν_1 is a bounded input disturbance and ν_2 is an output disturbance which is bounded and differentiable [Fig. 3]. The choice of the reference model, as well as the controller structure are identical to those in the ideal case. The overall system equations are given by [36]

$$\dot{x} = Ax + b(\phi^T \omega + r + \bar{\nu}(t)); \qquad y_p = h^T x \tag{15}$$

where $x^T = [x_p^T, \omega^T], h^T(sI - A)^{-1}b = W_m(s)$, $\phi \triangleq \theta - \theta^*$ and $\bar{\nu}$ is an equivalent input disturbance due to ν_1 and ν_2.

4.1 Robustness Without Persistent Excitation

Under the same assumptions on the plant transfer function as in the ideal case, for the general case where $n^* \geq 2$, the adaptive law is based on the augmented error $\epsilon_1(t)$ defined as in section 3. However, when external disturbances are present, the corresponding error equation has the form

$$\phi^T(t)\varsigma(t) + \nu(t) = \epsilon_1(t) \tag{16}$$

where $\nu(t) \stackrel{\Delta}{=} W_m(s)\bar{\nu}(t)$ with $|\nu(t)| \leq \nu_0$. Equation (16) forms the starting point of all the results presented in this section. If the standard adaptive law (12b) is used in this case, it no longer follows directly that $\phi(t)$ will remain bounded. In fact $||\phi(t)||$ decreases only when $\text{sgn}[\phi^T\varsigma+\nu] = \text{sgn}[\phi^T\varsigma]$ and hence instability can result when $\phi^T\varsigma$ is small compared to $|\nu(t)|$.

(a) Use of a Dead-Zone: In [15] Peterson and Narendra suggested the use of a dead-zone in the adaptive law as follows:

$$\begin{aligned} \dot{\phi} &= -\frac{\epsilon_1\varsigma}{1+\varsigma^T\varsigma} \quad & |\epsilon_1| \geq \nu_0 + \delta \\ &= 0 & |\epsilon_1| < \nu_0 + \delta \end{aligned} \tag{17}$$

where δ is an arbitrary positive constant. According to this law, the parameter vector is not adjusted when the amplitude of the augmented error is smaller than $\nu_0 + \delta$. In [15] this is shown to result in the boundedness of all the signals in the adaptive system as follows:

(i) The boundedness of $||\phi||$ follows directly from the adaptive law. Since the structure of the adaptive system is unaltered, it follows from the same arguments as in the ideal case [7] that if the signals in the adaptive loop grow in an unbounded fashion,

$$\sup_{\tau\leq t}|y_p(\tau)| \sim \sup_{\tau\leq t}||\omega^{(2)}(\tau)|| \sim \sup_{\tau\leq t}||\omega(\tau)|| \sim \sup_{\tau\leq t}||\varsigma(\tau)||. \tag{13}$$

(ii) From the adaptive law (17) and the inequality

$$\frac{\delta}{\nu_0+\delta}|\phi^T\varsigma + \nu| \leq |\phi^T\varsigma| \leq \frac{2\nu_0+\delta}{\nu_0+\delta}|\phi^T\varsigma + \nu| \text{ when } |\epsilon_1(t)| > \nu_0 + \delta,$$

it also follows that $\dot{\phi} \in \mathcal{L}^2$. This implies that

$$||\omega^{(2)}(\tau)|| = o\left[\sup_{\tau\leq t}||\omega(t)||\right] + \bar{y}(t) \tag{18}$$

where $\bar{y}(t)$ is a bounded signal.

(iii) Equation (18) contradicts (13) and hence all signals are bounded.

With this approach, adaptation takes place for only a finite time. This implies that in practice, the system will converge to a linear time-invariant system in a finite time after which the output error will lie entirely in the dead-zone and hence adaptation ceases. Further, it is clear, that to implement the adaptive law, it is necessary to know a bound ν_0 on the disturbance $\nu(t)$. In practice, ν_0 may be quite conservative. Further, if an adaptive law using a dead-zone based on ν_0 is used, the output and parameter errors may not tend to zero even when the disturbance is removed. This has motivated research aimed at determining variable dead-zones in the adaptive law based on information available regarding the magnitude of the disturbance. [Ref. section 5]

(b) Bound on $||\theta^*||$: An alternate approach to the robustness problem in the presence of bounded disturbances was taken by Kreisselmeier and Narendra in [16]. Instead of stopping adaptation inside a "dead-zone", it is assumed in [16] that the desired vector θ^* has a norm less than a specified value $||\theta^*||_{max}$. Hence the search procedure for the control parameter vector can be confined essentially to the set $S : \triangleq \{\theta| \, ||\theta|| \leq ||\theta^*||_{max}\}$. The adaptive law in this case is given by

$$\begin{aligned} \dot{\phi} &= -\frac{\epsilon_1 \varsigma}{1+\xi^T \xi} && \text{if } ||\theta|| < ||\theta^*||_{max} \\ &= -\frac{\epsilon_1 \varsigma}{1+\xi^T \xi} - \theta(1 - \frac{||\theta||}{||\theta^*||_{max}})^2 && \text{elsewhere} \end{aligned} \tag{19}$$

where $\xi^T \triangleq [\omega^T, \varsigma^T]$. In [16] it is shown that such a scheme results in the boundedness of all signals in the system and is briefly outlined below:

The overall system equations can be written from (15) and (12) as

$$\dot{z} = \overline{A}z + \overline{b}(\phi^T \omega + \overline{\nu} + r) \tag{20}$$

where $z^T \triangleq [x^T, \varsigma^T]$ and $\overline{A}$ is asymptotically stable. Assume that signals in the adaptive system grow in an unbounded fashion so that $\lim_{t\to\infty} \sup_{\tau \leq t} ||\xi(\tau)|| = \infty$.

(i) It directly follows from the adaptive law (19) that $||\theta||$ is bounded and hence the state of the system, x, as well as ξ can grow at most exponentially. Then it can be shown that unbounded sequences $\{t_i\}, \{a_i\}$ exist with

$$||\xi(t)|| \geq a_i \qquad \forall t \in [t_i, t_i + a_i]. \tag{21}$$

(ii) From the adaptive law (19) it also follows that over an interval $[t_1, t_2]$,

$$\int_{t_1}^{t_2} \frac{\epsilon_1^2}{1+\xi^T \xi} dt \leq \Delta_1 + \Delta_2 \int_{t_1}^{t_2} \frac{1}{[1+\xi^T \xi]^{1/2}} dt \tag{22}$$

where Δ_1 and Δ_2 are finite positive constants. Since ν is bounded, from (21) we obtain that $\int_{t_i}^{t_i+a_i} \frac{(\phi^T \varsigma)^2}{1+\xi^T \xi} dt \leq \Delta$ where Δ is bounded. In addition,

$$|\frac{d}{dt} \frac{(\phi^T \varsigma)^2}{1+\xi^T \xi}| \leq c \text{ where } c \text{ is a finite constant}$$

since ξ can grow at most exponentially. We partition the interval $T_i \triangleq [t_i, t_i + a_i]$ into two sets $T_{i1} \triangleq [t| \frac{(\phi^T \varsigma)^2}{1+\xi^T \xi} < \epsilon]$ and $T_{i2} \triangleq T_i - T_{i1}$. If L_{i1} and L_{i2} are the measures of the sets T_{i1} and T_{i2} respectively, since ν is bounded, we obtain that $L_{i2} \leq \frac{2c\Delta}{\epsilon^2}$ and $L_{i1} \geq a_i - \frac{2c\Delta}{\epsilon^2}$.

(iii) If $||z|| \geq a$ and $\frac{|\phi^T \omega|}{||z||} < \varepsilon$, from (20) it follows that $||z||$ decays exponentially, where a and ε are chosen appropriately. It can therefore be shown that when $t \in T_{i1}$, $||z||$ decays exponentially and can grow at most exponentially when $t \in T_{i2}$. From the discussions in (ii), it follows that by making a_i arbitrarily large, the measure L_{i1} can be made arbitrarily large compared to L_{i2}. This contradicts (21) and hence all signals in the adaptive loop are bounded.

The differences between the schemes suggested in [15] and [16], are worth emphasizing. These can be stated in terms of the prior information assumed in the two cases as well as the mathematical proofs used in establishing the boundedness of all the signals in the system.

1. In [15] ν_0, an upper bound on $|\nu(t)|$ is assumed to be known while implementing the adaptive law. In [16] such knowledge is not needed. However, information regarding $||\theta^*||_{max}$ is required in this case. While the adaptive law with a dead-zone cannot be u.a.s., the adaptive law (19) assures u.a.s. (as in the ideal case) when the disturbance tends to zero and the reference input is persistently exciting.

2. The proof of boundedness in [15] uses limiting arguments as $t \to \infty$ to show that $\phi \in \mathcal{L}^2$. Such a procedure cannot be used in [16] since $\phi(t)$ need not tend to any limit as $t \to \infty$. Hence all arguments are based on the analysis of the behavior of the system over a finite time interval. As shown later in this paper, the proof of [16] is applicable to other adaptive control problems as well. This is due to the fact that in most problems of practical interest, the output error rarely tends to zero as $t \to \infty$ so that adaptation never ceases entirely.

(c) The σ-modification scheme [37]: In approaches (a) and (b), prior information in the form of the bound on the disturbance or the desired parameter vector was assumed. In contrast to this, a scheme suggested by Ioannou and Kokotovic [37] assures boundedness of all signals in the system, without such assumptions. The motivation for this method may be described briefly as follows: If $V(e,\phi)$ is a quadratic Lyapunov function candidate, the time-derivative $\dot{V}(e,\phi)$ has the general form $-e^TQe + e^T\alpha\bar{\nu}$ where $Q = Q^T > 0, \alpha$ is a constant vector and $\bar{\nu}$ is a bounded disturbance. Since $\dot{V}$ is indefinite very little can be concluded regarding the boundedness of e and ϕ and accounts for the modifications suggested in [15] and [16]. In [37], an additional term $-\sigma\theta$ is used in the adaptive law, as a result of which $\dot{V}(e,\phi)$ becomes negative definite outside a bounded region in the (e,ϕ) space. From this it follows that all signals in the system are bounded. To the authors' knowledge, the method has been shown to result in global boundedness only for the special case when the reference model is strictly positive real.

While the use of the additional term $-\sigma\theta$ in the adaptive law is certainly a novel one, the approach suffers to some extent due to the very presence of this term. Since $-\sigma\theta = \sigma(\theta^* + \phi)$ where θ^* is the desired parameter vector, the error equations can never be u.a.s. if the plant being controlled is not stable to begin with (i.e. if $\theta^* \neq 0$). Further, the magnitude of the bias term is proportional to θ^* and hence qualitatively to the degree of adaptation needed. Many of these weaknesses can be compensated for by the proper choice of σ but in no case can the bias be entirely eliminated if prior information regarding the plant is limited to that available in the ideal case. The possibility of modifying the adaptive law when a bound on $||\theta||$ is known, as in [16], was recently suggested by Ioannou and Tsakalis [39] for the general case when $n^* \geq 2$. In such a case, the adaptive law does not significantly differ from that suggested in [16].

d. A New Adaptive Law: Recently a new adaptive law motivated by that given in [37], was proposed by Narendra and Annaswamy in [38]. With the same error model (16), the new adaptive law proposed is given by

$$\dot{\phi}(t) = -\frac{\epsilon_1 \zeta}{1+\xi^T\xi} - \gamma\frac{|\epsilon_1|\theta}{1+\xi^T\xi} \qquad \gamma > 0. \tag{23}$$

It is shown in [38] that the adaptive law (23) leads to

(i) boundedness of all signals in the system in the ideal case,

(ii) u.a.s. of equation (23) when ζ is sufficiently persistently exciting, and

(iii) boundedness of all signals in the system when a bounded disturbance $\nu(t)$ is present.

The above results make the scheme superior to all others known in this category. Like the adaptive law in [37] (i) and (iii) are obtained without any additional prior information. (ii) makes the scheme particularly attractive for tracking problems in the presence of disturbances. The proofs of (i) and (iii) follow along the same lines as in [16]. Since the new algorithm has not yet appeared in the published literature, we include here a more detailed discussion.

Boundedness of Signals in The Ideal Case:($\nu(t) \equiv 0$)

(i)Choosing a quadratic function $V(\phi) = \frac{1}{2}\phi^T\phi$, the time-derivative $\dot{V}$ is given by

$$\dot{V} = -\frac{|\epsilon_1|}{1+\xi^T\xi}\left[|\epsilon_1| + \gamma\phi^T\theta^* + \gamma\phi^T\phi\right]. \tag{24}$$

Since $\dot{V} \leq 0$ when $\|\phi\| \geq \|\theta^*\|$, it follows that $\|\phi(t)\|$ is bounded. Assuming that signals in the adaptive system grow in an unbounded fashion, it follows from (24) that the state of the system,z, (ref. (20)) as well as ξ can grow at most exponentially. Hence, as in [16], it can be shown that unbounded sequences $\{t_i\}, \{a_i\}$ exist with

$$\|\xi(t)\| \geq a_i \ \forall t \in [t_i, t_i + a_i]. \tag{21}$$

(ii) Integrating (24) over a finite interval $[t_1, t_2]$, and using the fact that $\|\phi(t)\|$ is bounded, we obtain that

$$\int_{t_1}^{t_2} \frac{\epsilon_1^2}{1+\xi^T\xi}dt \leq \Delta_1 + \Delta_2 \int_{t_1}^{t_2} \frac{1}{[1+\xi^T\xi]^{1/2}}dt$$

where Δ_1 and Δ_2 are appropriate positive constants. Again, since ν is bounded, it follows that

$$\int_{t_i}^{t_i+a_i} \frac{(\phi^T\varsigma)^2}{1+\xi^T\xi}dt \leq \Delta \qquad \text{and } \left|\frac{d}{dt}\frac{(\phi^T\varsigma)^2}{1+\xi^T\xi}\right| \leq c$$

where Δ, c are positive constants.

(iii) Partitioning the interval $T_i \triangleq [t_i, t_i + a_i]$ into two sets $T_{i1} \triangleq [t| \frac{(\phi^T\varsigma)^2}{1+\xi^T\xi} < \epsilon]$ and $T_{i2} \triangleq T_i - T_{i1}$, as in [16], it can be shown that when $t \in T_{i1}$, $\|z\|$ decays exponentially and can grow at most exponentially when $t \in T_{i2}$. From the discussions in (ii), it follows that by making a_i arbitrarily large, the measure of T_{i1} can be made arbitrarily large compared to that of T_{i2}. This contradicts the assumption that signals grow in an unbounded fashion and therefore boundedness follows.

Uniform Asymptotic Stability: We now show that, in the ideal case, if the persistent excitation of ς is sufficiently large, the parameter error ϕ and hence ϵ_1 and e_1 tend to zero.

The problem under consideration is the u.a.s. of equation (23) which can be rewritten as

$$\dot{\phi} = -uu^T\phi - \alpha(t)(\theta^* + \phi)|u^T\phi| \tag{25}$$

where $u(t) \triangleq \frac{\varsigma}{(1+\xi^T\xi)^{1/2}}$ and $\alpha(t) \triangleq \frac{\gamma}{(1+\xi^T\xi)^{1/2}}$. Equation (25) is of the form $\dot{x} = A(t)x + f(x,t)$ and hence can be shown to be u.a.s. if $\dot{x} = A(t)x$ is u.a.s. and $f(x,t)$ is sufficiently small, by the proper choice of γ.

Theorem 4: In the differential equation (25), if u is persistently exciting, there exists a constant γ^* such that the equilibrium state is u.a.s.l. for all $0 < \gamma < \gamma^*$.

Proof: Since u is persistently exciting $\dot{\phi} = -uu^T\phi$ is exponentially stable. Hence positive constants σ and k exist such that

$$\|p(t,t_0)\| \leq k\, exp\,[-\sigma(t-t_0)]$$

where $p(t,\tau)$ is the transition matrix of $\dot{\phi} = -uu^T\phi$. Further, all the solutions of (25) are bounded and hence a constant M exists such that they tend to the invariant set $\mathcal{D} \triangleq \{\phi| \ \|\phi\| < M\}$. The following arguments show that any solution starting in $\mathcal{D}$ tends to the origin as $t \to \infty$ by choosing $\gamma^* \triangleq \frac{\sigma}{k^2(M+\|\theta^*\|)}$:

Defining $f(\phi,t) \triangleq \alpha(t)(\theta^* + \phi)|u^T\phi|$, we have, for all $\phi \in \mathcal{D}$, $\|f(\phi,t)\| \leq \gamma(M + \|\theta^*\|)\|\phi(t)\|$. Following the main result in [40], if

$$V = \phi^T C(t)\phi \qquad \text{where } C(t) = \int_t^\infty \|p(\tau,t)\|^2 d\tau \tag{26}$$

it follows that

$$\dot{V} \leq -||\phi||^2 \left[1 - \frac{\gamma k^2(M+||\theta^*||)}{\sigma}\right].$$

Choosing $\gamma < \gamma^*$, we have $\dot{V}$ negative definite so that the origin of (25) is u.a.s.l. Since ς is bounded, from equation (16), it follows that ϵ_1 tends to zero.

Remarks

1. The main advantage of the adaptive law (23) over that suggested in [37] is that the error equations are u.a.s.l. in the ideal case when ς is sufficiently persistently exciting. From a theoretical viewpoint, this is attractive since the usual arguments related to total stability with respect to small perturbations can be applied to such an adaptive system [5].
2. The condition given in the above theorem is quite stringent and is expressed in terms of the convergence parameters σ and k which are unknown. Hence theorem 4 merely assures that a sufficiently large degree of PE of $u(t)$ will assure u.a.s.l. Work is currently in progress to obtain less conservative bounds.

Boundedness of Signals in The Perturbed system:($|\nu(t)| \leq \nu_0$)

When a bounded disturbance $\nu(t)$ is present in (16), the signals of the adaptive system continue to remain bounded. The arguments follow along very similiar lines to that of the ideal system and are briefly outlined below.

(i) The quadratic function $V = \frac{1}{2}\phi^T\phi$ yields

$$\dot{V} \leq -\epsilon_1^2/(1+\xi^T\xi) - |\epsilon_1|/(1+\xi^T\xi)\left[-\nu_0 + \gamma\phi^T\theta^* + \gamma\phi^T\phi\right].$$

Hence $||\phi(t)||$ is uniformly bounded.

(ii) Integrating $\dot{V}$ over an interval $[t_i,\, t_i + a_i]$, as in the ideal system, we obtain

$$\int_{t_i}^{t_i+a_i} \frac{\epsilon_1^2}{1+\xi^T\xi}dt \leq \Delta_3 + \Delta_4 \int_{t_i}^{t_i+a_i} \frac{1}{[1+\xi^T\xi]^{1/2}}dt \tag{27}$$

where Δ_3 and Δ_4 are finite positive constants.

(iii) Since $||\dot{\xi}|| \leq M_1||\xi|| + M_2$ for some constants $M_1 > 0$ and $M_2, \nu(t) = W_m(s)\bar{\nu}(t)$ and $||\dot{\phi}||$ is bounded, it follows that a constant c_1 exists such that

$$\left|\frac{d}{dt}\frac{\epsilon_1^2}{1+\xi^T\xi}\right| \leq c_1. \tag{28}$$

(iv) Defining T_{i1} and T_{i2} as before, it follows that

$$\begin{aligned} \dot{z} &\leq -kz & \quad t \in T_{i1} \\ \text{and} &\leq \lambda z & \quad t \in T_{i2} \end{aligned}$$

where $k, \lambda > 0$.

(v) Assuming that the signals grow without bound, by choosing ϵ sufficiently small, the measure of T_{i1} can be made small relative to that of T_{i2}, leading to a contradiction.

Hence all solutions of the perturbed system (20) are bounded.

4.2 Robustness With Persistent Excitation

It is now generally realized that the concept of persistent excitation plays an important role in a variety of situations in adaptive systems. It first arose in the context of adaptive identification where the persistent excitation of some signals of the system was needed to assure the convergence of the identification parameters to their true values. More recently it recurred in adaptive control problems where the persistent excitation of some of the internal signals of the system was needed to demonstrate the convergence of the control parameters.

The fact that these internal signals are dependent variables motivated further research in the area of persistent excitation. In both identification and control problems, persistent excitation is found to be a necessary and sufficient condition for the u.a.s. of related time-varying differential equations. The problems become considerably more difficult when external disturbances are present, since the arguments tend to become circular when the standard adaptive law is used. Persistent excitation of the relevant signals cannot be demonstrated (using currently known definitions - ref. section 2) unless they are bounded and the latter in turn can be demonstrated only by using arguments related to persistent excitation. These and similar questions that arise while using the notion of persistent excitation in robust adaptive systems are discussed in this section.

Statement of the Problem: The error differential equations of the adaptive control problem in the presence of bounded disturbances can be written as:

Case: $n^* = 1$ Error Equation: $\dot{e} = Ae + b(\phi^T\omega + \bar{\nu})$, $e_1 = h^T e$ (29a)
Adaptive Law: $\dot{\phi} = -e_1\omega$

Case: $n^* \geq 2$ Error Equation: $\epsilon_1 = \phi^T\varsigma + \nu$ (29b)
Adaptive Law: $\dot{\phi} = -\dfrac{\epsilon_1\varsigma}{1+\varsigma^T\varsigma}$

where $e : \mathcal{R}^+ \to \mathcal{R}^{3n}$, $\omega : \mathcal{R}^+ \to \mathcal{R}^{2n}, \varsigma = W_m(s)I\omega$, $\epsilon_1 = e_1 + \theta^T\varsigma - W_m\theta^T\omega$ and $\nu = W_m(s)\bar{\nu}$. The aim is to determine conditions under which the solutions remain bounded with the above adaptive laws.

From section 3, it follows that the equilibrium state of the error equations

$$\begin{aligned} \dot{e} &= Ae + b\phi^T\omega; e_1 = h^T e \\ \dot{\phi} &= -e_1\omega \end{aligned}$$

is uniformly asymptotically stable if ω is persistently exciting. Hence it is tempting to conclude that a bounded disturbance will result in bounded outputs e and ϕ if ω is persistently exciting. Since the boundedness of signals has not been established, ω cannot be assumed to be bounded and proving that it is persistently exciting becomes specious. The two approaches that have been suggested in the literature to study this problem are discussed below.

4.2.1 Local Stability

From Malkin's theorem [5], it is known that when an unperturbed system is u.a.s., a related perturbed system has bounded solutions provided that the initial perturbations are small. Many results have been reported in the literature [26-27,41-42] where the robustness of adaptive control is discussed in the presence of such small disturbances. Most of these are based on the notion of total stability and can be considered as extensions or modifications of Malkin's theorem. Using such an approach, it appears possible to treat nonlinearities, disturbances and time-varying plant parameters in a unified manner [41]. For instance, the error equation (29a) is u.a.s. when the disturbance $\bar{\nu}(t) \equiv 0$. Therefore, Malkin's theorem implies that given $\epsilon > 0$ there exist δ_1 and δ_2 such that for $\|x(0)\| \leq \delta_1$ and

$|\bar{\nu}(t)| \leq \delta_2$, $\|x(t)\| \leq \epsilon$ where $x(t) \triangleq [e^T, \phi^T]^T$. Using this approach, Anderson and Johnstone treated robust adaptive control of time-varying plants in [42], where they relaxed the magnitude constraint on the initial condition at the expense of a stronger condition on the perturbation.

4.2.2 Global Stability

In contrast to the above, efforts have been underway at Yale University to use the concept of persistent excitation to ensure global boundedness of all the signals [36]. The principal outcome of these efforts is considered briefly in this section and is based on the analysis of a set of nonlinear rather than linear differential equations. The main result, contained in theorem 5 assures the boundedness of all the signals in the system if the degree of persistent excitation of the reference input is sufficiently large compared to the magnitude of the disturbance. The results are given for a system with $n^* = 1$ using the error equations (29a).

The circularity involving persistent excitation of ω is avoided in the following manner. If ω^* is the signal in the model corresponding to ω in the plant, then we have $\omega = \omega^* + Ce$ where C is a known constant matrix. Hence, it is more appropriate analytically to require ω^*, which is a bounded signal, rather than ω, to be persistently exciting. This results in the nonlinear differential equation

$$\begin{aligned} \dot{e} &= Ae + b\phi^T(\omega^* + Ce) + b\bar{\nu};\, e_1 = h^T e \\ \dot{\phi} &= -e_1(\omega^* + Ce). \end{aligned} \tag{30}$$

Hence the problem is to determine sufficient conditions under which the output of a nonlinear time-varying system is bounded for a given bounded input $\bar{\nu}(t)$ with $|\bar{\nu}(t)| \leq \nu_0$. This problem is addressed in [36] and the principal result is expressed in theorem 5.

Since $W_m(s) = h^T(sI - A)^{-1}b$ is strictly positive real, given a symmetric positive definite matrix Q, a matrix $P = P^T > 0$ exists such that $A^T P + PA = -Q,\ \ Pb = h$. Let λ_Q be the minimum eigenvalue of Q and $E_0 \triangleq \frac{2\|h\|\nu_0}{\lambda_Q} = \gamma\nu_0$.

Theorem 5: Let the $2n$-dimensional vector ω^* be persistently exciting and satisfy condition (5) for constants T_0, δ_0, and ϵ_0. Then all solutions of (30) are bounded if

$$\epsilon_0 \geq \gamma\nu_0 + \delta \tag{31}$$

where δ is an arbitrary positive constant.

Proof: Let $z^T \triangleq [e^T\sqrt{P},\ \phi^T]$. Then a quadratic function $W(z) = e^T Pe + \phi^T\phi$ has a time-derivative $\dot{W}(z) = -e^T Qe + 2e_1\bar{\nu}$. Defining the region $D \triangleq \{z|\ \|e\| \leq E_0\}$, it follows that $W(z)$ can increase only in D and the maximum increase in $W(z)$ over any period T_1 is given by

$$W(z(t_0 + T_1)) - W(z(t_0)) \leq \frac{E_0^2}{4}\lambda_Q T_1. \tag{32}$$

We consider the following two cases:

$$\text{Case (i)} \quad \frac{1}{T_0}\left|\int_{t_2}^{t_2+\delta_0} (\omega^*(\tau) + Ce(\tau))^T \mathrm{w}\, d\tau\right| \geq k\epsilon_0,\ k = \frac{\delta}{E_0 + \delta}: \tag{33}$$

If condition (33) is satisfied for any unit vector $\mathrm{w} \in \mathcal{R}^{2n}$ for any $t_0 \geq 0$, then $\omega = \omega^* + Ce$ is persistently exciting and (30) represents an exponentially stable system [12] with a bounded input $\bar{\nu}$. Hence the system

$$\dot{z} = J(t)z + \mu(t) \tag{34}$$

is also exponentially stable with a bounded input μ where

$$J(t) \triangleq \begin{bmatrix} \sqrt{P}A\sqrt{P}^{-1} & \sqrt{P}b\omega^T(t) \\ -\omega(t)h^T\sqrt{P}^{-1} & 0 \end{bmatrix}, \quad \mu(t) \triangleq \begin{bmatrix} \sqrt{P}b \\ 0 \end{bmatrix}\bar{\nu}(t)$$

From the results of [13] it follows that for the unforced system $\dot{z} = J(t)z, W(z(t_0 + T_0)) \leq c^2 W(z(t_0))$ for all t_0, where $c < 1$. Equivalently, if $\Phi(t_0 + T_0, t_0)$ is the transition matrix of $\dot{z} = J(t)z, ||\Phi(t_0 + T_0, t_0)|| \leq c$ for all t_0. Hence for the forced system (34), $||z(t_0 + T_0)|| \leq c||z(t_0)|| + k_1\nu_0$ where k_1 is a constant. If $c_1 \triangleq \frac{k_1\nu_0}{1-c}$, then $||z(t_0)|| \geq c_1 \Rightarrow ||z(t_0 + T_0)|| < ||z(t_0)||$ or

$$W(z(t_0 + T_0)) < W(z(t_0)) \text{ if } W(z(t_0)) \geq c_1^2/2. \tag{35}$$

$$\text{Case (ii)} \quad \frac{1}{T_0}\left|\int_{t_2}^{t_2+\delta_0} (\omega^*(\tau) + Ce(\tau))^T \mathrm{w} d\tau\right| < k\epsilon_0 : \tag{36}$$

If condition (33) is not satisfied, then (36) is satisfied for some constant unit vector $\mathrm{w} \in \mathcal{R}^{2n}$, i.e. ω is not persistently exciting over the interval $[t_0, t_0 + T_0]$ in some direction w. As shown below, this once again assures that $||z(t_0 + T_0)|| < ||z(t_0)||$ for any initial condition.

From the inequality (31) and the choice of k, it follows that $(1 - k)\epsilon_0 \geq E_0$. Since ω^* satisfies inequality (5), we have

$$\left|\int_{t_2}^{t_2+\delta_0} [Ce(\tau)]^T \mathrm{w} d\tau\right| \geq (1 - k)\epsilon_0 T_0.$$

Since $\dot{W} = -e^T Qe + 2e_1\bar{\nu}$, the decrease in $W(z)$ over the interval $[t_2, t_2 + \delta_0]$ satisfies the inequality

$$\begin{aligned} W(z(t_2)) - W(z(t_2 + \delta_0)) &\geq \lambda_Q \textstyle\int_{t_2}^{t_2+\delta_0} ||e(\tau)||^2 d\tau - 2||h||\nu_0 \int_{t_2}^{t_2+\delta_0} ||e(\tau)|| d\tau \\ &\geq \frac{\lambda_Q}{\delta_0} \textstyle\int_{t_2}^{t_2+\delta_0} ||e(\tau)|| d\tau \{\int_{t_2}^{t_2+\delta_0} ||e(\tau)|| d\tau - \frac{2||h||\nu_0\delta_0}{\lambda_Q}\} \\ &\geq E_0^2(T_0 - \delta_0)\lambda_Q \qquad \text{by the choice of } k. \end{aligned}$$

Since the minimum decrease in $W(z(t))$ over the interval $[t_2, t_2 + \delta_0]$ is greater than the maximum increase $\frac{E_0^2}{4}(T_0 - \delta_0)\lambda_Q$ over a period $(T_0 - \delta_0), W(z(t_0 + T_0)) < W(z(t_0))$ for any $W(z(t_0))$. Since cases (i) and (ii) represent the only two possibilities over any period $[t_0, t_0 + T_0]$, we obtain that

$$W(z(t_0)) \geq c_1^2/2 \Rightarrow W(z(t_0 + T_0)) < W(z(t_0))$$

which implies that all solutions of (30) are uniformly bounded. □

Comments

1. If condition (33) holds for any arbitrary interval $[t_0, t_0 + T_0], \omega$ is persistently exciting and the boundedness of $||z||$ follows trivially. The conditions given in the literature dealing with local stability are generally expressed in terms of the PE of ω.
2. When ω does not satisfy (33), it follows from the analysis of case (ii) that the error e is persistently exciting in a subspace over the interval $[t_0, t_0 + T_0]$ so that $W(e, \phi)$ decreases. It is worth noting that e is not required to be persistently exciting over the entire space in the interval $[t_0, t_0 + T_0]$.
3. Sufficient conditions for boundedness of solutions are stated in the above theorem in terms of the degree of persistent excitation of $\omega^*(t)$. For design purposes, it is more desirable to express them in terms of the reference input $r(t)$. From the results of the previous section, it follows that by choosing a reference input $r(t)$ with n distinct frequencies, ω^* can be made persistently exciting. The results of the above theorem imply that a reference input $r(t)$ must be chosen so that ω^* is persistently exciting with a degree of PE large compared to the amplitude of the disturbance. Very little is currently known regarding the relationship between the degree of persistent excitation of inputs and outputs of linear systems. Such relationships have to be better understood before the above theorem can be directly used in design.

4.3 Hybrid Adaptive Control - Improved Robustness

Recently Narendra, Khalifa and Annaswamy [43] presented several new adaptive algorithms which can be applied to continuous or discrete systems. The common feature of such algorithms is that control parameters are adjusted at rates significantly slower than the rates at which the system operates. In discrete systems this corresponds to infrequent adjustment of the control parameters based on the information collected over several instants of time. The resulting algorithms involve reduced computational effort and assure the u.a.s. of the ideal system when the input is persistently exciting. One such algorithm is briefly discussed below.

If the parameter error and the output error are denoted as $\phi(.)$ and $e_1(.)$ respectively, the error equations for a discrete adaptive control problem can be expressed as

$$\begin{aligned} \phi^T(k)\omega(l) &= e_1(l) \quad k,l \in N, l \in [kT,(k+1)T-1] \quad \text{(ideal case)} \\ \phi^T(k)\omega(l)+\nu(l) &= e_1(l) \quad \text{(with bounded perturbations)} \end{aligned}$$

Throughout the interval $[kT,(k+1)T-1]$, the error e_1 and the signal $\omega(.)$ are measured while the parameter vector θ (and hence ϕ) remains constant. At the end of the interval, using one of the adaptive laws suggested in [43], given by

$$\Delta\theta(k) = \Delta\phi(k) = -\frac{1}{T}\sum_{l=kT}^{(k+1)T-1} \frac{e_1(l)\omega(l)}{1+\omega^T(l)\omega(l)}$$

the parameter vector $\theta(k)$ is updated. The resulting equation has the form

$$\phi(k+1) = [I-R(k)]\phi(k) + S(k)$$

where

$$R(k) = \frac{1}{T}\sum_{l=kT}^{(k+1)T-1} \frac{\omega(l)\omega^T(l)}{1+\omega^T(l)\omega(l)}, \quad \text{and} \quad S(k) = \frac{1}{T}\sum_{l=kT}^{(k+1)T-1} \frac{\omega(l)\nu(l)}{1+\omega^T(l)\omega(l)}.$$

Qualitatively, the above control can be said to improve the degree of persistent excitation while decreasing the effective magnitude of the disturbance. Hence, by the results of the previous section, the robustness of the overall system is significantly improved. While simulation results have confirmed this, no theoretical proof currently exists for the general case when the relative degree n^* of the plant is greater than unity. In the authors' opinion hybrid adaptive control is preferable in all situations over instantaneous adjustment of the control parameters.

4.4 Comments on Adaptive Algorithms for Bounded Disturbances

1. All the algorithms discussed in section 4 with the exception of those in section 4.2.1 assure global boundedness of the signals in the adaptive loop when the disturbances are bounded. In section 4.1 this is accomplished by modifying the adaptive laws while the result in section 4.2.2 assures the same by using a sufficiently large degree of PE of the reference input.
2. The schemes (a) and (c) in section 4.1 are not u.a.s. when no external disturbances are present. Scheme (b) assures u.a.s. if the reference input is persistently exciting and scheme (d) assure u.a.s. if the reference input has a sufficiently large degree of PE. Further prior information regarding bounds on the external disturbance and the parameter vector θ^* is needed in schemes (a) and (b) respectively. In contrast to this, schemes (c) and (d) do not need such information. The authors feel that the state-dependent term $-|e_1|\theta$ in the new adaptive law may have wider applications in adaptive control.

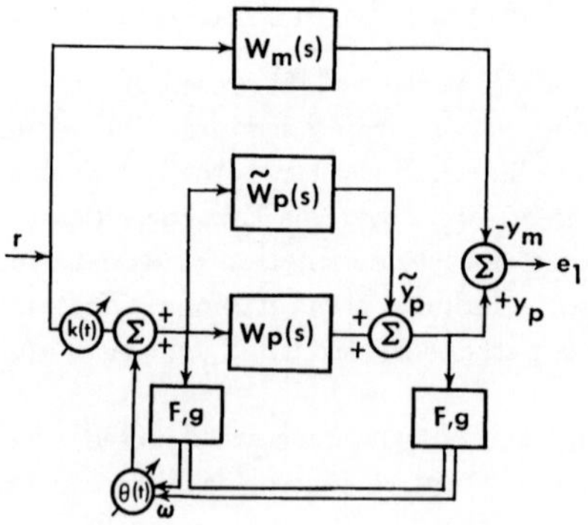

Figure 4: Robust Adaptive Control in the Presence of State-Dependent Disturbances.

3. The scheme in [15] using a dead-zone is a conservative approach in which adaptation is stopped when the error becomes small. This assures the boundedness of the parameter vector. If a time-varying bound can be specified for the disturbance, a time-varying dead-zone may be incorporated in the adaptive law.
4. The sufficient condition for the boundedness of the solutions in the presence of bounded disturbances given in section 4.2 is expressed in terms of the degree of PE of ω^*. This in turn is determined by the reference input $r(.)$ as well as the frequency responses of the reference model and the state-variable filters F in the adaptive loop. Hence the result indicates that all the above are important factors in the design of efficient adaptive controllers.
5. Given the initial conditions, the local stability result in [42] assures the existence of a sufficiently small perturbation so that all solutions are bounded. Hence this can be considered as an example of total stability. In contrast to this, the result in section 4.2 is more along the lines of practical stability. Given the bounds on the disturbance, it is shown that for a sufficiently large degree of PE, the system has bounded solutions for arbitrary initial conditions.

5. Robust Adaptive Control In The Presence of State - Dependent Disturbances

In contrast to the problems considered in section 4, those discussed in this section are concerned with disturbances that cannot be a priori assumed to be bounded. Such a situation occurs in cases in which the plant is only modeled partially. The problem of adaptively controlling such a plant using a reduced order controller is perhaps the most important one in the field at the present time, a fact which is reflected by much of the published literature.

Briefly stated, the problem is to control a given plant of order $\bar{n}$ adaptively, using a controller of dimension $2n < 2\bar{n}$. In such a case the problem can be posed as one in which $(\bar{n} - n)$ states of the plant are unmodeled so that the output resulting from them would constitute a state-dependent disturbance. While different formulations of the problem have been suggested in the literature, it is our opinion that the theoretical questions that arise and the nature of prior information needed in all of them are quite similar. We present below, a formulation of the problem which is best suited for the discussions that follow:

The unknown plant $\overline{P}$ consists of two linear time-invariant parts P and $\tilde{P}$ with transfer functions $W_p(s)$ and $\widetilde{W}_p(s)$ connected in parallel so that the overall transfer function of the plant is $\overline{W}_p(s) = W_p(s) + \widetilde{W}_p(s)$ [Fig. 4]. While both of them are unknown, the prior information assumed regarding them is different. P is referred to as the modeled part of the plant and $\tilde{P}$ the unmodeled part. The entire design of the controller is carried out assuming that $W_p(s)$ represents the true transfer function of the plant. If P is of dimension n, the controller of dimension $2n$ is sufficient to stabilize it as shown in section 3, provided $W_p(s)$ satisfies the assumptions (I) made in the ideal case. Based on the prior information (and in particular the relative degree n^* of P) a reference model with an asymptotically stable transfer function $W_m(s)$ is chosen. In the absence of $\tilde{P}$ it is known that for some constant value θ^* of $\theta(t)$, the plant together with the controller will match the transfer function $W_m(s)$ of the reference model exactly. To make the problem analytically tractable in the presence of $\tilde{P}$, we further make the important assumption where necessary that, when $\theta(t) \equiv \theta^*$ the overall system has a transfer function $\overline{W}_m(s)$ which is asymptotically stable. This assumption assures the existence of a constant control parameter vector which results in the boundedness of all the signals in the system. Given the above formulation of the problem, the aim is to determine stable adaptive laws for adjusting $\theta(t)$ so that the error between plant and model output satisfies some performance criterion, while all the other signals remain bounded.

The analysis of the above problem can be carried out either in terms of the transfer function $\overline{W}_m(s)$ or $W_m(s)$. While $\overline{W}_m(s)$ can only be assumed to exist, $W_m(s)$ can be chosen a priori by the designer and hence the two approaches lead to different error models and consequent theoretical questions. In the following we discuss the problem in two stages. In the first stage $\overline{W}_m(s)$ is assumed to be SPR while in the second stage $\overline{W}_m(s)$ is not assumed to satisfy this condition. The latter problem in turn can be conveniently divided into two parts. In the first part several local results which have recently appeared in the literature are briefly presented. They provide sufficient conditions under which the transfer function $\overline{W}_m(s)$ behaves almost like an SPR transfer function. In the second part the more interesting problem, of global boundedness of the signals in the system, is discussed. Here the same methods, which were suggested in section 4 for the case of bounded disturbances, are used to derive conditions on the output $\tilde{y}_p$ of $\widetilde{W}_p(s)$ to assure global boundedness of all the signals in the system. In every case it is found that these requirements on $\tilde{y}_p$ are the same, indicating that the results obtained are not fortuitous. How these relate to recent results proposed by other authors [44,45,46,39] are discussed towards the end of the section. The implications of these developments for future research on the reduced order model problem are included in section 6.

5.1 Case (i) $\overline{W}_m(s)$ Strictly Positive Real

While in our opinion the assumption that $\overline{W}_m(s)$ is SPR is quite unrealistic, it represents a special case in which some general results can be obtained. It is also helpful in providing a frame of reference for assessing solutions obtained using more realistic assumptions.

If $\theta(t) = \theta^* + \phi(t)$, the output of the plant can be described by the equation [1]

$$y_p(t) = \overline{W}_m(s)[r(t) + \phi^T(t)\omega(t)].$$

Since the output of the reference model $y_m(t)$ is given by $y_m(t) = W_m(s)r(t)$, it follows that the output error $e_1(t) \triangleq y_p(t) - y_m(t)$ is described by

$$e_1(t) = \overline{W}_m(s)\phi^T(t)\omega(t) + \nu(t) \tag{37}$$

where ν is the bounded signal $\nu(t) \triangleq [\overline{W}_m(s) - W_m(s)]r(t)$ with $|\nu(t)| \leq \nu_0$. If $\overline{W}_m(s)$ is strictly positive real, it follows that any one of the approaches suggested in section 4 can be used, provided the corresponding assumptions are made. For exmple, if ν_0, the upper bound on $\nu(t)$ is known, a dead-zone can be used in the adaptive law. The modification suggested by Kreisselmeier and Narendra [16] can be resorted to if ν_0 is not known, but a bound on $\|\theta^*\|$ is assumed. This was suggested by Kosut and Friedlander in [47]. The authors however feel that the new adaptive law given in section 4.1 is more attractive in this case [38], since no additional information regarding $\overline{W}_m(s)$ is needed. In such a case the adaptive law takes the form

$$\dot{\phi}(t) = -e_1(t)\omega(t) - \gamma|e_1(t)|\theta(t).$$

The reader is referred to [38,46] for details.

An alternate condition based on persistent excitation is also included in [47] where it is shown that if $\omega(t)$ is persistently exciting, the signals in the system will be bounded. As mentioned in section 4.2 this suffers from the weakness that $\omega(t)$ is the output of the nonlinear system under consideration. In contrast to the above, the results due to the authors in [36] discussed in section 4.2 based on persistent excitation (Theorem 5) provide better insights regarding the stability of the adaptive system and is briefly stated as follows:

> Let $\omega(t) \triangleq \omega^*(t)$ when $\theta(t) \equiv \theta^*$ in the controller, with a degree of PE ϵ_0. If $\bar{\nu}(t) \triangleq [1 - \overline{W}_m^{-1}(s)W_m(s)]r(t)$ with $|\bar{\nu}(t)| \leq \bar{\nu}_0$, then all the signals in the adaptive system are bounded if
>
> $$\epsilon_0 \geq \gamma\bar{\nu}_0 + \delta$$
>
> where γ is a constant that depends on the system parameters as defined in (31) and δ is an arbitrary positive constant.

It is worth noting that since both ϵ_0 and $\bar{\nu}_0$ are proportional to the amplitude of the reference input, the condition has to be satisfied based on the frequency response of the transfer functions $\overline{W}_m(s)$ and $W_m(s)$ and the spectrum of $r(\cdot)$. Hence the choice of the reference input, the transfer function of the reference model, and the state variable filters of the controller, is critical in assuring the boundedness of all the signals in the overall system. The condition however is independent of the degree of PE of the reference input. The result is particularly appealing, since it indicates that the adaptive system will be well behaved if the frequency response $\overline{W}_m(j\omega)$ is large compared to $[\overline{W}_m(j\omega) - W_m(j\omega)]$ in the frequency range of interest.

5.2 Case (ii) $\overline{W}_m(s)$ Is Not Strictly Positive Real

The simplification that results from the assumption that $\overline{W}_m(s)$ is SPR is immediately evident while determining the adaptive law from the equation (37). When $\overline{W}_m(s)$ is not SPR, an augmented error has to be generated (ref. section 3) before a stable adaptive law can be obtained. However, this is not possible in the present case since $\overline{W}_m(s)$ is unknown. Hence several attempts were made to determine conditions under which $\overline{W}_m(s)$ could deviate from an SPR transfer function without making the adaptive system unstable. While several papers have appeared in this area, all of them contain stability results which are local in nature. We discuss below a few representative papers indicating the approach used and the results obtained.

1. Local Stability: The starting point for most of the above approaches is the error equation

$$e_1(t) = \overline{W}_m(s)\phi^T(t)\omega(t) \tag{38a}$$

where $\overline{W}_m(s)$ is not strictly positive real. When $\overline{W}_m(s)$ is SPR, the standard adaptive law

$$\dot{\phi}(t) = -\epsilon e_1(t)\omega(t) \qquad \epsilon > 0 \tag{38b}$$

assures exponential stability if the reference input is persistently exciting. Hence the problem posed here is to find the conditions under which (38) is exponentially stable when $\overline{W}_m(s)$ is not SPR.

Method of Averaging: As mentioned in section 2, the stability of a nonautonomous differential equation can under certain conditions be analyzed by considering an autonomous system obtained by the method of averaging. In the adaptive control problem, since the parameters generally vary slowly compared to the states of the system, such a method can be applied. The method of averaging was introduced by Åström [48] in an effort to analyze instability resulting from unmodeled dynamics.

In [49] Kokotovic et al. consider the equations (38a,b) and using a small parameter in the adaptive law and the averaging method, proved the following theorem.

Theorem 6: Let ω be a bounded, almost periodic signal, which is persistently exciting. Then there exists an ϵ^* such that for all $\epsilon \in (0, \epsilon^*]$, the system (38) is exponentially stable if

$$\sum_{k=-\infty}^{\infty} Re[\overline{W}_m(j\nu_k)]\ Re[\Omega(j\nu_k)\overline{\Omega}^T(j\nu_k)] > 0 \tag{39}$$

where $\omega(t) = \sum_{k=-\infty}^{\infty} \Omega(j\nu_k) exp(j\nu_k t)$ and $\overline{\Omega}(j\nu_k)$ is the complex-conjugate of $\Omega(j\nu_k)$.

Qualitatively the above theorem states that the solution of the adaptive control problem (38a,b) is critically dependent on the spectrum of the excitation in relation to the frequency response $\overline{W}_m(j\nu)$. In particular, if ϵ is sufficiently small and ϕ varies slowly and ω is pesistently exciting with dominant frequencies for which $Re[\overline{W}_m(j\nu)] > 0$, the solutions of (38a,b) will be exponentially stable. Results of a similar nature have also been obtained by Fu et al. [22], Kosut et al.[50] and Anderson et al. [51]. In [22] the results are extended to stationary reference inputs and applied to the ideal case. In [50], Kosut et al. extend the results to signals in $\mathcal{P}_{[0,\infty)}$. In [51], similar results are derived using the small gain theorem [25].

<u>**Comments:**</u> (i) It should be noted that in [49,50] and [51], it is assumed that ω is bounded and the stability conditions are then expressed in terms of the freequency spectrum of ω. As pointed out in [49] this restricts the validity of the stability results to a neighborhood of the tuned solutions. While the above results provide valuable insights regarding the stability and instability mechanisms in the adaptive control problem, they are nonetheless expressed in terms of signals which are outputs of a nonlinear system. As pointed out earlier, establishing the boundedness of the parameter vector $\theta(t)$ and deducing from it the boundedness of all the signals in the adaptive system are still by far the most difficult questions encountered in adaptive control problems. It is worth comparing these results to those given in section 5.1 when $r(.)$ is persistently exciting.

(ii) It is stated in [47] that "... (The stability result in [47]) is achieved by requiring $\overline{W}_m(s) \in$ SPR, a condition which is difficult to maintain in normal circumstances ... It would appear then, that a more valid approach to providing a system theoretic setting for adaptive control is to develop local stability conditions, which hopefully do not require that $\overline{W}_m(s) \in$ SPR." The authors however do not subscribe to this viewpoint and believe that global stability results can be achieved in most problems of interest without invoking an SPR type condition.

<u>**2. Global Stability:**</u> In section 4, global boundedness of all the signals in the system in the presence of bounded disturbances was demonstrated using two distinct approaches, where the first involved modification of the standard adaptive law while the second was based on the arguments involving the PE of the reference input. Based on the results obtained earlier in this section, the authors are of the opinion that the same two approaches can also be used for studying the robustness with respect to state-dependent disturbances. While work is in progress using both approaches, preliminary results are currently available using only the first approach. We first present the extension of the methods described in section 4.1 to the problem at hand.

The error model considered in case (i) was expressed in terms of $\overline{W}_m(s)$, which is a stable closed loop transfer function of the plant including the unmodeled part. In contrast to this, we now express the error model in terms of $W_m(s)$ and the output $\tilde{y}_p$ of the unmodeled part as a disturbance, where $\tilde{y}_p \triangleq \widetilde{W}_p(s)u$. With the controller structure as in (10), the output of the plant can be expressed as

$$y_p(t) = W_m(s)[r(t) + \phi^T(t)\omega(t)] + \eta(t)$$

where $\eta(t)$ is the effect of the disturbance $\tilde{y}_p(t)$ at the output of the plant when $\theta(t) \equiv \theta^*$ (ref. derivation in [15].). An augmented error is generated (eq. (12)) and therefore the error equation is given by

$$\phi^T \varsigma + \eta = \epsilon_1. \tag{40}$$

In contrast to ν in equation (16), the disturbance η in equation (40), which is due to unmodeled dynamics, cannot be assumed to be bounded a priori. However, we apply the adaptive laws (a)-(d) mentioned in section 4 to this problem and derive conditions on η under which the signals in the adaptive system can be shown to be bounded.

(a) Use of a Dead-Zone: Since the disturbance η cannot be assumed to be bounded, it is not possible to determine a fixed dead-zone in this problem as in section 4. Assuming that an instantaneous upper bound $\eta_0(t)$ on $\eta(t)$ is available, an adaptive scheme is proposed as follows where adaptation ceases when the error ϵ_1 becomes smaller than $\eta_0(t)$:

$$\begin{aligned} \dot{\phi} &= -\tfrac{\epsilon_1 \varsigma}{1+\xi^T \xi} & & |\epsilon_1(t)| \geq (1+\delta)\eta_0(t) \\ &= 0 & & |\epsilon_1(t)| < (1+\delta)\eta_0(t) \end{aligned} \tag{41}$$

where $|\eta(t)| \leq \eta_0(t) \quad \forall t \geq t_0$ and δ is an arbitrary positive constant. As in the bounded disturbance case, the adaptive law (41) assures the boundedness of $\|\phi\|$ and hence the parameter vector $\theta(t)$. Further $\dot{\phi} \in \mathcal{L}^2 \cap \mathcal{L}^\infty$. Hence the state of the overall system can grow at most exponentially. However, as in the bounded disturbance case, the boundedness of all the signals in the system cannot be concluded directly since $\epsilon_1(t)$ may grow in an unbounded fashion without becoming larger than $(1+\delta)\eta_0(t)$. Hence additional conditions have to be imposed on the growth rate of $\eta_0(t)$ to assure the boundedness of all the signals. This is expressed by

$$\eta_0(t) \leq \mu[1 + \xi^T(t)\xi(t)]^{1/2} \qquad \text{where } \mu > 0 \tag{42}$$

where ξ is the state of the system as defined in section 4.1. The proof of boundedness can be given by considering the following three cases, assuming that the signals grow in an unbounded fashion.

(i) If the trajectory lies inside the dead-zone for ever, the feedback signal to the plant $|\phi^T \omega + \bar{\eta}| \leq \mu[1 + \xi^T \xi]^{1/2}$ where $W_m(s)\bar{\eta}(t) \stackrel{\Delta}{=} \eta(t)$. Hence if μ is sufficiently small, this leads to a contradiction if we assume that $\lim_{t\to\infty} \sup_{\tau \leq t}[\|\xi(\tau)\|] = \infty$.

(ii) If the trajectory lies outside the dead-zone, contradiction follows as in the bounded disturbance case, since $\epsilon_1 = \beta[1 + \xi^T \xi]^{1/2}$ where $\beta \in \mathcal{L}^2$.

(iii) If the trajectory alternately lies in and out of the dead-zone, the proof of boundedness follows along similar lines as in [16] if μ is sufficiently small, by showing that the measure of $\{t | \frac{|\phi^T \omega|}{\|z\|} > \epsilon\}$ is much smaller than the measure of $\{t | \frac{|\phi^T \omega|}{\|z\|} \leq \epsilon\}$.

Hence in all three cases global boundedness can be assured by choosing μ to be sufficiently small and using the same arguments as those used in sections 4.1-(b) and 4.1-(d).

(b) Bound on $\|\theta\|^*$: The above arguments can also be used if the adaptive law (19) is applied directly to the present problem assuming that a bound on $\|\theta^*\|$ is known a priori as shown below. Let

$$|\eta(t)| \leq \mu[1 + \xi^T(t)\xi(t)]^{1/2}, \qquad \mu > 0.$$

The proof can be briefly given as follows assuming that $\lim_{t\to\infty} \sup_{\tau \leq t}[\|\xi(\tau)\|] = \infty$.

(i) Boundedness of $\|\phi\|$ follows immediately from the adaptive law.

(ii) Since $\int_0^\infty \dot{V} dt < \infty$, we have

$$\int_{t_1}^{t_2} \frac{\epsilon_1^2}{1+\xi^T \xi} dt \leq \Delta_1 + \mu \Delta_2 [t_2 - t_1]; \qquad \frac{d}{dt}\left[\frac{\epsilon_1^2}{1+\xi^T \xi}\right] \leq \text{ a constant if } |\dot{\eta}| \leq k_1|\eta| + k_2.$$

(iii) For a sufficiently small μ, the measure of $\{t| \frac{|\phi^T\omega|}{||z||} > \epsilon\}$ can be shown to be much smaller than the measure of $\{t| \frac{|\phi^T\omega|}{||z||} \leq \epsilon\}$.

Hence a contradiction follows.

(c) Other Adaptive Laws: Following along similar lines as above, it can be shown that the two other adaptive laws given section 4.1 can be modified to apply in the present context as follows:

$$\dot{\phi} = -\frac{\epsilon_1 \zeta}{1+\xi^T\xi} - \frac{\sigma\eta_0^2(t)}{1+\xi^T\xi}\theta \tag{43}$$

where $\sigma > 0$, and

$$\dot{\phi} = -\frac{\epsilon_1 \zeta}{1+\xi^T\xi} - \frac{|\epsilon_1|\eta_0(t)}{1+\xi^T\xi}\theta. \tag{44}$$

Both adaptive laws lead to global boundedness of all signals, provided

$$|\eta(t)| \leq \eta_0(t) \leq \mu[1+\xi^T(t)\xi(t)]^{1/2}$$

for a sufficiently small μ. The reader is referred to [46] for details.

Comments: 1. We note that in all the above proofs, it is assumed that $|\eta(t)| \leq \mu[1+\xi^T(t)\xi(t)]^{1/2}$ and the boundedness of all the signals in the system is derived assuming that the parameter μ is sufficiently small. This assumption results in the proofs in all four cases parallelling those in section 4.1 for the bounded disturbance case. While the choice of μ assures the output of the unmodeled part to be small relative to the magnitude of the state of the system for all inputs, it is the opinion of the authors that less restrictive and practically more interesting results can be derived based on the frequency responses of the modeled and unmodeled parts of the plant and the spectral content of the reference input, as shown in section 5.1.

2. In [52] a normalization factor $m(t)$ is introduced, where $|\eta(t)| \leq \mu|m(t)|$ and $m(t)$ is generated using signals in the system. The state of the system is then augmented to be $\overline{\xi} = [\xi^T,\ m]^T$ so that condition (42) is satisfied by replacing ξ by $\overline{\xi}$ in the adaptive law. In [44], Praly shows that such a normalization together with a projection of the control parameter into a convex compact set, leads to robustness of the plant with respect to a graph topology. In [45] Kreisselmeier and Anderson derive a similar result by using a combination of the dead-zone scheme in [15] and a bound on $||\theta^*||$ as in [16]. It is not clear to the authors as to why such a combination was used in the adaptive law since it follows from the above discussions that either the dead-zone or the projection method is adequate to solve this problem, provided condition (42) is satisfied. A recent report by Ioannou and Tsakalis consider the same problem in [39] by using a combination of the $\sigma-$modification scheme and a bound on $||\theta^*||$.

6. Conclusions

Recent developments in robust adaptive control theory have been treated briefly in the preceding sections. While all of them provide sufficient conditions under which the signals in the system will remain bounded, our primary interest has been in those results which are global in character. As a consequence, greater emphasis has been given to such results in the paper. In contrast to the local results discussed in section 5, which impose conditions on the internal signals of the nonlinear system, the global results given in the paper are expressed in terms of parameters and signals of the system which can be chosen at the discretion of the designer.

Global boundedness of all the signals is assured in cases where the disturbance is bounded by either modifying the adaptive law or by requiring that the reference input have a sufficiently large degree of PE. Comments regarding the relative advantages of such schemes were given at the end of section 4. As shown in section 5, these methods also assure boundedness of signals when state dependent disturbances are present, if either the overall system transfer function satisfies a positive realness condition, or if the output due to the unmodeled plant is sufficiently small compared to the state of the system. However, the above results represent only a preliminary stage in the resolution of the general reduced order model problem.

That considerable prior information is necessary regarding the plant to be controlled to achieve stable adaptive control is necessary has been generally recognized for a long time. This is confirmed by the results in section 5 which indicate that when state-dependent disturbances are present, information concerning the spectral content of the reference input, and the frequency characteristics of the unmodeled part, the state variable filters and the reference model, are all important to assure the boundedness of all the signals.

In section 4, it was argued that hybrid adaptation in which parameters are adjusted intermittently, outperforms continuous adjustment when external disturbances are present. Recent work by Elliott et al. [53] and Goodwin et al. [54] which use block processing of data, can also be considered to be hybrid algorithms. Since they have demonstrated that by the proper choice of the interval over which the parameters remain constant, nonminimum phase plants can be controlled adaptively, the method appears to be particularly attractive for practical implementation. The averaging method discussed in section 5, which can also be considered as an approximate version of hybrid adaptation, was shown to assure local stability in the presence of unmodeled dynamics when the frequency content of $\omega(t)$ was such that $\overline{W}_m(s)$ was a strictly positive real transfer function. All these results indicate that intermittent adaptation is a desirable feature to assure robustness in adaptive systems.

It is well known that the adaptive control problem in the ideal case can be divided into an algebraic part and an analytic part. The former is concerned with the existence of a time-invariant controller which can achieve the desired objective, while the latter deals with the asymptotic convergence of the time-varying controller, using the adaptive law, to a time-invariant controller. A similar division can also be made in all the robust adaptive control problems described in this paper, though the limiting behavior of the controller may not always be time-invariant. Recent work on relaxing the assumptions on the plant transfer function [55]-[59] has shown that none of the conditions in (I) may be necesssary to control the plant adaptively. This provides the theorist considerable flexibility in posing the adaptive control problem. This is particularlly true of the reduced order model problem where the existence of a controller which stabilizes the modeled part P and the entire plant $\overline{P}$ simultaneously, may be all that is required. This in turn has brought the algebraic part of adaptive control theory closer in spirit to the problems currently being discussed in the area of robust linear control. The discussions in this paper also reveal that there is a wide choice of approaches while addressing the analytic part of the problem. It is the confluence of these different ideas that leads the authors to believe that some significant developments can be anticipated in the near future in the resolution of the reduced order model problem.

Acknowledgments

The work reported here was supported by the National Science Foundation under Grant ECS-8300223. The second author would like to thank the IBM Corporation for a post-doctoral fellowship.

References

[1] K.S. Narendra, Y.H. Lin and L.S. Valavani, "Stable Adaptive Controller Design - Part II:Proof of Stability," *IEEE Trans. on Autom. Control*, AC-25, pp. 440-448, June 1980.

[2] A.S. Morse, "Global Stability of Parameter Adaptive Control Systems," *IEEE Trans. on Autom. Control*, AC-25, pp. 433-439, June 1980.

[3] K.S. Narendra and Y.H. Lin, "Stable Discrete Adaptive Control," *IEEE Trans. on Autom. Control*, AC-25, pp. 456-461, June 1980.

[4] G.C. Goodwin, P.J. Ramadge and P.E. Caines, "Discrete time Multivariable Adaptive Control," *IEEE Transactions on Automatic Control*, vol. 25, pp. 449-456, June 1980.

[5] I.G. Malkin, "Theory of Stability of Motion," Tech. Report Tr. 3352, U.S. Atomic Energy Commission, English ed. 1958.

[6] A.N. Krylov and N.N. Bogoliuboff, Introduction to Nonlinear Mechanics, (English Translation), Princeton University Press, Princeton, New Jersey, 1943.

[7] K.S. Narendra, A.M. Annaswamy and R.P. Singh, "A General Approach to the Stability Analysis of Adaptive Systems," *International Journal of Control*, vol. 41, No. 1, 1985.

[8] K.S. Narendra and A.M. Annaswamy, "Persistent Excitation and Stability Properties of Adaptive Systems," *Proceedings of the Fourth Yale Workshop on Applications of Adaptive Systems Theory*, New Haven, May 1985.

[9] R.E. Kalman and J.E. Bertran, "Control Systems Analysis and Design via the 'Second Method' of Lyapunov," *Journal of Basic Engineering*, pp. 371-392, June 1960.

[10] J.P. LaSalle, "Some Extensions of Liapunov's Second Method," *IRE Transactions on Circuit Theory*, pp. 520-527, Dec. 1960.

[11] J.P. LaSalle, "Asymptotic Stability Criteria," *Proceedings of Symposium on Appl. Math.*, Hydrodynamic Instability, Providence, R.I., vol. 13, pp. 299-307, 1962.

[12] A.P. Morgan and K.S. Narendra, "On the Uniform Asymptotic Stability of Certain Linear Nonautonomous Differential Equations" *SIAM Journal of Control and Optimization*, Vol. 15, pp. 5-24, Jan. 1977.

[13] A.P. Morgan and K.S. Narendra, "On the Stability of Nonautonomous Differential Equations $\dot{x} = [A + B(t)]x$ with Skew-Symmetric Matrix $B(t)$" *SIAM Journal of Control and Optimization*, Vol. 15, pp. 163-176, Jan. 1977.

[14] J.S.-C Yuan and W.M. Wonham, "Probing Signals for Model Reference Identification," *IEEE Transactions on Automatic Control*, AC-22, pp. 530-538, August 1977.

[15] B.B. Peterson and K.S. Narendra, "Bounded Error Adaptive Control," *IEEE Trans. on Autom. Control*, AC-27, pp. 1161-1168, Dec. 1982.

[16] G. Kreisselmeier and K.S. Narendra, "Stable Model Reference Adaptive Control in the Presence of Bounded Disturbances," *IEEE Trans. on Autom. Control*, AC-27, pp. 1169-1175, Dec. 1982.

[17] J.P. LaSalle and S. Lefschetz, Stability by Liapunov's Direct Method, Academic Press, New York, 1961.

[18] W. Hahn, Theory and Application of Liapunov's Direct Method, Prentice-Hall International Series in Applied Mathematics, Englewood Cliffs, New Jersey, 1963.

[19] N.N. Bogoliuboff and Yu. A. Mitropolskii, Asymptotic Methods in the Theory of Nonlinear Oscillators, Gordon & Breach, New York, 1961.

[20] J.K. Hale, Ordinary Differential Equations, Kreiger, Molaban, Florida, 1980.

[21] V.I. Arnold, Geometric Methods in the Theory of Differential Equations, Springer-Verlag, New York, 1982.

[22] L.-C. Fu, M. Bodson and S.S. Sastry, "New Stability Theorems for Averaging and Their Applications to the Convergence Analysis of Adaptive Identification and Control Schemes," Tech. Report UCB/ERL M85/21, UC at Berkeley, March 1985.

[23] D. Hill and P. Moylan, "Connection Between Finite-Gain and Asymptotic Stability," *IEEE Transactions on Automatic Control*, vol. AC-25, pp. 931-936, Oct. 1980.

[24] C.A. Desoer, R. Liu and L.V. Auth, Jr., "Linearity and Nonlinearity and Asymptotic Stability in the Large," *IEEE Transactions on Circuit Theory*, pp. 117-118, March 1965.

[25] C.A. Desoer and M. Vidyasagar, Feedback Systems: Input-Output Properties, Academic Press, 1985.

[26] R.L. Kosut and C.R. Johnson, Jr., "An Input-Output View of Robustness in Adaptive Control," Special Issue on Adaptive Control, *Automatica*, September 1984.

[27] R.L. Kosut and B.D.O. Anderson, "Robust Adaptive Control: Conditions for Local Stability," *Proceedings of the 23rd IEEE CDC*, Las Vegas, 1984.

[28] K.S. Narendra and L.S. Valavani, "Stable Adaptive Controller Design-Direct Control," *IEEE Trans. on Autom. Control*, AC-23, pp.570-583, Aug. 1978.

[29] K.S. Narendra and P. Kudva, "Stable Adaptive Schemes for System Identification and Control - Parts I&II" *IEEE Trans. on Systems, Man and Cybernetics*, SMC-4, pp. 541-560, Nov. 1974.

[30] K.S. Narendra and Y.H. Lin, "Design of Stable Model Reference Adaptive Controllers," Applications of Adaptive Control, pp. 69-130, Academic Press, New York, 1980.

[31] S. Dasgupta, B.D.O. Anderson and A.C. Tsoi, "Input Conditions for Continuous Time Adaptive Systems Problems," Tech. Report, Australian National University, 1983.

[32] K.S. Narendra and A.M. Annaswamy, "Persistent Excitation and Robust Adaptive Algorithms," *Proceedings of the Third Yale Workshop on Applications of Adaptive Systems Theory*, Yale University, June 1983.

[33] S. Boyd and S.S. Sastry, "On Parameter Convergence in Adaptive Control," *Systems and Control Letters"*, vol. 3, pp. 311-319, 1983.

[34] B. Egardt, Stability of Adaptive Controllers, Springer- Verlag, Berlin, 1979.

[35] K.S. Narendra and B.B. Peterson, "Bounded Error Adaptive Control - Part I," Tech. Report No. 8005, Center for Systems Science, Yale University, New Haven, Dec. 1980.

[36] K.S. Narendra and A.M. Annaswamy, "Robust Adaptive Control in the Presence of Bounded Disturbances," Tech. Report 8406, Center for Systems Science, Yale University, New Haven, Ct., 1984. (Also presented at the 1984 ACC conference, San Diego, 1984.); To appear in the *IEEE Transactions on Automatic Control.*

[37] P.A. Ioannou and P.V. Kokotovic, "Adaptive Systems with Reduced Models," Springer-Verlag, New York, 1983.

[38] K.S. Narendra and A.M. Annaswamy, "A New Adaptive Law for Robust Adaptive Control Without Persistent Excitation," Tech. Report No. 8510, Center for Systems Science, Yale University, New Haven, September 1985.

[39] P.A. Ioannou and K. Tsakalis, "A Robust Direct Adaptive Controller," Tech. Report 85-07-1, Univ. of Southern California, May 1985.

[40] J.S.W. Wong, "Remarks on Global Asymptotic Stability of Certain Quasi-linear Differential Equations," *Proc. Amer. Math. Soc.* , vol. 17, No. 4, August 1966.

[41] M. Bodson and S.S. Sastry, "Small Signal I/O Stability of Nonlinear Control Systems - Applications to the Robustness of a MRAC Scheme," Tech. Report, UC at Berkeley, September 1985.

[42] B.D.O. Anderson and R.M. Johnstone, "Adaptive Systems and Time-varying Plants," *International Journal of Control*, vol. 37, No. 2, pp. 367-377, Feb. 1983.

[43] K.S. Narendra, I.H. Khalifa and A.M. Annaswamy, "Error Models for Stable Hybrid Adaptive Systems - Part II," *IEEE Transactions on Automatic Control*, AC-30, No. 4, pp. 339-347, April 1985.

[44] L. Praly, "Global Stability of a Direct Adaptive Control Scheme Robust with Respect to a Graph Topology," this volume, Plenum Press, 1985.

[45] G. Kreisselmeier and B.D.O. Anderson, "Robust Model Reference Adaptive Control," DFVLR Tech. Report, Oberpfaffenhofen, W. Germany, 1984.

[46] K.S. Narendra and A.M. Annaswamy, "Robust Adaptive Control Using a Reduced Order Model," Tech. Report, New Haven, Ct. (In preparation.).

[47] R.L. Kosut and B. Friedlander, "Robust Adaptive Control: Conditions for Global Stability," *IEEE Transactions on Automatic Control*, vol. AC-30, pp. 610-624, April 1985.

[48] K.J. Åström , "Analysis of Rohrs Counter-example to Adaptive Control," *Proceedings of the 22nd IEEE CDC*, San Antonio, Texas, pp. 982-987, Dec. 1983.

[49] P. Kokotovic, B. Riedle and L. Praly, "On a Stability Criterion for Continuous Slow Adaptation," *Systems & Control Letters*, vol. 6, pp. 7-14, June 1985.

[50] R.L. Kosut, B.D.O. Anderson and I. Mareels, "Stability Theory for Adaptive Systems: Methods of Averaging and Persistent Excitation," Preprint, 1985.

[51] B.D.O. Anderson, R.R. Bitmead, C.R. Johnson, Jr., and R.L. Kosut, "Stability Theorems for the Relaxation of the Strictly Positive Real Condition in Hyperstable Adaptive Schemes," *Proceedings of the 23rd IEEE CDC*, Las Vegas, 1984.

[52] L. Praly, "Robustness of Model Reference Adaptive Control," *Proceedings of the Third Yale Workshop on Applications of Adaptive Systems Theory*, Yale University, New Haven, Ct., 1983.

[53] H. Elliott, R. Cristi and M. Das, "Global Stability of Adaptive Pole Placement Algorithms," *IEEE Transactions on Automatic Control*, AC-30, pp. 348-356, April 1985.

[54] G.C. Goodwin and E.K. Teoh, "Persistency of Excitation in the Presence of Possibly Unbounded Signals," *IEEE Transactions on Automatic Control*, AC-30, pp. 595-597, June 1985.

[55] A.S. Morse, "Recent Problems in Parameter Adaptive Control," Proc. CNRS colloquium on Development and Utilization of Mathematical Models in Automatic Control, Belle-Isle, France, Sept. 1982.

[56] R.D. Nussbaum, "Some Remarks on a Conjecture in Parameter Adaptive Control," *Systems and Control Letters*, vol. 3, pp. 243-246, 1983.

[57] D.R. Mudgett and A.S. Morse, "Adaptive Stabilization of Linear Systems with Unknown High-Frequency Gains," *IEEE Transactions on Automatic Control*, AC-30, pp. 549-554, June 1985.

[58] T.H. Lee and K.S. Narendra, "Stable Discrete Adaptive Control with Unknown High Frequency Gain," Tech. Report No. 8408, Center for Systems Science, Yale University, New Haven, Ct., Nov. 1984.(Also submitted to the IEEE Trans. on Autom. Control).

[59] J. Willems and C. Byrnes, "Global Adaptive Stabilization in the Absence of Information on the Sign of the High Frequency Gain," *Proc. INRIA Conf. on Analysis and Optimization of Systems, Springer Lecture Notes in Control and Information Sciences*, vol. 62, pp. 49-57, June 1984.

Methods of Averaging for Adaptive Systems

Robert L. Kosut
Integrated Systems, Inc.
101 University Avenue
Palo Alto, Ca. 94301

Abstract

A summary of methods of averaging analysis is presented for continuous-time adaptive systems. The averaging results of Riedle and Kokotovic [1] and of Ljung [2] are examined and are shown to be closely related. Both approaches result in a sharp stability-instability boundary which can be tested in the frequency domain and interpreted as a signal dependent positivity condition.

1. Introduction

For a large class of adaptive systems, as well as for some output error identification schemes, a stability analysis in the neighborhood of the desired behavior leads to investigating the stability of the *linearized adaptive system* described by an equation of the form,

$$\dot{\theta} = -\epsilon z H(z'\theta) \tag{1}$$

where $\theta(t) \epsilon \mathcal{R}^p$ is the adaptation parameter vector, $z(t) \epsilon \mathcal{R}^p$ is the regressor, and $\epsilon > 0$ is the adaptation gain. The theory developed in [3,4] shows that the stability of adaptive systems in the neighborhood of the equilibrium trajectories is dependent on the stability of this system of linear time-varying equations. System (1) for example, can be obtained as a result of linearization of the adaptive system in the neighborhood of a "tuned" system, i.e., a system where the adaptive parameters are set to a constant value $\theta_* \epsilon \mathcal{R}^p$ and whose behavior is deemed acceptable. Hence, in (1), $\theta(t)$ is the vector of parameter errors between the parameter estimate at time t and the tuned value θ_*, $z(t)$ is the regressor vector from the tuned system (e.g., filtered revisions of measured signals), and the scalar ϵ is the magnitude of the adaptation gain which essentially controls the rate of adaptation. The operator H depends on the actual system being controlled or identified and also on the tuned parameter setting θ_*.

It is shown in [3] that if the zero solution of (1) is uniformly asymptotically stable (u.a.s.), then the adaptive system is locally stable, i.e., the adaptive system behavior will remain in a neighborhood of the desired behavior provided the initial parameter error $\theta(0)$ and the effect of external disturbances are sufficiently small. Although these results were arrived at using input-output properties, local stability properties can also be obtained from the results on "total" stability [5].

1.1 Unmodeled Dynamics and Slow Adaptation

In the ideal case there are a sufficient number of adaptive parameters (the number p) such that the tuned parameter setting results in $H(s)$ being strictly positive real (SPR), i.e., $H(s)$ proper and stable, and $Re\ H(j\omega) > 0, \forall \omega \epsilon \mathcal{R}_+$. Under these conditions, we have the following known results: (i) the zero solution of (1) is stable, i.e., $\theta(t)$ is bounded but not necessarily constant; (ii) if, in addition, $z(t)$ is persistently exciting, then the zero solution is u.a.s., thus, $\theta(t) \rightarrow 0$ exponentially fast as $t \rightarrow \infty$. The trouble starts when there are an insufficient number of parameters to obtain $H(s) \epsilon SPR$, as is the case in adaptive control when the plant has unmodeled dynamics.

In this paper we will examine the stability of (1) when ϵ is small, $z(t)$ is persistently exciting, and $H(s)$ is not necessarily SPR but only stable. We will refer to this case as *slow adaptation.*

1.2 Approaches Based on Averaging

In a recent paper by Riedle and Kokotovic [1], a classical method of averaging as described by Hale [6] was applied to the linearized adaptive system. The result is a sharp stability-instability boundary determined by a signal dependent positivity condition which asserts that the zero solution of (1) is u.a.s. if

$$\lambda(\sum_{\omega\epsilon\Omega}[\alpha(\omega)\alpha(\omega)^*]Re\ H(j\omega)) > 0 \tag{2}$$

where Ω and $\{\alpha(\omega), \omega\epsilon\Omega\}$ are, respectively, the Fourier exponents and coefficients of $u(t)$. Condition (2) can be considered as a ***signal dependent positivity condition***, but unlike the SPR condition $Re\ H(j\omega)$ is not required to be positive at *all* frequencies. Thus, this result is significantly weaker than the SPR condition required in the proof of stability of adaptive systems, e.g., [7,8]. In order to apply the averaging theory to obtain this result, the linearized system has first to be decoupled into slow (parameter) states and fast states. It is this transformation which is essential to the averaging approach and is a major contribution in the Riedle-Kokotovic method.

Averaging has also been applied to the counter-example of Rohrs et al. [9] by Åström [10,11]. In this analysis, by "freezing" the parameters, the parameter and state equations are decoupled thereby obtaining the asymptotic trajectories. Both of these averaging analyses assume that the system is periodic or almost periodic, an assumption that can be dispensed with by introducing the notion of a sample average [12].

In [13], the averaging approach is extended to nonlinear systems by introducing the *integral manifold* which completely separates the parameter and state equations. This latter approach is valid for the nonlinear adaptive system, and not just the linearized part. Related results can also be found in [14].

Averaging methods for adaptive systems have appeared in earlier work, the most notable of these being the averaging method developed by Ljung [2] for use in discrete-time recursive parameter estimation. The analysis shows that the convergence properties of the estimates can be determined from the stability properties of a related set of ordinary differential equations; the method usually referred to as the ODE analysis.

In this paper we summarize the results obtained by Riedle and Kokotovic [1,13] and show (heuristically) how they are related to the local stability analysis in [3,4] and the ODE averaging approach of Ljung in [2].

2. Adaptive Error System

Although it is unlikely that a truly generic adaptive error system can be formed to capture all the nuances of adaptive systems, the SISO adaptive system shown in Figure 1 is offered as a good representation for the purposes of analysis. The system equations are:

$$e = e_* - H_{ev}v \tag{3a}$$
$$z = z_* - H_{zv}v \tag{3b}$$
$$v = z'\theta \tag{3c}$$
$$\dot{\theta} = e\Gamma z \tag{3d}$$

The development of (3) can be found in [15,16] and in [17]. In (3), $e(t) \in \mathcal{R}$ is a measured error signal which drives the parameter update (3d), $z(t) \in \mathcal{R}^p$ is the regressor, and $\theta(t) \in \mathcal{R}^p$ is the parameter error between the current estimate at t and a *tuned* parameter setting $\theta_* \in \mathcal{R}^p$.

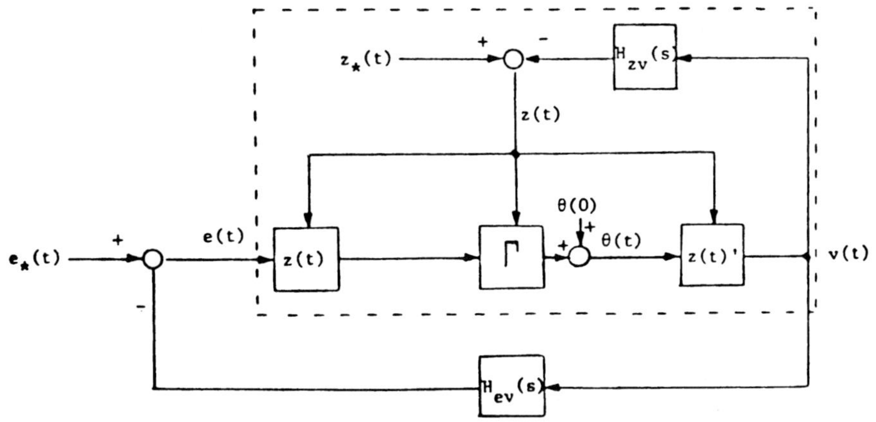

Figure 1: Adaptive error system

The selection of θ_* is based on complete knowledge of the actual plant and disturbances. The system corresponding to this setting is referred to as the ***tuned system***. The signals $e_*(t) \in \mathcal{R}$ and $z_*(t) \in \mathcal{R}^p$ are outputs of the tuned system, and are referred to as the ***tuned error*** and ***tuned regressor***, respectively. The signal $v(t) \in \mathcal{R}$ can be regarded as the adaptive control error.

The operators H_{ev} and H_{zv} are dependent on θ_* and describe how v effects the error and regressor signals. We assume here that H_{ev} and H_{zv} are linear-time-invariant (LTI) with stable proper transfer functions $H_{ev}(s)$ and $H_{zv}(s)$. This would arise, for example, when the plant to be controlled is LTI and the adaptive controller is linear in the adaptive parameters. The stability of H_{ev} and H_{zv} is a consequence of the definition of θ_* as the tuned parameter setting.

The operation Γ depends on the choice of parameter update algorithm. We will restrict attention here to the following representatives:

Gradient

$$\begin{aligned} &(\Gamma z)(t) = \epsilon z(t) \\ &\epsilon > 0 \end{aligned} \tag{4}$$

Recursive Least Squares

$$\begin{aligned} &(\Gamma z)(t) = P(t)z(t) \\ &\tfrac{d}{dt}P^{-1}(t) = z(t)z(t)' \\ &P(0) = P(0)' > 0 \end{aligned} \tag{5}$$

3. Global Stability and Passivity

It is of interest to determine under what conditions the adaptive error system (3,5) produces bounded outputs (θ, e, v, z) for *all* bounded initial parameter errors $\theta(o) \in \mathcal{R}^p$. This is what is meant here by "global" stability. As it turns out, it is possible to prove such a result provided that:

(i) $H_{ev}(s) \in SPR$ with gradient (6)

(ii) $H_{ev}(s) - \frac{1}{2} \in SPR$ with least squares (7)

(iii) $z_*, \dot{z}_* \in L_\infty^p$ and either (8)

a) $e_*, \dot{e}_* \in L_2 \cap L_\infty$ (9)

b) $e_*, \dot{e}_* \in L_\infty$ and
$z \in PE$ (persistently exciting) (10)

Parameter convergence to a constant in $\mathcal{R}^p$ or to a well defined subset in $\mathcal{R}^p$, requires that (9) be strengthened to:

$$e_*, \dot{e}_* \in L_2 \cap L_\infty, \; z_* \in PE \tag{11}$$

The above results can be found in [7,15,16] and in [18]. Although of theoretical significance, they are not feasible to obtain in practice. In the first place, due to unmodeled dynamics [9], $H_{ev}(s) \in SPR$ is practically impossible to achieve in adaptive feedback and even in some output error identification. (This is not the case in equation error identification.) Secondly, when $e_*, \in L_\infty$ as in (10), it is required that $z \in PE$ which cannot be guaranteed in advance since z is *inside* the adaptive loop. Case (11) which requires $z_* \in PE$ - which is feasible to establish - conflicts with $e_*, \dot{e}_* \in L_2 \cap L_\infty$. The latter implies $e_*(t) \to 0$ which can only occur for $z_* \in PE$ - and where there are no unmodeled dynamics which we argue is not possible.

With these impossible to satisfy theoretical requirements, it is doubtful that a global stability theory can be attained which relies on passivity, i.e., condition (6,7). On the practical side, however, there is substantial evidence of well engineered algorithms that work without SPR [10]. These do not work for all $\theta(o)$ and for all e_*, z_* in L_∞, but rather, for restricted magnitudes and signal spectrums. For example, if $H_{ev}(s)$ is SPR for $\omega \le \omega_{Bw}$ then it is expected that the adaptive system will be well behaved provided there is insignificant excitation above ω_{Bw}. The following example illustrates some of this phenomena.

Example: Consider the model reference adaptive control (MRAC) system studied by Rohrs et al. [9] with plant

$$P(s) = \frac{2}{s+1} \; \frac{229}{(s+15)^2+4}$$

reference model

$$H_{ref}(s) = \frac{3}{s+3}$$

and adaptive control law

$$u = -\hat{\theta}_1 y + \hat{\theta}_2 r$$

The adaptive parameters are obtained from the gradient algorithm,

$$\begin{aligned} \dot{\hat{\theta}}_1 &= ye \\ \dot{\hat{\theta}}_2 &= -re \\ e &= y - H_{ref} r \end{aligned}$$

For this example we have the tuned error given by

$$e_* = H_{er} r$$

with

$$H_{er}(s) = \frac{458\theta_{*2}}{s^3+31s^2+259s+229(1+2\theta_{*1})} - \frac{3}{s+3}$$

We also have

$$H_{ev}(s) = \frac{458}{s^3+31s^2+259s+229(1+2\theta_{*1})}$$

Observe that $H_{er}(s)$ and $H_{ev}(s)$ are stable provided that

$$\theta_{*1} \in [0, 17.03)$$

Since $H_{ev}(s)$ has a relative degree of three, it follows that $H_{ev}(s)$ is not SPR, and so global stability is not guaranteed.

Figure 2 shows $\theta_1(t)$ vs. $\theta_2(t)$ for simulations corresponding to two selected inputs:

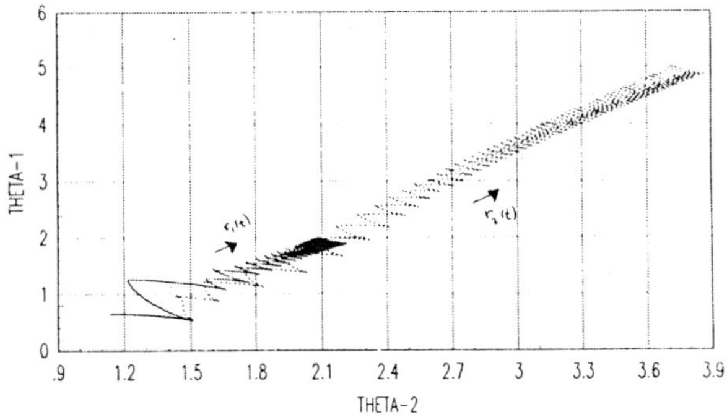

Figure 2: Parameter drift to inputs $R1$ and $R2$.

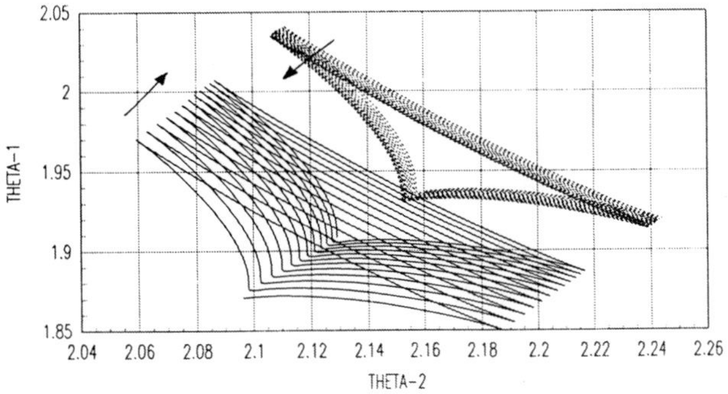

Figure 3: Blow-up of drift to input $R1$.

$$\begin{aligned} r_1(t) &= 1 + sin\ 3t \\ r_2(t) &= 1 + sin\ 5t \end{aligned}$$

with initial conditions $\widehat{\theta}_1(0) = .65, \widehat{\theta}_2 = 1.15$ which satisfy the DC tracking requirement. The response to $r_1(t)$ undergoes a transient and then drifts down a line in $\mathcal{R}^2$ to an apparent stable orbit. Figure 2 shows a blow-up near the stable orbit as well as a trajectory which starts just below it and, drifts upward. The response to $r_2(t)$, however, is unstable in the sense that the parameters continue to drift and eventually $\widehat{\theta}_1(t)$ will exceed 17.03 and the system becomes unstable.

Most adaptive control systems show the characteristic behavior illustrated in our example. The parameter first exhibit a transient followed by a steady-state drifting. The papers by Åström [10,11] contain many examples. In the example here, the drifting appears to occur along a line in $\mathcal{R}^2$. In one case (input r_1) the drift stops and the parameters settle into a periodic orbit. With an apparently modest change in the input spectrum (r_2) the parameters now drift into the instability region. Therefore, either the orbital center has drastically changed and is now outside the constant parameter stability set, or else there is no stable orbit at all anywhere along the $\mathcal{R}^2$ line of drift.

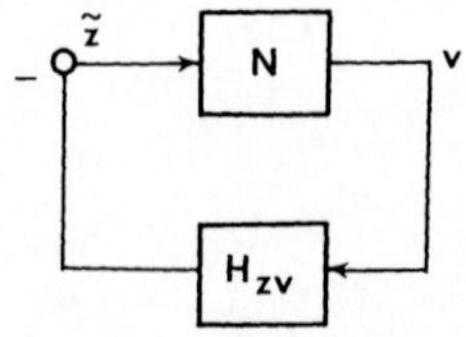

Figure 4: Equivalent feedback representation of adaptive error system.

In the forthcoming sections we will establish conditions under which the qualitative properties of the drifting phenomena can be predicted under slow adaptation. Our analysis is local and based on the classical methods of linearization and averaging for nonlinear systems.

4. Local Stability: Small Gain Theory and Averaging

Another way to view the system (3) under ideal conditions (6)-(11) is to arrange the system in the form shown in Fig. 4. Here, the forward path operator is defined by the map $N = \tilde{z} \to v$ such that

$$z = z_* + \tilde{z} \tag{12a}$$
$$\dot{\theta} = \epsilon[ze_* - zH_{ev}(z'\theta)] \tag{12b}$$
$$v = z'\theta \tag{12c}$$

with $\tilde{z}$ obtained from the feedback path,

$$\tilde{z} = -H_{zv}(\phi'\theta) \tag{13}$$

Observe that N is in effect the *linear* adaptive system that we mentioned earlier. Clearly $\tilde{z}$ is the amount by which the regressor $z(t)$ differs from the tuned regressor $z_*(t)$.

Now, let $T : \tilde{z} \to \tilde{\xi}$ denote the *loop-gain operator* defined by (12) and

$$\tilde{\xi} = -H_{zv}(\phi'\theta) \tag{14}$$

Small Gain Theory asserts [19] that if, for some $p \in [1, \infty]$, the L_p-gain of T is less than one, then the system is L_p-stable. Let $\gamma_p(T)$ denote the L_p-gain of T, i.e.,

$$\gamma_p(T) = inf\{k : \exists b \geq 0 \; s.t. ||T\tilde{z}||_p \leq k||\tilde{z}||_p + b, \forall \tilde{z} \in L_p\} \tag{15}$$

when $p = 2$, it is possible to show that for all $\tilde{z} \in L_2$,

$$||v||_2 \leq c|\theta(0)| \tag{16}$$

where c is a constant *independent* of $\tilde{z}$ [16]. Hence,

$$\begin{aligned} ||T\tilde{z}||_2 &= ||H_{zv}v||_2 \\ &\leq \gamma_2(H_{zv})c|\theta(0)| \end{aligned} \tag{17}$$

Comparing this result with the definition of gain (15), we see that $\gamma_2(T) = 0$. Thus, under ideal conditions, *the loop-gain of the adaptive error system is zero*!

Now, suppose that $H_{ev}(s)$ is not SPR and $e_*(t) \neq 0$. One would expect that small deviations in the SPR'ness of $H_{ev}(s)$ and small non-zero magnitudes of $e_*(t)$ could be tolerated without trouble. Unfortunately, this is not quite the case. In the first place (16) holds without persistent excitation. This means that system N (12) is only uniformly stable (in the sense of Lyapunov). Recall that uniform stability is not robust to typical perturbations. Uniform asymptotic stability of N (equivalently exponential asymptotic stability, since N

is linear) is robust to a large class of perturbations. Thus, a basic idea behind the use of various forms of linearization theorems in the analysis of adaptive systems, is to insure that the system N is u.a.s. (uniformly asymptotically stable) which necessitates that $z_*(t)$ be persistently exciting. Since space limitations do not permit us to elaborate on linearization theory here, the interested reader is referred to [3,4]. We will, however, see that averaging imposes a natural linearization.

By restricting the magnitude of $\theta(o)$ and the magnitude and spectrum of $z_*(t)$ and $e_*(t)$, it is possible to obtain conditions to prove local stability [3,4]. The local stability property hinges on two premises: (i) the error system trajectories are in a (not necessarily small) neighborhood of the tuned solution, and (ii) the linear time varying system which maps $w \rightarrow \theta$ as given by

$$\dot{\theta} = -(\Gamma z_*)(t) H_{ev}(z_*'(\cdot)\theta(\cdot)) + (\Gamma w)(t) \tag{18}$$

is L_∞-stable, i.e., there exists constants k and b $s.t. \|\theta\|_\infty \leq k\|w\|_\infty + b$. The choice of Γ comes from (4) or (5) and $z_*(t)$ is the tuned regressor. We can regard (18) as a linearization of the update algorithm. There are several ways to establish the L_∞ stability of (18).

4.1 Gradient Algorithm

We first consider the case when Γ represents the gradient algorithm, i.e., $(\Gamma z_*)(t) = \epsilon z_*(t)$ with $\epsilon > 0$.

In [20], it is shown that if $H_{ev}(s) \epsilon SPR$ and $z_* \epsilon PE$, then for all $\epsilon > 0, w \rightarrow \theta$ is exponentially stable, and hence, L_∞-stable. In [21], if $H_{ev}(s) = \overline{H}_{ev}(s) + s\Delta(s)$, $\overline{H}_{ev}(s) \epsilon SPR$, $\Delta(s)$ is stable, and $z_* \epsilon PE$ then for sufficiently small ϵ and $\|\dot{z}_*\|_\infty, w \rightarrow \theta$ is still exponentially stable, and hence is L_∞-stable. This latter method relies on loop-transformations and application of small gain theory.

Another approach is to use averaging. In [1] it is shown that if $z_* \epsilon PE$ with the Fourier series representation

$$z_*(t) \sim \sum_k \alpha(\omega_k) e^{j\omega_k t} \tag{19}$$

and if the eigenvalues of the real matrix

$$B = \sum_k \alpha(\omega_k)\alpha(-\omega_k)' H_{ev}(-j\omega_k) \tag{20}$$

all have positive real parts, then for all sufficiently small $\epsilon > 0, w \rightarrow \theta$ is exponentially stable, and hence, L_∞-stable. Moreover, if any one eigenvalue of B has a negative real part, then $w \rightarrow \theta$ is exponentially unstable. Hence, there exists $w \epsilon L_\infty s.t. |\theta(t)| \rightarrow \infty$ as $t \rightarrow \infty$ exponentially fast. It is obvious then when $H_{ev}(s)$ is *not* SPR, but only approximately so, then the Riedle-Kokotovic result provides a sharp stability-instability boundary. Note that when $H_{ev}(s)$ is SPR and $z_* \epsilon PE$ we have from [20] that $w \rightarrow \theta$ is exp. stable for *all* $\epsilon > 0$. At the present time, averaging theory as applied here, does not hold for all $\epsilon > 0$ even when $H_{ev}(s)$ is SPR. On the other hand, the result in [21] remains valid for $H_{ev}(s) \epsilon SPR (\Delta(s) = o)$ because then $\epsilon > 0$ is bounded above by infinity.

Example In this example we illustrate what happens when $Re\lambda(B) > 0$ but ϵ is too large. Consider the scalar system

$$\dot{\theta} = -\epsilon z_* H_{ev}(z_* \theta)$$

with $z_*(t) = sin(.35t)$ and $H_{ev}(s) = 1/(s^2 + 2s + 2)$. In this case B is a scalar and it is easily verified that $B > 0$. The simulations in Figure 5 with $\theta(0) = 1$ show that the zero solution is u.a.s. for $\epsilon = 4$ (and for all $\epsilon \leq 4$), but is completely unstable for $\epsilon = 8$ (and for all $\epsilon \gtrsim 8$).

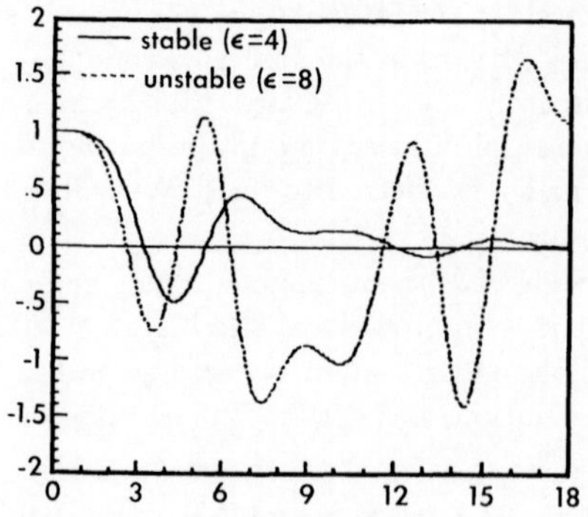

Figure 5: Parameter trajectories with varying gain

4.2 Recursive Least Squares Algorithm

In this case we have from (5) that

$$(\Gamma z_*)(t) = P_*(t) z_*(t)$$
$$\tfrac{d}{dt} P_*(t)^{-1} = z_*(t) z_*(t)', P_*(0) > 0.$$

When $z_* \epsilon PE$ there exists $\alpha > 0$ such that

$$P_*(t)^{-1} = P_*(o)^{-1} + \int_o^t z_*(\tau) z_*(\tau)' d\tau \geq \alpha t \cdot I.$$

Thus, it is convenient to define $R(t) = \frac{1}{t} P_*(t)^{-1}$ for $t > 0$. Hence $R(t)^{-1} = t P_*(t) \leq \frac{1}{\alpha} I$ and we can write (18) and (5) as,

$$\begin{aligned} \dot{\theta} &= \tfrac{1}{t} R^{-1} [w - z_* H_{ev}(z_*' \theta)] \\ \dot{R} &= \tfrac{1}{t} (z_* z_*' - R) \end{aligned} \tag{21}$$

When $H_{ev}(s) - \frac{1}{2}$ is not SPR we can now follow [2] and for $t \geq s$ and s sufficiently large, approximate the right hand side by its average. Letting "overbar" denote average (assuming it exists) we have:

$$\begin{aligned} \dot{\theta}(t) &\approx \tfrac{1}{t} \overline{R}^{-1} (\overline{w} - (\overline{z_* H_{ev} z_*'}) \overline{\theta}) \\ \dot{\overline{R}}(t) &\approx \tfrac{1}{t} (\overline{z_* z_*'} - \overline{R}) \end{aligned} \tag{22}$$

Integrating from s to $s + T, T > 0$, gives

$$[\theta(s+T) - \theta(s)] / \int_s^{s+T} dt/t \approx \overline{R}^{-1} (w - (\overline{z_* H_{ev} z_*'}) \overline{\theta}) \tag{23}$$

$$[R(s+T) - R(s)] / \int_s^{s+T} dt/t \approx \overline{z_* z_*'} - \overline{R} \tag{24}$$

Now change time scales $s + T \to \tau + \Delta\tau, \Delta\tau = \int_s^{s+T} dt/t$ and letting $s \to \infty$ gives the differential equations:

$$\dot{\theta}_A(\tau) = R_A(\tau)^{-1} [\overline{w} - B \theta_A(\tau)] \tag{25}$$

$$\dot{R}_A(\tau) = \overline{z_* z_*'} - R_A(\tau) \tag{26}$$

with $B = \overline{z_* H_{ev} z_*'}$ given by (20). These equations actually describe the asymptotic behavior of (18) in just the same way as they do for discrete-time [2]. In order to validate the approximations in each of the steps leading to (25) and (26), it is necessary to introduce various regularity conditions. A complete proof can be found in [2,22]. Here, as warned, we offer only heuristics.

Observe that in (26) as $\tau \to \infty, R_A(\tau) \to \overline{z_* z_*'}$. Thus, when $H_{ev}(s) - \frac{1}{2}$ is not SPR and $z_* \epsilon PE$ with Fourier representation (19) the asymptotes are stable if

$$Re\ \lambda(L) > 0 \tag{27}$$

where

$$\begin{aligned} L &= (\overline{z_* z_*'})^{-1} B \\ &= (\textstyle\sum_k \alpha(\omega_k)\alpha(-\omega_k)')^{-1} \sum_k \alpha(\omega_k)\alpha(\omega_k)' H_{ev}(-j\omega_k) \end{aligned} \tag{28}$$

If $Re\ H(j\omega_k) \geq \rho > 0$ at low frequencies, and if $|\alpha(\omega_k)|$ is small at frequencies where $Re\ H(j\omega_k) \leq 0$, then $Re\ \lambda(L) = \rho$. Thus, all parameter asymptotes have a uniform rate of convergence which is not the case for the gradient algorithm with a time-invariant gain.

5. Averaging: A More General Approach

In this section we will establish a general form of the adaptive error system (3,5) which is useful for application of averaging methods. The first step is to transform (3,5) into a set of nonlinear time-varying differential equations. To do this observe that if $H_{ev}(s)$ and $H_{zv}(s)$ are strictly proper functions (a convenient illustrative, but not necessary, assumption) then we can write

$$\begin{aligned} H_{ev}(s) &= c'(sI - A)^{-1}b \\ H_{zv}(s) &= D(sI - A)^{-1}b \end{aligned} \tag{29}$$

where $A \in \mathcal{R}^{nxn}, b \in \mathcal{R}^n, D \in \mathcal{R}^{pxn}$, with $(A, b, [c\ D'])$ a minimal representation. Also, $Re\ \lambda(A) < 0$ reflecting the fact that H_{ev} and $H_{zv}(s)$ are stable. The error system (3) is then equivalently expressed as

$$\begin{aligned} e &= e_* - c'x \\ z &= z_* - D_x \\ \dot{x} &= Ax + bz'\theta \\ \dot{\theta} &= (\Gamma z)e \end{aligned} \tag{30}$$

By eliminating the variables e and z we can reduce (30) to the coupled state-space description:

$$\dot{\phi} = \gamma(t) f(t, \phi, x) \tag{31}$$

$$\dot{x} = Ax + g(t, \phi, x) \tag{32}$$

With the gradient algorithm (4), let

$$\begin{aligned} \phi(t) &= \theta(t) \\ \gamma(t) &= \epsilon \end{aligned} \tag{33}$$

and

$$\begin{aligned} f(t, \theta, x) &= z_*(t)e_*(t) - Q_*(t)x + c'xDx \\ Q_*(t) &= z_*(t)c' + e_*(t)D \\ g(t, \theta, x) &= b(z_*(t) - Dx)'\theta \end{aligned} \tag{34}$$

With the recursive least squares algorithm (5), define:

$$R(t) = \frac{1}{t} P(t)^{-1} \tag{35}$$

and let

$$\phi(t) = \begin{pmatrix} \theta(t) \\ col\{R(t)\} \end{pmatrix}, \gamma(t) = \frac{1}{t} \tag{36}$$

where the operator $col\{R\}$ stacks up the columns of the matrix R to form a vector. Thus,

$$f(t,\phi,x) = \begin{pmatrix} R^{-1}(z_*(t)e_*(t) & -Q_*(t)x + c'xDx) \\ col\{z_*(t)z_*(t)' & -z_*(t)(Dx)' - Dxz_*(t)' \\ & +Dx(Dx)' - R\} \end{pmatrix} \tag{37}$$

$$g(t,\phi,x) = b(z_*(t) - Dx)'\theta \tag{38}$$

The $col\{\cdot\}$ operator was used by Ljung in [2] to develop the discrete-time version of (31,32).

5.1 The Integral Manifold

The basic idea in the application of averaging methods to (31,32) is to see what happens when $\gamma(t)$ is small. Essentially, $\phi(t)$ slows down and we would expect to be able to approximate the right hand side of (31) with its average, i.e.,

$$\dot{\phi} \approx \gamma(t)\overline{f}(\phi) \tag{39}$$

where

$$\overline{f}(\phi) = \lim_{T\to\infty} \frac{1}{T}\int_o^T f(t,\phi,\overline{x}(t,\phi))dt \tag{40}$$

assuming the limit exists. (Such is the case, for example, when $f(t,\phi,x)$ and $g(t,\phi,x)$ are periodic in t for all bounded ϕ and x). The function $\overline{x}(t,\phi)$ is referred to as the state of the *frozen parameter* system, i.e., $\overline{x}(t,\phi)$ solves (32) whenever ϕ is a fixed vector. To emphasize this point we may express $\overline{x}(t,\phi)$ as the solution to the partial differential equation.

$$\frac{\partial}{\partial t}\overline{x} = A\overline{x} + g(t,\phi,\overline{x}), \overline{x}(0,\phi) = x(0)$$

The frozen parameter system was introduced in the averaging analysis proposed by Åström [11].

In order to remove the approximation in (39) we introduce the *integral manifold* as suggested by [13] [see [6] for discussion of the integral manifold]

The integral manifold M of (31,32) is the set,

$$M = \{t,\phi,x : x(t_o) = h(t_o,\phi(t_o)) \text{ implies } x(t) = h(t,\phi(t)), \forall t \geq t_o\} \tag{41}$$

By substituting $x = h(t,\phi)$ into (31,32), the *manifold function* $h(t,\phi)$ is seen to satisfy the partial differential equation

$$\frac{\partial h}{\partial t} + \gamma(t)\frac{\partial h}{\partial \phi}f(t,\phi,h) = Ah + g(t,\phi,h) \tag{42}$$

Whenever $\gamma(t)$ is sufficiently small, a reasonable approximation to $h(t,\phi)$ is given by $h_o(t,\phi)$ which is the solution to

$$\begin{aligned} \tfrac{\partial h_o}{\partial t} &= Ah_o + g(t,\phi,h_o) \\ &= F(\theta)h_o + G(\theta)z_*(t) \end{aligned} \tag{43}$$

where the last line follows from (34) with

$$F(\theta) = A - b\theta'D,\ G(\theta) = b\theta' \tag{44}$$

In (42), θ and t are regarded as independent variables and, hence, we can define the stabilizing parameter set

$$D_s = \{\theta \in \mathcal{R}^p : Re\ \lambda(F(\theta)) < 0\} \tag{45}$$

Thus, for $\gamma(t)$ sufficiently small, we can refer to $h(t,\phi)$ with $\theta\epsilon D_s$ as the *stable manifold*, which we will approximate by $h_o(t,\theta), \theta\epsilon D_s$.

An important observation to make at this point is that the approximate manifold function $h_o(t,\phi)$ satisfies the same partial differential equation as the frozen parameter system state $\overline{x}(t,\phi)$. The only difference is in initial conditions. However, if $\theta \epsilon D_s$ then as $t \to \infty$ we have $h_o(t,\phi) - \overline{x}(t,\phi) \to 0$ exponentially.

The final transformation on (31,32) is obtained by examining the behavior of (ϕ, x) in the neighborhood of the stable manifold. Introduce the error state,

$$\xi = x - h(t,\phi) \tag{46}$$

Using (46), and (31,32), we have

$$\dot{\phi} = \gamma(t) f(t,\phi,h(t,\phi)+\xi) \tag{47}$$

$$\dot{\xi} = F(\theta)\xi - \gamma(t) h_\phi(t,\phi) f(t,\phi,h(t,\phi)+\xi) \tag{48}$$

where we have used $h_\phi(t,\phi)$ to denote $\frac{\partial h}{\partial \phi}(t,\phi)$. If $\gamma(t)$ is sufficiently small it can be shown that under suitable regularity conditions we can approximate $h(t,\phi)$ with the frozen parameter state $\overline{x}(t,\phi)$ and obtain the approximate system,

$$\dot{\phi} = \gamma(t) f(t,\phi,\overline{x}(t,\phi)+\xi)$$
$$\dot{\xi} = F(\theta)\xi - \gamma(t)\overline{x}_\phi(t,\phi) f(t,\phi,\overline{x}(t,\phi)+\xi)$$

Moreover, if $\gamma(t)$ is sufficiently small and θ remains (moving slowly) in D_s then $\xi(t) \to 0$ exp. fast. As a result, by the same reasoning as in Section 4, the stability of the asymptotic system:

$$\dot{\phi}_A(\tau) = \overline{f}(\phi_A(\tau)) \tag{49}$$

where

$$\overline{f}(\phi) = \lim_{T\to\infty} \frac{1}{T} \int_o^T f(t,\phi,\overline{x}(t,\phi))dt \tag{50}$$

assuming the limit exists. The stability of (49) is given as follows. The proof is in [6].

Theorem

Let ϕ^o denote a solution of

$$\overline{f}(\phi^o) = 0$$

and define the matrix,

$$G = \frac{\partial \overline{f}}{\partial \phi}(\phi^o)$$

Then, provided $Re\ \lambda(G) \neq 0$, the equilibrium solution $\phi_A(\tau) = \phi^o$ of the asymptotic system (49) is:

(i) u.a.s. if $\max_i\ Re\ \lambda_i(G) < 0$.

(ii) unstable if $\max_i\ Re\ \lambda_i(G) > 0$.

5.2 Application to Gradient Algorithm

Applying this result to (33,34) with the gradient algorithm and with $z_* \epsilon PE$ and $\overline{z_* e_*} = 0$, gives $G = -B$ from (20). Since $\gamma(t) = \epsilon > 0$, we can only conclude that if $Re\ \lambda(B) > 0$ and ϵ is sufficiently small, then $\theta(t) \epsilon D_s$ long enough for transients to die out, which is unprovable as yet in general.

Observe that $\theta_* \in \mathcal{R}^p$ such that $\overline{z_* e_*} = 0$ does not define an equilibrium of the actual system. All we can say is that with $\epsilon > 0$ small, there is a $\hat{\theta}(t)$ which orbits near (to order ϵ) the equilibrium of the asymptotic system. We can also choose to consider $\overline{z_* e_*} = 0$ as a defining equation in a *candidate* tuned setting. Other conditions would also have to hold (e.g. small $e_*(t)$, etc.) which may be obtainable with proper input selection. In other words, the signals present *during adaptation* should be similar to those present during tracking or disturbance rejection. Otherwise, the algorithms choice of the tuned setting ($\overline{z_* e_*} = 0$) may be undesirable.

5.3 Application to Recursive Least Squares Algorithm

Under the same conditions and with the same provisions as above, $G = -L$ with L from (28). This time, since $\gamma(t) = 1/t \rightarrow 0$ as $t \rightarrow \infty$, we can conclude that if $Re\ \lambda(L) > 0$, then $\theta(t) \rightarrow 0$ as $t \rightarrow \infty$ at a rate $1/t$. In this case, due to the presence of $1/t$, the parameters $\widehat{\theta}(t)$ asymptotically approach the solution of (49).

6. Concluding Remarks

The averaging theory described here, as well as averaging theory in general, has its uses and limitations for adaptive system. In the first place, the theory requires slow adaptation which can be counter-productive because performance can be below par for the long period of time it takes for the parameters to readjust. Secondly, averaging theory is a form of linearization so that the (nonlinear) adaptive system must be initialized in a (not necessarily small) neighborhood of the tuned system. On the positive side, however, we do obtain frequency domain conditions which explain the system behavior near the tuned solutions. In this sense, we can consider the results of averaging theory to be necessary conditions for good performance of adaptive systems.

To obtain the heralded goal of frequency-domain stability conditions, it may be inevitable to encounter linearization. Somewhat less intuitively appealing results can be obtained without resorting to direct linearization or averaging, e.g., in [4,17,21]. These results arise from a combination of small gain theory and perturbation theory.

Acknowledgments

The work reported here was sponsored by the Air Force Office of Scientific Research (AFSC), under contracts F49620-84-C-0054 and F49620-85-C-0094. The author is indebted to the above average work of B. Riedle and P. Kokotovic on methods of averaging and for many enlivening discussions on these and related matters.

References

[1] B.D. Riedle and P.V. Kokotovic, "A Stability-Instability Boundary for Disturbance-Free Slow Adaptation," Proc. 23rd IEEE CDC, Las Vegas, Nevada, 1984.

[2] L. Ljung, "Analysis of Recursive Stochastic Algorithms," *IEEE Trans. on Automatic Control*, vol. AC-22, pp. 551-575, 1977.

[3] R.L. Kosut and B.D.O. Anderson, "Robust Adaptive Control: Conditions for Local Stability," Proc. 23rd IEEE CDC, Las Vegas, Nevada, 1984.

[4] R.L. Kosut and B.D.O. Anderson, "A Local Stability Analysis for a Class of Adaptive Systems," *IEEE Trans. on Automatic Control*, to appear.

[5] J.P. LaSalle and S. Lefschetz, Stability by Liapunov's Direct Method, Academic Press, New York, 1961.

[6] J.K. Hale, Ordinary Differential Equations, Krieger, Molaben, Florida, 1980.

[7] K.S. Narendra, Y.H. Lin and L.S. Valavani, "Stable Adaptive Controller Design, Part II: Proof of Stability," *IEEE Trans. on Automatic Control*, vol. AC-25, pp. 440-448, 1980.

[8] Y.D. Landau, Adaptive Control: The Model Reference Approach, Marcel Dekker, 1979.

[9] C.E. Rohrs, L. Valavani, M. Athans and G. Stein, "Robustness of Adaptive Algorithms in the Presence of Unmodeled Dynamics," Proc. 21st IEEE CDC, Orlando, Florida, 1982.

[10] K.J. Åström , "Analysis of Rohr's Counter Example to Adaptive Control," Proc. 22nd IEEE CDC, San Antonio, Texas, 1983.

[11] K.J. Åström , "Interactions Between Excitation and Unmodeled Dynamics in Adaptive Control," Proc. 23rd IEEE CDC, Las Vegas, Nevada, 1984.

[12] R.L. Kosut, B.D.O. Anderson and I. Mareels, "Stability Theory for Adaptive Systems: Methods of Averaging and Persistency of Excitation," submitted 24th CDC and *IEEE Trans. on Automatic Control*, 1985.

[13] B.D. Riedle and P.V. Kokotovic, "Integral Manifold Approach to Slow Adaptation," CSL Report DC-80, University of Illinois, Urbana, 1985.

[14] L.C. Fu, M. Bodson and S. Sastry, "New Stability Theorems for Averaging and Their Applications to the Convergence Analysis of Adaptive Identification and Control Schemes," Memo No. UCB/ERL M85/21, Univ. of California, Berkeley, 1985.

[15] R.L. Kosut and B. Friedlander, "Performance Robustness Properties of Model Reference Adaptive Control," Proc. 21st IEEE CDC, Orlando, Florida, 1982.

[16] R.L. Kosut and B. Friedlander, "Robust Adaptive Control: Conditions for Global Stability," to appear, *IEEE Trans. on Automatic Control*, vol. AC-30, No. 7, 1985.

[17] R.L. Kosut and C.R. Johnson, Jr., "An Input-Output View of Robustness in Adaptive Control," *Automatica*, vol. 20, No. 5, pp. 569-681, 1984.

[18] S. Boyd and S. Sastry, "Necessary and Sufficient Conditions for Parameter Convergence in Adaptive Control," Memo No. UCB/ERL M84/25, Univ. of California, Berkeley, 1984.

[19] C.A. Desoer and M. Vidyasagar, Feedback Systems: Input-Output Properties, Academic Press, New York, 1975.

[20] B.D.O. Anderson, "Exponential Stability of Linear Equations Arising in Adaptive Identification," *IEEE Trans. on Automatic Control*, vol. AC-22, pp. 83-88, 1977.

[21] B.D.O. Anderson, R. Bitmead, C.R. Johnson, Jr. and R.L. Kosut, "Stability Theorems for the Relaxation of the Strictly Positive Real Condition in Hyperstable Adaptive Systems," Proc. 23rd IEEE CDC, Las Vegas, Nevada, 1984.

[22] L. Ljung, Theory and Practice of Recursive Identification, MIT Press, Cambridge, Mass., 1983.

[23] K.J. Åström , "Theory and Applications of Adaptive Control - A Survey," *Automatica*, vol. 19, pp. 471-486, 1983.

A Robust Indirect Adaptive Control Approach*

Gerhard Kreisselmeier
University of Kassel, Dept. of Electrical Engineering,
Wilhelmshöher Allee 73, D-3500 Kassel, West Germany

Abstract

This paper considers the robust design of an indirect adaptive control approach, which is applicable when the unknown parameters of a linear, time invariant plant lie in a known convex set throughout which no unstable pole-zero cancellation occurs. In order to achieve the robustness, the use of a relative dead zone in the adaptive law is proposed. It is shown that, with a suitably designed relative dead zone, the adaptive control system is (globally) stable, even in the presence of small, unmodeled plant uncertainties.

1. Introduction

In practical control applications the plant to be controlled is typically of higher order than the plant model used to design the controller. It was shown recently, that such model-plant mismatch, even when the associated modeling errors are small, may cause the instability of an adaptive control system, which otherwise would have been stable [1,2]. Hence there is a need to establish some kind of robustly stable adaptive control, i.e. adaptive control with guaranteed stability even in the presence of (small) unmodeled plant uncertainties.

First results in this direction have been reported by several authors. For example, in [3] the robust local stability of an adaptive regulator was obtained as an additional property of a scheme, which had been designed without robustness considerations. In [2] the robust local stability of a model reference adaptive controller was established on the basis of a singular perturbation analysis, and by means of an additional, linear feedback term in the adaptive law (i.e. $\dot{\theta} = -\sigma\theta - \cdots$) and the assumption that the external reference input has a uniformly bounded time derivative. The latter two means aimed at reducing the high frequency content of the signals in the closed loop system and thus at not exciting the neglected high frequency modes.

An alternate approach to the robustness problem is pursued in [4,5] and [6]. Here the basic idea is to use the external reference signal so as to ensure persistence of excitation in the adaptive system, thereby to get exponential stability, and to obtain robustness as a consequence of the exponential stability. Obviously, this kind of robustness does not follow as a structural property of the adaptive system but rather as a property attached to the system through the exciting signal. The important question, how to ensure the persistence of excitation in the presence of the plant uncertainty, seems, however, not to have been completely resolved as yet.

Different from the above mentioned approaches, the present paper proposes the concept of a relative dead zone in the adaptive law as a general tool for compensating the effect of unmodeled plant uncertainties and for achieving robust stability in adaptive control systems. The idea of using a dead zone in the adaptive law is not new. It has been used, for example, to obtain (global) stability of model reference adaptive control systems in the presence of disturbances, which are uniformly bounded (e.g. [7]). The effect of an unmodeled plant uncertainty can also be interpreted in terms of a disturbance acting on the plant, but this disturbance depends on the plant input and therefore, as long as the stability of the adaptive

* Reprinted by permission of Taylor and Francis Ltd. from International Journal of Control, Vol. 43, pp. 161-167, 1986

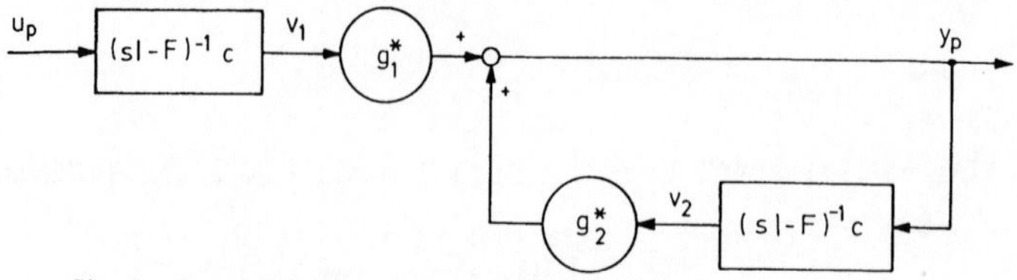

Fig. 1a Plant Model

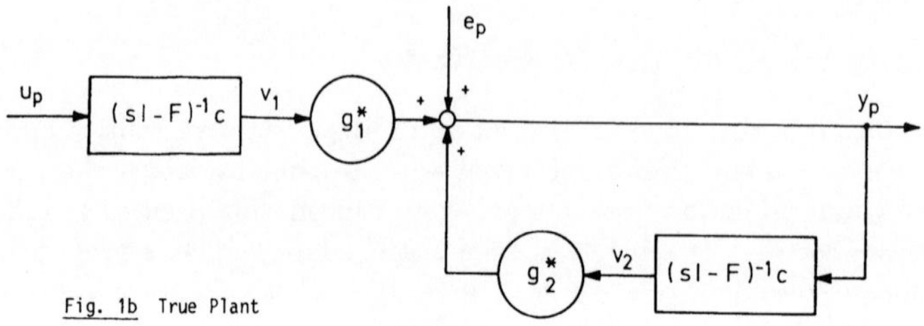

Fig. 1b True Plant

Figure 1

system has not been established, the possibility exists that this disturbance grows without bound. Hence, results as in [7] cannot be applied directly. To overcome the difficulties arising from the potentially unbounded disturbance, we propose to take the dead zone function not of the identification error itself but rather of the identification error, suitably normalized relative to the signals in the adaptive system (relative dead zone). Thereby the strict parameter descent property of the adaptive scheme is preserved, even in the presence of small plant uncertainties. This together with the inherent robustness properties of linear time invariant and almost time invariant systems finally allows us to conclude the robust (global) stability of the adaptive control system.

Although the concept of the relative dead zone is likely to be of more general applicability, it is developed here in the context of a particular indirect adaptive control approach for the sake of clarity and ease of exposition.

2. Problem Statement

a) The Plant Model

Consider a plant model with a rational, strictly proper transfer function $T_M(s)$. It is straight forward to see that such a transfer function can be represented in the form [8]

$$T_M(s) = \underline{g}_1^{*T}(s\underline{I} - \underline{F})^{-1}\underline{c}/(1 - \underline{g}_2^{*T}(s\underline{I} - \underline{F})^{-1}\underline{c}), \tag{1}$$

where $\underline{F}$ is an arbitrary, strictly stable matrix and such that $(\underline{F}, \underline{c})$ is controllable. The vectors $\underline{g}_1^*$ and $\underline{g}_2^*$ assign the numerator and denominator polynomials of $T_M(s)$, and thus represent the plant parameters.

Equation (1) gives rise to the (nonminimal) state space representation of the plant model.

$$\begin{aligned} \dot{\underline{v}}_1 &= \underline{F}\underline{v}_1 + \underline{c}u_p \\ \dot{\underline{v}}_2 &= \underline{F}\underline{v}_2 + \underline{c}y_p \\ y_p &= \underline{g}^{*T}\underline{v}, \end{aligned} \tag{2a}$$

where u_p and y_p denote the input and output of the plant, which can be measured, $\underline{g}^{*T} = [\underline{g}_1^{*T}, \underline{g}_2^{*T}]$ represents the unknown plant parameters, and $\underline{v}^T = [\underline{v}_1^T, \underline{v}_2^T]$.

In practice, the plant parameters $\underline{g}^*$ are often not completely arbitrary. Therefore, some a priori knowledge about $\underline{g}^*$ is assumed to be available in the form

$$\underline{g}^* = \underline{G}\underline{\theta}^* + \underline{g}^o \tag{2b}$$

$$\underline{\theta}_{min} < \underline{\theta}^* < \theta_{max}, \tag{2c}$$

where $\underline{\theta}^* = [\theta_1^*, \ldots, \theta_\ell^*], (1 \le \ell \le dim(\underline{g}^*))$ is an unknown parameter vector, and the inequality in (2c) is to be understood componentwise.

Relations (2b,c) say that $\underline{g}^*$ is known to lie in an ℓ-dimensional subspace (with origin at $\underline{g}^o$) of the parameter space, and that $\underline{g}^*$ is also known to lie in a convex subset of this subspace. So, we need to deal only with a reduced number of ℓ unknown parameters $\underline{\theta}^*$ in a known convex set. (Note, that the general case with $\ell = 2n, \underline{G} = \underline{I}, \underline{g}^o = 0$, i.e. $\underline{g}^* = \underline{\theta}^*$ is also covered).

Given the plant model of the form (1), (2a), we assume that

(A1) The model order n ($\triangleq$ dimension of $\underline{g}_1^*, \underline{g}_2^*$, and $\underline{v}_1, \underline{v}_2$) is known,

(A2) $\underline{G}, \underline{g}^o$, and $\underline{\theta}_{min}, \underline{\theta}_{max}$ are finite and known,

(A3) There exists $\epsilon > 0$ such that for all $\underline{\theta} \in [\underline{\theta}_{min}, \underline{\theta}_{max}]$ we have the following: Let $\underline{g} = \underline{G}\underline{\theta} + \underline{g}^o$, and let z_i and λ_j denote the zeros and poles of the transfer function

$$\underline{g}_1^T(s\underline{I} - \underline{F})^{-1}\underline{c}/(1 - \underline{g}_2^T(s\underline{I} - \underline{F})^{-1}\underline{c}).$$

Then $|z_i - \lambda_j| > \epsilon$ whenever $Re\, z_i > -\epsilon$.

In summation, the plant model has a transfer function of known order n, its parameters lie in a known convex set, and throughout this set an unstable pole-zero cancellation must not occur. Otherwise the plant model is arbitrary, for example, it may have a minimum as well as a nonminimum phase transfer function.

The above plant model is to serve as a basis for synthesizing a suitable adaptive controller for the plant, which we want to control.

b) The Plant

The plant, which we want to control, may slightly differ from the assumed model (2), and is defined by the equations

$$\begin{aligned} \dot{\underline{v}}_1 &= \underline{F}\underline{v}_1 + \underline{c}u_p \\ \dot{\underline{v}}_2 &= \underline{F}\underline{v}_2 + \underline{c}y_p \\ y_p &= \underline{g}^{*T}\underline{v} + e_p, \end{aligned} \tag{3a}$$

(see also Fig. 1b), and

$$\underline{g}^* = \underline{G}\underline{\theta}^* + \underline{g}^o \tag{3b}$$

$$\underline{\theta}_{min} < \underline{\theta}^* < \theta_{max}. \tag{3c}$$

The only distinction between (2) and (3) is the quantity e_p in (3a), which represents the effect of modeling errors. Regarding these modeling errors, which are also referred to as unmodeled plant uncertainty, we make the following assumption.

(A4) Let $0 < \sigma_m < -Re\lambda_i(\underline{F})$ and $\gamma > 0$ be fixed constants, and let m be defined by the equation

$$\dot{m} = -\sigma_m m + |u_p| + |y_p| \tag{3d}$$

with arbitrary initial condition $m_o > 0$. Then there exists a finite $\mu > 0$ such that e_p satisfies the relation

$$|e_p| < \mu(\gamma + m). \tag{3e}$$

While the quantity m in (3d) is related to the magnitude of the plant input and output signals, the quantity μ in (3e) rates the magnitude of the uncertainty. To illustrate assumption (A4), we consider two relevant examples of plant uncertainties.

Example 1. Let the plant be given by

$$Y_p = (T_M(s) + T_u(s)) \cdot U_p$$

where $T_u(s)$ represents the unmodeled plant uncertainty. Taking the Laplace transform of (3a) we get

$$Y_p = \frac{(\underline{g}_1^{*T}(s\underline{I} - \underline{F})^{-1}\underline{c}U_p + E_p)}{(1 - \underline{g}_2^*(s\underline{I} - \underline{F})^{-1}\underline{c})},$$

and comparing this equation with the preceding one, we obtain

$$E_p = (1 - \underline{g}_2^{*T}(s\underline{I} - \underline{F})^{-1}\underline{c}) \cdot T_u(s) \cdot U_p.$$

Note that $\underline{F}$ has all its eigenvalues to the left of $-\sigma_m$. If $T_u(s)$ is also more stable than $-\sigma_m$, then it follows that the impulse response of the transfer function E_p/U_p is bounded in modulus by $\mu\, exp\{\sigma_m t\}$ for some finite $\mu > 0$. This implies

$$\begin{aligned} |e_p| &< \int_o^t \mu\, exp\{-\sigma_m(t-\tau)\}|u_p(\tau)|d\tau \\ &< \mu(\gamma + m). \end{aligned}$$

As a result, assumption (A4) is satisfied, if the unmodeled dynamics of the plant are more stable than $-\sigma_m$.

Example 2. Let the plant be given by

$$Y_p = \tilde{T}_M(s) U_p$$

where $\tilde{T}_M(s)$ differs from $T_M(s)$ only by having parameters $\underline{g}^* + \mu\underline{\tilde{g}}$ instead of $\underline{g}^*$. Suppose further that there does not exist a $\underline{\tilde{\theta}}^*$ such that $\underline{G}\underline{\tilde{\theta}}^* + \underline{g}^o$ equals $\underline{g}^* + \mu\underline{\tilde{g}}$, for example, because $\underline{\tilde{g}}$ is not in the range space of $\underline{G}$. Then the plant is also not in the model set, and an unavoidable modeling error arises.

It is seen that this kind of uncertainty is closely connected with the use of a priori parameter information. Specifically, this uncertainty includes inaccurate modeling of low frequency modes and even of unstable modes of the plant.

Using the parameters $\underline{g}^* + \mu\underline{\tilde{g}}$ instead of $\underline{g}^*$ in (2a) we obtain the (nonminimal) state space representation of the plant

$$\begin{aligned} \dot{\underline{v}}_1 &= \underline{F}\underline{v}_1 + \underline{c}u_p \\ \dot{\underline{v}}_2 &= \underline{F}\underline{v}_2 + \underline{c}y_p \\ y_p &= (\underline{g}^* + \mu\underline{\tilde{g}})^T \underline{v}, \end{aligned}$$

which is already in the form (3a), where

$$e_p = \mu\underline{\tilde{g}}^T \underline{v}.$$

It follows that

$$\begin{aligned} |e_p| &< \mu \int_o^t |\underline{\tilde{g}}_1^T exp\{\underline{F}(t-\tau)\}\underline{c}\; u_p(\tau) + \underline{\tilde{g}}_2^T exp\{\underline{F}(t-\tau)\}\underline{c}\; y_p(\tau)|d\tau \\ &< \mu\,\text{const} \cdot \int_o^t exp\{-\sigma_m(t-\tau)\}(|u_p(\tau)| + |y_p(\tau)|)d\tau \\ &< \mu\,\text{const} \cdot (\gamma + m) \end{aligned}$$

and again assumption (A4) is satisfied.

c) The Robust Adaptive Control Problem

Knowing the plant model and the structure of the plant uncertainty, the problem is to devise an adaptive controller such that the resulting closed loop adaptive control system is robustly stable.

By robust stability we mean that there exists a $\mu' > 0$ such that for all $0 < \mu < \mu'$ and all possible plant uncertainties satisfying the above assumptions, the closed loop system is stable.

We say that the closed loop system is stable if for arbitrary initial conditions and arbitrary, piecewise continuous, uniformly bounded external reference inputs $r(t)$, all of its states remain uniformly bounded. This kind of stability is usually also referred to as global stability.

3. The Adaptive Controller

a) Controller Structure

Based on the above plant model (2) a suitable controller for the plant is defined by the equations

$$\begin{aligned} \dot{\underline{x}}_c &= \underline{K}_o(\underline{\theta})\underline{x}_c + \underline{k}_1(\underline{\theta})y_p + \underline{k}_2(\underline{\theta})r && (4a) \\ u_p &= \underline{k}_3(\underline{\theta})\underline{x}_c + k_4(\underline{\theta})y_p + k_5(\theta)r && (4b) \end{aligned}$$

where $\underline{x}_c$ is the state vector of the controller, $r(t)$ is a bounded external reference input, and $\underline{\theta}$ is an estimate of $\underline{\theta}^*$. It is assumed that the dimension of $\underline{x}_c$ and the controller gains $\underline{K}_o(\cdot), \ldots, k_5(\cdot)$ are designed such that

(A5) $\underline{K}_o(\cdot), \underline{k}_1(\cdot), \ldots, k_5(\cdot)$ are functions of $\underline{\theta}$, defined and continuous for all $\|\underline{\theta}\| < \infty$.

(A6) There exists $\sigma_c > 0$ such that for all constant vectors $\underline{\theta}\epsilon[\underline{\theta}_{min}, \underline{\theta}_{max}]$ we have the following: Equation (2a) with $\underline{g}^*$ replaced by $\underline{g}(\underline{g} = \underline{G\theta} + \underline{g}^o)$, together with the controller (4), results in a closed loop system all eigenvalues of which have real parts less than $-\sigma_c$.

Equations (4) represent a general (adjustable) compensator with feedforward and feedback compensation. It is designed so that a plant model with parameters $\underline{\theta}$ together with the controller, tuned with the same parameters, results in a stable closed loop system, and further that small changes of $\underline{\theta}$ result in small changes of the controller gains. Assumption (A3) assures that a compensator, which is of the form (4) and satisfies (A5), (A6), can always be designed. (Of course this requires a control strategy which does not attempt undue control of poorly controllable/observable (stable) subsystems. Note also that the controller gains $k_i(\underline{\theta})$ need only be designed for $\underline{\theta}\epsilon[\underline{\theta}_{min}, \underline{\theta}_{max}]$. To also define the controller gains outside this domain we may, for example, redefine them as $k_i(\underline{\theta} - \underline{f}(\underline{\theta}))$, where $\underline{\theta} - \underline{f}(\underline{\theta})$ projects $\underline{\theta}$ onto the boundary of $[\underline{\theta}_{min}, \underline{\theta}_{max}]$ when $\underline{\theta}$ is outside. $\underline{f}(\underline{\theta})$ is defined in eqn. (5h) below). In fact, assumptions (A5), (A6) leave much freedom for the controller design both with respect to the controller structure and with respect to satisfying various design objectives beyond stabilization.

b) Adaptive Scheme

In order to adaptively adjust the controller, we need an adaptive scheme, which generates an estimate $\underline{\theta}$ of the unknown parameter vector $\underline{\theta}^*$. Based on the plant representation (3), we set up the following scheme.

Define

$$\begin{aligned} \dot{\underline{v}}_1 &= \underline{F}\underline{v}_1 + \underline{c}u_p \\ \dot{\underline{v}}_2 &= \underline{F}\underline{v}_2 + \underline{c}y_p && (5a) \\ y &= \underline{g}^T(\underline{\theta})\underline{v}, \end{aligned}$$

where y is an estimated plant output and

$$\underline{g}(\underline{\theta}) = \underline{G}\underline{\theta} + \underline{g}^o \tag{5b}$$

is an estimated plant parameter vector. Further define the identification error

$$e = y - y_p \tag{5c}$$

and define the relative identification error

$$e_1 = e/(\gamma + m), \tag{5d}$$

where γ and m are as defined in assumption (A4), i.e. $\gamma > 0$ is arbitrary,

$$\dot{m} = -\sigma_m m + |u_p| + |y_p|, \quad m_o > 0 \tag{5e}$$

and $0 < \sigma_m < -Re\ \lambda_i(\underline{F})$. Then the adaptive law is chosen as

$$\dot{\underline{\theta}} = -\underline{\Gamma}\left[\frac{\underline{G}^T \underline{v} D(e_1)}{\gamma + m} + \underline{f}(\underline{\theta})\right], \tag{5f}$$

Here $\underline{\Gamma} = \underline{\Gamma}^T > \underline{0}$ is an arbitrary adaptive gain, $D(\cdot)$ is the dead zone function defined by

$$D(e_1) = \left\{ \begin{array}{ccc} 0 & \text{if} & |e_1| < d \\ e_1 - d & \text{if} & e_1 > d \\ e_1 + d & \text{if} & e_1 < -d \end{array} \right\} \tag{5g}$$

d is the size of the dead zone, which can be chosen arbitrarily in an interval $[0, d_o)$, where d_o depends on the plant model and the controller design and is specified later, and $\underline{f}(\underline{\theta})$ is defined by

$$f_i(\underline{\theta}) = \left\{ \begin{array}{ll} 0 & \text{if } [\underline{\theta}_{min}] < \theta_i < [\underline{\theta}_{max}]_i \\ \theta_i - [\underline{\theta}_{max}]_i & \text{if } \theta_i > [\underline{\theta}_{max}]_i \\ \theta_i - [\underline{\theta}_{min}]_i & \text{if } \theta_i < [\underline{\theta}_{min}]_i \end{array} \right\} \tag{5h}$$

The idea behind the adaptive law (5f) is illustrated as follows. Using (5a-c) and (3a,b) the identification error may be written in the form

$$e = (\underline{\theta} - \underline{\theta}^*)^T \underline{G}^T \underline{v} - e_p. \tag{6}$$

In (6) the plant modeling error e_p is an unknown disturbance, the magnitude of which depends on the plant input and/or output signals (e_p may even grow without bound if u_p and/or y_p grow without bound). Such a disturbance is difficult to deal with directly. Normalization of e with respect to $\gamma + m$, i.e. with respect to the signal level, gives the relative identification error in the form

$$e_1 = [(\underline{\theta} - \underline{\theta}^*)^T \underline{G}^T \underline{v} - e_p]/(\gamma + m), \tag{7}$$

and from (3e) we obtain

$$|e_p|/(\gamma + m) < \mu. \tag{8}$$

Hence the disturbance, caused by e_p in the relative identification error e_1, is bounded by a constant independent of the plant input and output signals.

As $\partial e_1^2/\partial \underline{\theta}$ is given by $2\underline{G}^T \underline{v} e_1/(\gamma + m)$, it is seen that the first term in the adaptive law (5f) is essentially a steepest descent term, but with the error e_1 replaced by $D(e_1)$, thus applying the known dead zone idea for bounded disturbances (e.g. [7]).

The $\underline{f}(\underline{\theta})$-term in the adaptive law (5f) is zero whenever $\underline{\theta}$ is in the set $[\underline{\theta}_{min}, \underline{\theta}_{max}]$ and becomes nonzero only to drive $\underline{\theta}$ back to this set, when $\underline{\theta}$ gets outside it.

4. The Closed Loop Adaptive Control System

The overall adaptive closed loop system is defined by equations (3)-(5). Substituting (5c) into (5a) gives

$$\begin{aligned} \dot{\underline{v}}_1 &= \underline{F}\underline{v}_1 + \underline{c}u_p \\ \dot{\underline{v}}_2 &= \underline{F}\underline{v}_2 + \underline{c}y_p \\ y_p &= \underline{g}^T(\underline{\theta})\underline{v} - e. \end{aligned} \tag{9}$$

The plant, represented by (9), together with the controller (4), gives rise to a representation of the main control loop in the form

$$\dot{\underline{x}} = \underline{A}(\underline{\theta})\underline{x} - \underline{b}_1(\underline{\theta})e + \underline{b}_2(\underline{\theta})r \tag{10}$$

where $\underline{x} = [\underline{v}^T, \underline{x}_c^T]^T$ and

$$\underline{A}(\underline{\theta}) = \begin{bmatrix} \underline{F} + \underline{c}k_4(\underline{\theta})\underline{g}_1^T(\underline{\theta}) & \underline{c}k_4(\underline{\theta})\underline{g}_2^T(\underline{\theta}) & \underline{c}\underline{k}_3^T(\underline{\theta}) \\ \underline{c}\underline{g}_1^T(\underline{\theta}) & \underline{F} + \underline{c}\underline{g}_2^T(\underline{\theta}) & \underline{0} \\ \underline{k}_1(\underline{\theta})\underline{g}_1^T(\underline{\theta}) & \underline{k}_1(\underline{\theta})\underline{g}_2^T(\underline{\theta}) & \underline{K}_o \end{bmatrix} \tag{11}$$

$$\underline{b}_1(\underline{\theta}) = \begin{bmatrix} \underline{c}k_4(\underline{\theta}) \\ \underline{c} \\ \underline{k}_1(\underline{\theta}) \end{bmatrix}, \underline{b}_2(\underline{\theta}) = \begin{bmatrix} \underline{c}k_5(\underline{\theta}) \\ \underline{0} \\ \underline{k}_2(\underline{\theta}) \end{bmatrix}.$$

Observe that the above equations hold for arbitrary $\mu > 0$, i.e. whether or not unmodeled plant uncertainty is present. In particular, if $\mu = 0$, then (3) represents the plant model, and so does (9). Moreover, a comparison of (2a) and (9) shows that, except for the error signal e, equations (9) represent a plant model with parameters $\underline{g}(\underline{\theta})$ in place of $\underline{g}(\underline{\theta}^*)$. Hence the closed loop system matrix $\underline{A}(\underline{\theta})$ in (10) arises from a plant model with parameters $\underline{\theta}$ and a controller tuned with the *same* parameters. As a consequence, assumption (A6) assures that if $\underline{\theta}\epsilon[\underline{\theta}_{min}, \underline{\theta}_{max}]$ then all eigenvalues of $\underline{A}(\underline{\theta})$ have real parts less than $-\sigma_c$. Also, from assumptions (A2) and (A5) it follows that $\underline{A}(\underline{\theta}), \underline{b}_1(\underline{\theta})$, and $\underline{b}_2(\underline{\theta})$ are bounded for all $\underline{\theta}\epsilon[\underline{\theta}_{min}, \underline{\theta}_{max}]$.

The first step of the analysis is to establish certain properties of the adaptive scheme as it operates within the overall adaptive control system.

Lemma 1

Consider the overall adaptive closed loop system (3)-(5) and let the size d of the dead zone be arbitrary $(0 < d < \infty)$. Further consider an arbitrary plant uncertainty satisfying assumption (A4) with $\mu < d$. Then for arbitrary initial conditions and arbitrary, piecewise continuous reference inputs $r(t)$,

(i) $\underline{\theta}(t)$ is uniformly bounded,

(ii) $\int_0^\infty |D(e_1)|dt = \Delta < \infty$,

(iii) $\lim_{t\to\infty} \underline{f}(\underline{\theta}) = 0$,

(iv) $\lim_{t\to\infty} \underline{\theta}(t) = \overline{\underline{\theta}}$, $\overline{\underline{\theta}}\epsilon[\underline{\theta}_{min}, \underline{\theta}_{max}]$.

The proof of Lemma 1, which is given in [9], shows that whether the overall adaptive system is stable or not, the signal e_p which is due to the modeling errors (and which would grow without bound if the overall system were unstable!) is within the relative dead zone, and therefore the adaptive law is able to make the norm of the parameter error decrease monotonically as if no modeling errors were present. From this we get the desirable results that $\underline{\theta}(t) \to \overline{\underline{\theta}}$ as $t \to \infty$, and $\overline{\underline{\theta}}\epsilon[\underline{\theta}_{min}, \underline{\theta}_{max}]$.

However, from these facts the stability of the overall adaptive system does not immediately follow: Due to the relative dead zone, the adaptive law is unable to reduce the parameter error $\underline{\theta} - \underline{\theta}^*$ to zero. Therefore the controller will not be exactly adapted to the plant, even in the limit. Also, we do not have a property like $e_1 \to 0$ as $t \to \infty$ (or e_1 being "small" when $\|x\|$ is "large"), which had been the basis of stability proofs when no modeling errors were present. We have only $\int_0^\infty |D(e_1)|dt = \Delta < \infty$, i.e. except for a finite integral of excursions, e_1 will be less than the size d of the dead zone.

Based on the above Lemma, one obtains the following robust stability property of the adaptively controlled plant.

Theorem (Robust Stability)

There exist $d_o > 0$, such that if the size of the relative dead zone in the adaptive law is chosen so that $0 < d < d_o$, then the adaptive control system (3)-(5) is robustly stable for all plant uncertainties satisfying (A4) with $\mu < d$.

The proof of robust stability of the adaptive control system is given in [9]. This proof is constructive in the sense that it specifies an interval $[0, d_o)$ within which the size d of the relative dead zone can be chosen arbitrarily. Roughly, d_o depends on the stability margin of the ideally adapted control system.

Since robust stability will be achieved for all modeling uncertainties of magnitude $\mu > d$, the choice of d will typically reflect the maximally expected unmodeled plant uncertainty. Hence, the more accurate the plant model is, the smaller values of d can be chosen, and the more accurately will the adaptive scheme be able to adapt. The converse is also true.

5. Conclusions

A robust indirect adaptive control approach is presented. The approach can be applied in practical situations, where the unknown parameters of the plant lie in a known convex set throughout which no unstable pole-zero cancellation occurs. Also, the approach enables a fairly general design of the controller, which is then tuned adaptively using appropriate plant parameter estimates.

As a means to achieve robustness of the adaptive control system with respect to unmodeled plant uncertainties, we propose the use of a relative dead zone in the adaptive law, from which the plant parameter estimates are generated. With the relative dead zone in the adaptive law, adaptation takes place only when the identification error, suitably normalized relative to the signals in the adaptive system, is within the dead zone. Thereby the parameter descent property of the adaptive law is preserved in the presence of small plant uncertainties, and robustness is obtained, but the estimated plant parameters will, in general, converge to some $\overline{\underline{\theta}} \neq \underline{\theta}^*$. It is shown in [9] that, nevertheless, the closed loop adaptive system operates as an ideal system in the sense of the design (but at design point $\overline{\underline{\theta}}$ instead of $\underline{\theta}$, and with remaining modeling uncertainty of size d instead of μ), and has the desirable potential of adaptive performance improvements.

Finally we mention that the relative dead zone idea has also been useful to establish robustly stable model reference adaptive control [10].

Acknowledgments

This work was carried out while the author was with the DFVLR - Institute for Flight Systems Dynamics, Oberpfaffenhofen.

References

[1] C.E. Rohrs, L. Valavani, M. Athans and G. Stein, "Robustness of Adaptive Control Algorithms in the Presence of Unmodeled Dynamics," Proc. 21st IEEE Conference on Decision and Control, Orlando, Florida, 1982, pp. 3-11.

[2] P.A. Ioannou and P.V. Kokotovic, "Adaptive Systems with Reduced Models," in Lecture Notes in Control and Information Sciences, vol. 47, Springer, New York, 1983.

[3] G. Kreisselmeier, "On Adaptive State Regulation," *IEEE Trans. Autom. Contr.*, vol. AC-27, pp. 3-7, 1982.

[4] B.D.O. Anderson and R.M. Johnstone, "Robust Lyapunov Stability Results and Adaptive Systems," Proc. 20th IEEE Conference on Decision and Control, San Diego, Ca., 1981, pp. 510-515.

[5] B.D.O. Anderson, "Exponential Convergence and Persistent Excitation," Proc. 21st IEEE Conference on Decision and Control, Orlando, Florida, 1982, pp. 12-17.

[6] K.S. Narendra and A.M. Annaswamy, "Persistent Excitation and Robust Adaptive Algorithms," Proc. 3rd Workshop on Applications of Adaptive Systems Theory, Yale University, 1983, pp. 11-18.

[7] B.B. Peterson and K.S. Narendra, "Bounded Error Adaptive Control," *IEEE Trans. Autom. Contr.*, vol. AC-27, pp. 1161-1168, 1982.

[8] B.D.O. Anderson, "Adaptive Identification of Multiple-Input, Multiple-Output Plants," Proc. 13th IEEE Conference on Decision and Control, 1974, pp. 273-281.

[9] G. Kreisselmeier, "A Robust Indirect Adaptive Control Approach," DFVLR-Institut fuer Dynamik der Flugsysteme, 1983, also *Int. J. Contr.*, to appear 1985/86.

[10] G. Kreisselmeier and B.D.O. Anderson, "Robust Model Reference Adaptive Control," *IEEE Trans. Autom. Contr.*, to appear, 1985/86.

[11] C.A. Desoer and M. Vidyasagar, Feedback Systems: Input-Output Properties, Academic Press, New York, 1975.

[12] R.L. Kosut and B. Friedlander, "Performance Robustness Properties of Adaptive Control Systems," Proc. 21st IEEE Conference on Decision and Control, Orlando, Florida, 1982, pp. 18-23.

[13] R.L. Kosut, C.R. Johnson and B.D.O. Anderson, "Robustness of Reduced-Order Adaptive Model Following," Proc. 3rd Workshop on Applications of Adaptive Systems Theory, Yale University, 1983, pp. 1-6.

[14] G. Kreisselmeier and K.S. Narendra, "Stable Model Reference Adaptive Control in the Presence of Bounded Disturbances," *IEEE Trans. Autom. Contr.*, vol. AC-27, pp. 1169-1175, 1982.

[15] L.A. Zadeh and C.A. Desoer, Linear System Theory, McGraw Hill, New York, 1963.

Global Stability of a Direct Adaptive Control Scheme With Respect to a Graph Topology

L. Praly
CAI Ecole des Mines
35 Rue Saint Honore
77305 Fontainebleau Cedex
France

Abstract

We study the stability given by a modified model reference adaptive controller. Modifications are projections of the adapted parameters into a convex compact set and normalization of the signals entering the adaptation law by a weighted ℓ_2-norm of the I/O signals. The plant is assumed to satisfy properties which are proved to be robust with respect to a graph topology based on μ-exponential stability. Global stability is established and the mean square tracking error is shown to converge to zero linearly with the unmodeled effects.

1. Introduction

Global stability and convergence are now well established properties for model reference adaptive controllers in a closed loop with an ideal plant Π_0. For such a plant Π_0, it is assumed that:

A1: The input $u(t)$ and the output $y(t)$ are related by

$$A_0(q^{-1})y(t) = q^{-d}B_0(q^{-1})u(t) \tag{1}$$

where A_0, B_0 are polynomials in the unit delay operator q^{-1}.

A2: The plant Π_0 is minimum phase.

A3: The delay d is known.

A4: n_A, n_B are known upperbounds of the degree of A_0, B_0 respectively.

This result can be easily extended to a plant described by

$$G_0(q^{-1})A_0(q^{-1})y(t) = q^{-d}G_0(q^{-1})B_0(q^{-1})u(t) \tag{2}$$

where $G_0(q^{-1})$ represents cancellable exponentially stable poles-zeros.

Unfortunately, in practice, the relation between plant input and plant output is at best a small perturbation of (2). Among the causes of perturbation, one can find the presence of a bounded disturbance $v(t)$:

$$G_0(q^{-1})A_0(q^{-1})y(t) = q^{-d}G_0(q^{-1})B_0(q^{-1})u(t) + v(t) \tag{3}$$

or the presence of unmodeled dynamics, i.e. we have for example:

$$\begin{aligned}(G_0(q^{-1}) &+ \epsilon_c G_1(q^{-1}))(A_0(q^{-1}) + \epsilon_p A_1(q^{-1}))y(t) = \\ (G_0(q^{-1}) &+ \epsilon_c G_2(q^{-1}))(\epsilon_u B_1(q^{-1}) + q^{-d}B_0(q^{-1}) + \epsilon_z q^{-d}B_2(q^{-1}))u(t)\end{aligned} \tag{4}$$

with G_1, G_2, A_1, B_1, B_2 polynomials. In this case neglecting ϵ_c corresponds to neglecting nearly cancellable exponentially stable poles-zeros, neglecting ϵ_p or ϵ_z corresponds to neglecting fast stable poles or zeros and neglecting ϵ_u corresponds to neglecting fast unstable zeros.

The case of a bounded disturbance has been investigated by Egardt [1]. He has shown that even infinitely small bounded disturbance can lead to unbounded signals; however, if the adapted parameters are bounded then so are the signals. As a consequence standard theoretical adaptive controllers have to be modified. For example, at each step, one can project the adapted parameters into a bounded convex set.

In the case of unmodeled dynamics, the problem is to show that for sufficiently small $\epsilon_c, \epsilon_p, \epsilon_z, \epsilon_u$ (but not zero), global stability is preserved. More generally we would be interested in the robustness of this global stability. This means (see [2]) that if global stability holds for the ideal plant Π_0 (for which the adaptive controller has been designed), it holds also for all plants in an open neighborhood of Π_0. For this to make sense, a topology on the set of plants is needed. To prove the advantage of adaptive control, this topology should be for example the weakest topology in which classical linear feedback exponential stability is robust, i.e. the graph topology of Vidyasagar [3].

To study this problem, Gawthrop and Lim [4] have applied the input-output operator techniques to the error model. For the case $\epsilon_u = 0$, they have shown that if the plant in feedback with a linear time invariant controller (lying among those reachable by the adaptation law) is strictly inside a cone, then global stability holds. But no proof is available for standard theoretical adaptive controllers that this cone is not reduced to one point (i.e. the ideal case). A new modification helps in avoiding this difficulty. To motivate this, we express the effects of unmodeled dynamics as an open loop disturbance and rewrite (4) in the form (3):

$$G_0(q^{-1})A_0(q^{-1})y(t) = q^{-d}G_0(q^{-1})B_0(q^{-1})u(t) + w(t) \tag{5}$$

where $w(t)$ depends on the signals and therefore is potentially unbounded. Nevertheless it can be easily seen that for any set of polynomials G_1, G_2, A_1, B_1, B_2 and for any $\mu > 0$, there exists γ such that:

$$w(T)^2 \leq \gamma^2(\epsilon_c^2 + \epsilon_p^2 + \epsilon_z^2 + \epsilon_u^2) \sum_{t=0}^{T} \mu^{2(T-t)}(u(t)^2 + y(t)^2) \tag{6}$$

This shows that the effects of unmodeled dynamics $w(t)$, when "normalized" by $n(t)$ defined as

$$n(t)^2 = \mu^2 n(t-1)^2 + u(t)^2 + y(t)^2$$

are bounded and even small if $\epsilon_c, \epsilon_p, \epsilon_z, \epsilon_u$ are small. This remark and Egardt's result is the motivation for normalizing the signals entering the adaptation law (but not the control law).

In [5], we have shown that both this normalization and the projection, proposed by Egardt, allow the global stability to be proved in the case $\epsilon_u = 0$ and $\epsilon_p, \epsilon_z, \epsilon_c$ sufficiently small (see also [6]). A more interesting result is obtained by applying the technique of Gawthrop and Lim to decide as to how $w(t)$ is transformed by feedback. This has been done for the case $\epsilon_u = 0$ by Ortega, Praly and Landau [7]. They have shown also that due to the normalization, the cone mentioned above is not reduced to a point. This result has been extended to the case when ϵ_u is not zero and when a bounded disturbance is present in [8].

In this paper, we reestablish the result of [8] for a more realistic controller and using a more compact (though more conservative) proof. Also, we state the robustness of the global stability with respect to a graph topology based on μ-exponential stability (the poles are inside a disk of radius μ).

To shorten this introduction, we have mentioned only some results of the discrete time case. Very interesting work has also been done in the continuous time case for both the bounded disturbance case ([9-11]) and the unmodeled dynamics case ([12-14]).

Following our definition of robustness, we first need to define our adaptive controller, which is done in section 2. In section 3, we give a new set of assumptions which define non-ideal plants which are robust. Finally in section 4, we prove that these assumptions are sufficient to guarantee global stability and we establish that the mean square tracking error goes to zero as the unmodeled effects disappear.

2. A Model Reference Adaptive Controller

The objective of a model reference adaptive controller is to minimize the tracking error

$$e^m(t) = C(q^{-1})(y(t) - y^m(t)) \tag{7}$$

where $y^m(t)$ is a uniformly bounded reference output and $C(q^{-1})$ is a monic polynomial of degree n_c with exponentially stable roots. For the ideal plant Π_0 defined by (1), the following adaptive controller solves the problem.

Let $\phi(t), \theta(t)$ be vectors of $\mathcal{R}^n$ defined as

$$\phi(t) = (u(t)\ldots u(t-n_s)y(t)\ldots y(t-n_r))' \tag{8}$$

$$\theta(t) = (s_0(t)\ldots s_{n_s}(t)r_0(t)\ldots r_{n_r}(t))' \tag{9}$$

with

$$n_s = n_B + d - 1,\ n_r = \max(n_A - 1, n_c - d),\ n = n_s + n_r + 2 \tag{10}$$

The standard updating algorithm is specified by the following equations:

$$\begin{aligned} e(t) &= C(q^{-1})y(t) - \theta(t-1)'\phi(t-d) \\ g(t) &= \tfrac{1}{1+\overline{\phi}(t-d)'F(t-1)\overline{\phi}(t-d)} \\ \theta'(t) &= \theta(t-1) + g(t)F(t-1)\overline{\phi}(t-d)\overline{e}(t) \\ F'(t) &= F(t-1) - g(t)F(t-1)\overline{\phi}(t-d)\overline{\phi}(t-d)'F(t-1) \end{aligned} \tag{11}$$

where

$$F(t) = (1 - \frac{\lambda_0}{\lambda_1})F'(t) + \lambda_0 I \qquad \text{(matrix regularization)} \tag{12}$$

and

$$\theta''(t) = \theta'(t) + \max(0, \sigma_0 - s_0'(t))\frac{F_{.1}(t)}{F_{11}(t)} \qquad \text{(leading coefficient regularization)} \tag{13}$$

and

$$\theta(t) = \theta_c + (\theta''(t) - \theta_c)\min(1, \frac{R}{||\theta''(t) - \theta_c||}) \quad \text{(projection into the sphere}(\theta_c, R)) \tag{14}$$

The control law (implicit in $u(t)$) is given by :

$$C(q^{-1})y^m(t+d) = \theta(t)'\phi(t) \tag{15}$$

Here $F_{.1}(t)$ is the first column of $F(t)$, $F_{11}(t)$ is the first entry of $F(t)$;$\overline{\phi}(t-d), \overline{e}(t)$ are normalized signals defined in (17). In the following, we refer to equations (11) to (14) as the adaptation law.

Normalization Procedure: Before specifying the adaptation law, the signals are normalized as follows: Let $\rho(t)^2$ be the output of a first order filter with $\phi(t-d)'\phi(t-d)$ as input, or more precisely with:

$$\rho(t)^2 = \mu^2\rho(t-1)^2 + \max(||\phi(t-d)||^2, (1-\mu^2)\rho^2) \tag{16}$$

where $0 < \mu < 1,\ \ \rho(0) = \rho$. We define

$$\overline{e}(t) = \frac{e(t)}{\rho(t)}, \qquad \overline{\phi}(t-d) = \frac{\phi(t-d)}{\rho(t)} \tag{17}$$

Note that $1/\rho(t)$ and $\overline{\phi}(t)$ are bounded.

Comment: This algorithm is a least square version of the "DSA-algorithm with projection" proposed by Egardt (p. 69 [1]). In [7],[8], a d-interlaced version of it has been studied. Since our goal is to deal with nonideal plants, it incorporates three modifications in the standard theoretical least square adaptive controller:
1. Monitoring of the adapted parameters using projections (13), (14)- this is an efficient remedy to the problem of bounded disturbances (see [1]).
2. Normalization procedure - As mentioned in the introduction, this causes the effects of unmodeled dynamics to appear as bounded disturbances in the adaptation law.
3. Matrix Regularization (12): This causes the eigenvalues of $F(t)$ to stay in $[\lambda_0, \lambda_1]$ resulting in a non-decaying gain, a desirable property in applications.

Implementation considerations:
1. The ratio $\frac{\lambda_0}{\lambda_1}$ is equivalent to a forgetting factor. It can be made time varying according to the "quality" of the information contained in $\overline{\phi}(t-d)$ compared to $F(t-1)$.
2. λ_1 allows a trade-off between the rate of convergence and the reduction of noise through smoothing. It can be made time varying according to the plant time variations.
3. A dead zone on the (unnormalized) estimation error can be introduced. This improves convergence in the presence of bounded (numerical) noise and takes care of some "burst" phenomena. In this case we can take:

$$g(t) = \frac{\max(1 - \frac{V(t)}{|e(t)|}, 0)}{1 + \overline{\phi}(t-d)' F(t-1) \overline{\phi}(t-d)} \tag{18}$$

where $V(t)$ defines the dead zone.
4. θ_c may be (slowly) time varying. For example, it can be given by a gain scheduling. In this case R, also time varying, characterizes the confidence we have in θ_c.
5. The update of $F(t)$ is made using a $U-D$ or square root algorithm.

All the results presented below can be extended to cases which include these modifications.

Let us denote by Θ, the intersection in $\mathcal{R}^n$ of the closed sphere $(\theta_c,\ R\sqrt{\frac{\lambda_0}{\lambda_1}})$ and the closed half space $(s_0 \geq \sigma_0)$. Due to projections, our algorithm looks for a parameter vector lying in a set containing Θ. For any fixed element θ in Θ, we can define a sequence $e_\theta(t)$ as:

$$e_\theta(t) = C(q^{-1})y(t) - \theta' \phi(t-d). \tag{19}$$

Clearly $e_\theta(t)$ is the estimation error using a fixed vector θ. The technical interest in the above three modifications is that the following properties hold:

The adapted parameter vector $\theta(t)$ satisfies:

(i) $\|\theta(t)\| \leq R' = R + \|\theta_c\|, \quad s_0(t) \geq \sigma_0$ (20)

(ii) $\|\theta(t) - \theta(t-1)\| \leq \sqrt{\lambda_1}(1 + \sqrt{\frac{\lambda_0}{\lambda_1}})\overline{e}(t)$ (21)

(iii) For any $\theta \in \Theta$, we have

$$e(t)^2 \leq (V_\theta(t-1) - V_\theta(t))\rho(t)^2 + (1 + \lambda_1)e_\theta(t)^2 \tag{22}$$

where $V_\theta(t)$ satisfies:

$$0 \leq V_\theta(t) \leq V_5 \tag{23}$$

Proof: The arguments of this proof are standard. The proof is omitted here due to space limitations. Please see [15] for details.

3. Robust Assumptions

Assumptions A1 to A4 from which the adaptive controller of section 2 was designed, are not robust with respect to any practically interesting topology. In this section, we propose a new set of assumptions:

Let $\Im$ be the ring of proper rational functions $F(q)$ whose poles are all in the open disk of radius μ (they are said to be μ-exponentially stable). For example, $A_0(q^{-1})$, $B_0(q^{-1})$, $C(q^{-1})$, as functions of q, are elements of $\Im$. $\Im$ is equipped with the norm:

$$\gamma(F) = \sup_{|q|\geq\mu} |F(q)|. \tag{24}$$

For a sequence $x(t)$, we define its $\ell_2(\mu)$-norm as:

$$\|x\|_T^2 = \sum_{t=0}^{T} \mu^{-2t} x(t)^2 \tag{25}$$

we have:

$$\mu^{-T}|x(T)| \leq \|x\|_T. \tag{26}$$

Given an element F of $\Im$, we define a sequence $z(t)$ by

$$z(t) = F(q)x(t) \tag{27}$$

where q is the forward shift operator. We have (see [16]):

$$\|z\|_T \leq \gamma(F)\|x\|_T. \tag{28}$$

In the following, we consider plants Π which can be described by:

A1′: $A(q)y(t) = B(q)u(t-1) + v(t)$ (29)

where A, B are coprime elements of $\Im$ such that B/A is a proper rational function and

$$A(0) = 1 \tag{30}$$

and $v(t)$, appearing as an external signal, may incorporate the effects of bounded disturbances, some distributed parameters, some nonlinearities, or some time variations, but is restricted to satisfy:

$$|v(t+1)| \leq \mu_0|v(t)| + \nu(|u(t)| + |y(t)|) + V \tag{31}$$

for some constants μ_0, ν, V such that

$$0 \leq \mu_0 < \mu, \quad 0 \leq \nu \leq 1, \quad 0 \leq V. \tag{32}$$

With the presence of $v(t)$ in (29), we can assume:

$$u(t) = 0, \;\; y(t) = 0, \;\; v(t) = 0 \qquad \forall t \leq 0 \tag{33}$$

Now since $B(q)$, given by the plant Π and $C(q^{-1})$, given by the controller, are analytic functions outside the disk of radius μ, we can write a Laurent's series:

$$B(q)C(q^{-1}) = \sum_{i=0}^{\infty} h_i q^{-i} \tag{34}$$

and define a polynomial $P(q^{-1})$ of degree $d-2$ and an element $D(q)$ of $\Im$ given by:

$$P(q^{-1}) = \sum_{i=0}^{d-2} h_i q^{-i} \tag{35}$$

$$D(q) = \sum_{i=0}^{\infty} h_{i+d-1} q^{-i} \tag{36}$$

We have the identity:

$$B(q)C(q^{-1}) = P(q^{-1}) + q^{1-d}D(q) \tag{37}$$

For the ideal plant Π_0, we remark that:

$$P_0(q^{-1}) = 0 \tag{38}$$
$$D_0(q) = B_0(q^{-1})C(q^{-1}) \tag{39}$$

A consequence of A2 is that $D_0(q)$ is stably invertible. For the plant Π, we assume:

A2′ $D(q)$ is an invertible element (i.e. a unit) of $\Im$.

It follows that $\gamma(D^{-1})$ is finite.

This assumption is satisfied by Π_0 if $C(q^{-1})$ and $B_0(q^{-1})$ are units of $\Im$. This implies that μ, a controller parameter, should be chosen larger than the modulus of the desired poles given by C and of the plant zeros given by B_0.

A consequence of the knowledge of n_A, n_B given by A4 is that, with $Q_0(q^{-1})$ a polynomial vector defined by:

$$Q_0(q^{-1}) = (A_0(q^{-1}) \cdots q^{-n_s} A_0(q^{-1})\ q^{-d}B_0(q^{-1}) \cdots q^{-(d+n_r)}B_0(q^{-1}))' \tag{40}$$

there exists a vector θ_0 satisfying (Bezout's theorem):

$$1 = \frac{Q_0(q^{-1})'\theta_0}{C(q^{-1})B_0(q^{-1})} \tag{41}$$

and using this particular fixed θ_0 in the control law (15), we achieve:

$$\lim_{t\to\infty} e^m(t) = 0 \tag{42}$$

In the case of Π, we define $Q(q)$ as:

$$Q(q) = (A(q) \cdots q^{-n_s} A(q)\ q^{-1}B(q) \cdots q^{-(n_r+1)}B(q))' \tag{43}$$

And for any θ, with D a unit of $\Im$ (given by A2′), we define a new element of $\Im$ by:

$$H_\theta(q) = 1 - D(q)^{-1}Q(q)'\theta \tag{44}$$

Clearly $\gamma(H_\theta)$ is a continuous function of θ. Hence we can choose θ as an argument of $\min_{\theta\in\Theta}[\gamma(H_\theta)]$. This vector is denoted as θ_* and the subscript $*$ in e_*, v_*, H_* refer to this vector. We assume

A4′ $\gamma(H_*) < \gamma_h$ (45)

where

$$\gamma_h^{-1} = \sqrt{1+\lambda_1}(1 + k_1(d-1)) \tag{46}$$

with k_1 a positive constant given in Table 1.

This assumption shows how λ_1 allows a trade-off between the rate of adaptation and the amount of allowed unmodeled dynamics. In the case $d = 1$, we have exactly the conicity condition given in [7],[8] (for a d-interlaced recursion algorithm).

This assumption is satisfied by Π_0 if θ_0 belongs to Θ. This means that the sign and a lower bound of the leading coefficient, and an upperbound of the norm of $\theta_0 - \theta_c$ are assumed to be known in the design of our adaptive controller.

Finally a consequence of A3 is that $P_0(q^{-1})$ is equal to zero. For the plant Π, we assume: A3′:

$$\gamma(P) < \frac{\gamma_h - \gamma(H_*)}{\gamma(D^{-1})\left(k_2\gamma(D^{-1}) + k_3 + k_4(\gamma_h - \gamma(H_*))\right)} \tag{47}$$

with k_2, k_3, k_4 strictly positive constants given in table 1.

Let us now prove that A2$'$ to A4$'$ are robust. Following the approach in [3], we consider a graph topology constructed from the set $\Im$. Since $F(q) \rightarrow F(\mu q)$ is an isomorphism on the field of rational functions, all the properties of [3] can be rederived here. In particular we obtain a topology which is the weakest one in which feedback μ-exponential stability is robust. Compared with the graph topology of [3], the main loss is that we do not allow pole-zero cancellation of roots whose modulus is between μ and 1. Since this topology on the set of plants Π is obtained from the topology on $\Im \times \Im$, our robustness result follows from Theorem 1.

Theorem 1:
The set of elements (A, B) which satisfy A2$'$ to A4$'$ is open.
Proof: Part 1: The set of triple (F, H, G) which satisfy

$$\gamma(H) < \gamma_h \tag{48}$$

$$\gamma(G) < \frac{\gamma_h - \gamma(H)}{\gamma(F)\,(k_2\gamma(F) + k_3 + k_4(\gamma_h - \gamma(H)))} \tag{49}$$

is clearly an open set of $\Im^3$. As mentioned in [3], the application $F^{-1} \rightarrow F$ is continuous on the set $\mathcal{U}$ of units of $\Im$ and the set $\mathcal{U}$ is an open set of $\Im$. Hence the set of triple (F^{-1}, H, G) satisfying (48),(49) is open.
Part 2: Assume for the time being that the application $(A, B) \rightarrow P$ is continuous. Then clearly the application $(A, B) \rightarrow D$ is continuous. This implies that the set of pairs (A, B) such that D belongs to $\mathcal{U}$ is open. Moreover on this set, the application $(A, B) \rightarrow H_\theta$ is continuous for any fixed θ. Hence we have established that given a fixed θ, the application $(A, B) \rightarrow (D, H_\theta, P)$ is continuous. And with part 1, the set $\mathcal{V}_\theta$ of pairs (A, B) such that D belongs to $\mathcal{U}$ and

$$\gamma(H) < \gamma_h \tag{50}$$

$$\gamma(P) < \frac{\gamma_h - \gamma(H_\theta)}{\gamma(D^{-1})\,(k_2\gamma(D^{-1}) + k_3 + k_4\,(\gamma_h - \gamma(D^{-1})))} \tag{51}$$

is open.

We now prove that $(A, B) \rightarrow P$ is continuous. With (34),(35) we have:

$$\gamma(P) = \sup_{|q|\geq\mu} |\sum_{i=0}^{d-2} h_i q^{-i}| \leq \sum_{i=0}^{d-2} |h_i|\mu^{-i} \tag{52}$$

But with the Cauchy-Schwarz inequality, we have

$$\sum_{i=0}^{d-2} |h_i|\mu^{-i} \leq \sqrt{d-1}(\sum_{i=0}^{d-2} h_i^2\mu^{-2i})^{\frac{1}{2}} \tag{53}$$

$$\leq \sqrt{d-1}(\sum_{i=0}^{\infty} h_i^2\mu^{-2i})^{\frac{1}{2}} \tag{54}$$

Using Plancherel's theorem:

$$\sum_{i=0}^{\infty} h_i^2\mu^{-2i} = \frac{1}{2\pi}\int_0^{2\pi} |B(\mu e^{i\theta})C(\mu^{-1}e^{-i\theta})|^2 d\theta \tag{55}$$

$$\leq \sup_{|q|\geq\mu} |B(q)C(q^{-1})|^2 \tag{56}$$

The property follows since the mapping is linear and we have shown:

$$\gamma(P) \leq \sqrt{d-1}\,\gamma(BC) \leq \sqrt{d-1}\,\gamma(C)\gamma(B) \tag{57}$$

Part 3: The conclusion follows noticing that the set mentioned in the theorem is:

$$\cup_{\theta\in\Theta} \mathcal{V}_\theta$$

4. Global Stability and Convergence

Theorem 2:
Under assumptions A1′ to A4′, if ν in (31) is sufficiently small (see (115)), then the adaptive controller (11)-(15) in feedback with the plant Π yields globally bounded signals and:

$$\lim_{T\to\infty}\sup\left[\frac{1}{T}\sum_{t=0}^{T}\left(\frac{e^m(t)}{\rho(t)}\right)^2\right]^{\frac{1}{2}}(\gamma_h-\gamma(H^*)) \le \gamma(D^{-1})(k_5\gamma(P)+k_6\nu+k_7V) + k_8\left[\frac{1}{T}\sum_{t=0}^{T}|H_*(q)C(q^{-1})y^m(t)|^2\right]^{\frac{1}{2}} \tag{58}$$

This shows that the mean square tracking error converges at least linearly to zero with $\gamma(H_*), \gamma(P), \nu$ and V. Moreover if

$$\gamma(P)=\nu=V=0 \tag{59}$$

and if there exists θ in Θ such that A3′, A4′ are satisfied and a perfect tracking is possible, i.e.

$$H_\theta(q)C(q^{-1})y^m(t)=0 \qquad \forall\, t \tag{60}$$

then we have an ℓ_2-convergence of the tracking error.

Remark: In the general case, since the convergence is only for the mean square error, some "burst" phenomena can appear.

Before embarking on the proof of this theorem, let us prove some technical results. We will use positive constants $\alpha_i, \beta_i, \gamma_i, \delta_i, V_i$, given in Table 1, which depend only on the norms $\gamma(A), \gamma(B)$, on V, on

$$Y=\sup_t |C(q^{-1})y^m(t)| \tag{61}$$

and on the controller parameters $n_s, n_r, d, \lambda_0, \lambda_1, \mu, \rho, R, \sigma_0, \theta_c, \gamma(C), \gamma(C^{-1})$.

Lemma 4.1: The definitions (8),(16) imply:

(i) $\mu^{-t}\rho(t) \ge \mu^{-(t-1)}\rho(t-1)$ (62)

(ii) $\|\phi\|_T \le \gamma_1\mu^{-(T+d)}\rho(T+d) \le \|\phi\|_T + V_1\mu^{-(T+d)}$ (63)

(iii) $\|u\|_T + \|y\|_T \le \|\phi\|_T \le \gamma_2\|u\|_T + \gamma_3\|y\|_T$ (64)

Proof: From (16) we have

$$\mu^{-2t}\rho(t)^2 - \mu^{-2(t-1)}\rho(t-1)^2 = \mu^{-2t}\max\left(\|\phi(t-d)\|^2,\ \rho^2(1-\mu^2)\right) \tag{65}$$

This proves (i). Summing from $t=1$ to $t=T+d$, we get:

$$\mu^{-2d}\|\phi\|_T^2 \le \mu^{-2(T+d)}\rho(T+d)^2-\rho^2 \tag{66}$$

$$\le \mu^{-2d}\|\phi\|_T^2+\rho^2\mu^{-2(T+d)} \tag{67}$$

This proves (ii).

The left inequality of (iii) is trivial. For the right inequality, we have

$$\|\phi\|_T^2 = \sum_{t=0}^{T}\mu^{-2t}u(t)^2+\cdots\mu^{-2n_s}\mu^{-2(t-n_s)}u(t-n_s)^2 \tag{68}$$

$$+\sum_{t=0}^{T}\mu^{-2t}y(t)^2+\cdots\mu^{-2n_r}\mu^{-2(t-n_r)}y(t-n_r)^2 \tag{69}$$

Lemma 4.2: From assumption (31) we have:

$$\mu^{-T}|v(T)| \leq \|v\|_T \leq \alpha_1 \nu \mu^{-(T+d-1)} \rho(T+d-1) + \delta_1 V \mu^{-T} \tag{70}$$

Proof: Since we have

$$|v(t+1)| \leq \mu_0 |v(t)| + \nu(\|u\|_T + \|y\|_T) + V \tag{71}$$

taking the $\ell_2(\mu)$-norm, we get:

$$\mu\|(1-\mu_0 q^{-1})|v|\,\|_{T+1} \leq \nu(\|u(t)\| + \|y(t)\|) + \frac{V\mu^{-T}}{(1-\mu^2)^{\frac{1}{2}}} \tag{72}$$

But

$$(\mu - \mu_0)\|v\|_{T+1} \leq \mu\|(1-\mu_0 q^{-1})|v|\|_{T+1}. \tag{73}$$

Then the conclusion follows with lemma 4.1.

Lemma 4.3: From (15), (20) and (29) we have:

$$\rho(t+1) \leq \gamma_4 \rho(t) \tag{74}$$

Proof: Let us rewrite (29) as:

$$y(t+1) - v(t+1) = (1 - A(q))y(t+1) + B(q)u(t) \tag{75}$$

Then with (30), the $\ell_2(\mu)$-norm gives:

$$\begin{aligned}
\mu^{-t}|y(t+1) - v(t+1)| &\leq \gamma(q(A-1))\|y\|_t + \gamma(B)\|u\|_t && (76)\\
&\leq \mu\gamma(A-1)\|y\|_t + \gamma(B)\|u\|_t && (77)
\end{aligned}$$

Hence with lemma 4.1:

$$|y(t+1)| \leq (\mu\gamma(A-1) + \gamma(B))\mu^{-d}\gamma_1\rho(t+d) + |v(t+1)| \tag{78}$$

To obtain the same type of inequality for $u(t)$, let us denote:

$$\begin{aligned}
\phi_r(t) &= (u(t-1),\ldots,u(t-n_s),y(t),\ldots,y(t-n_r))' && (79)\\
\theta_r(t) &= (s_1(t),\ldots,s_{n_s}(t),r_0(t),\ldots,r_{n_r}(t))' && (80)
\end{aligned}$$

The control law (15) can be rewritten as:

$$u(t) = \frac{-\theta_r(t)'\phi_r(t) + C(q^{-1})y^m(t+d)}{s_0(t)} \tag{81}$$

Since with (16)

$$\|\phi_r(t)\|^2 \leq \rho(t+d-1)^2 + y(t)^2 \tag{82}$$

we obtain with (20):

$$u(t)^2 \leq 2\left(\frac{R'}{\sigma_0}\right)^2 \left(\rho(t+d-1)^2 + y(t)^2\right) + \frac{2}{\sigma_0^2}Y^2 \tag{83}$$

The result follows since (16) implies:

$$\rho^2 \leq \rho(t+1)^2 \leq (1+\mu^2)\rho(t)^2 + u(t-d+1)^2 + y(t-d+1)^2 + \rho^2(1-\mu^2) \tag{84}$$

Lemma 4.4: From (29),(37) and assumption A2′, we have

$$\|u\|_{T-d} \leq \mu^d\gamma(D^{-1})(\gamma(A)\|Cy\|_T + \gamma(C)\|v\|_T + \mu^{-1}\gamma(P)\|u\|_{T-1}) \tag{85}$$

Proof: With (29), (37), we can obtain:

$$D(q)u(t-d) = A(q)C(q^{-1})y(t) - C(q^{-1})v(t) - P(q^{-1})u(t-1) \tag{86}$$

The conclusion follows clearly.

Lemma 4.5: From (11), (15), (21), we have:

$$\|Cy\|_T \leq \|Cy^m\|_T + \gamma_5\|e\|_T \tag{87}$$

where

$$\gamma_5 = 1 \qquad \text{if } d = 1 \tag{88}$$

Proof: From (11),(15), we obtain

$$C(q^{-1})y(t) = C(q^{-1})y^m(t) + e(t) + (\theta(t-1) - \theta(t-d))'\phi(t-d). \tag{89}$$

But with (21), we have:

$$\|\,(\theta(t) - \theta(t-d))'\,\phi(t-d)\| \leq \sqrt{\lambda_1}(1 + \sqrt{\frac{\lambda_0}{\lambda_1}})\sum_{i=1}^{d-1}|\bar{e}(t-i)|\|\bar{\phi}(t-d)\|\rho(t) \tag{90}$$

Clearly (16) gives:

$$\|\bar{\phi}(t-d)\|^2 \leq 1 \tag{91}$$

Also with lemma 4.3, we get:

$$\frac{\rho(t)}{\rho(t-i)} \leq \gamma_4^i \tag{92}$$

All these inequalities lead to:

$$\begin{aligned}\|(\theta(t) - \theta(t-d))'\phi(t-d)\|_T &\leq \gamma_4^{d-1}\sqrt{\lambda_1}(1 + \sqrt{\frac{\lambda_0}{\lambda_1}})\|\sum_{i=1}^{d-1}|\bar{e}(t-i)|\|_T && (93)\\ &\leq \left(\frac{\gamma_4}{\mu}\right)^{d-1}\sqrt{\lambda_1}(1 + \sqrt{\frac{\lambda_0}{\lambda_1}})(d-1)\|e\|_{T-1} && (94)\end{aligned}$$

The conclusion follows.

Lemma 4.6: From (19), (29), (37), and assumptions A2′ to A4′, we have:

$$\|e_*\|_T \leq \gamma(H_*)\|Cy\|_T + \gamma(D^{-1})(\alpha_2\|v\|_T + \beta_1\gamma(P)\|\phi\|_{T-1}) \tag{95}$$

Proof: From (29), and the definitions (8), (43) of the vectors $\phi(t), Q(q)$, we have:

$$\theta_*' Q(q)y(t) = \theta_*'(B(q)\phi(t-1) + \sqcup(q^{-1})v(t)) \tag{96}$$

where $\sqcup(q^{-1})$ is a polynomial vector:

$$\sqcup(q^{-1}) = (1 \cdots q^{-n_s}\ 0 \cdots 0)' \tag{97}$$

On the other hand (19) is:

$$e_*(t) = C(q^{-1})y(t) - q^{1-d}\theta_*'\phi(t-1) \tag{98}$$

Hence using the identity (37), we get:

$$D(q)e_*(t) = (D(q) - \theta_*'Q(q))C(q^{-1})y(t) + \theta_*' \sqcup (q^{-1})C(q^{-1})v(t) + P(q^{-1})\theta_*'\phi(t-1) \tag{99}$$

Since D is a unit of $\mathfrak{S}$ (assumption A2′) and the initial conditions are zero ((33)), we obtain:

$$e_*(t) = H_*(q)C(q^{-1})y(t) + D(q)^{-1}\theta_*' \sqcup (q^{-1})C(q^{-1})v(t) + D(q)^{-1}P(q^{-1})\theta_*'\phi(t-1) \tag{100}$$

The conclusion follows.

Proof of the first part:
Clearly it is sufficient to prove that $\rho(t)$ is bounded. We will need two key inequalities.
First Inequality: We derive an upper bound of $||e||_T$ in terms of $\mu^{-T}\rho(T)$: The summation of (22) gives:

$$||e||_T^2 \leq \sum_{t=1}^{T}(V_*(t-1) - V_*(t))\mu^{-2t}\rho(t)^2 + (1+\lambda_1)||e_*||_T^2 \tag{101}$$

To simplify the notations, we introduce:

$$\Delta(T)^2 = \max\left(0, \sum_{t=1}^{T}(V_*(t-1) - V_*(t))\mu^{2(T-t)}\left(\frac{\rho(t)}{\rho(T)}\right)^2\right) \tag{102}$$

We obtain:

$$||e||_T \leq \Delta(T)\mu^{-T}\rho(T) + \gamma_6||e_*||_T \tag{103}$$

$||e_*||_T$ is given by lemma 4.6, with $||v||_T$ and $||\phi||_{T-1}$ given by leammas 4.2, 4.1 respectively (i.e. with lemma 4.3):

$$||e_*||_T \leq \gamma(H_*)||Cy||_T + \gamma(D^{-1})(\alpha_3\nu + \beta_2\gamma(P))\mu^{-T}\rho(T) + \gamma(D^{-1})\delta_2 V\mu^{-T} \tag{104}$$

But $||Cy||_T$ is related to $e(t)$ by lemma 4.5. hence we have established:

$$(1-\gamma(H_*)\gamma_5\gamma_6)||e||_T \leq \left(\Delta(T) + \gamma(D^{-1})(\alpha_4\nu + \beta_3\gamma(P))\right)\mu^{-T}\rho(T) + (\gamma(H_*)V_2 + \gamma(D^{-1})\delta_3 V)\mu^{-T} \tag{105}$$

Second Inequality: We derive an upperbound of $\mu^{-T}\rho(T)$ in terms of $||e||_T$: From lemma 4.1, we have:

$$\mu^{-T}\rho(T) \leq \frac{\gamma_2}{\gamma_1}||u||_{T-d} + \frac{\gamma_3}{\gamma_1}||y||_{T-d} + \frac{V_1}{\gamma_1}\mu^{-T} \tag{106}$$

But $||u||_{T-d}$ is given by lemma 4.4, with help of lemmas 4.1 and 4.2:

$$||u||_{T-d} \leq \gamma(D^{-1})(\gamma_7||Cy||_T + (\alpha_5\nu + \beta_4\gamma(P))\mu^{-T}\rho(T) + \delta_4 V\mu^{-T}) \tag{107}$$

Also $||y||_{T-d}$ is obtained from $||Cy||_{T-d}$ since the roots of $C(q^{-1})$ are μ-exponentially stable.

$$||y||_{T-d} \leq \gamma(C^{-1})||Cy||_{T-d} \tag{108}$$

This yields (since $||Cy||_T$ is increasing in T):

$$\begin{aligned}\mu^{-T}\rho(T) &\leq (\tfrac{\gamma_2\gamma_7}{\gamma_1}\gamma(D^{-1}) + \tfrac{\gamma_3}{\gamma_1}\gamma(C^{-1}))||Cy||_T \\ &+\tfrac{\gamma_2}{\gamma_1}\gamma(D^{-1})(\alpha_5\nu + \beta_4\gamma(P))\mu^{-T}\rho(T) + (\tfrac{V_1}{\gamma_1} + \tfrac{\gamma_2}{\gamma_1}\gamma(D^{-1})\delta_4 V)\mu^{-T}\end{aligned} \tag{109}$$

Hence finally with lemma 4.5

$$\begin{aligned}[1-\gamma(D^{-1})(\alpha_6\nu + \beta_5\gamma(P))]\mu^{-T}\rho(T) &\leq (\gamma(D^{-1})\gamma_8 + \gamma_9)||e||_T \\ &+ (V_3 + \gamma(D^{-1})(V_4 + \delta_5 V))\mu^{-T}\end{aligned} \tag{110}$$

End of the Proof: From assumption A4′, we know that there exists $\epsilon > 0$ defined by:

$$1 - \gamma(H_*)\gamma_5\gamma_6 = \epsilon \tag{111}$$

Hence our two inequalities give:

$$\left[1 - \gamma(D^{-1})(\alpha_6\nu + \beta_5\gamma(P)) - \frac{\gamma(D^{-1})\gamma_8 + \gamma_9}{\epsilon}\left(\Delta(T) + \gamma(D^{-1})(\alpha_4 v + \beta_3\gamma(P))\right)\right]\rho(T) \leq M_0 \tag{112}$$

with

$$M_0 = V_3+\gamma(D^{-1})V\left(\delta_5+\delta_3\frac{\gamma(D^{-1})\gamma_8+\gamma_9}{\epsilon}\right)+\gamma(D^{-1})V_4+\frac{\gamma(D^{-1})\gamma_8+\gamma_9}{\epsilon}\gamma(H_*)V_2 \quad (113)$$

Also from assumption A3′, we know that:

$$1-\gamma(P)\gamma(D^{-1})(\beta_5+\beta_3\frac{\gamma(D^{-1})\gamma_8+\gamma_9}{\epsilon})>0 \quad (114)$$

Hence if ν satisfies

$$\nu<\frac{1-\gamma(P)\gamma(D^{-1})(\beta_5+\beta_3\frac{\gamma(D^{-1})\gamma_8+\gamma_9}{\epsilon})}{\gamma(D^{-1})(\alpha_6+\alpha_4\frac{\gamma(D^{-1})\gamma_8+\gamma_9}{\epsilon})} \quad (115)$$

there exists $\eta>0$ such that:

$$1-\gamma(D^{-1})(\alpha_6\nu+\beta_5\gamma(P)+\frac{\gamma(D^{-1})\gamma_8+\gamma_9}{\epsilon}(\alpha_4\nu+\beta_3\gamma(P)))=\eta \quad (116)$$

It follows that for each time T for which:

$$\Delta(T)\leq\delta<\frac{\eta\epsilon}{\gamma(D^{-1})\gamma_8+\gamma_9} \quad (117)$$

We have:

$$\rho(T)\leq M \quad (118)$$

with:

$$M=\frac{\epsilon M_0}{\epsilon\eta-(\gamma(D^{-1})\gamma_8+\gamma_9)\delta} \quad (119)$$

Hence let us consider time intervals generically denoted as (T_0,T_1) such that

$$\begin{array}{lll}\rho(T_0) & \leq M & \\ \rho(T) & > M & \forall T\in(T_0,T_1) \\ \rho(T_1) & \leq M & \end{array} \quad (120)$$

T_1-T_0 may be infinite or zero. On such intervals, we necessarily have

$$\Delta(T)>\delta \qquad \forall T\in(T_0,T_1) \quad (121)$$

Let us study the consequences of this inequality. From the definition of $\Delta(T)$, we have

$$\Delta(T)>\delta\Rightarrow\sum_{t=1}^{T}(V_*(t-1)-V_*(t))\frac{\mu^{-2t}\rho(t)^2}{\mu^{-2T}\rho(T)^2}>\delta^2 \quad (122)$$

But

$$\sum_{t=1}^{T}(V_*(t-1)-V_*(t))\frac{\mu^{-2t}\rho(t)^2}{\mu^{-2T}\rho(T)^2}=V_*^{av}(T-1)-V_*(T) \quad (123)$$

where we have defined

$$\begin{aligned} V_*^{av}(T) &= \sum_{t=1}^{T}V_*(t)\frac{\mu^{-2(t+1)}\rho(t+1)^2-\mu^{-2t}\rho(t)^2}{\mu^{-2(T+1)}\rho(T+1)^2}+\frac{V_*(0)\rho^2}{\mu^{-2(T+1)}\rho(T+1)^2} && (124)\\ &= \frac{\mu^{-2T}\rho(T)^2}{\mu^{-2(T+1)}\rho(T+1)^2}V_*^{av}(T-1)+\left(1-\frac{\mu^{-2T}\rho(T)^2}{\mu^{-2(T+1)}\rho(T+1)^2}\right)V_*(T) && (125)\end{aligned}$$

We remark that since $V_*(T)$ is bounded and $\mu^{-T}\rho(T)$ is increasing, $V_*^{av}(T)$ is also bounded:

$$V_*^{av}(T)\leq V_5 \quad (126)$$

Now if

$$V_*(T) < V_*^{av}(T-1) - \delta^2 \tag{127}$$

we obtain (with lemma 4.3):

$$V_*^{av}(T-1) - V_*^{av}(T) \;>\; \delta^2 \frac{\mu^{-2(T+1)}\rho(T+1)^2 - \mu^{-2T}\rho(T)^2}{\mu^{-2(T+1)}\rho(T+1)^2} \tag{128}$$

$$\;>\; \left(\frac{\delta\mu}{\gamma_4}\right)^2 \frac{\mu^{-2(T+1)}\rho(T+1)^2 - \mu^{-2T}\rho(T)^2}{\mu^{-2T}\rho(T)^2} \tag{129}$$

Summation of such inequalities on the interval (T_0, T_1) gives $\forall T \in (T_0, T_1)$:

$$\sum_{t=T_0+1}^{T} \frac{\mu^{-2(t+1)}\rho(t+1)^2 - \mu^{-2t}\rho(t)^2}{\mu^{-2t}\rho(t)^2} < (\frac{\gamma_4}{\delta\mu})^2 V_5 \tag{130}$$

Since $\mu^{-t}\rho(t)$ is increasing, this implies:

$$\log \frac{\mu^{-2T}\rho(T)^2}{\mu^{-2(T_0+1)}\rho(T_0+1)^2} < (\frac{\gamma_4}{\delta\mu})^2 V_5 \tag{131}$$

Hence, with the definition of $\rho(T_0)$ and lemma 4.3, we have obtained $\forall T \in (T_0, T_1)$:

$$M < \rho(T) < \mu^{T-T_0} exp(\frac{1}{2}(\frac{\gamma_4}{\delta\mu})^2 V_5)\gamma_4 \frac{M}{\mu} \tag{132}$$

This implies that $\rho(T)$ is bounded.

Proof of the second part: Let us first prove a technical result: For a sequence $x(t)$, we denote

$$\overline{||x||}_T^2 = \sum_{t=0}^{T} (\frac{x(t)}{\rho(t)})^2 \tag{133}$$

Let F be an element of $\Im$ and $x(t), z(t)$ be two sequences related by:

$$z(t) = F(q)x(t) \tag{134}$$

we have:

Lemma 4.7:

$$\overline{||z||}_T \leq \gamma(F)\overline{||x||}_T \tag{135}$$

Proof: We have:

$$\overline{||z||}_T^2 = \sum_{t=0}^{T} \frac{\mu^{-2t}z(t)^2}{\mu^{-2t}\rho(t)^2} \tag{136}$$

And we remark that

$$\mu^{-2t}z(t)^2 = ||z||_t^2 - ||z||_{t-1}^2 \tag{137}$$

We know that for any $t \geq 0$ (see (28)):

$$||z||_t^2 \leq \gamma(F)^2 ||x||_t^2 \tag{138}$$

Hence

$$\overline{||z||}_T^2 = \sum_{t=0}^{T-1} ||z||_t^2 \left(\frac{1}{\mu^{-2t}\rho(t)^2} - \frac{1}{\mu^{-2(t+1)}\rho(t+1)^2}\right) + \frac{||z||_T^2}{\mu^{-2T}\rho(T)^2} \tag{139}$$

Then since $\mu^{-t}\rho(t)$ is increasing, we use (138):

$$\overline{||z||}_T^2 \;\leq\; \gamma(F)^2 \left[\sum_{t=0}^{T-1} ||x||_t^2 (\frac{1}{\mu^{-2t}\rho(t)^2} - \frac{1}{\mu^{-2(t+1)}\rho(t+1)^2}) + \frac{||x||_T^2}{\mu^{-2T}\rho(T)^2}\right] \tag{140}$$

$$\;=\; \gamma(F)^2 \overline{||x||}_T^2 \tag{141}$$

Now as in lemma 4.5, 4.6, we have

$$\begin{aligned} e^*(t) &= H_*(q)[C(q^{-1})y^m(t) + e(t) + (\theta(t-1) - \theta(t-d))'\phi(t-d) \\ &\quad + D(q)^{-1}\theta_*' \sqcup (q^{-1})C(q^{-1})v(t) + D(q)^{-1}P(q^{-1})\theta_*'\phi(t-1) \end{aligned} \tag{142}$$

Hence with lemma 4.7 (using (21) as in lemma 4.5):

$$\overline{\|e^*\|_T} \le \overline{\|H_*Cy^m\|_T} + \gamma_5\gamma(H_*)\overline{\|e\|_T} + \gamma(D^{-1})(\alpha_2\overline{\|v\|_T} + \beta_1\gamma(P)\overline{\|\phi\|_{T-1}}) \tag{143}$$

Now with lemma 4.3, we have:

$$\frac{\|\phi(t)\|}{\rho(t)} = \frac{\rho(t+d)}{\rho(t)}\|\overline{\phi}(t)\| \le \gamma_4^d \tag{144}$$

It follows that:

$$\overline{\|\phi\|_{T-1}} \le \gamma_4^d\sqrt{T} \tag{145}$$

Similarly, we have with lemma 4.2:

$$\frac{|v(t)|}{\rho(t)} \le \alpha_1\nu\mu^{-(d-1)}\gamma_4^{d-1} + \frac{\delta_1 V}{\rho(t)} \tag{146}$$

It follows that

$$\alpha_2\overline{\|v\|_T} \le \alpha_3\nu\sqrt{T+1} + \delta_2 V\left(\sum_{t=0}^{T}\frac{1}{\rho(t)^2}\right)^{\frac{1}{2}} \tag{147}$$

Finally summation of (22) gives:

$$\overline{\|e\|_T} \le \sqrt{V_*(0)} + \gamma_6\overline{\|e_*\|_T} \tag{148}$$

We have obtained:

$$\begin{aligned} (1 - \gamma_5\gamma_6\gamma(H_*))\overline{\|e\|_T} &\le \sqrt{V_5} + \gamma_6\overline{\|H_*Cy^m\|_T} + \gamma(D^{-1})(\alpha_3\nu + \beta_1\gamma_4^d\gamma(P))\sqrt{T+1} \\ &\quad + \delta_2 V\left(\textstyle\sum_{t=0}^{T}\frac{1}{\rho(t)^2}\right)^{\frac{1}{2}} \end{aligned} \tag{149}$$

The result follows since:

$$e^m(t) = e(t) + (\theta(t-1) - \theta(t-d))'\phi(t-d) \tag{150}$$

which implies:

$$\frac{e^m(t)}{\rho(t)} \le \bar{e}(t) + \sum_{i=1}^{d-1}|\bar{e}(t-i)|\sqrt{\lambda_1}\left(1 + \sqrt{\frac{\lambda_1}{\lambda_0}}\right) \tag{151}$$

and:

$$\overline{\|e^m\|_T} \le \left(1 + (d-1)\sqrt{\lambda_1}\left(1 + \sqrt{\frac{\lambda_1}{\lambda_0}}\right)\right)\overline{\|e\|_T} \tag{152}$$

Table 1

$$
\begin{aligned}
\alpha_1 &= \frac{\gamma_1}{\mu-\mu_0} \\
\alpha_2 &= \gamma(C)\mu^{-n_s}R'\sqrt{n_s+1} \\
\alpha_3 &= \alpha_1\alpha_2(\tfrac{\gamma_4}{\mu})^{d-1} \\
\alpha_4 &= \alpha_3\gamma_6 \\
\alpha_5 &= \gamma(C)\alpha_1\mu\gamma_4^{d-1} \\
\alpha_6 &= \frac{\alpha_5\gamma_2}{\gamma_1} \\
\beta_1 &= \frac{R'}{\mu} \\
\beta_2 &= \beta_1\gamma_1(\tfrac{\gamma_4}{\mu})^{d-1} \\
\beta_3 &= \beta_2\gamma_6 \\
\beta_4 &= \gamma_1\gamma_4^{d-1} \\
\beta_5 &= \frac{\beta_4\gamma_2}{\gamma_1} \\
\gamma_1 &= \mu^d \\
\gamma_2 &= \mu^{-n_s}\sqrt{n_s+1} \\
\gamma_3 &= \mu^{-n_r}\sqrt{n_r+1} \\
\gamma_4^2 &= 1+(1+2(\tfrac{R'}{\sigma_0})^2)\left[1+\tfrac{2\mu^2}{(\mu-\mu_0)^2}\tfrac{2V^2}{\rho^2}+2+(\mu-\mu_0)^2(\mu\gamma(A-1)+\gamma(B))^2\right] \\
&\quad +\tfrac{2}{\sigma_0^2}\tfrac{Y^2}{\rho^2} \\
\gamma_5 &= 1+(d-1)(\tfrac{\gamma_4}{\mu})^{d-1}\sqrt{\lambda_1}(1+\sqrt{\tfrac{\lambda_1}{\lambda_0}}) \\
\gamma_6 &= \sqrt{1+\lambda_1} \\
\gamma_7 &= \mu^d\gamma(A) \\
\gamma_8 &= \frac{\gamma_2\gamma_5\gamma_7}{\gamma_1} \\
\gamma_9 &= \frac{\gamma_3\gamma_5\gamma(C^{-1})}{\gamma_1} \\
\gamma_{10} &= \frac{1+(d-1)\sqrt{\lambda_1}(1+\sqrt{\frac{\lambda_1}{\lambda_0}})}{\gamma_5\gamma_6} \\
\gamma_h &= \frac{1}{\gamma_5\gamma_6} \\
\delta_1 &= \frac{\mu}{\mu-\mu_0}\sqrt{\frac{1}{1-\mu^2}} \\
\delta_2 &= \alpha_2\delta_1 \\
\delta_3 &= \delta_2\gamma_6 \\
\delta_4 &= \mu^d\gamma(C)\delta_1 \\
\delta_5 &= \frac{\gamma_2}{\gamma_1}\delta_4 \\
k_1 &= (\tfrac{\gamma_4}{\mu})^{d-1}\lambda_1\left(1+\sqrt{\tfrac{\lambda_1}{\lambda_2}}\right) \\
k_2 &= \frac{\beta_3\gamma_8}{\gamma_5\gamma_6} \\
k_3 &= \frac{\beta_3\gamma_9}{\gamma_5\gamma_6} \\
k_4 &= \beta_5 \\
k_5 &= \beta_1\gamma_4^d\gamma_{10} \\
k_6 &= \alpha_3\gamma_{10}
\end{aligned}
$$

$$
\begin{aligned}
k_7 &= \frac{\delta_2\gamma_{10}}{\rho} \\
k_8 &= \frac{\gamma_{10}}{\rho} \\
V_1 &= \rho\mu^d\sqrt{2} \\
V_2 &= \frac{\gamma_6 Y}{\sqrt{1-\mu^2}} \\
V_3 &= \frac{V_1}{\gamma_1} + \frac{\gamma_3\gamma(C^{-1})}{\gamma_1}\frac{Y}{\sqrt{1-\mu^2}} \\
V_4 &= \frac{\gamma_2\gamma_7}{\gamma_1}\frac{Y}{\sqrt{1-\mu^2}} \\
V_5 &= 4R^2\frac{1+\lambda_1}{\lambda_0}
\end{aligned}
$$

References

[1] B. Egardt, Stability Analysis of Adaptive Controllers, Lecture Notes in Control and Information Sciences, No. 20, Springer-Verlag, New York, 1979.

[2] B.A. Francis, "Robustness of the Stability of Feedback Systems," *IEEE Transactions on Automatic Control*, August 1980.

[3] M. Vidyasagar, "The Graph Metric for Unstable Plants and Robustness Estimates for Feedback Stability," *IEEE Transactions on Automatic Control*, May 1984.

[4] P.J. Gawthrop, and K.W. Lim, "Robustness of Self-tuning Controllers," *Proceedings of the IEE*, vol. 129, p. 21-29, 1982.

[5] L. Praly, "Robustness of Model Reference Adaptive Control," *Proceedings of the Third Yale Workshop on Applications of Adaptive Systems Theory*, June 1983.

[6] R. Cristi, "A Globally Stable Adaptive Algorithm for Discrete Time Systems with Fast Unmodeled Dynamics," Tech. Report, Dept. of Elec. Engg., University of Michigan, Dearborn, Sept. 1983.

[7] R. Ortega, L. Praly, I.D. Landau, "Robustness of Discrete-Time Adaptive Controllers," *IEEE Transactions on Automatic Control*, vol. 30, pp. 1179-1187, Dec. 1985.

[8] L. Praly, "Robust Model Reference Adaptive Controllers, Part I: Stability Analysis," *Proceedings of the IEEE CDC*, Dec. 1984.

[9] B.B. Peterson and K.S. Narendra, "Bounded Error Adaptive Control," *IEEE Transactions on Automatic Control*, Dec. 1982.

[10] G. Kreisselmeier and K.S. Narendra, "Stable Model Reference Adaptive Control in the Presence of Bounded Disturbances," *IEEE Transactions on Automatic Control*, Dec. 1982.

[11] K.S. Narendra and A.M. Annaswamy, "Robust Adaptive Control in The Presence of Bounded Disturbances," To appear in the *IEEE Transactions on Automatic Control.*

[12] R.L. Kosut, B. Friedlander, "Robust Adaptive Control: Conditions for Global Stability," *IEEE Transactions on Automatic Control*, July 1985.

[13] P.A. Ioannou and P.V. Kokotovic, Adaptive Systems with Reduced Models, Lecture Notes in Control and Information Sciences, Springer-Verlag, New York, 1983.

[14] P.A. Ioannou and K. Tsakalis, "A Robust Model Reference Adaptive Controller," Tech. Report, Dept. of Elec. Engg., University of Southern California, May 1985.

[15] L. Praly, "Robustness of Indirect Adaptive Control based on Pole Placement Design," submitted for publication in *Automatica*; see also *IFAC workshop on adaptive systems in control and signal processing*, June 1983.

[16] C.A.Desoer and M. Vidyasagar, Feedback Systems: Input-Output Properties, Academic Press, 1975.

Robust Discrete-Time Adaptive Control

Petros Ioannou and Kostas Tsakalis
University of Southern California
EE-Systems, SAL 300
Los Angeles, Ca. 90089-0781

Abstract

This paper proposes a discrete-time model reference adaptive control algorithm which is robust with respect to additive plant uncertainties. The algorithm employs the same controller structure as in [1] but a different adaptive law for adjusting the controller parameters. If the plant uncertainty is "small" the algorithm guarantees the boundedness of all signals in the adaptive loop and "small" residual tracking errors for any bounded initial conditions. In the absence of plant uncertainties the algorithm guarantees zero residual tracking errors.

1. Introduction

Recently, several attempts have been made to analyze the robustness properties of adaptive control algorithms with respect to bounded disturbances and unmodeled dynamics. In [2-9] it is shown that unmodeled dynamics or even small bounded disturbances can cause most of the adaptive control algorithms to go unstable. The outcome of these instability studies is that most of the adaptive controllers need to be modified or redesigned in order to counteract instability and improve robustness.

Several different approaches for robustness have been recently proposed in the literature. In the case of bounded disturbances, the basic idea of most of the modifications is to prevent instability by eliminating the integral action of the adaptive laws. This can be achieved by the use of a dead zone [2,4,10-13], σ-modification [6,9] and similar other techniques [3,10]. In another approach [14] for handling disturbances the reference input signal is required to have enough frequencies for the measurement vector to be persistently exciting and be large relative to the disturbance. Persistent excitation guarantees exponential stability and therefore robustness with respect to disturbances.

When unmodeled dynamics are present, global stability cannot be guaranteed by simply eliminating the integral action of the adaptive laws. The unmodeled dynamics act as an external disturbance in the adaptive error equation and can no longer be assumed to be bounded. Despite this difficulty, however, several local results have been obtained in the literature [19,15,16] by using similar modifications as in the case of bounded disturbances. These results, however, are applicable to continuous-time plants whose dominant parts are minimum phase and of relative degree (n^*) one and whose unmodeled parts are stable and fast.

Global robustness results are recently obtained for discrete-time [17-19] and continuous-time [20,21] plants whose modeled parts may have arbitrary relative degree and whose unmodeled parts are stable. A normalizing signal, first used in [3] for the disturbance case, which bounds the modeling error signal, and a projection which keeps the parameter estimates bounded and within a chosen sphere are used in [17,18]. If the desired controller parameter vector is within the sphere, then global stability is guaranteed in the sense that for any bounded initial conditions all the signals in the adaptive loop are bounded. In another approach [19], the idea of normalization together with a relative dead zone and a projection are used in the adaptive law to achieve robustness for a discrete-time algorithm. The approach of [19] requires bounds on the unknown plant and controller parameters as well as on the plant zeros. A similar approach as in [19] is used to prove robustness in the case of an indirect continuous-time adaptive controller in [20]. It is shown that if the unknown parameters of the plant lie in a known convex set throughout which no unstable pole-zero cancellation occurs, then the use of normalization together with a suitably designed relative dead zone guarantees stability in the presence of small plant uncertainties. In [21] a direct adaptive control algorithm which is applicable to continuous-time plants with arbitrary relative degree is proposed. The algorithm is designed for the reduced order plant which is assumed to be minimum phase and of known relative degree and order, but is shown to stabilize the full order plant which, due to the unmodeled fast dynamics, may be nonminimum phase and of unknown order. The algorithm uses the idea of normalization and σ-modification. The only apriori information required for the implementation of the algorithm is an upper bound for the norm of the desired controller parameter vector as in [10,15].

In this paper the results of [21] are extended to discrete-time plants with additive plant uncertainties. The structure of the controller is similar to that in [1] but is different from that of [17-19,22]. The adaptive laws for adjusting the controller parameters employ the idea of normalization and σ-modification and are new. It is shown that the proposed algorithm guarantees boundedness and small residual tracking errors for any bounded initial conditions. In the absence of plant uncertainties the residual tracking error reduces to zero. The only apriori information required for implementation of the algorithm is an upper bound for the norm of the desired controller parameter vector.

2. Plant and the Control Objective

Consider a single-input single-output (SISO) plant represented by

$$y_k = G_o(z)u_k + \mu\eta_k \tag{1}$$
$$\eta_k = \Delta G(z)u_k \tag{2}$$

where

$$G_o(z) = k_p \frac{N(z)}{D(z)} \tag{3}$$

is the nominal transfer function of the plant, $\mu\Delta G(z)$ is an additive plant uncertainty rated by the scalar $\mu > 0$ and giving rise to the disturbance term $\mu\eta_k$ in the nominal plant equation (1).

For the nominal part of the plant we make the following assumptions:

(i) $N(z)$ is a monic m^{th} order polynomial in z with all roots inside the unit circle.

(ii) $D(z)$ is a monic n^{th} order polynomial in z with $n > m$.

(iii) k_p is an unknown constant with known sign.

(iv) The coefficients of the polynomials $N(z)$, $D(z)$ are unknown.

The disturbance term $\mu\eta_k$ is assumed to be bounded from above by the normalizing signal m_k (as in [17-21]) generated by the equation

$$m_{k+1} = \delta_0 m_k + \delta_1(1 + |u_k| + |y_k|), m_0 \geq \frac{\delta_1}{1-\delta_0} \tag{4}$$

i.e.,

$$\frac{|\eta_k|}{m_k} \leq N_0 + g_k \tag{5}$$

where δ_0, δ_1 are positive design parameters such that $\delta_0 < 1, N_0$ is a finite positive constant and g_k is a geometrically decaying term.

Remark 2.1 It should be noted that (5) does not guarantee that η_k is bounded unless u_k and y_k are bounded.

Remark 2.2 If $\Delta G(z)$ is a rational proper transfer function with poles strictly inside the unit circle, then δ_0 can be chosen so that (5) is satisfied. In this case, δ_0 needs to be such that

$$|p_j| < \delta_0 - \delta_2, \qquad j = 1, 2, \ldots, \tag{6}$$

where p_j is the jth pole of $\Delta G(z)$ and $0 < \delta_2 < \delta_0 < 1$, is satisfied. Thus, the only apriori information required about $\Delta G(z)$ in order to satisfy (6) and, consequently (5), is the maximum magnitude of the poles of $\Delta G(z)$.

The plant output y_k is required to track the output y_k^m of the reference model

$$y_k^m = W_m(z) r_k \tag{7}$$

where r_k is a bounded reference signal,

$$W_m(z) = k_m \frac{1}{D_m(z)}, \tag{8}$$

$D_m(z)$ is a monic $(n-m)th$ order polynomial with roots inside the unit circle and k_m is a constant of the same sign as k_p.

In the absence of plant uncertainties, i.e., $\mu = 0$, the adaptive controller in [1] guarantees boundedness of all signals in the closed loop and asymptotic convergence of the tracking error $e_k = y_k - y_k^m$ to zero. When $\mu \neq 0$, however, the stability properties of the algorithm in [1] can no longer be guaranteed no matter how small μ is.

The control objective is to devise an adaptive controller for the plant (1) so that the closed loop system is globally stable and the tracking error is as small as possible for a class of plant uncertainties defined by (2),(5).

3. The Design of the Adaptive Controller ($k_p = 1$)

The input u_k and output y_k are used to generate a $(2n-1)$ dimensional auxiliary vector ω_k as

$$\omega_{k+1}^{(1)} = F\omega_k^{(1)} + g u_k \tag{9}$$

$$\omega_{k+1}^{(2)} = F\omega_k^{(2)} + g y_k \tag{10}$$

where the eigenvalues of $F \in \mathcal{R}^{(n-1)\times(n-1)}$ are strictly within the unit circle, (F, g) is a controllable pair, and $\omega_k = [\omega_k^{(1)^T}, \omega_k^{(2)^T}, y_k]^T$. For ease of presentation we will assume that $k_p = k_m = 1$ and use as input to the plant the sequence

$$u_k = \theta_k^T \omega_k + r_k \tag{11}$$

where $\theta_k = [\theta_{1k}^T, \theta_{2k}^T, \theta_{3k}]^T$ is a $(2n-1)$ dimensional adjustable control parameter vector. The controller structure (9)-(10) is the same as the one used in [1]. However, the adaptive law for adjusting the controller parameter θ_k is different from that of [1] and is given below.

The following auxiliary signals are generated:

$$\varsigma_k = W_m(z)\omega_k \tag{12}$$

$$\nu_k = W_m(z)\theta_k^T \omega_k \tag{13}$$

$$\epsilon_k = y_k - y_k^m + \theta_k^T \varsigma_k - \nu_k \tag{14}$$

$$m_{k+1} = \delta_0 m_k + \delta_1(1 + |u_k| + |y_k|), \; m_0 \geq \frac{\delta_1}{1-\delta_0} \tag{15}$$

Then the adaptive law is given by

$$\theta_{k+1} = (1-\sigma_s)\theta_k - \gamma \frac{\epsilon_k \varsigma_k}{m_k^2 + \varsigma_k^T \varsigma_k} \tag{16}$$

$$\sigma_s = \begin{cases} 0 & \text{if } ||\theta_k|| < M_0 \\ \sigma_0 & \text{if } ||\theta_k|| \geq M_0 \end{cases} \tag{17}$$

where $\delta_0, \delta_1, \delta_2, \sigma_0, M_0, \gamma$ are positive design constants such that

$$\gamma + 2\sigma_0 < 1, \qquad M_0 \geq 2||\theta^*||, \tag{18}$$

$$0 < \delta_2 < \delta_0 < 1, \ \max_i |\lambda_i| < \delta_0 - \delta_2; \ i = 1, 2, \ldots n^* + n - 1 \tag{19}$$

are satisfied with λ_i being the ith pole of $W_m(z)$ and $(zI - F)^{-1}g$ and $\theta^* = [\theta_1^{*T}, \theta_2^{*T}, \theta_3^*]^T$ is the desired controller parameter vector defined in [1].

Remark 3.1 The only apriori information used in the adaptive law (12)-(19) is an upper bound for the norm of the unknown desired controller parameter vector θ^*. Our analysis allows M_0 to be relatively large, therefore, a conservative bound for $||\theta^*||$ can be safely used.

Remark 3.2 No apriori information about the plant is required in choosing the design parameters $\gamma, \delta_0, \delta_1, \delta_2$ and σ_0 because (19) can always be satisfied by properly choosing $W_m(z)$ and matrix F.

The two terms in the adaptive law (16) which are crucial for robustness are the normalizing signal m_k and the $\sigma_s\theta_k$ term. The $\sigma_s\theta_k$ term retards adaptation whenever θ_k grows to large values and, therefore, counteracts parameter drift [9], a form of instability, which may arise even when η_k is bounded and small. The normalizing signal bounds the disturbance term η_k and, therefore, guarantees that the input to the adaptive law is bounded. A key property of the normalizing signal, which is crucial for the analysis of the algorithm (9)-(17), is given by the following lemma.

Lemma 3.1 Let $x_k \in R^n$ be the state of the following system

$$x_{k+1} = Ax_k + B\overline{u}_k \tag{20}$$

where $|\lambda_i(A)| < \delta_0 - \delta_2$, $i = 1, 2, \ldots, n$, and $||\overline{u}_k|| \leq (|u_k| + |y_k|)$. Then there exists a positive constant N_1 such that

$$\frac{||x_k||}{m_k} < N_1 + g_k \tag{21}$$

where g_k is a geometrically decaying term which depends on the initial conditions.

Proof From (20) we have

$$x_k = A^k x_0 + \sum_{i=0}^{k-1} A^{k-i-1} B\overline{u}_i \tag{22}$$

Since $0 < \delta_0 - \delta_2 < 1$, then there exists a constant $c_0 > 0$ such that

$$||A_k|| \leq c_0 \delta_0^k \tag{23}$$

Therefore,

$$||x_k|| \leq ||x_0|| c_0 \delta_0^k + c_0 ||B|| \sum_{i=0}^{k-1} \delta_0^{k-i-1} (|u_i| + |y_i|) \tag{24}$$

From (15) we have

$$m_k = \delta_0^k m_0 + \delta_1 \sum_{i=0}^{k-1} \delta_0^{k-i-1} (1 + |y_i| + |u_i|) \tag{25}$$

Hence

$$m_k \geq m_0 \geq \frac{\delta_1}{1-\delta_0} > 0 \tag{26}$$

and, therefore, $\frac{1}{m_k}$ is a well-defined positive quantity for $k \geq 0$. Taking $N_1 \geq \frac{c_0\|B\|}{\delta_1}$, (21) follows directly from (24), (25).

4. Robustness Analysis ($k_p = 1$)

Let us now apply the algorithm (9)-(17) to the actual plant (1), (2) and ask the following question: Is there a $\mu^* > 0$ such that for each $\mu \in [0, \mu^*)$ and all η_k satisfying (2), (5) the closed-loop plant is stable? We answer this question by the following theorem.

Theorem 4.1 There exists a $\mu^* > 0$ such that for each $\mu \in [0, \mu^*)$ all the signals in the closed loop plant (1), (9)-(17) are bounded for any bounded initial conditions. Furthermore, the tracking error e_k converges to the residual set

$$D_e = \{e_k : \lim_{N\to\infty} \sup_{N>0} \frac{1}{N} \sum_{k=0}^{N-1} |e_k| \leq q_6\sqrt{\epsilon_0} + q_7\mu\} \tag{27}$$

where q_6, q_7 are positive constants and $\epsilon_0 > 0$ is an arbitrarily small number.

Corollary 4.1 In the absence of plant uncertainties, i.e., $\mu = 0$, the controller (9)-(17) guarantees zero residual tracking error, in addition to stability.

Proof of Theorem 4.1

The proof of Theorem 4.1 is completed in the following three steps: In Step 1 we prove that θ_k is bounded and $\frac{(\phi_k^T \varsigma_k)^2}{\overline{m}_k^2} + \sigma_s \theta_k^T \phi_k$ is small in the mean for small μ, where $\overline{m}_k^2 = m_k^2 + \varsigma_k^T \varsigma_k$ and $\phi_k = \theta_k - \theta^*$. The result of Step 1 is used to show that $\frac{|\phi_k^T \omega_k|}{\overline{m}_k}$ is also small in the mean for small μ in Step 2. In Step 3 the results of Step 1 and 2 are used to prove boundedness for all signals in the closed loop and the smallness of the tracking error e_k for small μ.

Step 1: Boundedness of θ_k and "Smallness" of $\left(\frac{\phi_k^T \varsigma_k}{\overline{m}_k}\right)^2 + \sigma_s \theta_k^T \phi_k$

Using the definition of θ^* given in [1] the tracking error e_k can be expressed as

$$e_k = W_m(z)\phi_k^T \omega_k + \mu\overline{\eta}_k \tag{28}$$

where $\overline{\eta}_k = \eta_k + W_m(z)[\theta_2^{*T}(zI - F)^{-1}g + \theta_3^*]\eta_k$. Then from (14) and (28) we have

$$\epsilon_k = \phi_k^T \varsigma_k + \mu\overline{\eta}_k \tag{29}$$

and, therefore, (16) may be written as

$$\phi_{k+1} = \phi_k - \gamma\frac{\phi_k^T \varsigma_k \varsigma_k}{\overline{m}_k^2} - \sigma_s\theta_k - \mu\gamma\frac{\overline{\eta}_k \varsigma_k}{\overline{m}_k^2}. \tag{30}$$

We consider the following positive definite function

$$V_k = \phi_k^T \phi_k \tag{31}$$

Then $\Delta V_k = V_{k+1} - V_k$ may be written as

$$\Delta V_k \leq -\frac{\gamma}{2}\frac{(\phi_k^T \varsigma_k)^2}{\overline{m}_k^2} - \frac{1}{4}\sigma_s\theta_k^T \phi_k + \beta_0\mu^2 \tag{32}$$

where β_0 is a positive constant, by using (5), Lemma 3.1, the fact that $\frac{\|\varsigma_k\|}{\overline{m}_k} < 1$ and by completing the squares. Since $\sigma_s\theta_k^T\phi_k \geq 0$ for all $k \geq 0$, (32) implies that $\Delta V_k < 0$ whenever $V_k \geq V_0$ where V_0 is a finite constant. Therefore, ϕ_k is a uniformly bounded sequence. Furthermore, there exist positive constants Δ_0, Δ_1 such that

$$\frac{1}{N}\sum_{k=k_o}^{k_o+N-1}\frac{(\phi_k^T\varsigma_k)^2}{\overline{m}_k^2} + \sigma_s\theta_k^T\phi_k \leq \frac{\Delta_o}{N} + \Delta_1\mu^2,\ k_o \geq 0. \tag{33}$$

Hence $\frac{(\phi_k^T\varsigma_k)^2}{\overline{m}_k^2} + \sigma_s\theta_k^T\phi_k$ is bounded, and small in the mean for small μ.

Step 2: The "Smallness" of $\frac{|\phi_k^T\omega_k|}{\overline{m}_k}$

In order to prove that (33) implies that $\frac{|\phi_k^T\omega_k|}{\overline{m}_k}$ is small in the mean for small μ we need the following lemmas:

Lemma 4.1: The sequence $\frac{|\phi_k^T\omega_k|}{\overline{m}_k}$ is bounded.

Proof of Lemma 4.1: Since $\omega_k = [\omega_k^{(1)^T}, \omega_k^{(2)^T}, y_k]^T$ we have

$$\frac{|\phi_k^T\omega_k|}{\overline{m}_k} \leq \|\phi_k\|\left[\frac{\|\omega_k^{(1)}\|}{m_k} + \frac{\|\omega_k^{(2)}\|}{m_k} + \frac{|y_k|}{m_k}\right] \tag{34}$$

and from (1), (9)-(11) we can show that

$$y_k = W_m(z)u_k - W_m(z)r_k - \theta_1^{*^T}W_m(z)(zI-F)^{-1}gu_k - \theta_2^{*^T}W_m(z)(zI-F)^{-1}gy_k - \theta_3^*W_m(z)y_k\,. \tag{35}$$

Hence from Lemma 3.1 and the boundedness of r_k the result follows from (34), (35).

Lemma 4.2: The sequence $\varsigma_{k+i}, i = 0, 1, \ldots, n^*$, where $n^* = n - m$ is the relative degree of $G_o(z)$ satisfies (36), i.e.,

$$\frac{\|\varsigma_{k+i}\|}{\overline{m}_k} < \bar{c}_i,\ i = 0, 1, \ldots, n^* \tag{36}$$

where the $\bar{c}_i$'s are positive finite constants.

Proof of Lemma 4.2: For $i = 0$, (36) follows from (9), (10), (12) and Lemma 3.1. Let us now consider the following representation for $\varsigma_k = [\varsigma_k^{(1)^T}, \varsigma_k^{(2)^T}, \varsigma_k^{(3)^T}]^T$, i.e.,

$$v_{k+1}^{(1)} = A_m v_k^{(1)} + b_m u_k \tag{37}$$

$$v_{k+1}^{(2)} = A_m v_k^{(2)} + b_m y_k \tag{38}$$

$$\varsigma_{k+1}^{(1)} = F\varsigma_k^{(1)} + gc_m^T v_k^{(1)} \tag{39}$$

$$\varsigma_{k+1}^{(2)} = F\varsigma_k^{(2)} + gc_m^T v_k^{(2)} \tag{40}$$

$$\varsigma_k^{(3)} = c_m^T v_k^{(2)} \tag{41}$$

where (A_m, b_m, c_m) is in the observable canonical form and is a minimal state representation of $W_m(z)$. Let c_i be a vector of dimension n^* whose $i-th$ element is 1 and the rest are zeros. Then

$$c_m^T v_{k+i}^{(j)} = c_{i+1}^T v_k^{(j)},\ i = 0, 1, \ldots, n^* - 1;\ j = 1, 2 \tag{42}$$

and

$$c_m^T v_{k+n^*}^{(2)} = c_{n^*}^T A_m v_k^{(2)} + b_m y_k. \tag{43}$$

Hence

$$\frac{||\varsigma_{k+i}^{(j)}||}{\overline{m}_k} \leq ||F||\frac{||\varsigma_{k+i-1}^{(j)}||}{\overline{m}_k} + ||g||||c_i||\frac{||v_k^{(j)}||}{\overline{m}_k},\ i = 1,2,\ldots,n^*;\ j = 1,2 \quad (44)$$

$$\frac{|\varsigma_{k+i}^{(3)}|}{\overline{m}_k} \leq ||c_{i+1}||\frac{||v_k^{(2)}||}{\overline{m}_k},\ i = 1,2,\ldots,n^* - 1 \quad (45)$$

$$\frac{|\varsigma_{k+n^*}^{(3)}|}{\overline{m}_k} \leq ||c_{n^*}^T A_m||\frac{||v_k^{(2)}||}{\overline{m}_k} + ||b_m||\frac{|y_k|}{\overline{m}_k} \quad (46)$$

From Lemma 3.1 and (35) it follows that the right hand sides of (44)-(46) are bounded; therefore, (36) holds.

Lemma 4.3: If

$$\frac{(\phi_k^T\varsigma_k)^2}{\overline{m}_k^2} + \sigma_s\theta_k^T\phi_k \leq \epsilon_o \quad (47)$$

is satisfied for some $\epsilon_o \in [0,1]$ and some $k > 0$, then there exist positive constants c_3, c_4 such that

$$||\phi_{k+1} - \phi_k|| \leq c_3\sqrt{\epsilon_o} + c_4\mu. \quad (48)$$

Proof of Lemma 4.3: From (30) we obtain

$$||\phi_{k+1} - \phi_k|| \leq \gamma\frac{|\phi_k^T\varsigma_k|}{\overline{m}_k}\frac{||\varsigma_k||}{\overline{m}_k} + \gamma\mu\frac{||\overline{\eta}_k||}{\overline{m}_k}\frac{||\varsigma_k||}{\overline{m}_k} + \sigma_s||\theta_k|| \quad (49)$$

and from (47) we have

$$\frac{|\phi_k^T\varsigma_k|}{\overline{m}_k} \leq \sqrt{\epsilon_o}, \quad \sigma_s\theta_k^T\phi_k \leq \epsilon_o \quad (50)$$

for some $k > 0$. In view of (50) we can write

$$\sigma_s||\theta_k|| \leq \frac{\epsilon_o}{M_o - ||\theta^*||}. \quad (51)$$

Substituting (50), (51) in (49) and using the fact that $||\varsigma_k||/\overline{m}_k < 1$, we obtain

$$||\phi_{k+1} - \phi_k|| \leq \gamma\sqrt{\epsilon_o} + \gamma\mu(N_o + g_k) + \frac{\epsilon_o}{||\theta^*||} \quad (52)$$

and therefore (48) follows.

Lemma 4.4: Assume that

$$\frac{(\phi_k^T\varsigma_k)^2}{\overline{m}_k^2} + \sigma_s\theta_k^T\phi_k \leq \epsilon_o,\ k = n, n+1, \ldots, n+n^* \quad (53)$$

is satisfied for some $\epsilon_o \in [0,1]$. Then

$$\frac{|\phi_k^T\omega_k|}{\overline{m}_k} \leq c_5\sqrt{\epsilon_o} + c_6\mu \quad (54)$$

holds for $k = n, n+1, \ldots, n+n^*$ and some positive constants c_5, c_6.

Proof of Lemma 4.4: The sequence ω_k can be written as

$$\omega_k = W_m^{-1}(z)\varsigma_k = \sum_{i=0}^{n^*} d_i\varsigma_{k+i} \quad (55)$$

for some scalars d_i, $i = 0, 1, \ldots, n^*$. Hence

$$\begin{aligned} \frac{|\phi_k^T \omega_k|}{\overline{m}_k} &\leq \sum_{i=0}^{n^*} |d_i| \frac{|\phi_k^T \varsigma_{k+i}|}{\overline{m}_k} \\ &\leq d_m \sum_{i=0}^{n^*} \frac{|\phi_{k+i}^T \varsigma_{k+i}|}{\overline{m}_k} + d_m \sum_{i=1}^{n^*} \sum_{j=1}^{i} \|\phi_{k+j} - \phi_{k+j-1}\| \frac{\|\varsigma_{k+i}\|}{\overline{m}_k} \end{aligned} \tag{56}$$

where $d_m = \max_i |d_i|$. Using (53) and the results of Lemmas 4.2, 4.3 in (56), (54) follows directly.

Let us now consider a time interval consisting of N samples and define the following sets

$$\Omega_1 = \{k : \frac{(\phi_k^T \varsigma_k)^2}{\overline{m}_k^2} + \sigma_s \theta_k^T \phi_k \leq \epsilon_o\} \tag{57}$$

$$\Omega_2 = \{N\} \backslash \Omega_1 \tag{58}$$

$$\Omega_3 = \{k : \frac{(\phi_i^T \varsigma_i)^2}{\overline{m}_k^2} + \sigma_s \theta_i^T \phi_i \leq \epsilon_o;\ i = k, k+1, \ldots, k+n^*\} \tag{59}$$

$$\Omega_4 = \{N\} \backslash \Omega_3 \tag{60}$$

Let M_i be the number of elements in Ω_i. Then the following relationships hold:

$$\Omega_3 \subset \Omega_1 \text{ and } M_3 \leq M_1 \leq N \tag{61}$$

$$\Omega_2 \subset \Omega_4 \text{ and } M_2 \leq M_4 \leq N \tag{62}$$

The inequality (33) together with the fact that

$$M_2 \epsilon_o \leq \sum_{k=k_o}^{k_o+N-1} \frac{(\phi_k^T \varsigma_k)^2}{\overline{m}_k^2} + \sigma_s \theta_k^T \phi_k \tag{63}$$

for any $k_o \geq 0$ imply that

$$M_2 \leq \frac{\Delta_o}{\epsilon_o} + \mu^2 \frac{\Delta_1}{\epsilon_o} N \tag{64}$$

Since each point in Ω_2 can prevent at most $n^* + 1$ points of Ω_1 from belonging in Ω_3 it follows that

$$M_4 \leq M_2(n^* + 1) \tag{65}$$

and therefore

$$M_3 \geq N - M_2(n^* + 1) \geq N - \frac{(n^* + 1)}{\epsilon_o}(\mu^2 N \Delta_1 + \Delta_o). \tag{66}$$

Using Lemmas 4.1, 4.4 we obtain that

$$\frac{|\phi_k^T \omega_k|}{\overline{m}_k} \leq c_5 \sqrt{\epsilon_o} + c_6 \mu \qquad \text{for } k \in \Omega_3 \tag{67}$$

and

$$\frac{|\phi_k^T \omega_k|}{\overline{m}_k} \leq c_7 \qquad \text{for } k \in \Omega_4 \tag{68}$$

and some $c_7 > 0$. Hence

$$\begin{aligned} \sum_{k=k_o}^{k_o+N-1} \frac{|\phi_k^T \omega_k|}{\overline{m}_k} &\leq M_3(c_5 \sqrt{\epsilon_o} + c_6 \mu) + M_4 c_7 \\ &\leq N(c_5 \sqrt{\epsilon_o} + c_6 \mu) + \mu^2 \frac{N \Delta_1}{\epsilon_o}(n^* + 1) c_7 \end{aligned} \tag{69}$$

Therefore there exists positive constants f_i, $i = 1, 2, 3, 4$ such that

$$\frac{1}{N} \sum_{k=k_o}^{k_o+N-1} \frac{|\phi_k^T \omega_k|}{\overline{m}_k} \leq f_1 \sqrt{\epsilon_o} + \mu f_2 + \mu^2 \frac{f_3}{\epsilon_o} + \frac{f_4}{N \epsilon_o} \tag{70}$$

which implies that $\frac{|\phi_k^T \omega_k|}{\overline{m}_k}$ is small in the mean for small μ and ϵ_o.

Step 3: Boundedness of Signals and "Smallness" of e_k

It can be shown as in [1] that e_k may be expressed as

$$e_k = W_m(z)\phi_k^T\omega_k + \mu\overline{\eta}_k \,. \tag{71}$$

Consider the following non-minimal state representation of (71), i.e.,

$$\overline{e}_{k+1} = A_c\overline{e}_k + b_c\theta_k^T\omega_k \tag{72}$$

$$e_k = h_c^T\overline{e}_k + \mu\overline{\eta}_k \tag{73}$$

where A_c is stable and $h_c^T(zI - A_c)^{-1}b_c = W_m(z)$. Similarly, ς_k can be represented as the state of the system

$$\varsigma_{k+1} = A\varsigma_k + B\begin{pmatrix} u_k \\ y_k \end{pmatrix} \tag{74}$$

where A is a stable matrix.

Let us now choose the positive definite function

$$V_k = \alpha\overline{e}_k^T P\overline{e}_k + \overline{m}_k^2 + \beta\varsigma_k^T P_1\varsigma_k \tag{75}$$

where α, β are positive constants and $P = P^T > 0$, $P_1 = P_1^T > 0$ satisfy the Lyapunov equations

$$A_c^T P A_c - P = -I \tag{76}$$

$$A^T P_1 A - P_1 = -I \tag{77}$$

respectively. Then

$$\begin{aligned} V_{k+1} - V_k \leq & -\alpha\|\overline{e}_k\|^2 - (1-\delta_o^2)m_k^2 - \beta\|\varsigma_k\|^2 \\ & +\alpha\tau_1\|e_k\||\phi_k^T\omega_k| + \alpha\tau_2|\phi_k^T\omega_k|^2 + \tau_3(1+|u_k|+|y_k|)^2 + \tau_4 m_k(1+|u_k|+|y_k|) \\ & +\tau_5\|\varsigma_k\|^2 + \tau_6\|\varsigma_k\|(|u_k|+|y_k|) + \tau_7(|u_k|+|y_k|)^2 + \beta\tau_8\|\varsigma_k\|(|u_k|+|y_k|) \\ & +\beta\tau_9(|u_k|+|y_k|)^2 \end{aligned} \tag{78}$$

where τ_1 to τ_9 are positive constants. Noting that $|u_k|+|y_k| \leq \tau_o\|\overline{e}_k\|+\tau_{10}$, for some constants τ_o, τ_{10}, and choosing α, β such that

$$\beta > \frac{\tau_9}{\delta_o(2-\delta_o)} \tag{79}$$

and

$$\alpha > max\left[(\tau_3+\tau_7+\beta\tau_9)\frac{\tau_o^2}{\delta_o(2-\delta_o)}, 8\frac{(\tau_4\tau_o)^2}{(1-\delta_o)^4}, \frac{(\tau_6\tau_o)^2}{2\beta} + \tau_6\tau_8\tau_o^2 + \beta\frac{\tau_8^2\tau_o^2}{2}\right] \tag{80}$$

then (78) can be rewritten as

$$\begin{aligned} V_{k+1} & \leq V_k\left[\gamma_o + \alpha\frac{|\phi_k^T\omega_k|}{\sqrt{V_k}}\left(\frac{\tau_1\|\overline{e}_k\|+\tau_2|\phi_k^T\omega_k|}{\sqrt{V_k}}\right)\right] + \gamma_2 \\ & \leq V_k\left[\gamma_o + \gamma_1\frac{|\phi_k^T\omega_k|}{\overline{m}_k}\right] + \gamma_2 \end{aligned} \tag{81}$$

where $\gamma_o = 1 - \frac{(1-\delta_o)^2}{2}$ and γ_1, γ_2 are positive constants. In order to analyze the stability properties of (81), we consider the discrete-time system

$$\overline{V}_{k+1} = \overline{V}_k\left[\gamma_o + \gamma_1\frac{|\phi_k^T\omega_k|}{\overline{m}_k}\right] \tag{82}$$

From (82) we have that

$$\overline{V}_{k+1} = \Pi_{i=0}^k\left[\gamma_o + \gamma_1\frac{|\phi_i^T\omega_i|}{\overline{m}_i}\right]\overline{V}_o. \tag{83}$$

Following the same steps as in the proof of Lemma 4.4, it can be shown that

$$\Pi_{i=0}^{k}\left(\gamma_o+\gamma_1\frac{|\phi_k^T\omega_k|}{\overline{m}_k}\right)\leq(\gamma_o+\mu\gamma_3+\gamma_4\sqrt{\epsilon_o})^{M_3}(1+\gamma_5)^{M_4} \tag{84}$$

for some positive constants $\gamma_3, \gamma_4, \gamma_5$. Choosing ϵ_o such that

$$0<\epsilon_o\leq\left(\frac{1-\gamma_o}{4\gamma_4}\right)^2 \tag{85}$$

and μ_1 as

$$\mu_1=\frac{1-\gamma_o}{4\gamma_3}>0 \tag{86}$$

it follows that for each $\mu\in[0,\mu_1]$

$$0<\gamma_o+\mu\gamma_3+\gamma_4\sqrt{\epsilon_o}\leq\frac{1+\gamma_o}{2}<1 \tag{87}$$

and (84) becomes

$$\Pi_{i=0}^{k}\left(\gamma_o+\gamma_1\frac{|\phi_k^T\omega_k|}{\overline{m}_k}\right)\leq\left(\frac{(1+\gamma_o)}{2}\right)^{M_3}(1+\gamma_5)^{M_4} \tag{88}$$

Using the inequalities (64-66) for M_3, M_4, (88) can be rewritten as

$$\Pi_{i=0}^{k}\left(\gamma_o+\gamma_1\frac{|\phi_k^T\omega_k|}{\overline{m}_k}\right)\leq\Delta_2^{1/\epsilon_o}\left[\frac{(1+\gamma_o)}{2}\Delta_3^{\mu^2/\epsilon_o}\right]^k \tag{89}$$

for some positive constants $\Delta_2, \Delta_3>1$. Hence from (89) and (83) we have

$$\overline{V}_{k+1}\leq\overline{V}_o\Delta_2^{1/\epsilon_o}\left[\frac{(1+\gamma_o)}{2}\Delta_3^{\mu^2/\epsilon_o}\right]^k. \tag{90}$$

Choosing μ_2 such that $\left(\frac{1+\gamma_o}{2}\right)\Delta_3^{\mu_2^2/\epsilon_o}=\frac{3+\gamma_o}{4}$, i.e.,

$$\mu_2=\sqrt{\epsilon_o}\left[\frac{\ell n(3+\gamma_o)-\ell n2(1+\gamma_o}{\ell n\Delta_3}\right] \tag{91}$$

and

$$\mu^*=min[\mu_1,\ \mu_2] \tag{92}$$

we have that for each $\mu\in[0,\mu^*]$

$$\overline{V}_{k+1}=\overline{V}_o\Delta_2^{1/\epsilon_o}\left(\frac{3+\gamma_o}{4}\right)^k. \tag{93}$$

Since $0<\gamma_o<1$, it follows that $\overline{V}_k\to 0$ geometrically fast as $k\to\infty$. Since (82) is the homogeneous part of (81) with the equal sign, and is geometrically stable, it follows from the comparison principle [23] and bounded-input-bounded-output stability that V_k is bounded. The boundedness of V_k implies that all the signals in the closed-loop are bounded.

Using (72), (73) we can write

$$e_k=h_c^TA_c^k\overline{e}_o+h_c^T\sum_{j=0}^{k-1}A_c^{k-j-1}b_c\phi_j^T\omega_j+\mu\overline{\eta}_k. \tag{94}$$

Since A_c is stable, then there exists positive constants q_1, q_2 and $\lambda_o<1$ such that

$$|e_k|\leq q_1\lambda_o^k\|\overline{e}_o\|+q_2\sum_{j=0}^{k-1}\lambda_o^{k-j-1}|\phi_j^T\omega_j|+\mu|\overline{\eta}_k|. \tag{95}$$

Therefore

$$\sum_{k=k_o}^{k_o+N-1} |e_k| \le q_3 + q_4 \sum_{j=k_o}^{k_o+N-1} |\phi_j^T \omega_j| + \mu \sum_{j=k_o}^{k_o+N-1} |\overline{\eta}_k| \tag{96}$$

for some positive constants q_3, q_4 and any $k_o \ge 0$. Using (70), (5), Lemma 3.1 and the boundedness of $\overline{m}_k$, it follows from (96) that

$$\frac{1}{N} \sum_{k=k_o}^{k_o+N-1} |e_k| \le \frac{q_5}{N} + q_6\sqrt{\epsilon_o} + \mu q_7 \tag{97}$$

for some positive constants q_5, q_6, q_7. Letting $N \to \infty$, (27) follows.

Proof of Corollary 4.1: Setting $\mu = 0$ in (33) we have that

$$\sum_{k=0}^{N-1} \frac{(\phi_k^T \varsigma_k)^2}{\overline{m}_k^2} + \sigma_s \theta_k^T \phi_k \le \Delta_o, \qquad \forall\, N \ge 1. \tag{98}$$

Since $\overline{m}_k$ is a bounded sequence for any $\mu \in [0, \mu^*]$, (98) implies that

$$\lim_{k \to \infty} (\phi_k^T \varsigma_k) = 0, \qquad \lim_{k \to \infty} \sigma_s \theta_k^T \phi_k = 0. \tag{99}$$

Using (99) and following Step 2 of the proof of Theorem 4.1, it can be shown that for $\mu = 0$

$$\lim_{k \to \infty} (\phi_k^T \omega_k) = 0. \tag{100}$$

Then setting $\mu = 0$ in (71) and using (100) we have that $\lim_{k \to \infty} e_k = 0$ and the proof is complete.

5. Adaptive Controller for k_p Unknown

When k_p is unknown, two additional estimated parameters need to be introduced: one for estimating k_p and one for estimating $1/k_p$. The control law is given by

$$u_k = \theta_k^T \omega_k + c_{o_k} r_k \tag{101}$$

where θ_k, ω_k are as defined in section 3 and c_{o_k} is the estimate of k_p at instant k. The following auxiliary signals are generated.

$$\xi_k = \theta_k^T \varsigma_k - W_m(z) u_k + c_{o_k} y_k^m \tag{102}$$

$$\overline{\epsilon}_k = y_k - y_k^m + \psi_k \xi_k \tag{103}$$

where ξ_k is the estimate of $1/k_p$ at instant k and ς_k is given by (12). Defining the estimated parameter vector $\overline{\theta}_k$ as

$$\overline{\theta}_k = [\theta_k^T, c_{o_k}, \psi_k]^T \tag{104}$$

the update law is chosen as

$$\overline{\theta}_{k+1} = (1 - \sigma_s)\overline{\theta}_k - \gamma \frac{\overline{\epsilon}_k \overline{\varsigma}_k}{\hat{m}_k} \tag{105}$$

$$\sigma_s = \begin{cases} 0 & \text{if } \|\overline{\theta}_k\| < \overline{M}_0 \\ \sigma_0 & \text{if } \|\overline{\theta}_k\| \ge \overline{M}_0 \end{cases} \tag{106}$$

where $\overline{\varsigma}_k = [\varsigma_k^T, y_k^m, \xi_k]^T, \hat{m}_k = m_k^2 + \overline{\varsigma}_k^T \overline{\varsigma}_k, \overline{M}_o \ge 2\|\overline{\theta}^*\|$, $\overline{\theta}^* = [\theta^{*T}, k_p, 1/k_p]^T$ and m_k, γ are as defined in section 3.

The stability properties of the controller (101)-(106) when applied to the actual plant (1), (2) are similar to those of Theorem 4.1 and Corollary 4.1. Furthermore the proof of these properties follows very closely from the proof of Theorem 4.1 and Corollary 4.1 and is omitted.

Remark 5.1 The σ-modification used in (17), (106) switches σ from 0 to σ_o whenever $||\theta_k||, ||\bar{\theta}_k||$ exceed $M_o, \overline{M}_o$ respectively. An alternative approach is to use a different σ for each element of θ_k or $\bar{\theta}_k$. Such an approach will not change the stability results and proofs of the algorithm, but it may have a beneficial effect on the transient of the signals in the closed loop.

Remark 5.2 In this paper it is assumed that $n > m$, i.e., the relative degree of the nominal transfer function of the plant is $n^* \geq 1$. When $n = m$ two possible robust adaptive controllers can be designed with similar properties as those given by Theorem 4.1 and Corollary 4.1.

The first one may employ a delay term z^{-1} at the plant and reference model output. The relative degree of the augmented plant will then be increased by one and the results of section 4, 5 are directly applicable.

The second one may choose $W_m(z) = 1$ and the same form of filters as in (9), (10) but with F being an $n \times n$ matrix instead of $(n-1) \times (n-1)$. In this case $\varsigma_k = \omega_k$ and the proof of stability and robustness is much simpler than that of Theorem 4.1 since Step 2 of the proof is no longer needed.

6. Conclusion

The paper presents a discrete-time model reference adaptive control algorithm which is robust with respect to additive plant uncertainties. The crucial features of the algorithm is a normalizing signal which bounds the disturbance term due to the plant uncertainty and the σ-modification [15,21] which counteracts parameter drift and guarantees bounded adaptation. If the plant uncertainty is small, the algorithm guarantees boundedness for all signals in the closed loop and small residual tracking errors. In the absence of any modeling errors, the residual tracking error is zero.

Acknowledgments

This work was supported by the National Science Foundation under Grant ECS-8312233.

References

[1] K.S. Narendra and Y.H. Lin, "Stable Discrete Adaptive Control," *IEEE Trans. on Automatic Control*, AC-25, No. 3, pp. 456-461, 1980.

[2] B. Egardt, "Stability Analysis of Adaptive Control Systems with Disturbances," *Proc. JACC*, San Francisco, Ca., 1980.

[3] B. Egardt, Stability of Adaptive Controllers, Springer-Verlag, Berlin, 1979.

[4] B.B. Peterson and K.S. Narendra, "Bounded Error Adaptive Control," *IEEE Trans. on Automatic Control*, vol. AC-27, No. 6, pp. 1161-1168, 1982.

[5] C.E. Rohrs, L. Valavani, M. Athans and G. Stein, "Robustness of Adaptive Control Algorithms in the Presence of Unmodeled Dynamics," Preprints 21st IEEE CDC, Orlando, Fl., 1982.

[6] P.A. Ioannou and P.V. Kokotovic, Adaptive Systems with Reduced Models, Springer-Verlag, Berlin, 1983.

[7] K.J. Åström , "Analysis of Rohrs Counterexamples to Adaptive Control," Dept. of Automatic Control, Lund Inst. of Tech. Report, 1983.

[8] B.D. Riedle and P.V. Kokotovic, "Disturbance Instabilities in an Adaptive System," *IEEE Trans. on Automatic Control*, AC-29, pp. 822-824.

[9] P.A. Ioannou and P.V. Kokotovic, "Instability Analysis and Improvement of Robustness of Adaptive Control," *Automatica*, vol. 20, No. 5, Sept. 1984.

[10] G. Kreisselmeier and K.S. Narendra, "Stable Model Reference Adaptive Control in the Presence of Bounded Disturbances," *IEEE Trans. on Automatic Control*, vol. AC-27, No. 6, pp. 1169-1176, 1982.

[11] C. Samson, "Stability Analysis of Adaptively Controlled System Subject to Bounded Disturbances," *Automatica*, vol. 19, pp. 81-86, 1983.

[12] S.S. Sastry, "Model Reference Adaptive Control: Stability, Parameter Convergence and Robustness," *I.M.A. Journal of Control and Information*, vol. 1, pp. 27-66, 1984.

[13] J.M. Martin-Sanches, "A Globally Stable APCS in the Presence of Bounded Noises and Disturbances," *IEEE Trans. on Automatic Control*, vol. AC-29, pp. 461-464, 1984.

[14] K.S. Narendra and A.M. Annaswamy, "Persistent Excitation and Robust Adaptive Algorithms," Proc. 3rd Yale Workshop on Applications of Adaptive Systems Theory, Yale University, New Haven, Ct., June 15-17, 1983, pp. 11-18.

[15] P.A. Ioannou, "Robust Adaptive Controller with Zero Residual Tracking Errors," The 24th IEEE Conf. on Decision and Control, Fort Lauderdale, Florida, 1985.

[16] R.L. Kosut and C.R. Johnson, Jr., "An Input-Output View of Robustness in Adaptive Control," *Automatica*, vol. 20, No. 5, Sept. 1984.

[17] L. Praly, "Robustness of Model Reference Adaptive Control," Third Yale Workshop on Applications of Adaptive Systems Theory, Yale University, New Haven, Ct., June 15-17, 1983.

[18] L. Praly, "Robust Model Reference Adaptive Controllers, Part 1: Stability Analysis," Proc. 23rd IEEE Conf. on Decision and Control, 1984.

[19] G. Kreisselmeier and B.D.O. Anderson, "Robust Model Reference Adaptive Control," DFVLR - Institut fur Dynamik der Flugsysteme Report, May 1984.

[20] G. Kreisselmeier, "A Robust Indirect Adaptive Control Approach," The Fourth Yale Workshop on Theory and Applications of Adaptive Control, May 1985.

[21] P.A. Ioannou and K. Tsakalis, "A Robust Direct Adaptive Controller," University of Southern California, EE-Systems, Report 85-07-1, June 1985.

[22] G. Goodwin, P. Ramadge and P. Caines, "Discrete-Time Multivariable Adaptive Control," *IEEE Trans. on Automatic Control*, AC-25, June 1980.

[23] K.S. Narendra, Y.H. Lin and L.S. Valavani, "Stable Adaptive Controller Design, Part II: Proof of Stability," *IEEE Trans. on Automatic Control*, AC-25, June 1980.

Identification and Adaptive Control of Linear Stochastic Systems

P. R. Kumar
Department of Electrical and Computer Engineering
and
Coordinated Science Laboratory
University of Illinois at Urbana-Champaign

Abstract

We provide an account of some of the recent progress on identification and adaptive control of linear stochastic systems. This complements a recent survey [1].

1. Identification

We will begin with the problem of identification. Consider a stationary process $\{y(t)\}$ with a rational spectral density

$$\Phi_{yy}(z) = \frac{C(z)C(z^{-1})}{A(z)A(z^{-1})}.$$

It can be represented as the output of the ARMA (auto-regressive moving average) system

$$y(t) = \sum_{i=1}^{p}(a_i y(t-i) + c_i w(t-i)) + w(t) \tag{1}$$

where $\{w(t)\}$ is a white noise process. The polynomial

$$A(z) := 1 - \sum_{i=1}^{p} a_i z^i \tag{2}$$

has all its roots strictly outside the unit circle, and

$$C(z) := 1 + \sum_{i=1}^{p} c_i z^i \tag{3}$$

has all its roots on or outside the unit circle. We interpret z as the *backward* shift operator, i.e.,

$$zy(t) \triangleq y(t = 1).$$

It should be noted that the ARMA representation (1) can also be used to represent some non-stationary processes. For example, if $A(z)$ has roots inside the unit circle then $\{y(t)\}$ "blows" up, and is non-stationary. Furthermore, if a control input $\{u(t)\}$ is also present, then we can generalize the system (1) to an ARMAX (i.e., ARMA with exogenous inputs) representation of the form:

$$y(t) = \sum_{i=1}^{p}(a_i y(t-i) + b_i u(t-i) + c_i w(t-i)) + w(t) \tag{4}$$

We shall, therefore, take the ARMAX representation (4) as the model of a linear stochastic system and pose the identification problem as follows. Given observations, on line, of the processes $\{y(t)\}$ and $\{u(t)\}$, how may we estimate the *unknown* parameters $(a_1, \ldots, a_p, b_1, \ldots, b_p, c_1, \ldots, c_p)$ describing the system (4)?

Let $\theta(t)$ be an *estimate* of $\theta^o \triangleq (a_1, \ldots, a_p, b_1, \ldots, b_p, c_1, \ldots, c_p)^T$ made on the basis of observations $\{y(i), u(i) : i \leq t\}$. We would like to have an estimation procedure which is (i) *recursive*, i.e. we can update $\theta(t)$ to $\theta(t+1)$ by storing only a finite number of "states" and (ii) *consistent*, i.e., $\lim_{t \to \infty} \theta(t) = \theta^o$ and (iii) reasonably *efficient*, i.e., $cov(\theta(t) - \theta^o)$ is small.

2. The Least Squares Method for ARX Systems

The main difficulty in the identification problem arises in the estimation of the coefficients $(c_1, \ldots, c_p)$. It has its source in the fact that in contrast to $y(t)$ and $u(t)$, the variables $w(t)$ are not available as observations.

Let us therefore begin by considering the identification problem for the *simpler* ARX system:

$$y(t) = \sum_{i=1}^{p} (a_i y(t-i) + b_i u(t-i)) + w(t). \tag{5}$$

Defining

$$\theta(t-1) \triangleq (y(t-1), \ldots, y(t-p), u(t-1), \ldots, u(t-p))^T \tag{6}$$

$$\theta^o \triangleq (a_1, \ldots, a_p, b_1, \ldots, b_p)^T \tag{7}$$

the system (5) can be written as

$$y(t) = \phi^T(t-1)\theta^o + w(t).$$

Consider the *Least Squares Estimate* $\theta(t)$ defined by

$$\theta(t) \triangleq \arg\min_{\theta} \sum_{i=1}^{t} [y(i) - \phi^T(i-1)\theta]^2$$

Simple calculations, say differentiating and setting the derivative to 0, show that

$$\theta(t) = [\sum_{i=1}^{t} \phi(i-1)\phi^T(i-1)]^{-1} [\sum_{i=1}^{t} \phi(i-1) y(i)]. \tag{8}$$

Moreover, it can also be seen [2], that $\theta(t)$ satisfies the recursions:

$$\theta(t+1) = \theta(t) + R^{-1}(t)\phi(t)[y(t+1) - \phi^T(t)\theta(t)] \tag{9}$$

$$R(t+1) = R(t) + \phi(t+1)\phi^T(t+1). \tag{10}$$

Hence the Least Squares Estimates can be recursively obtained.

To analyze the asymptotic behavior, we can substitute $y(i) = \phi^T(i-1)\theta^o + w(i)$ in (8) to obtain

$$\theta(t) = \theta^o + [\sum_{i=1}^{t} \phi(i-1)\phi^T(i-1)]^{-1} [\sum_{i=1}^{t} \phi(i-1) w(i)]. \tag{11}$$

Hence, if $E(w(i)w(j)) = \sigma^2 \delta_{ij}$, then

$$E \quad (\theta(t)) = \theta^o$$

$$E \quad \|\theta(t) - \theta^o\|^2 = E(\text{trace } R^{-1}(t-1)).$$

By Chebyshev's inequality [3], it follows that

$$Prob(\|\theta(t) - \theta^o\| \geq \epsilon) \leq \frac{E(\text{trace } R^{-1}(t-1))}{\epsilon^2}.$$

Hence, if

$$\lim_{t\to\infty} E(\text{trace } R^{-1}(t)) = 0 \tag{12}$$

then

$$\theta(t) \to \theta^o \text{ in probability.} \tag{13}$$

The condition (12) clearly imposes some restrictions on the input sequence $\{u(t)\}$. For example, as a degenerate case, if

$$u(t) = 0 \text{ for all } t$$

then $R(t)$ is not even invertible. One may refer to (12) as a *Persistency of Excitation* condition on the input sequence $\{u(t)\}$.

In a similar vein one can also analyze the covariance of the parameter error and prove the efficiency of the scheme under certain additional assumptions.

To prove *strong* consistency, i.e., $\lim_{t\to\infty} \theta(t) = \theta^o$ a.s., rather than convergence in probability as in (13), we need to analyze the last term on the right hand side in (8).

Let $R_{nn}(t)$ be the n-th diagonal element of the matrix $R(t)$ and let $\phi_n(t)$ be the n-th element of the vector $\epsilon(t)$. Clearly, $\{\sum_{i=1}^{t} \frac{\phi_n(i-1)}{R_{nn}(i-1)} w(i)\}$ is *martingale*, i.e.,

$$E[\sum_{i=1}^{t} \frac{\phi_n(i-1)}{R_{nn}(i-1)} w(i) | \mathcal{F}_{t-1}] = \sum_{i=1}^{t-1} \frac{\phi_n(i-1)}{R_{nn}(i-1)} w(i)$$

where $\mathcal{F}_t$ is the σ-algebra generated by the past up till time t. Moreover, this is a square integrable martingale [4], and so it converges a.s. [5]. Hence

$$\sum_{i=1}^{\infty} \frac{\phi_n(i-1)}{R_{nn}(i-1)} w(i) \text{ exists a.s.}$$

By Kronecker's Lemma, [3], it follows that if

$$\lim_{t\to\infty} R_{nn}(t) = +\infty \text{ a.s.}$$

then

$$\lim_{t\to\infty} \frac{1}{R_{nn}(t-1)} \sum_{i=1}^{t} w(i) = 0 \text{ a.s.} \tag{14}$$

Hence if

$$\lim_{t\to\infty} \lambda_{min}(R(t)) = +\infty \text{ a.s.} \tag{15i}$$

$$\frac{\lambda_{min}(R(t))}{\lambda_{max}(R(t))} \geq \epsilon > 0 \text{ for all large } t \text{ a.s.} \tag{15ii}$$

where λ_{min} and λ_{max} denote the minimum and maximum eigenvalues, then clearly (14) implies

$$\lim_{t\to\infty} R^{-1}(t-1) \sum_{i=1}^{t} w(i) = 0 \text{ a.s.}$$

Hence (11) implies that

$$\lim_{t\to\infty} \theta(t) = \theta^o \text{ a.s.}$$

Hence (15i,ii) are sufficient conditions for strong consistency. The similarity between (12) and (15i,ii) may be noted. The above proof of strong consistency is due to Ljung [4].

3. The ARMAX Case

Let us now return to the ARMAX system (4) which includes the term $\sum_{i=1}^{p} c_i w(t-i)$ on the right hand side, in contrast to (5).

Motivated by the good properties of the Least Squares Estimates of $(a_1, \ldots, a_p, b_1, \ldots, b_p)$ in the ARX case, one is tempted to use the same estimate (8) even for the ARMAX case, with $\phi(t)$ defined as in (6). Unfortunately, such a scheme will not converge to the true values of the parameters, as the following example shows.

3.1 Example of Bias of Least Squares Estimates in ARMAX Case

Consider the simple system

$$y(t) = \theta^o y(t-1) + c_1 w(t-1) + w(t) \tag{16}$$

with $\theta^o = 0$. In this special case, $\phi(t) \triangleq y(t)$, and so the Least Squares Estimate (8) is

$$\theta(t) = \frac{\frac{1}{t}\sum_{i=1}^{t} y(i-1)y(i)}{\frac{1}{t}\sum_{i=1}^{t} y^2(i-1)}. \tag{17}$$

Since (16) is a simple MA process, it is easy to compute that $E(y(i-1)y(t)) = c_1\sigma^2, E(y^2(i-1)) = (1+c_1^2)\sigma^2$ where σ^2 is the variance of $w(t)$. Moreover, the numerator of (17) is the *lag one autocorrelation* of y along a *sample*, while the denominator is the *sample variance*. By the strong law of large numbers, it follows that

$$\lim_{t\to\infty} \theta(t) = \frac{c_1}{1+c_1^2} \text{ a.s.}$$

and so the parameter estimates are *biased*. □

However, a slight modification of the Least Squares Estimates, which goes by the name of the Instrumental Variables Method allows us to eliminate the bias in the estimates of $(a_1, \ldots, a_p, c_1, \ldots, c_p)$.

3.2 The Instrumental Variables Method

The system (4) can be written as

$$y(t) = \phi^T(t-1)\theta^o + v(t) \text{ where } v(t) \triangleq w(t) + \sum_{i=1}^{p} c_i w(t-i) \tag{18}$$

with $\phi(t)$ and θ^o defined as in (6) and (7). Let $\psi(t)$ be some undefined vector for the time being, and consider the estimate

$$\theta(t) \triangleq [\sum_{i=1}^{t} \psi(i-1)\phi^T(i-1)]^{-1}[\sum_{i=1}^{t} \psi(i-1)y(i)]. \tag{19}$$

It should be noted that the choice of $\psi(t) \triangleq \phi(t)$ will yield the Least Squares Estimate (8), but we will not make such a choice. Substituting $y(i) = \phi^T(i-1)\theta^o + v(i)$ in (19) gives

$$\theta(t) = \theta^o + [\frac{1}{t}\sum_{i=1}^{t} \psi(i-1)\phi^T(i-1)]^{-1}[\frac{1}{t}\sum_{i=1}^{t} \psi(i-1)v(i)]. \tag{20}$$

In order to have strong consistency, we want the last turn on the right hand side of (20) to converge to 0. Hence if one chooses $\{\psi(i)\}$ so that

(i) $\psi(i-1)$ is uncorrelated with $v(i)$ so that $\lim_{t\to\infty} \frac{1}{t}\sum_{i=1}^{t} \psi(i-1)w(i) = 0$

(ii) $\psi(i-1)$ is sufficiently correlated with $\phi(i-1)$ so that $\limsup_{t\to\infty}[\frac{1}{t}\sum_{i=1}^{t} \psi(i-1)\phi^T(i-1)]^{-1}$ exists

then the parameter estimates will be strongly consistent. For objective (i) to be satisfied, $\psi(i-1)$ may be chosen so that it depends only on system variables up till time $(i-p)$, for then $\psi(i-1)$ will be uncorrelated with each of $w(i-p), w(i-p+1), \ldots, w(i)$ and hence also with $v(i)$. Such an appropriate choice of $\psi(i)$ is called an Instrumental Variable.

As an illustration, in the preceding example of system (16), one may choose $\psi(t) \triangleq y(t-1)$. For more on the Instrumental Variables Method the reader is referred to Ljung and Söderström[6].

3.3 Adaptive Spectral Factorization

It should be noted that the Instrumental Variables Method only gives us estimates of $(a_1, \ldots, a_p, b_1, \ldots, b_p)$ and not $(c_1, \ldots, c_p)$. Until very recently, there was *no proof* of strong consistency of any recursive algorithm for estimating $(c_1, \ldots, c_p)$ which works even when the polynomial $C(z)$ of (3) is ***not*** positive real. Recently, however, Solo [7] has shown that the Instrumental Variable Method used in conjunction with an *adaptive spectral factorization* scheme does give strongly consistent estimates of all the parameters.

To illustrate this scheme it is sufficient to consider ARMA systems and drop the input sequence $\{u(t)\}$. So consider the system (1) and let us also assume that the roots of the polynomials $A(z)$ and $C(z)$ of (2) and (3) are strictly outside the unit circle.

Let

$$\begin{aligned}\phi(t) &\triangleq (y(t-1), \ldots, y(t-p))^T \\ \theta^o &\triangleq (a_1, \ldots, a_p)^T\end{aligned}$$

and suppose that by the Instrumental Variables Method we have an estimate $\theta(t)$ of θ^o which is strongly consistent, i.e.,

$$\lim_{t\to\infty} \theta(t) = \theta^o \quad \text{a.s.} \tag{21}$$

Note that due to (18), we can form *estimates* of $v(t)$, called say $\hat{v}(t)$, from

$$\hat{v}(t) \triangleq y(t) - \phi^T(t-1)\theta(t).$$

Due to (21) and the asymptotic stationarity of the system it follows that

$$\lim_{t\to\infty} \frac{1}{t}\sum_{i=1}^{t} \hat{v}(i)\hat{v}(i-\ell) = \lim_{t\to\infty} \frac{1}{t}\sum_{i=1}^{t} v(i)v(i-\ell) \quad a.s. \tag{22}$$

for every integer ℓ. Let

$$P_\ell \triangleq \lim_{t\to\infty} \frac{1}{\sigma^2 t}\sum_{i=1}^{t} v(i)v(i-\ell) = \frac{1}{\sigma^2}E(v(i)v(i-\ell))$$

be the correlations of lag ℓ, and it is clear that since $v(i) = C(z)w(i)$,

$$\Phi(z) \triangleq \sum_{\ell=-p}^{p} P_\ell z^\ell = C(z)C(z^{-1}). \tag{23}$$

Define

$$\begin{aligned}P_\ell(t) &\triangleq \tfrac{1}{t}\textstyle\sum_{i=1}^{t} \hat{v}(i)\hat{v}(i-\ell) \\ \hat{\Phi}(t,z) &\triangleq \textstyle\sum_{\ell=-p}^{p} \hat{P}_\ell(t)z^\ell\end{aligned}$$

and we have

$$\lim_{t\to\infty} \hat{\Phi}(t,z) = \Phi(z) = C(z)C(z^{-1}) \quad \text{a.s. for every } z.$$

Presently, we will exhibit the details of a recursive spectral factorization scheme of Wilson [8], but just suppose that we have such a recursive scheme

$$\hat{C}_n(z) \triangleq f(\Phi(z), \hat{C}_{n-1}(z)) \tag{24}$$

which has the property that $\hat{C}_n(z) \to C(z)$. Then it is reasonable to suspect that if we use

$$\hat{C}(t,z) \triangleq f(\hat{\Phi}(t,z), \hat{C}(t-1,z)) \tag{25}$$

then we should also have

$$\lim_{t\to\infty} \hat{C}(t,z) = C(z) \quad \text{a.s. for every } z. \tag{26}$$

The difference between (25) and (24) is that in (25) we use an estimate $\hat{\Phi}(t,z)$ in place of the unknown $\Phi(z)$.

Solo [7] has proved that (26) does hold when the function $f(\cdot)$ in (25), defining the adaptive spectral factorization algorithm, is the recursion due to Wilson [8]. We will not go into Solo's proof, but we will present Wilson's algorithm below because it has several interesting features.

3.4 Wilson's Algorithm for Spectral Factorization

The problem is this. Give $\{P_\ell\}$ and $\Phi(z)$ as in (23), we want a recursive scheme such as (24) with the property that

$$\lim_{n\to\infty} \hat{C}_n(z) = C(z)$$

where $C(z)$ has all its roots (say) inside the unit circle. (So the $C(z)$ of the previous section is $C(z^{-1})$ of the present section.)

Note that this just means that we have to solve the simultaneous nonlinear equations

$$\sum_{k=0}^{p-j} \hat{c}_k \hat{c}_{k+j} = P_j \qquad \text{for } j = 0, \ldots, p \tag{27}$$

with the constraint that all roots of

$$\hat{C}(z) \triangleq \sum_{i=0}^{P} \hat{c}_i z^i$$

are strictly inside the unit circle. Solving (27) is equivalent to solving

$$h_j(\hat{c}_o, \ldots, \hat{c}_p) = 0 \qquad \text{for } j = 0, \ldots, p \tag{28}$$

where

$$h_j(\hat{c}_o, \ldots, \hat{c}_p) \triangleq \sum_{k=0}^{p-j} \hat{c}_k \hat{c}_{k+j} - P_j \tag{29}$$

Wilson's algorithm is just a Newton-Raphson method for solving (28). Defining

$$T_{jk}(\hat{c}_o, \ldots, \hat{c}_p) \triangleq \frac{\partial h_j(\hat{c}_o, \ldots, \hat{c}_p)}{\partial \hat{c}_k}$$

it is clear that

$$T = T_1 + T_2$$

where

$$T_1(\hat{c}_0, \ldots, \hat{c}_p) := \begin{bmatrix} \hat{c}_0 & \cdots & \cdots & \hat{c}_k \\ \hat{c}_1 & \cdots & \hat{c}_k & 0 \\ \cdot & & & \\ \cdot & & & \\ \cdot & & & \\ \hat{c}_k & \cdots & 0 & 0 \end{bmatrix} ; \quad T_2(\hat{c}_0, \ldots, \hat{c}_p) := \begin{bmatrix} \hat{c}_0 & \cdots & \cdots & \hat{c}_k \\ 0 & \hat{c}_0 & \cdots & \hat{c}_{k-1} \\ \cdot & & & \\ \cdot & & & \\ \cdot & & & \\ 0 & \cdots & \cdots & \hat{c}_0 \end{bmatrix}$$

Let

$$\hat{C}_n \triangleq (\hat{c}_o(n), \ldots, \hat{c}_p(n))^T$$

be the iterate at step n, and the Newton Raphson scheme gives

$$\hat{C}_{n+1} = \hat{C}_n - T^{-1}(\hat{C}_n) h(\hat{C}_n) \tag{30}$$

where $h(\hat{C}_n) \triangleq (h_1(\hat{C}_n), \ldots, h_p(\hat{C}_n))^T$.

Now we will exhibit a remarkable property of the scheme (30). Note that (30) is equivalent to

$$T(\hat{C}_n)\hat{C}_{n+1} = T(\hat{C}_n)\hat{C}_n - h(\hat{C}_n). \tag{31}$$

Using (29) and the property that

$$T(\hat{C}_n)\hat{C}_n = 2\hat{P}(n)$$

where

$$\text{j}-\text{th element of } \hat{P}(n) \triangleq \sum_{k=0}^{p-j} \hat{c}_k\hat{c}_{k+j}(n)$$

it follows that (31) can be written as

$$T(\hat{C}_n)\hat{C}_{n+1} = \hat{P}(n) + P$$

with $P \triangleq (P_o, \ldots, P_p)^T$. Using $T = T_1 + T_2$ we therefore get

$$T_1(\hat{C}_n)\hat{C}_{n+1}T_2(\hat{C}_n)\hat{C}_{n+1} = \hat{P}(n) + P. \tag{32}$$

In polynomial terms, (32) is equivalent to

$$\hat{C}_n(z)\hat{C}_{n+1}(z^{-1}) + \hat{C}_n(z^{-1})\hat{C}_{n+1}(z) = \hat{C}_n(z)\hat{C}_n(z^{-1}) + C(z)C(z^{-1})$$

which is the same as

$$\hat{C}_{n+1}(z)\hat{C}_{n+1}(z^{-1}) = [\hat{C}_{n+1}(z) - \hat{C}_n(z)][\hat{C}_{n+1}(z^{-1}) - \hat{C}_n(z^{-1}) + C(z)C(z^{-1}). \tag{33}$$

Hence we have

$$|\hat{C}_{n+1}(z)|^2 = |\hat{C}_{n+1}(z) - \hat{C}_n(z)|^2 + |C(z)|^2 \qquad \text{on } |z| = 1. \tag{34}$$

The remarkable property now is that by Rouche's Theorem [9], the polynomials $\hat{C}_{n+1}(z)$ and $\hat{(C)}_n(z)$ have the *same number of roots inside the unit circle.* Hence if the initial iterate is chosen as $\hat{(C)}_o(z) \equiv 1$, i.e., has no roots inside the unit circle, then *every iterate* $\hat{C}_n(z)$ *also has this property.*

One more property is also useful. If we can just show that $\{\hat{C}_n(z)\}$ converges, then clearly the first term on the right hand side of (33) will vanish, and so the left hand side converges to $C(z)C(z^{-1})$, proving that $\lim_{n\to\infty} \hat{C}_n(z) = C(z)$.

To prove convergence of $C_n(z)$ note that by dividing (33) by $C_n(z)C_n(z^{-1})$ we get

$$2\, Re\left[\frac{\hat{C}_{n+1}(z)}{\hat{C}_n(z)}\right] = 1 + \left|\frac{C(z)}{\hat{C}_n(z)}\right|^2 \qquad \text{on } |z| = 1$$

and since the last term on the right hand side is less than or equal to 1 from (33), an application of the Maximum Principle [9], shows that

$$\frac{1}{2} \leq \frac{\hat{C}_{n+1}(z)}{\hat{C}_n(z)} \leq 1 \qquad \text{for } z \in [-1, 1].$$

Hence $\{\hat{C}_n(z)\}$ is a monotone decreasing sequence for every $z \in [-1, 1]$, bounded below by 0, and so converges.

3.5 The Pseudo-Linear Regression Method

Let us return to the problem of identifying $(a_1, \ldots, a_p, c_1, \ldots, c_p)$ for the system (1)

$$y(t) = \phi^T(t-1)\theta^o + w(t)$$

with

$$\begin{aligned}\phi(t-1) &\triangleq (y(t-1), \ldots, y(t-p), w(t-1), \ldots, w(t-p))^T \\ \theta^o &\triangleq (a_1, \ldots, a_p, c_1, \ldots, c_p)^T.\end{aligned}$$

The Least Squares Estimate

$$\theta(t+1) \triangleq \theta(t) + R^{-1}(t)\phi(t)[y(t+1) - \phi^T(t)\theta(t)]$$

is good, but since $\phi(t)$ contains $w(t), \ldots, w(t-p+1)$ as components, it cannot be implemented.

However,

$$\hat{w}(t) \triangleq y(t) - \phi^T(t-1)\theta(t)$$

can be regarded as an estimate of $w(t)$. One can use this instead of $w(t)$. The resulting scheme is called a Pseudo-Linear Regression scheme.

It has been analyzed by Solo [10] and shown to be convergent to the true parameter values provided

$$Re\, C^{-1}(e^{i\omega}) \geq \frac{1}{2} \qquad \text{for all } \omega \in [0, 2\pi]$$

This is a restrictive condition on the noise and can be rewritten as

$$|C(e^{i\omega}) - 1| < 1 \qquad \text{for } \omega \in [0, 2\pi]$$

and so amounts to requiring that the moving average noise

$$v(t) \triangleq w(t) + \sum_{i=1}^{p} c_i w(t-i)$$

be not too "colored."

3.6 The Recursive Prediction Error Method

The drawback to the Instrumentation Variables cum Wilson's Adaptive Spectral Factorization scheme of Solo [7] is that the Instrumental Variables Method is not generally efficient. Moreover, the computation of the inverse in (19) can lead to numerical problems when $\psi(i)$ and $\phi(i)$ are weakly correlated.

A possibly superior algorithm in this regard is the Recursive Prediction Error Method [6]. There is currently, however, to the author's knowledge, *no* proof of convergence of this algorithm.

To motivate the algorithm consider the following Off-line *Prediction Error Method.* Consider the family of systems

$$\begin{aligned}y(t) &= \phi^T(t-1)\theta + w(t) \\ \phi^T(t-1) &\triangleq (y(t-1), \ldots, y(t-p), w(t-1), \ldots, w(t-p))\end{aligned}$$

indexed by θ. For each θ, given observations of the past $y(s), s \leq t-1)$ one can form a prediction

$$\hat{y}_{t|t-1}(\theta) \triangleq E^\theta[y(t)|y(s), s \leq t-1].$$

One can then compare this prediction with the *observed* value $y(t)$ and form the *prediction error*

$$\tilde{y}_{t|t-1}(\theta) \triangleq y(t) - \hat{(y)}_{t|t-1}(\theta).$$

Clearly a model θ is a good fit for the observed data. Thus one may consider an estimate $\theta(t)$ which minimizes the predictor errors in the sense that

$$\theta(t) \triangleq \arg\min_{\theta} \sum_{i=1}^{t} (\tilde{y}_{i|i-1}(\theta))^2. \tag{35}$$

Such a $\theta(t)$ can be called a *Prediction Error Estimate.* Unfortunately, as the following example shows, the criterion (35) is difficult to minimize.

Example of Prediction Error Estimates

Consider the family of systems

$$y(t) = \theta_1 u(t-1) + \theta_2 w(t-1) + w(t) \tag{36}$$

with $w(0) = 0$, indexed by $\theta \triangleq (\theta_1, \theta_2)^T$. Note that

$$w(1) = y(1) - \theta_1 u(0) \tag{37}$$

$$w(t) = y(t) - \theta_1 u(t-1) - \theta_2 w(t-1) \qquad \text{for } t \geq 2. \tag{38}$$

Solving (38) with initial condition (37) gives

$$w(t-1) = \sum_{n=1}^{t-1} (-\theta_2)^{t-n-1} (y(n) - \theta_1 u(n-1)) \qquad \text{for } t \geq 2. \tag{39}$$

Substituting for $w(t-1)$ in (36) gives

$$y(t) = \theta_1 u(t-1) - \sum_{n=1}^{t-1} (-\theta_2)^{t-n} (y(n) - \theta_1 u(n-1)) + w(t).$$

Hence

$$E^{\theta}[y(t)|y(s), u(s); s \leq t-1] = \theta_1 u(t-1) - \sum_{n=1}^{t-1} (-\theta_2)^{t-n} (y(n) - \theta_1 u(n-1))$$

and so

$$\tilde{y}_{t|t-1}(\theta) = \sum_{n=1}^{t} (-\theta_2)^{t-n} (y(n) - \theta_1 u(n-1)).$$

Therefore,

$$\theta(t) \triangleq \arg\min_{\theta} \sum_{i=1}^{t} [\sum_{n=1}^{i} (-\theta_2)^{i-n} (y(n) - \theta_1 u(n-1))]^2. \tag{40}$$

The expression on the right hand side of (40) features arbitrarily large powers of θ_2 (as t becomes large) and is consequently not easily minimized. Hence it may not be possible to determine the Prediction Error Estimate on-line.

Nevertheless, the Prediction Error Estimates do serve as a benchmark because of their good asymptotic properties [6].

This motivates the goal of obtaining an approximation of the Prediction Error Estimates which can be *recursively* generated, the so-called *Recursive Prediction Error Method.*

Let $V_t(\theta)$ be the summation on the right hand side of (35). One then obtains estimates satisfying

$$\theta(t+1) = \theta(t) + R^{-1}(t)\psi(t)[y(t+1) - \hat{y}(t+1)] \tag{41}$$

where $R(t)$ is a recursively generated approximation of the Hessian of V_t, $\psi(t)$ is an approximation of the gradient, and $y(t+1)$ is a "prediction" of $\hat{y}(t+1)$. For more details on the nature of these approximations, we refer the reader to [6]. The reader will also find there details of the so-called "Projection" feature of the algorithm, which we have omitted to mention. This is basically a mechanism by which the algorithm is "reset" whenever it enters regions of the parameter space considered destabilizing. Some such mechanism, it appears, may be needed for convergence.

One advantage of this Recursive Prediction Error Method is that it leaves open the possibility that it may be asymptotically efficient.

This algorithm is very difficult to analyze directly and analysis has been mainly conducted using the *Ordinary Differential Equation* pioneered by Ljung [11,12]. As we will see below, this approach gives much valuable information and intuition regarding the behavior of the stochastic algorithm but unfortunately falls short of providing a conclusive statement regarding convergence.

3.7 The Ordinary Differential Equation Approach

Many recursive stochastic algorithms, the Pseudo-Linear Regression Method and the Recursive Prediction Error Method included, can be written as

$$\theta_{n+1} = \theta_n - \gamma_{n+1} f(\theta_n, x_{n+1}) \tag{42}$$

where x_n is the state of a time varying linear system

$$x_{n+1} = A_{n+1}(\theta_n)x_n + B_{n+1}(\theta_n)w_{n+1} \tag{43}$$

where $\{w_n\}$ is a "white" noise process. The equation (42) models the update equation for the pair (9),(10) say, while (43) models all the remaining dynamics including the ARMAX process itself. For details, see Ljung [11,12]. (Again, we have omitted the "Projection" feature found in some algorithms.)

Ljung [11,12] has shown that one can study the behavior of (42) by studying an associated ordinary differential equation. To provide an outline of this method, we follow the approach of Kushner and Clark [13] utilized by Metivier and Priouret [14].

In many algorithms the sequence $\{\gamma_n\}$ occurring in (42) is typically of the form

$$\gamma_n = \frac{1}{n}$$

Suppose now that the term $\gamma_{n+1} f(\theta_n, x_{n+1})$ can be further decomposed so that we can rewrite (42) as

$$\theta_{n+1} = \theta_n - \gamma_{n+1} h(\theta_n) + v_{n+1}$$

where $h(\cdot)$ is a continuous function. We are going to conduct a sample path analysis of $\{\theta_n(\omega), x_n(\omega)\}$ for fixed ω, and so we shall drop, for the time being, the dependence on ω. Define

$$\begin{aligned} V_n &\triangleq \textstyle\sum_{k=0}^{n} v_k \\ t_n &\triangleq \textstyle\sum_{k=1}^{n} \gamma_k. \end{aligned}$$

For any sequence $\{z_n\} \subseteq R^p$, one can consider the sequence $\{(z_n, t_n)\}$, respectively, by

$$\begin{aligned} z_s(t) &\triangleq z_n && \text{for } t_n \le t < t_{n+1} \\ z_\ell(t) &\triangleq \tfrac{(t-t_n)z_{n+1}+(t_{n+1}-t)z_n}{t_{n+1}-t_n} && \text{for } t_n \le t < t_{n+1}. \end{aligned}$$

It is then easy to check that the linearly interpolated function $\theta_\ell(\cdot)$ satisfies the integral equation,

$$\theta_\ell(t) = \theta_o - \int_o^t h(\theta_s(r))dr + V_\ell(t). \tag{44}$$

Moreover, if one defines

$$\begin{aligned} \theta_\ell^n(t) &\triangleq \theta_\ell(t + t_n) \\ (\Delta V)^n(t) &\triangleq V_\ell(t + t_n) - V_\ell(t_n) \end{aligned} \tag{45}$$

then $\theta_\ell^n(\cdot)$ satisfies the integral equation,

$$\theta_\ell^n(t) = \theta_\ell^n(0) - \int_o^t h(\theta_\ell^n(r))dr + (\Delta V)^n(t) + \epsilon_n(t) \tag{46}$$

where

$$\epsilon_n(t) \triangleq \int_o^t h(\theta_s(t_n + r))dr - \int_o^t h(\theta_\ell^n(r))dr. \tag{47}$$

Basically (44) and (46) are just rewritings of the difference equation (42) in interpolated form.

Note now that our goal is to study the asymptotic behavior of the sequence $\{\theta_n\}$, and from (45) it is clear that this is equivalent to studying the behavior of $\theta_\ell^n(t)$ for large n but fixed *finite* t. Hence consider the sequence of *functions* $\{\theta_\ell^n(\cdot)\}$. The limiting behavior of this sequence of *functions* evaluated at *fixed* t models the asymptotic behavior of the sequence $\{\theta_n\}$.

This is captured by the following result of Kushner and Clark [13].

Suppose

(i) $\{\theta_n\}$ is a bounded sequence.

(ii) $\lim_{t\to\infty} \sup_{|r|\le T} |V_\ell(t+r) - V_\ell(t)| = 0$ for every $T > 0$.

Then

(iii) every limit point of $\{\theta_\ell^n(\cdot)\}$, in the topology of uniform convergence on compacta, satisfies the *ordinary differential equation*

$$\frac{d\theta}{dt} = -h(\theta) \tag{48}$$

(iv) if θ^* is a locally asymptotically stable equilibrium point of (48) with domain of attraction D and if

$$\theta_n \in D \quad \text{i.o.} \tag{49}$$

then

$$\lim_{n\to\infty} \theta_n = \theta^*. \qquad \square$$

Thus, to show convergence of the sequence $\{\theta_n\}$ to θ^*, one has to prove that (i), (ii), and (49) hold. Condition (ii) is satisfied if

$$\lim_{n\to\infty} \Big(\sup_{n<k\le m(n,T)} \Big| \sum_{i=n+1}^{k} \gamma_i v_i \Big| \Big) = 0 \tag{50}$$

with

$$m(n,T) \triangleq max\{k : \sum_{i=n}^{k} \gamma_i \le T\}$$

when (i) holds, [14]; and (50) can in many cases be established by martingale arguments. To do this one usually chooses $h(\theta)$ as the mean value, in steady state, of $f(\theta, x_{n+1})$, when θ_n is identically set equal to θ in (43). For this steady state to exist, θ must of course be restricted to the set of θ's for which $A(\theta)$ in (43) is stable. Hence $h(\cdot)$ is defined only in this set. (This is linked to the Projection feature.)

Thus, the Ordinary Differential Equation approach gives valuable information, *provided* one can establish (i) and (49) by other independent techniques. This is the single biggest limitation to the power of the approach.

For more on the Ordinary Differential Equation approach, we refer the reader to Kushner and Shwartz [15], Metivier and Priouret [14], Ljung and Söderström[6], and Kushner [16].

4. Adaptive Control

Let us consider the ARX unit delay system

$$y(t) = \sum_{i=1}^{p}(a_i y(t-1) + b_i u(t-i)) + w(t) \quad ; \; b_1 \neq 0.$$

Suppose the goal is to minimize the steady state variance $E(y^2(t))$. Then the *minimum variance control law* is

$$u(t-1) = -\frac{1}{b_1}[\sum_{i=1}^{p} a_i y(t-i) + \sum_{i=2}^{p} b_i u(t-i)]. \tag{51}$$

It gives rise to an internally stable closed loop system if the polynomial $B(z) \triangleq \sum_{i=1}^{p} b_i z^{i-1}$ is minimum phase, see Åström [17]. If the coefficients $(a_1, \ldots, a_p, b_1, \ldots, b_p)$ are unknown, then it is natural to estimate them by the Least Squares Method (9,10) and use the estimates in place of the true values of the parameters. This sort of a scheme (with the slight modification of fixing b_1 instead of estimating it) was proposed by Åström and Wittenmark [18]. They analyzed this scheme and showed that even in the presence of colored noise, *if* the scheme converged, then the limiting regulator is a minimum variance one. The analysis was further pursued by Ljung [11,12] who used the Ordinary Differential Equation to identify a Positive Real condition on the spectrum of the colored noise as being important for convergence.

Unfortunately, there is still no conclusive analysis of the least squares based adaptive minimum variance control scheme, though modifications of the least squares scheme have been studied [19].

Rigorous convergence results have, however, been obtained for a related *gradient* algorithm for parameter estimation. Goodwin, Ramadge, and Caines [20] have analyzed such a scheme and shown that for both regulation and tracking problems, the resulting adaptive control algorithm does succeed in minimizing the variance of the tracking error. Subsequently, in [21] it has been shown that in the *regulation* problem, the parameter estimates also converge, but only to a *random* multiple of the true parameter and *not* the true parameter. However, since the minimum variance control law (51) is invariant under scaling of the parameter vector, the adaptive control law *self-tunes* to the optimal control law.

Recently, however, the *tracking* problem has also been analyzed in [22] and we shall outline these results below.

5. Adaptive Minimum Variance Tracking by the Stochastic Gradient Algorithm

Consider the ARMAX unit delay system,

$$y(t) = \sum_{i=1}^{p} a_i y(t-i) + \sum_{i=1}^{q} b_i u(t-i) + \sum_{i=1}^{s} c_i w(t-i) + w(t) \tag{52}$$

with $b_1 \neq 0$. Let us suppose that there is a given reference trajectory $\{y^*(t)\}$ and the goal is to minimize

$$\lim \frac{1}{N}\sum_{1}^{N}(y(t) - y^*(t))^2 \tag{53}$$

i.e., the variance of the tracking error. If $\{w(t)\}$ is observable, then the optimal control law is clearly

$$u(t-1) = -\frac{1}{b_1}\left[\sum_{i=1}^{p} a_i y(t-i) + \sum_{i=2}^{q} b_i u(t-i) + \sum_{i=1}^{s} c_i w(t-i) - y^*(t)\right]$$

and this results in

$$y(t) = y^*(t) + w(t) \quad or \quad w(t) = y(t) - y^*(t). \tag{54}$$

Since $\{w(t)\}$ cannot be observed one can consider substituting (54) in place. This gives

$$u(t-1) = -\frac{1}{b_1}\left[\sum_{i=1}^{pvs}(a_i + c_i)y(t-i) + \sum_{i=2}^{q} b_i u(t-2) - y^*(t) - \sum_{i=1}^{s} c_i y^*(t-i)\right] \tag{55}$$

and in fact if all the roots of $C(z) \triangleq 1 + \sum_{i=1}^{p} c_i z^i$ are strictly outside the unit circle, then this control is optimal with respect to (53). Moreover, when this control law is applied, the system (52) asymptotically behaves as

$$y(t) = -\sum_{i=1}^{pvs}(a_i + c_i)y(t-i) + \sum_{i=1}^{q} b_i u(t-i) - \sum_{i=1}^{s} c_i y^*(t-i) + w(t)$$

which can also be written as

$$y(t) - y^*(t) = \phi^T(t-1)\theta^o + w(t)$$

where

$$\phi^T(t-1) \triangleq (y(t-1), \ldots, y(t-pV_s), u(t-1), \ldots, u(t-1), -y^*(t), \ldots, -y^*(t-s))$$

$$\theta^o \triangleq (a_1 + c_1, \ldots, a_{pvs} + c_{pvs}, b_1, \ldots, b_q, 1, c_1, \ldots, c_s)^T.$$

This suggests the following *stochastic gradient* algorithm for estimating θ^o,

$$\theta(t+1) = \theta(t) + \frac{\mu\phi(t)}{r(t)}[y(t+1) - \phi^T(t)\theta(t)] \tag{56}$$

$$r(t+1) = r(t) + \phi^T(t+1)\phi(t+1) \quad ; \mu > 0. \tag{57}$$

Moreover, since the optimal control law (55) can be written as $\phi^T(t)\theta^o = 0$, an adaptive version of it is simply

$$\phi^T(t)\theta(t) = 0. \tag{58}$$

The algorithm (56),(57),(58) now gives us an adaptive control law for the minimum variance tracking problem.

There is one nonstandard feature of the above algorithm. Note that the $(pvs + q + 1)$-th component of θ^o is 1, a known quantity. However, the algorithm ignores this information and estimates it anyway through (56). Thus the dimension of the parameter estimator is one-dimension larger than the corresponding algorithm in [20].

However, for the above "unnormalized" algorithm, [21] proves convergence of the parameter estimates to the true parameters and also self-tuning of the adaptive control law (58) to the optimal control law, provided that

$$\{y^*(t)\} \text{ is sufficiently rich of order at least } (q + s). \tag{59}$$

In many interesting tracking problems the condition (59) is not satisfied. For example, consider the problem of maintaining the output at a fixed constant value, i.e.,

$$y^*(t) = y^*(t-1) = \text{constant}. \tag{60}$$

Such a $\{y^*(t)\}$ is only sufficiently rich of order 1.

This motivates the following problem. Consider $\{y^*(t)\}$ generated by a *linear model*, i.e., one of the form

$$y^*(t) = \sum_{i=1}^{\ell} h_i y^*(t-i). \tag{61}$$

Note that the constant reference trajectory (60) is a special case of (61).

In [22] it is proved that when $\{y^*(t)\}$ satisfies (61), it is enough for the order of sufficient richness to be greater than s for the parameter estimates to be strongly consistent. The point here is that the regularity of the model cuts down on the order of sufficient richness (a reduction from $(q+s)$ to $(s+1)$) needed for strong consistency.

However, the sufficient richness of $\{y^*(t)\}$ may in applications be even less than s. In such cases one can reduce the dimension of the parameter estimator, as follows.

Suppose that $\{y^*(t)\}$ satisfies (61) and it is sufficiently rich of order ℓ, with $\ell \leq s$. Let us redefine

$$\phi(t) \triangleq (y(t-1),\ldots,y(t-pvs),u(t-1),\ldots,u(t-q),-y^*(t),\ldots,-y^*(t+1-\ell))^T$$

and employ the algorithm (56),(57),(58). Note that we have reduced the dimension of $\theta(t)$ and $\phi(t)$ by $(s+1-\ell)$ components. For example, if $\ell = 1$ as in (60), then there is a saving of s components. Effectively then we are only using ℓ components more than in the minimum variance regulation problem.

In [22] it is proved that with this reduced dimension adaptive controller, one still has convergence of the parameter estimates, but not strong consistency. More importantly the adaptive control law is *self-tuning*. For full details, see [22].

There is now therefore a unified treatment of the regulation and tracking problems, when the stochastic gradient algorithm is used for parameter estimation.

The full treatment of least squares based parameter estimation algorithms in adaptive control is however, still elusive.

6. Adaptive LQG

A disadvantage with the minimum variance control law (51) is that it leads to an internally stable closed loop system only when the polynomial $B(z)$ is of minimum phase. One can ask for the control law which minimizes the variance *subject* to the constraint that the resulting closed loop system is internally stable. This is a subset of the set of all control laws and optimization of the variance can be performed over this subset. The single-input, single-output case is treated in Peterka [23] and the multi-input, multi-output case in [24]. One can design an adaptive control scheme based on such a control law, but the resulting closed-loop adaptive control system has yet to be analyzed.

Let us turn instead to the quadratic cost criterion

$$\lim_{N\to\infty} \frac{1}{N} \sum_{t=1}^{N} [y^2(t) + \rho u^2(t)] \tag{62}$$

with $\rho > 0$. It appears, however, that one may not obtain self-optimality with respect to such a cost criterion by using a straightforward certainty equivalent adaptive control law [25].

We will recount below some of the arguments of [25]. Consider the first order system

$$y(t) = ay(t-1) + bu(t-1) + w(t). \tag{63}$$

The optimal control law for (63) with the cost criterion (62) is given by

$$u(t) = k(a,b)y(t)$$

where the optimal gain $k(a,b)$ can be obtained through standard LQG theory.

Since (a,b) are unknown one can use a stochastic gradient algorithm

$$\begin{aligned}
a(t+1) &= a(t) + \tfrac{\mu y(t)}{t\bar{r}(t)}[y(t+1) - a(t)y(t) - b(t)u(t)] \\
b(t+1) &= b(t) + \tfrac{\mu u(t)}{\bar{r}(t)}[y(t+1) - a(t)y(t) - b(t)u(t)] \\
\bar{r}(t+1) &= \bar{r}(t) + \tfrac{1}{t+1}[y^2(t+1) + u^2(t+1) - \bar{r}(t)]
\end{aligned}$$

to estimate (a, b). The the control input may be chosen as

$$u(t) = k(a(t), b(t))y(t).$$

The Ordinary Differential Equations corresponding to the above have been analyzed in [25]. They show that there is a manifold of equilibrium points all of which attract the solutions of the O.D.E.'s. However, except for one point on this manifold, all the other equilibria correspond to *strictly suboptimal* control laws.

This suggests that adaptive control laws of the above type will not generally self-tune to the optimal control law for the quadratic cost criterion (62). In this respect the minimum variance criterion (53) is fairly singular.

Acknowledgments

The research reported here has been supported in part by the National Science Foundation under Grant No. ECS-85-06628 and the Joint Services Electronics Program under Contract No. N00014-84-C-0149. The author is grateful to D. Connors, L. Praly, V. Solo and E. Trulsson for several useful discussions.

References

[1] P.R. Kumar, "A Survey of Some Results in Stochastic Adaptive Control," *SIAM Journal on Control and Optimization*, vol. 23, No. 3, pp. 329-380, May 1985.

[2] G. Goodwin and R.L. Payne, Dynamic System Identification, Academic Press, New York, 1977.

[3] K.L. Chung, "A Course in Probability Theory," Academic Press, New York, 1974.

[4] L. Ljung, "Consistency of the Least Squares Identification Method," *IEEE Transactions on Automatic Control*, vol. AC-21, pp. 779-781, 1976.

[5] J. Doob, Stochastic Processes, John Wiley, New York, 1953.

[6] L. Ljung and T. Söderström, Theory and Practice of Recursive Identification, M.I.T. Press, Cambridge, 1983.

[7] V. Solo, "Adaptive Spectral Factorization," Preprint, 1984.

[8] G. Wilson, "Factorization of the Covariance Generating Function of a Pure Moving Average Process," *SIAM Journal on Numerical Analysis*, vol. 6, pp. 1-7, 1969.

[9] L.V. Ahlfors, Complex Analysis, McGraw-Hill, New York, 1979.

[10] V. Solo, "The Convergence of AML," *IEEE Transactions on Automatic Control*, vol. AC-24, pp. 958-963, 1979.

[11] L. Ljung, "Analysis of Recursive Stochastic Algorithms," *IEEE Transactions on Automatic Control*, vol. AC-22, pp. 551-575, 1977.

[12] L. Ljung, "On Positive Real Transfer Functions and the Convergence of Some Recursive Schemes," *IEEE Transactions on Automatic Control*, vol. AC-22, pp. 539-551, 1977.

[13] H. Kushner and D. Clark, "Stochastic Approximation for Constrained and Unconstrained Systems," *Appl. Math. Sci.*, vol. 26, 1978.

[14] M. Metivier and P. Priouret, "Applications of a Kushner and Clark Lemma to General Classes of Stochastic Algorithms," *IEEE Transactions on Information Theory*, vol. IT-30, No. 2, pp. 140-150, 1984.

[15] H. Kushner and A. Shwartz, "An Invariant Measure Approach to the Convergence of Stochastic Approximations with State Dependent Noise," *SIAM Journal on Control and Optimization*, vol. 22, pp. 13-27, 1984.

[16] H. Kushner, Approximation and Weak Convergence Methods for Random Processes, M.I.T. Press, Cambridge, 1984.

[17] K.J. Åström , Introduction to Stochastic Control Theory, Academic Press, New York, 1970.

[18] K.J. Åström and B. Wittenmark, "On Self-tuning Regulators," *Automatica*, vol. 9, pp. 185-199, 1973.

[19] G.C. Goodwin and K.S. Sin, Adaptive Filtering Prediction and Control, Prentice Hall, Englewood Cliffs, 1984.

[20] G.C. Goodwin, P. Ramadge and P. Caines, "Discrete Time Stochastic Adaptive Control," *SIAM Journal on Control and Optimization*, vol. 19, pp. 829-853, 1981.

[21] A. Becker, P.R. Kumar and C.Z. Wei, "Adaptive Control with the Stochastic Approximation Algorithm: Geometry and Convergence," *IEEE Transactions on Automatic Control*, vol. AC-30, pp. 330-338, 1985.

[22] P.R. Kumar and L. Praly, "Self-tuning and Convergence of Parameter Estimates in Minimum Variance Tracking and Linear Model Following," University of Illinois, April 1985.

[23] V. Peterka, "On Steady State Minimum Variance Control Strategy," *Kybernetica*, vol. 8, pp. 218-231, 1972.

[24] U. Shaked and P.R. Kumar, "Minimum Variance Control of Multivariable ARMAX Systems," to appear in *SIAM Journal on Control and Optimization*.

[25] W. Lin, P.R. Kumar and T.I. Seidman, "Will the Self-tuning Approach Work for General Cost Criteria?" University of Maryland Baltimore County, 1983, to appear in *Systems and Control Letters*.

ADAPTIVE CONTROL APPLICATIONS

The great tragedy of Science -
the slaying of a beautiful hypothesis by an ugly fact.

Thomas Henry Huxley

It is common knowledge that the rapid and revolutionary progress in microelectronics in the last decade has had a profound impact on all aspects of technology. In the context of adaptive control, the computing capability and low cost of microprocessors have made the implementation of complex control algorithms in practical systems cost effective. In fact we are rapidly approaching a stage where the scope of specific applications will be limited only by the existing theory and the imagination of the designer. The six papers included in this section deal with recent successful efforts to apply adaptive concepts to practical problems. The theoretical analysis of adaptive systems is invariably based upon a number of assumptions such as the linearity of the plant being controlled, persistent excitation of the reference input signal and knowledge of the order and relative degree of the plant transfer function. These idealizations are rarely met in practice and hence various precautions must be taken to assure satisfactory performance when the algorithms are implemented.

The papers by Hang, Lim and Tay, Chesna and Ydstie, Unbehauen and Wiemer, and Bristol, are all concerned either explicitly or implicitly with process control. Industrial processes can be adequately represented by nonlinear and distributed models of high order. Further, sensors for measuring key variables are generally not available. Both these factors make the adaptive control of such processes considerably more difficult than those of electromechanical or power systems. But this is to some extent compensated for by performance requirements that are not too stringent and long time constants that permit efficient self-tuning based on on-line identification. All four papers discuss these aspects which are specific to process control problems. Contrary to what is believed by the novice regarding adaptive control, it is now generally realized that considerable prior information regarding both process and disturbance is needed to achieve tight control. This is clearly reflected in the last three papers which also reveal that the economic gains realized by using adaptation in process control are usually achieved by small but sustained improvements in overall efficiency.

Hang et al. discuss the use of adaptive Smith predictors in which on-line estimates of plant parameters are used to control the process. Chesna and Ydstie suggest modifications of well known adaptive algorithms by including a variable forgetting factor and using an extended horizon control policy for systems having a time-varying transport-delay and non-minimum phase dynamics. Unbehauen and Wiemer contend that adaptive control performs better than PID control or state feedback in distillation columns. Since sudden changes in flow and composition make it difficult to maintain the process near a specified operating point, they argue that adaptive control is needed. The use of an adaptive controller in such a case results in satisfactory disturbance rejection as well as considerable savings in cost.

The appearance of commercial products generally signifies a state of maturity in any field. The last few years have witnessed a significant increase in the number of feasibility studies, resulting in the introduction of several industrial adaptive regulators. The paper by Bristol describes one such, recently introduced by The Foxboro Company. This self-tuning controller is based on pattern recognition concepts rather than on recently developed adaptive control theory. While the advantages of this regulator over others can only be established by users in the future, it nevertheless serves to remind us that adaptation and learning may take place in a wide variety of forms. Bristol's comments regarding the pedagogy that is brought to bear on the problem of design provide an industrial perspective on the relation between theory and practice that the adaptive community should find most interesting.

The last two papers by Steinberg, and Neuman and Khosla, deal with the use of adaptive procedures in microwave radar imaging and robotics. In the former, the severe distortion due to position and phase errors in a large antenna is compensated for with an open-loop data-adaptive system. Using the measured signals to alter the weights in the antenna channels, a much sharper image is obtained. The first part of the paper by Neuman and Khosla presents a clear review of the properties of the Newton-Euler and Lagrange-Euler formulations of robot dynamics. Schemes for the on-line identification of dynamic parameters of robots which can be used for precise trajectory control, are discussed in the second part.

On Practical Implementations of Adaptive Control

C. C. Hang, K. W. Lim and T. T. Tay
Dept. of Electrical Engineering
National University of Singapore
Kent Ridge, Singapore 0511

Abstract

Practical aspects of adaptive control are discussed. Basic assumptions such as linearity, persistent excitation, known process order and dead time could be violated unless special precautions are taken. Implementation guidelines to avoid consequent problems such as estimator windup and parameter divergence are surveyed. Further findings on practical aspects are derived from extensive studies in the implementation of an adaptive Smith Predictor. In particular, the effects of d.c. levels and of faults such as static load disturbance and large parameter change are discussed. A new solution based on a two-level fault detection scheme with modified estimator is introduced and substantiated by simulation.

1. Introduction

The rapid development of adaptive control techniques in the seventies and eighties, in conjunction with the advent of low cost computing power, has attracted much interest in the industrial applications of adaptive control [1-5]. One simple and yet widely applicable way to use adaptive control laws is to make a control algorithm with automatic tuning [4-8]. The tuning may then be switched off when satisfactory performance is obtained. It can be used to solve industrial problems which arise due to lack of skilled field personnel, large number of controllers, and loops with long characteristic times especially when nonlinearities, variations in raw materials, or gradual changes in processing units require that loops be retuned intermittently. The tuning can be supervised by an operator and it is not necessary to have too much logic built into the algorithm.

When the process characteristics change frequently while tight control has to be maintained without operator intervention, the tuning of the control algorithm has to be switched on all the time. This is the more complete form of adaptive control and it requires a much more robust algorithm. Many feasibility studies and commercial installations indicate that this form of adaptive control can be used successfully to control industrial processes [1-6]. However, in all applications, different types of safety nets or special precautions in the form of 'jacketed software' have to be used. This is largely due to the gap between academic algorithms and practical algorithms. The basic academic design of adaptive controllers is based on several important assumptions such as linearity, persistent excitation, knowledge of process order and dead time. How these assumptions can be satisfied or relaxed in practice will be discussed in this paper. Further issues of practical significance such as reduction of switching-in transients, avoidance of estimator wind-up problem, etc., which are essentially refinements, will be addressed where appropriate.

A recent paper by Wittenmark and Åström [6] has given an excellent account of the state of the art and their experiences from practical work on adaptive control. These will be briefly surveyed to introduce the practical issues in adaptive control. The main contributions of this paper are: first, to introduce an important implementation issue arising from the occurrence of static load change and the on-line treatment of d.c. levels, which are not discussed by Wittenmark and Åström [6]. With load change, the estimated parameters would deviate sharply and would be erratic for a prolonged period during which control would be poor. Similar effects have been observed for large parameter changes. The main root of the problem is that an estimator well tuned for stationary operations in the presence of noise is not suitable for handling transient effects. A new solution, namely the two-level fault detection scheme with modified estimator will be introduced. Second, a different class of explicit adaptive controller, namely the adaptive Smith Predictor, is used to supplement the examples of self-tuning controllers commonly used by other authors [1-6].

2. Practical Issues and Implementation Guidelines

In this section we shall introduce the practical issues in adaptive control and summarize known, documentated solutions to the resulting problems. The basic assumptions made in adaptive control are those of linearity, persistent excitation, knowledge of process order and dead time, knowledge of noise properties, etc. They are necessary for the design and analysis of the estimator and controller. To ensure that these assumptions are not seriously violated, Wittenmark and Åström [6] have given the following implementation guidelines which are also in good agreement with other authors [2-5].

Robustness

A robust adaptive controller depends on both a robust controller design and on a robust estimator. For instance, a robust controller should provide high loop gain at low frequencies to eliminate low frequency disturbances; it should also have sufficient high frequency roll-off characteristics to eliminate the influence of high frequency disturbance or unmodeled dynamics. The model used should be accurate at least for frequencies where loop gain is around unity. To achieve this, the robust estimator should be provided with an input signal that has sufficient energy content around the cross-over frequency and is persistently exciting. Otherwise, an external perturbation signal should be injected or adaptation should be switched off when excitation is poor. The high frequency roll-off is obtained by signal filtering. Here the choice of sampling interval and the use of anti-aliasing filters are important considerations.

Parameter Tracking and Estimator Windup

To track variations in the process dynamics, exponential forgetting is usually used in the estimator. If there is a long period without sufficient excitation, exponential forgetting will cause the covariance matrix $P(k)$ to grow to a very large value. Then there may be large changes in the estimated parameters when new information comes into the system, for instance when the reference signal is given a step change. This phenomenon called 'estimator windup' or 'blow-up' may then cause a burst in the output of the process. Ways to avoid this problem include keeping the trace of $P(k)$ constant [8] and the use of variable forgetting factor [9,10]. A new method based on fault detection due to Hagglund [11,12] is found to be superior. Finally, poor excitation can aggravate numerical problems in the estimation algorithms but this can be alleviated by using the square rooting method or the U-D factorization method [3].

Start-up Procedures

To ensure closed-loop stability of a process which has natural non-linear characteristics, the adaptive controller should not be switched on until the estimator has converged reasonably. In the meantime, a safe but fixed controller has to be used. It can be advantageous to add a perturbation signal to speed up the convergence of the estimator. It may also be desirable to limit the control signal until better parameter estimates are obtained.

Reset Action and Anti-reset Windup

Reset action which gives the controller the ability to eliminate steady-state errors, may be introduced externally or internally included in the adaptive controller. Special precautions must be taken to avoid reset windup which can occur if the integrator loop is not closed due to signal saturation or malfunction of the process equipment or instrumentation.

Actuator Limits

The actuator has saturation limits and hence the controller output is usually limited likewise. The estimator should therefore be fed with the actual control signal that is sent out to the process. Otherwise, parameters like the gain of the process will be incorrectly estimated.

Tuning Parameters

The existing adaptive controllers still require certain process related and performance related parameters to be prespecified. While initial values of the estimator may be made insensitive by a proper start-up procedure, parameters such as sampling interval, dead time and order of the process model are more critical and may depend greatly on the type of adaptive controllers used.

3. Experience from Implementing Adaptive Smith Predictor [13-15]

The practical issues and implementation guidelines as outlined in the previous section are valid in any adaptive controller applications. In this section, we shall focus on the process control application, where reduced order models are essential and other practical factors, like frequent occurrence of static load disturbance and presence of d.c. levels in measurements introduce new problems to the estimator, which have not yet been given sufficient attention in the literature. Our experience comes mainly from the implementation of an adaptive Smith Predictor on process control computers.

Smith Predictors [14] are dead time compensators which have superior performance over conventional PID controllers. The adaptive Smith Predictor, which uses an estimator to generate the process model parameters on-line, solves the parameter sensitivity problem of the conventional Smith Predictor. The details of the adaptive Smith Predictor are well documented in the literature [13-15]. Suffice to say that it uses an explicit recursive parameter estimator with an exponential forgetting factor. As in the self-tuning regulator [1], the process dead time and order are assumed to be known. When measurement noise is serious, an unbiased estimator such as Recursive Instrumental Variables [13] should be used. However, for the purposes of illustration, we shall focus on the use of Recursive Least Squares (RLS) in this paper.

Process Dead Time and Order

In the process industry, prime candidates for adaptive control application are usually not simple, fast loops. Instead, they are typically sophisticated final quality control loops characterized by high order dynamics with dead time. This poses a severe problem to adaptive control because the full model would require too many parameters to be estimated or adapted on-line. Fortunately, the majority of these process dynamics exhibit overdamped characteristics and hence can be fairly accurately modelled by a dead time plus a low-order model [16,17]. In discrete parameter estimation and adaptive control, the above approximation can be reinforced by using a relatively large sampling interval T and appropriate filtering [3-6]. The determination of the dead time and order are then carried out using any suitable off-line system identification technique. A second or third order lag, and a dead time which is equal to one or more sampling intervals, are usually the final choices. The dead time is the more critical parameter to be pre-determined, as a wrong value will lead to serious estimation error resulting in poor control. If an external pseudo-random-binary-sequence (PRBS) test signal is used as perturbation signal, then a simple way to estimate the dead time [15] is to compute the cross-correlation of the test signal $x(k)$ and the process output $y(k)$ as follows:

$$\phi_{xy}(\tau) = \sum_{k-1}^{N} x(k)y(k+\tau) \tag{1}$$

where N is the number of data points used. As the autocorrelation of the PRBS test signal approximates a unit impulse, $\phi_{xy}(\tau)$ approximates the impulse response of the process and is quite insensitive to noise or other disturbance signals. Then the value of τ, where the magnitude of ϕ_{xy} starts to deviate consistently, is the pure dead time of the process. However, what we need is the pure dead time plus the apparent dead time due to reduced order modeling. This could be resolved by converting the impulse response into a step response using:

$$y(t) = \sum_{k=0}^{M-1} T\phi_{xy}(t-kT), t = MT \tag{2}$$

The process dead time for the reduced order model is then obtained as the intersection of the time axis with the straight line drawn through the point of inflexion along the rising step response [16]. (See Example 3.1).

If the test signal is present intermittently or permanently, then the dead time can be computed on-line and used to track drifting dead time. Another popular method of tracking unknown or varying dead time is by means of overparameterizations of the numerator polynomial in the process model. However, it has been found from simulations [14] that this method is not quite robust in the presence of measurement noise and static load disturbance.

Example 3.1

The transfer function of a process is $\frac{1}{(1+2s)^6}$. We shall model it by a second order lag plus a dead time of dT, where T is the sampling interval. Fig. 1(a) shows the crosscorrelation function $\phi_{xy}(\tau)$ obtained while Fig. 1(b) shows the converted step response using eqn. (2). The responses are not accurate as a short data length ($N = 64$) is used. However, the apparent dead time of $d = 6$ measured from Fig. 1(b) is found to be accurate for the purpose of reduced order modelling. The discrete reduced model obtained for $T = 1$ is:

$$G(z) = \frac{z^{-6}(0.011z^{-1} + 0.008z^{-2})}{1 - 1.76z^{-1} + 0.79z^{-2}}$$

Parameter Estimates

In order to track time-varying process characteristics, exponential weighting is used. The parameter estimates could hence be superimposed by noise-induced fluctuations as shown in Fig. 2. Two precautions are recommended here. One is to filter all parameter estimates obtained from the estimator before using them to update the controller even if a forgetting factor close to 1 is used. The other is to check the sign of the coefficients $\hat{b}_i$ of the numerator polynomial of the estimated model. The sign is usually known as it is required even for a fixed classical controller (commercially termed 'direct' or 'reverse' controller actions). As $\hat{b}_i$ are very small in magnitude and are proportional to the sampling interval, the fluctuations in estimates could cause them to change sign. An example is shown in Fig. 2. Large controller actions would be called upon if a wrong sign is updated to the model used to compute the controller action as in the case of the adaptive Smith Predictor. The natural

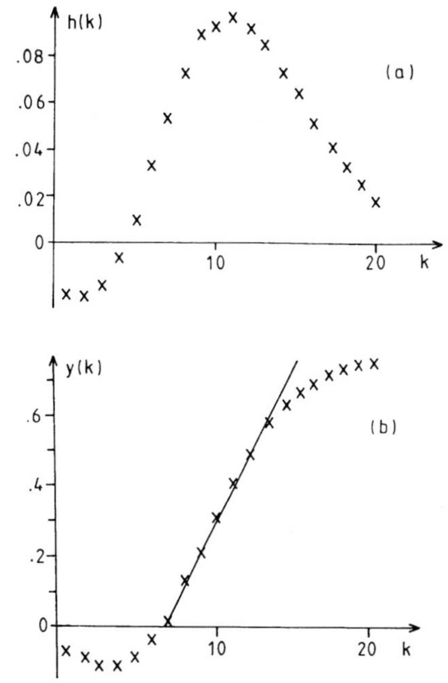

Figure 1: (a) Impulse response (b) Converted step response of a sixth order process without inherent dead time .

precaution is to check for sign changes and ignore estimated values which have the wrong sign instantaneously.

D.C. Levels in Measurements

In process control, inherent nonlinear process characteristics and the presence of unknown static load changes are quite common. To preserve the validity of linearity a small signal model has to be used [4,13] by taking deviations of the input $U(k)$ and output $Y(k)$ from their respective d.c. (or mean) levels, U_{DC} and Y_{DC}. The typical model is:

$$y(k) = -\sum_{i-1}^{n} a_i y(k-i) + \sum_{i-1}^{n} b_i u(k-d-i) + e(k) \tag{3}$$

where

$$\begin{aligned} u(k) &= U(k) - U_{DC} \\ y(k) &= Y(k) - Y_{DC} \end{aligned} \tag{4}$$

In general the d.c. levels, U_{DC} and Y_{DC} are unknown as they would change with the operating point as well as with the unmeasurable static load disturbances. There are several ways to handle this problem [4,13]. Two popular methods, the estimation of a *DC* constant and the use of high-pass filtering, have been compared in detail [13]. They are found to be similar in performance and they also share the same difficulties of causing large parameter deviations in the presence of static load disturbance to be discussed later. In this paper we shall only focus on the first method, estimation of a *DC* constant. The method simply requires the measurement vector $\underline{x}(k)$ and parameter vector $\hat{\underline{\theta}}(k)$ to be increased by one element:

$$\underline{x}^T(k) = [-Y(k-1), \ldots, -Y(k-n), U(k-d-1), \ldots, U(k-d-n), 1] \tag{5}$$

$$\underline{\theta}^T(k) = [\hat{a}_1(k), \ldots, \hat{a}_n(k),\ \hat{b}_1(k), \ldots, \hat{b}_n(k), \hat{c}(k)] \tag{6}$$

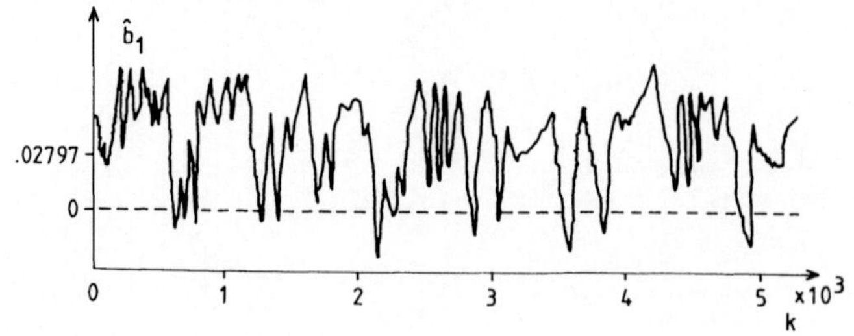

Figure 2: Estimate of $\hat{b}_1$ using noisy measurement ($\lambda = 0.99$; noise/signal = 40%).

The RLS estimation algorithm remains unchanged:

$$\underline{\hat{\theta}}(k) = \underline{\hat{\theta}}(k-1) + \underline{P}(k)\ \underline{x}(k)\{Y(k) - \underline{x}^T(k)\underline{\hat{\theta}}(k-1)\} \tag{7}$$

$$\underline{P}(k) = \{\underline{P}(k-1) - R(k)\underline{P}(k-1)\ \underline{x}(k)\underline{x}^T(k)\underline{P}(k-1)\}/\lambda \tag{8}$$

$$R(k) = 1/\{\lambda + \underline{x}^T(k)\underline{P}(k-1)\underline{x}(k)\} \tag{9}$$

The choice of the forgetting factor, λ, is more difficult than what is normally suggested in the literature. First let us consider the simpler case when there is no d.c. levels and no static load changes, i.e. $\hat{c}(k) = 0$ at all times. Fig. 3 and Fig. 4 show the effect of different λ on the covariance matrix $\underline{P}(k)$ and one of the parameter estimates $\hat{a}_1(k)$ respectively when there is a parameter change. Clearly, the convergence speed for $\lambda = 0.95$ is very fast but the estimates fluctuate too excessively and are not suitable for use in updating the controller even though the noise to signal ratio is only 5%. The fluctuations in parameter estimates for $\lambda = 0.99$ are much smaller but the convergence rate is too slow. This problem was not encountered by many authors in the literature as the noise level used by them was negligible. One excellent method of reconciliation between transient speed and stationary accuracy has been suggested recently by Hagglund [11,12] and we shall discuss it later in the next section. Another method is based on variable forgetting factor [9,10] and was originally proposed to avoid the problem of estimator windup. However, our experience indicates that it is not easy to use in practice and is not as robust as the method of Hagglund.

When there is either d.c. level or static load disturbance, the above problem is aggravated [13-15]. Fig. 5 shows the effect of a parameter change in the presence of d.c. level. The estimates would exhibit a large jump immediately upon the parameter change and it would take a long period (a few hundred samples for $\lambda = 0.99$) before they approach the correct values. The reason for the prolonged deviation can be traced to the large reduction of the trace of the covariance matrix $P(k)$ as shown earlier in Fig. 3. Similarly, Fig. 6 shows the effect of a static load change. In this case, irrespective of whether there is d.c. level, the estimates would exhibit a large jump immediately upon the presence of the static load change. This can be explained from the fact that there is not sufficient excitation to estimate $\hat{c}(k)$, as indicated in eqn. (5) and (6), although other states are sufficiently excited. Hence the parameter estimates for $\hat{b}_i$ and $\hat{a}_i$ would move in a certain direction to account for the load change. It would then take a fairly long period before $\hat{c}(k)$ is correctly estimated and the deviations in $\hat{b}_i$ and $\hat{a}_i$ are nulled. The estimates in these two cases, after a large parameter change or a load change, are clearly unsuitable for controller computation and we shall hence call these events "faults". Fortunately, the phenomenon that there is a jump in parameter estimates can be exploited to detect such "faults" and to trigger a recovery mechanism to be discussed in the next section.

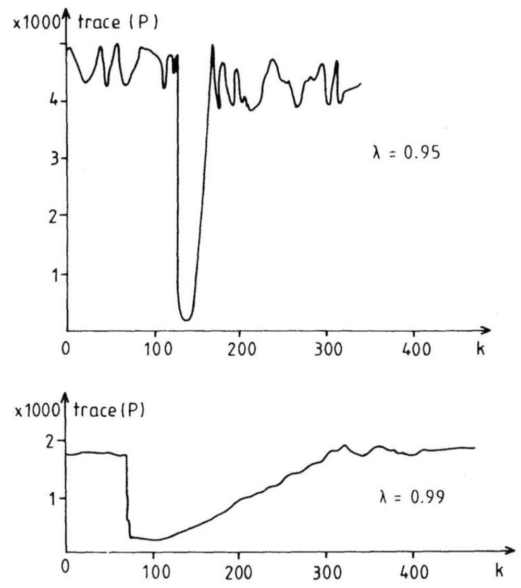

Figure 3: Trace of P matrix for different λ (noise/signal = 5%).

Actuator Nonlinearities

It is normally appropriate to use the signal sent to the actuator and the process measurement as the input-output data pair for parameter estimation. However, due to inherent actuator nonlinearities the estimator may produce a poor model and therefore wrong control. For a known nonlinearity such as a square law relationship between heating power and electric current [5], it is advantageous to cascade an inverse nonlinearity to make the process as linear as possible. Such compensation will greatly improve the robustness of the estimator. From our experience we would also like to recommend that other nonlinearities such as static friction, hysteresis and nonlinear gain of the actuator be diminished by a local feedback such as valve positioner control or flow control where possible. The main adaptive controller will then appear as a primary loop with the positioner or flow control as a secondary loop as in a cascade-control configuration. The estimator can then use the secondary loop setpoint as the input to the estimator. This approach has also been found to be beneficial by Hodgson [5].

4. Decoupling Transient and Stationary Properties of the Estimator

As discussed in the previous section, the choice of the forgetting factor, λ, in the recursive parameter estimator is difficult as it cannot reconcile the requirement of a smaller λ for fast transient speed and a larger λ for stationary accuracy. Furthermore, the presence of d.c. levels, large parameter changes and static load disturbances could contribute to the occurrence of "fault" resulting in poor parameter estimates. The solution to be presented in this section follows the principle of decoupling the transient and stationary properties of the estimator as suggested by Hagglund [11-12]. This is achieved by using a "fault detection" scheme to trigger the "transient modifications" of an otherwise fixed estimator which is optimal for stationary operations.

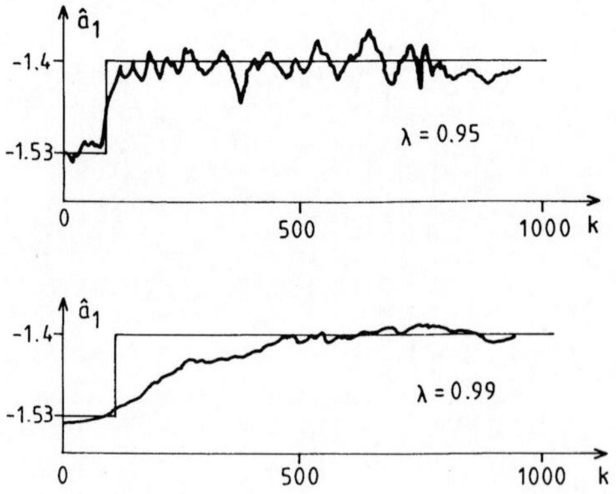

Figure 4: Estimate of $\hat{a}_1$ for different λ (noise/signal = 5%).

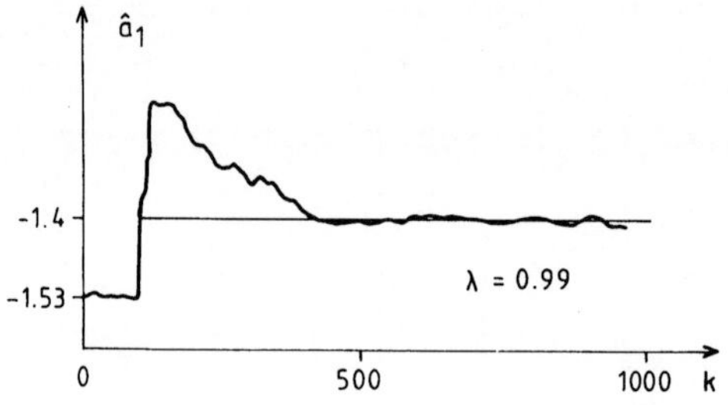

Figure 5: Effect of having a d.c. level.

Fault Detection

A good detector must satisfy two criteria. First, it must be able to detect a genuine "fault" such as parameter change as quickly as possible. Second, it must be robust enough not to be triggered by noise frequently. In this section, a two-level fault detection scheme is proposed. Simulations have shown that this two-level detection scheme gives quick detection of a fault and yet is robust enough not to give too many false alarms.

The first level is based on observations of the movement of the estimates during the occurrence of a fault. As discussed in the previous section, it is observed that when there is a sharp change in the parameters or when there is a large load change, the estimated parameters would jump immediately to some wrong values and remain there for a long period before converging to the correct values. Based on our experience, the following algorithm for fault detection is recommended: if $|\Delta\hat{a}_1(k-1) + \Delta\hat{a}_1(k)|/|\hat{a}_1(k-1)| > 0.15$ it is concluded that a fault has occurred. The parameter $\hat{a}_1$ is selected as a variable in the criterion for numerical reason as it always has a large magnitude relative to other $\hat{a}_i$ parameters. Normally for a process with unity gain, the magnitudes of the parameters, $\hat{b}_1$ and $\hat{b}_2$ are small, of the order of 10^{-2}. Therefore, when a ratio is taken, it is numerically more stable to choose a variable that has a large steady magnitude than one that is close to zero. Fig. 2 shows a plot of $\hat{b}_1$ over a period of more than five thousand samples with a noise to signal ratio of about 40%. Fairly often $\hat{b}_1$ crosses the zero level into the negative region. Thus at certain times, the estimate for b_1 is very close to zero. The criterion of 0.15 is optimal for $\lambda = 0.99$ and is chosen based on prolonged observations of the fluctuation of the estimates under noisy environment. Even with a noise level as high as 40%, the estimates do not make sudden jump of more than 15% over two successive samples. A typical trace of $\hat{a}_1$ over a span of five thousand samples is shown in Fig. 7.

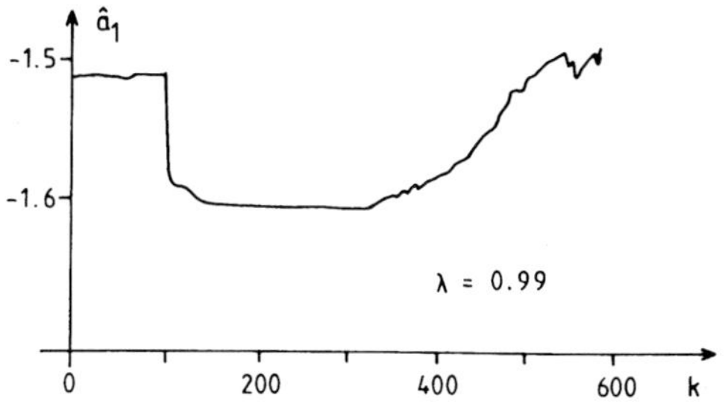

Figure 6: Parameter deviations due to a load change .

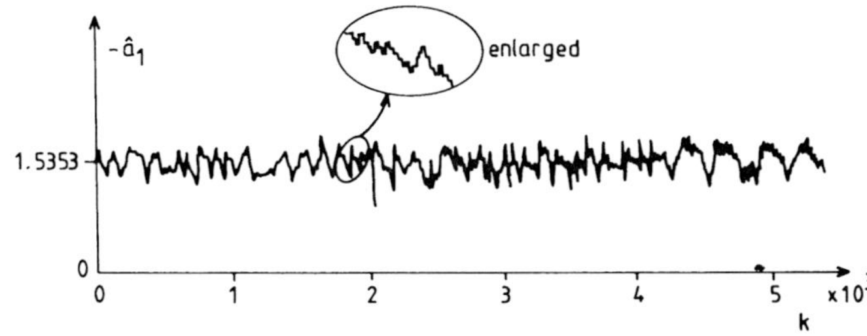

Figure 7: Estimate $\hat{a}_1$ using noisy measurements (noise/signal = 40%) .

The first level of detection is not designed to sense every fault that can occur. Rather it is designed to detect large and sudden fault that causes the parameters to deviate rapidly. However failure to detect a fault at this level is not disastrous since it will still be detected by a second level detection as a consistent drift of the parameter estimates.

The second level of detection is due to Hagglund [11,12] and is based on heuristic arguments. With forgetting factor equal to one, the linear least squares method is known to be the best linear unbiased estimator. This implies that there is no correlation between the increments of the parameter estimates in normal operation. Consequently, the probabilities for the estimate increments to have positive and negative $\underline{P}(k)\underline{x}(k)$ directions are equal. It is further assumed from intuition that this property will hold even for λ smaller than but very close to one, such as for $\lambda = 0.99$. The above argument implies that under normal operation,

$$P\{\Delta\underline{\hat{\theta}}(k)^T\Delta\underline{\hat{\theta}}(k-1) > 0\} \approx P\{\Delta\underline{\hat{\theta}}(k)^T\Delta\underline{\hat{\theta}}(k-1) < 0\} \tag{10}$$

where $P\{\}$ denotes probability.

However, when a fault occurs, $\underline{\hat{\theta}}$ is not close to its true value and the above arguments are no longer true. In this case,

$$P\{\Delta\underline{\hat{\theta}}(k)^T\Delta\underline{\hat{\theta}}(k-1) > 0\} > P\{\Delta\underline{\hat{\theta}}(k)^T\Delta\underline{\hat{\theta}}(k-1) < 0\} \tag{11}$$

The detection algorithm is given as follows:

$$\Delta\underline{\hat{\theta}}(k) = \underline{\hat{\theta}}(k) - \underline{\hat{\theta}}(k-1) \tag{12}$$

$$\underline{v}(k) = \gamma_1\underline{v}(k-1) + \Delta\underline{\hat{\theta}}(k) \qquad 0 < \gamma_1 < 1 \tag{13}$$

$$s(k) = sign\{\Delta\underline{\hat{\theta}}(k)^T\underline{v}(k-1)\} \tag{14}$$

$$r(k) = \gamma_2 r(k-1) + (1-\gamma_2)s(k) \qquad 0 < \gamma_2 < 1 \tag{15}$$

The difference between successive $\underline{\hat{\theta}}'s$ is filtered to give an indication of the trend $\underline{v}(k)$ in the movement of the estimates. $s(k)$ provides a measure of the extent of the correlation

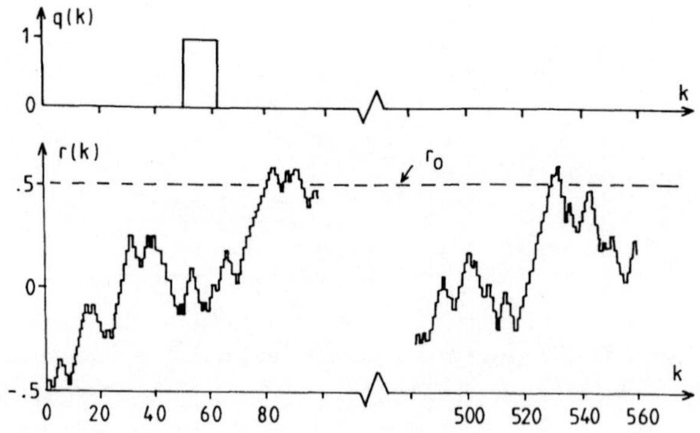

Figure 8: Fault detections using the two-level scheme.

between this trend and the instantaneous value of $\hat{\underline{\theta}}$. $r(k)$, the filtered value of $s(k)$ is the indicating variable for fault. During a fault, $\hat{\underline{\theta}}$ is expected to move in a persistent direction and it will cause $r(k)$ to tend to unity. A threshold r_o is set for $r(k)$. When $r(k)$ exceeds r_o, it is concluded that a fault has occurred. This detection scheme is found to be reliable for large parameter changes [12] and for load changes [17].

The selection of the parameters γ_1, γ_2 and r_o will depend on the sensitivity of the estimation scheme to parameter movement and the frequency and amplitude of the faults. Detailed guidelines are given by Hagglund [11]. For the following simulation runs, γ_1, γ_2 and r_o are chosen as suggested by Hagglund to be 0.85, 0.95 and 0.5 respectively.

Fig. 8 shows a plot of the variables of the fault detector. $q(k)$ is the fault indicator for first-level detection and a positive "one" indicates that a fault is detected. $r(k)$ is the fault indicator for the second-level detection and when $r(k)$ exceeds r_o, a fault is concluded. In the figure shown, a large load change occurs at $k = 50$. Two samples later this fault is detected by the first-level detector. The second-level detector did not sense the same fault until much (around thirty samples) later.

A small load change is injected at $k = 500$. This time the first-level detection has failed to sense the fault because there is only small drift in the parameter estimates. Hence, $q(k)$ remains at zero during this time. However, the fault is eventually detected by the second-level detector some thirty samples later, as indicated by $r(k)$ exceeding r_o.

Modifications of Estimator

The fault detector indicates when a fault has occurred. Action must then be taken to ensure that the fault is quickly corrected and disturbance to the controller and process is minimized. The following procedure is recommended after a fault is detected.

First, an option to suspend the updating of the controller parameters is provided. Most of the time, the estimates immediately following the occurrence of a fault would be poor. It is therefore more appropriate to freeze the updating of the controller until the parameters have converged than to risk the updating of poor estimates into the controller. However, if load changes are infrequent and the process parameters would change quite frequently, it may be advantageous to have continuous controller updating provided the process parameter estimates are filtered before they are used in control computation.

Second, the $\underline{P}$ matrix of the estimator is increased. The diagonal elements of the covariance matrix $\underline{P}$ are increased by fifty times. The inverse $\underline{P}$ matrix is an indication of the information contained in the estimator. When a fault occurs, we are no longer sure that the information contained in the inverse $\underline{P}$ matrix is correct. It is therefore necessary to decrease the information content of the estimator. This is done by a direct increase of the diagonal elements of the $\underline{P}$ matrix. With a larger $\underline{P}$ matrix, the estimator is capable of giving larger

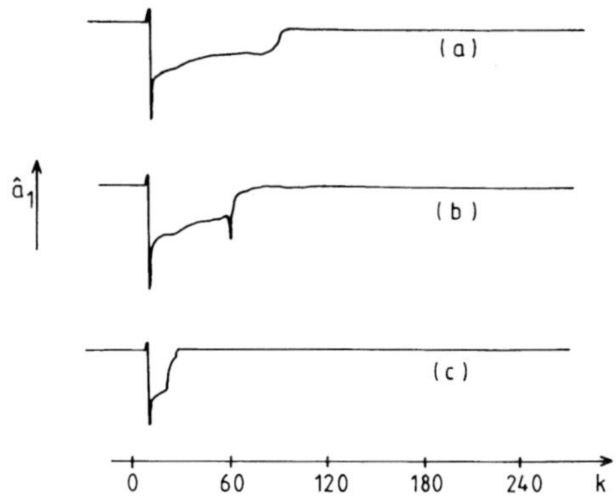

Figure 9: Effect of load change on parameter estimates (a) without fault detection; (b) with second-level fault detection; (c) with first-level fault detection.

corrections to the estimates. Therefore convergence will be much faster.

Third, the forgetting factor is decreased to a smaller value. In this case, the forgetting factor is set to 0.95 whenever a fault is detected. This then allows forgetting of the old information which is now useless and enables the estimator to identify the new parameters more quickly. The forgetting factor is set back to its original value when the parameters have converged.

Increasing the $\underline{P}$ matrix and decreasing the forgetting factor will cause the estimator to give much larger corrections to the estimates and assign larger weighting to more current information. Therefore it is important that the incoming data are more informative about the current process. The final action to take, if permissible by operations, is to increase the excitation to the process. It is recommended that when a fault is detected, the amplitude of any test signal being used is increased to two or three times the normal value. It is set back to its normal value when the parameters have converged.

Following the actions taken above, a convergence test is conducted every sample. When all the filtered parameters estimates do not move significantly, say more than five percent, from their immediate past values over a prespecified interval, the estimator is assumed to have converged. The forgetting factor is then reset to its previous value of 0.99. If the option of suspending controller updating has been chosen, the controller parameters will then be replaced by the new values; the input and output states, including those delayed output values should then be scaled according to the change in static gain to achieve a bumpless transfer before controller updating is resumed.

Simulation Results

The effectiveness of the fault detector and modified estimator will be demonstrated by simulation of an adaptive Smith Predictor [13-15] for the process $\frac{e^{-4s}}{(1+5s)(1+3s)}$ sampled at $T = 1$ sec. The design parameters are $\lambda = 0.99, \gamma_1 = 0.85, \gamma_2 = 0.95, r_o = 0.5$. The controller is a digital Smith Predictor with ringing poles removed. The adaptive controller is formed by updating the Smith Predictor using the estimator outputs. In the following simulations, the option of suspending controller updating during fault conditions is selected.

Fig. 9 shows a plot of $\hat{a}_1$ with and without the fault detector. A load change occurred at $k = 30$. Without the fault detector the parameter estimates moved to some wrong values almost immediately and it took the estimator a long time to resume the correct estimates. The corresponding process output therefore suffered from a poor load regulation as shown in Fig. 10. With the second-level detector plus the modified estimator, the parameter deviation period was vastly reduced thus resulting in much improved load regulation as shown. The

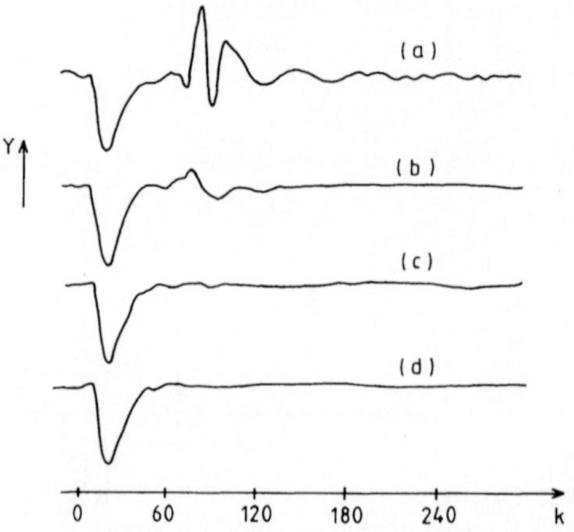

Figure 10: Effect of load change on process output (a) - (c) same as Fig. 9; (d) fixed controller.

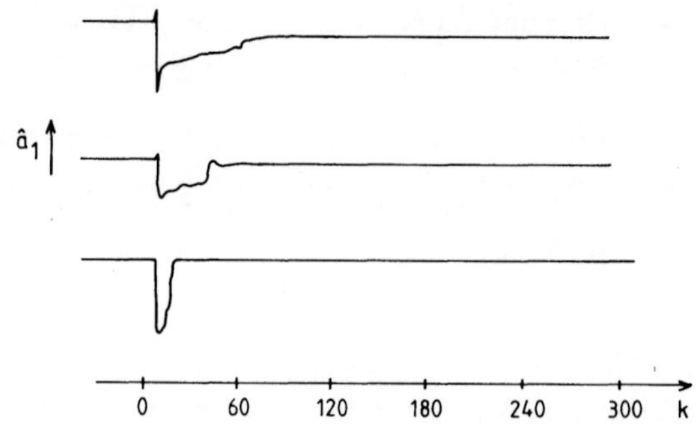

Figure 11: Effect of gain change on parameter estimates (a) without fault detection; (b) with second-level fault detection; (c) with first-level fault detection

application of the first-level detector further improved the load regulation due to the earlier detection of load and the performance was comparable to that of the fixed controller.

Fig. 11 and Fig. 12 show the effects of a large parameter change due to an increase in the process gain by three times. The performance improvement using the fault detector and modified estimator is evident. Without the fault detector the parameter estimates took a long time to track the new parameters and in the meantime, the process output response is extremely oscillatory. Whereas the second-level fault detector provided a large improvement in that the transient parameter adjustments converged much faster and the output response was in error for a shorter period. The first-level detector gave further improvement over the performance of the second-level detector.

5. Conclusions

It is indeed not straight forward to use the currently available adaptive control algorithms, especially for inexperienced users, as the assumptions of linearity, persistent excitation, exact knowledge of process dead time and order, etc. have to be satisfied or relaxed in practice. Many precautions, based largely on in-depth understanding of the algorithms and proven in extensive simulations, have been proposed in the literature and some of these are extended in this paper. In particular, the need to decouple the transient and stationary properties of the

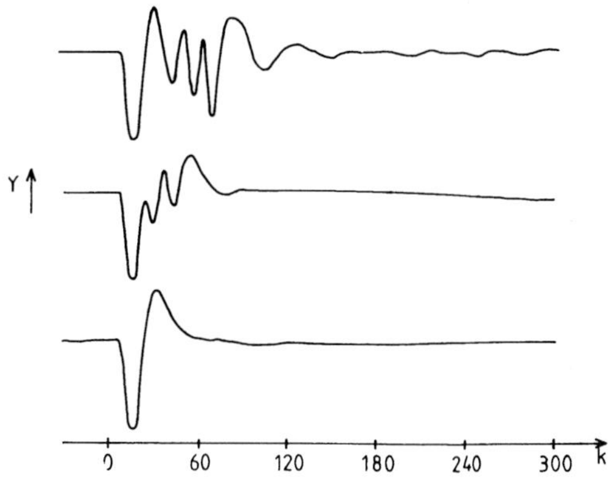

Figure 12: Effect of gain change on process output (a) - (c) same as Fig. 11.

estimator in the presence of faults, such as d.c. levels, large parameter changes and static load disturbances, is discussed. A two-level fault detection scheme, based on monitoring of sudden deviation of a parameter estimate plus the expected movement of the parameter estimates in a persistent direction, has been presented. Using the proposed fault detection scheme, plus modifications to the estimator such as resetting of covariance matrix and changing of forgetting factor, the performance of adaptive control in the presence of faults is significantly improved. These heuristic measures which provide practical robustness to the basic adaptive controller, can already be regarded as firm expert knowledge. Thus their implementation and further development will be simplified by the emerging expert system techniques [18].

References

[1] K.J. Åström , "Theory and Applications of Adaptive Control - a Survey," *Automatica*, vol. 19, pp. 471-486, 1983.

[2] G.C. Goodwin and K.S. Sin, Adaptive Filtering Prediction and Control, Prentice-Hall, 1984.

[3] C.J. Harris and S.A. Billings, "Self-tuning and Adaptive Control: Theory and Applications," *IEEE Publication*, 1981.

[4] R. Isermann, Digital Control Systems, Springer-Verlag, 1981.

[5] A. Hodgson, "Problems of Integrity in Applications of Adaptive Controllers," Ph.D. thesis, Dept. of Engrg. Sc., Oxford University, UK, 1982.

[6] B. Wittenmark and K.J. Åström , "Practical Issues in the Implementation of Self-tuning Control," *Automatica*, vol. 20, pp. 595-606, 1984.

[7] C.C. Hang, T.H. Lee and T.T. Tay, "Application of Recursive Parameter Estimation as an Auto-tuning Aid," to be presented at ISA Annual Conference, Oct. 1985.

[8] E. Irving, "Improving Power Network Stability and Unit Stress with Adaptive Generator Control," *Automatica*, vol. 15, pp. 31-46, 1979.

[9] T.R. Fortescue, L.S. Kershenbaum and B.E. Ydstie, "Implementation of Self-tuning Regulators with Variable Forgetting Factors," *Automatica*, vol. 17, pp. 831-835, 1981.

[10] P.E. Wellstead and S.P. Sanoff, "Extended Self-tuning Algorithm," *Int. J. Control*, vol. 34, pp. 433-455, 1981.

[11] T. Hagglund, "New Estimation Techniques for Adaptive Control," Ph.D. thesis, Dept. of Automative Control, Lund Institute of Technology, Sweden, 1984.

[12] T. Hagglund, "Adaptive Control of Systems Subject to Large Parameters Changes," Proc. IFAC World Congress, Budapest, July 1984, pp. 202-207.

[13] T.H. Lee and C.C. Hang, "The Effects of an Unmeasurable Step Disturbance on Recursive Parameter Estimation and Adaptive Control," Proc. Int. Workshop on Appl. of Adaptive Systems Theory, Yale University, June 1983, pp. 223-227.

[14] T.H. Lee and C.C. Hang, "A Performance Study of Parameter Estimation Schemes for Systems with Unknown Dead Time," Proc. American Control Conference, Boston, June 1985, pp. 512-516.

[15] T.T. Tay, "Computer Control of a Process with Significant Dead Time," Final Year Project Report, National Univ. of Singapore, March 1985.

[16] J.S. Moczek, R.E. Otto and T.J. Williams, "Approximation Models for the Dynamic Response of Large Distillation Columns," Proc. IFAC World Congress, Switzerland, August 1963, pp. 238-248.

[17] K.Y. Leong, C.C. Lim and C.C. Hang, "Auto-tuning of Cost Function Parameters for a Class of Explicit Stochastic Self-tuning Controllers," Report CI-84-6, National Univ. of Singapore, August 1984.

[18] K.J. Åström and J.J. Anton, "Expert Control," Proc. IFAC World Congress, Budapest, July 1984, pp. 245-250.

Self Tuning and Adaptive Control of Chemical Processes

Scott A. Chesna and B. Erik Ydstie
Department of Chemical Engineering
University of Massachusetts at Amherst
Amherst, Ma. 01003

Abstract

The paper reviews the results obtained from applying the self tuning regulator to several different chemical process control problems. This research has revealed that some minor modifications to the original theory lead to better performance and robustness characteristics in the face of nonlinear and time-varying process dynamics. This paper describes the results from using a variable in lieu of a fixed forgetting factor, and an extended horizon control policy for compensation of systems that have time-varying transport delay and non-minimum phase dynamics. We also describe how steady state design models can be used to compensate for some nonlinear process characteristics via nonlinear model compensation techniques.

1. Introduction

Chemical process applications confront the control engineer with challenges somewhat different from those met in electrical or mechanical engineering applications. They are usually best described by high order, nonlinear, distributed parameter equations involving parameters and states that are hard to determine and/or measure. The performance requirements are usually quite weak, and the control system is most often required to compensate for low frequency disturbances and low frequency changes in the set point. The processes are usually passive (they often involve diffusion processes) and they sometimes act as low pass filters with cut off frequencies in the order of several minutes, and sometimes even hours. In view of this, adaptive theory can be expected to work well since the adaptive regulators tune themselves on line to compensate for model uncertainties and nonlinearities. There is a second incentive to use the adaptive theory; expensive modeling and experimentation performed on the plant to obtain transfer functions can be bypassed. Because of these facts, adaptive control theory has been met with great enthusiasm by control engineers in the chemical industries.

One of the most well known adaptive controllers is the Self Tuning Regulator (STR) [2]. The STR was originally proposed for unknown, linear, time invariant systems. The systems need to have fixed, known delays and minimum phase characteristics. Some flexibility is allowed, however, since the STR can be made to work for a wide range of processes violating these assumptions if the process is stable, and sampling time and dead time are chosen judiciously [2,4]. The regulator can also be applied to mildly time varying and nonlinear systems if an exponential forgetting factor is included to discount old data.

Some extensions of STR theory have been proposed to make it applicable to more general systems. A control weighted algorithm has been proposed [5], and different versions of adaptive pole assignment strategies have also been considered [3,9]. These algorithms are more flexible than the STR and they can be used for control under less restrictive conditions. However, they also include more tuning parameters, and they are thus harder to initialize for someone who has not already gained a significant experience through practical adaptive control applications. Even the STR, which has only three important tuning parameters

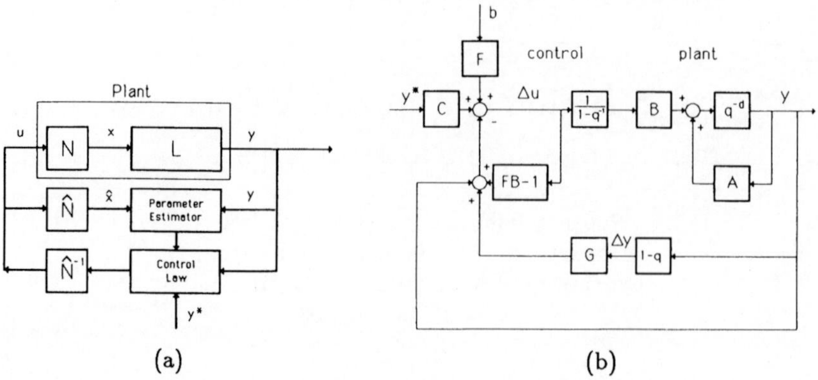

Figure 1: a) Nonlinear series compensation, b) Incremental controller.

(sampling time, forgetting factor and dead time), can be difficult to tune in practice. A recent review of applications to chemical process control problems is given in reference [12].

In our own research, we have tried to develop more robust algorithms that adhere closely to the intuitive simplicity of the STR and that have tuning parameters that are easy to choose. The purpose of this paper is to summarize our results and give a status report on our current research. While some of the results have been presented elsewhere [13,15], many are new and have not previously been published.

2. The Adaptive Control Algorithm

We consider the discrete, adaptive control of a nonlinear, continuous time system modeled as

$$\dot{x}(t) = Fx(t) + Gn(\psi, u(t), \tau), \text{ and} \tag{1}$$

$$y(t) = Cx(t) + b, \tag{2}$$

where $n(\psi, u(t), \tau)$ is a continuous differentiable, memoryless nonlinearity which may include a delay $\tau \geq 0$, ψ is a set of parameters, $u(t)$ and $y(t)$ are the process input and output respectively, b is a steady state bias or a low frequency disturbance, and F, G, C are the system matrices. By defining

$$m(t-\tau) = n(\psi, u(t), \tau) \tag{3}$$

system (1) becomes a linear system with the auxiliary variable $m(t)$ as an input. The simple form of the nonlinear system (1) motivates the use of nonlinear model compensation to cancel the process nonlinearities. If the nonlinear function $n(\psi, u(t), \tau)$ is known exactly and is invertible over its entire range, then a complete linearization can be achieved by the strategy shown in Figure 1. More complex nonlinear system can be handled in a similar way through the use of parallel models and output nonlinear compensation [1,7]. Only the simplest case with input nonlinearity is discussed here.

For the purpose of adaptive control it is convenient to replace the continuous time, state space model by a discrete, input-output model:

$$A(q^{-1})y(t) = B(q^{-1})m(t-d) + A(1)b, \quad \text{where} \tag{4}$$

$$m(t) = N(t)u(t). \tag{5}$$

The polynomials $A(q^{-1})$ and $B(q^{-1})$ are polynomials in the unit delay operator q^{-1}, defined so that

$$A(q^{-1}) = det(I - Aq^{-1})^{-1} = 1 + a_1 q^{-1} + a_2 q^{-1} + \cdots + a_n q^{-n},$$
$$B(q^{-1}) = C\ adj(I - Aq^{-1})^{-1}B = b_o + b_1 q^{-1} + b_2 q^{-2} + \cdots + b_{n+d-1} q^{-n-d+1},$$

120

where $A = exp(Fh)$, h is the sampling time, and $B = \int_o^h exp(F(h-\eta))Gd\eta$. The nonlinearity in equation (3) has been replaced by a time-varying gain $N(t)$, operationally equivalent to $n(\psi, \cdot)$, and $d = int(\tau/h) \geq 1$ is a time delay in integer units of the sampling time. Note that $B(q^{-1})$ has been expanded so that the first d coefficients are zero corresponding to the transport delay. The discretization step involves choosing a sampling time for the adaptive system. In our experience we have found that the algorithm we present here is less sensitive to choice of sampling interval than the STR. An additional advantage is that the Extended Horizon approach also allows fast sampling rates provided that the prediction horizon is chosen long enough. This may reduce intersample ripple.

In practice, the coefficients of models (4) and (5) are not known. However, we may have an estimate $\widehat{N}(t)$ of $N(t)$ based on experiments or steady state models of the process:

$$\widehat{m}(t) = \widehat{N}(t)u(t) \approx m(t). \tag{6}$$

The quality of this approximation is determined by the time-varying gain

$$\widetilde{N}(t) = N(t)\widehat{N}(t)^{-1}. \tag{7}$$

Exact matching of $m(t)$ and $\widehat{m}(t)$ is achieved when $\widetilde{N}(t) = 1$.

If $B(z) \neq 0$ for $|z| \leq 1$ and d is known, then the self tuning regulator can be applied. Convergence and stability is guaranteed if $\widetilde{N}(t) = 1$. As we shall see (Result 3), a certain amount of model mismatch is also allowed. Stability ensues when $\widetilde{N}$ is sector bounded (positive real). In the more general case, d is not known and there are, in addition to the model uncertainties and nonlinearities, non-minimum phase effects. This motivates the development of the Extended Horizon Controller (EHC) and the nonlinear model compensation technique described below.

Model (4) can be reformulated (by successive substitution) into a T-step ahead predictor model ($T \geq 0$)

$$y(t+T) = \alpha(q^{-1})y(t) + \beta(q^{-1})m(t+T-1) + \gamma, \tag{8}$$

where $\alpha(q^{-1}) = \alpha_o + \alpha_1 q^{-1} + \cdots + \alpha_{n-1}q^{-n-1}$, and $\beta(q^{-1}) = \beta_o + \beta_1 q^{-1} + \cdots + \beta_{n+d+T-1}q^{-n-d-T+1}$ with $\beta_i = 0$ for $i = 0, 1, \ldots, d-1$. γ is a transformed bias.

If d or a lower bound on d is known, then model (8) can be simplified by setting leading coefficients equal to zero. The exact number of zero coefficients of $B(q^{-1})$ is not assumed to be known; however, we have an upper bound.

The T-step ahead predictor model depends on T process inputs to be determined in the future. The EHC needs a current strategy for these future inputs. The 'heuristic' control strategies like the Dynamic Matrix [6] and the Model Algorithm [11] Controllers base these predicted inputs on the LQ optimal control theory. In this work we propose to use a simpler 'receding horizon control' law based on step inputs, i.e. set $u(t+1) = u(t)$ for $i = 1, 2, \ldots, T$ and recalculate at each step. This particular choice of strategy yields controllers that can be implemented almost as direct adaptive control laws. The receding horizon control law then becomes:

$$m(t) = (\Sigma_{i=0}^{T+d}\beta_i)^{-1}(y(t+T+d)^* - \overline{\beta}(q^{-1})m(t) - \alpha(q^{-1})y(t) - \gamma), \tag{9}$$

$$u(t) = \widehat{N}(t)^{-1}m(t), \tag{10}$$

where $\overline{\beta}(q^{-1}) = \beta_{d+T+1}q^{-1} + \cdots + \beta_{n+d+t-1}q^{-n}$, and $y(t+T)^*$ is the desired set point. This law reduces to the control law of the self tuning regulator for $T = 0$ when d is known exactly.

Result 1: Assume $(\Sigma_{i=0}^{T+d}\beta_i + \overline{\beta}(z)) = \gamma(z) \neq 0$ for $|z| \leq 1$, and $N = \widehat{N} = 1$, then for a constant setpoint $y(t)^* = y^*$:

$$\lim_{t\to\infty} |y(t) - y(t)^*| = 0, \ \lim_{t\to\infty} u(t) = c,$$

where c is a constant, and also $|u(t)| \leq K < \infty$ for all t.

Remarks: The stability condition reduces to the familiar non-minimum phase condition when $T = 0$ and d is exactly known. For larger T we have more flexibility since $\Sigma_{i=0}^{T+d}\beta_i$ grows as T grows. This tends to move the controller poles towards the origin reducing controller ringing. The tuning parameter T thus act in a similar way to the tuning parameter of the self tuning controller [5], however, its interpretation is different.

Proof of Result 1: Control law (10) and system (4) have the equivalent state space representations:

$$m(t-d) = -(CSB)^{-1}(CA^{T+1}x(t - y(t+T)^* + b), \tag{11}$$

where $S = \Sigma_{i=0}^{T} A^i$, and

$$x(t+1) = Ax(t) + Bm(t-d), x(o) = x_o, \tag{12}$$

$$y(t) = Cx(t) + b. \tag{13}$$

From (11) and (12) we have (with $y(t + T^* = y^*)$

$$x(t+1) = Ax(t) + B(CSB)^{-1}(y^* - CA^{T+1}x(t) - b)$$

By premultiplying with CS we obtain

$$CSx(t+1) = (CS - C)x(t) + y^* - b.$$

Now define a vector $\overline{x}$ so that $C\overline{x} = y^* - b$. By adding and subtracting $(CS - C)\overline{x}$, we then have

$$CS\tilde{x}(t+1) = (CS - C)\tilde{x}(t), \tag{14}$$

where $\tilde{x}(t) = x(t) - \overline{x}$. With $V(t) = \tilde{x}(t)'S'C'CS\tilde{x}(t) \geq 0$ we then have

$$V(t+1) - V(t) = -\tilde{x}(t)'C'C\tilde{x}(t) - \tilde{x}(t)'[(S-I)'C'C + C'C(S-I)]\tilde{x}(t).$$

By using the fact $(S - I) \geq 0$ (since $A \geq 0$), it follows from the lemma in the Appendix that the term $[\cdot]$ defines a non negative matrix. We must conclude $V(t+1) \leq V(t)$ and hence, since $V(t) \geq 0$ for all $t : \lim_{t\to\infty} \|C\tilde{x}(t)\|^2 = 0$. But from the definitions above we then have

$$\lim_{t\to\infty} |Cx(t) - y^* + b| = \lim_{t\to\infty} |y(t) - y(t)^*| = 0.$$

The boundedness of $u(t)$ then follows immediately since equation (10)

$$u(t) = (-\alpha(q^{-1}y(t) - b + y^*)(\Sigma_{i=0}^{T}\beta_i + \overline{\beta}(q^{-1})^{-1}$$

where the denominator is stable. Using the property above we also conclude

$$\lim_{t\to\infty} u(t) = \frac{1}{\beta(1)}(y^* - \alpha(1)y^* - \gamma) = c,$$

and the results follow.

We use a variation of the variable forgetting factor algorithm for the estimation of the parameters of equation (8). It is then convenient to rewrite equation (8) as:

$$y(t+T) = \phi(t)'\Theta, \tag{15}$$

where $\phi(t)$ is a regression vector:

$$\phi(t)' = (x(t+T), x(t+T-1), \ldots, x(t-m+1), y(t), \ldots, y(t-n+1), 1).$$

and Θ is a vector of parameters. In general $x(t)$ is not known exactly since $N(t)$ is not known accurately. $\phi(t)$ has to be replaced by an estimate:

$$\widehat{\phi}(t) = (\widehat{x}(t+T), \widehat{x}(t+T-1), \ldots, \widehat{x}(t-m), y(t), \ldots, y(t-n-1), 1)$$

Algorithm 1: Given $\{\widehat{\Theta}_o, P_o > 0, r, N_o, T, K, \lambda_{min}\}$ set $t = 1$ and compute:

1. $\lambda(t) = max\{\lambda_{min}, N_o(N_o + e(t)^2(1 + w(t-1))^{-1}r(t)^{-1})^{-1}\}$ where $e(t) = y(t) - \widehat{\phi}(t-T)^1\widehat{\Theta}(t-1), w(t-1) = \widehat{\phi}(t-T)'P(t-1)\widehat{\phi}(t-T)/r(t)$, and $r(t) = max\{r, \widehat{\phi}(t-T)'P(t-1)\widehat{\phi}(t-T)/K\}$
2. $P(t) = \lambda(t)^{-1}(P(t-1) - P(t-1)\widehat{\phi}(t-T)\widehat{\phi}(t-T)'P(t-1)/(\lambda(t) + w(t-1))r(t))$
3. $\widehat{\Theta}(t) = \widehat{\Theta}(t-1) + P(t)r(t)\widehat{\phi}(t-T)e(t)$
4. Solve $y(t+T)^* = \widehat{\phi}(t)'\widehat{\Theta}(t)$ so that $u(t+i) = u(t)$ for $i = 0, \ldots, T$
5. Set $t = t+1$ and go to step 1.

Comments: In the algorithm, $\lambda(t)$ and $r(t)$ are parameters adjusting the transient properties of the estimator. $\lambda(t)$ is the 'variable forgetting factor' chosen so that the memory length of the estimator is kept constant. It follows from the update formula that $0 < \lambda_{min} \leq \lambda(t) \leq 1$ for all t, where λ_{min} is a lower bound introduced to insure robustness during transients. $\lambda_{min} = 0.5$, for example, ensures that the memory length never goes below 2 sample periods since the minimum memory length is defined as:

$$N_{min} = 1/(1 - \lambda_{min}) = 2, \qquad \text{for } \lambda_{min} = 0.5. \tag{16}$$

In Algorithm 1 $\lambda(t)$ is chosen to keep $N(t) = N_o$, a fixed constant. A further discussion is given in [14,15]. $r(t)$ is supposed to be the variance of the innovation sequence. In practice $r(t) = r$ a fixed, arbitrary constant (often $r = 1$). In our algorithm, we have also had good success by using a moving average estimate of $r(t)$ so that

$$r(t) = r(t-1) - \frac{1}{N_o}(r(t-1) - e(t)^2), \text{ where } r(0) \text{ is given.}$$

We have also introduced an additional feature. By keeping $r(t) \geq \phi(t-T)'P(t-1)\phi(t-T)/K$, we assure that adaptation rates do not become arbitrarily high in a single step. By defining the aposteriori error:

$$\bar{e}(t) = y(t) - \phi(t-T)'\widehat{\Theta}(t),$$

from Algorithm 1 and by using the bounds on $\lambda(t)$ and $r(t)$:

$$\bar{e}(t) = \frac{\lambda(t)}{\lambda(t) + w(t-1)}e(t) \geq \frac{\lambda_{min}}{1+K}e(t).$$

Thus λ_{min} and K adjust an upper bound for the rate of decrease of the output error at each step. $0 < K << 1$ give adaptation rates $> \lambda_{min}$. Better smoothing then results when λ_{min} is close to unity. $K >> 1$ allows rapid adaptation, under some circumstances, at the cost of oscillating parameters and possible loss of robustness.

Algorithm 1 allows us to establish the following stability results:

Result 2: (Convergence) Suppose $\hat{x}(t) = x(t)$, $T = 0$, d known and $B(z) \neq 0$ for $|z| \leq 1$, then:

$$\lim_{t\to\infty} \lambda(t) = 1$$
$$\lim_{t\to\infty} |y(t) - y(t)^*| = 0$$

and $\{\phi(t), \widehat{\Theta}(t)\}$ is bounded uniformly.

Proof: See [14].

Result 3: (Robustness) Suppose $T = 0$, $d = 1$ and $B(z) \neq 0$ for $|z| \leq 1$ then $\{\widehat{\phi}(t), \widehat{\Theta}(t)\}$ is uniformly bounded and so is $\{y(t), u(t)\}$ provided that

$$(1 - \tilde{N}(t)^{-1}) < \lambda_{min}^{1/2}/((\lambda_{min} + K)^{1/2} \sup_{|z|=1} A(z)) \text{ for all } t \tag{17}$$

Proof: See [8].

Remark: Inequality (17) gives a sector bound on the model mismatch $\tilde{N}(t)$. This bound may be unrealistic in some applications, however the nonlinear compensation technique described above helps linearize the plant so that the bound can be more easily satisfied in regions of interest if not globally.

Result 4: (Time Varying Parameters) Suppose $T = 0$, $d = 1$, $\hat{x}(t) = x(t)$, and Θ is a time varying vector of system parameters so that $\Theta(t) = \Theta(t-1) + p(t)$. Suppose also the system is 'minimum-phase' in the sense defined in [16]. Then $\{\phi(t), \widehat{\Theta}(t)\}$ is bounded, and so is $\{y(t), u(t)\}$ provided that

$$p(t)'P^{-1}(t)p(t) \leq K \text{ for all } t$$

Proof: See [16] (Stochastic case).

Stability for a multistep version of the algorithm for $T > 0$ was established in [17].

Though the algorithm eventually gives zero offset, we have found it advantageous to modify the algorithm and use the delta operator $\Delta = 1 - q^{-1}$ rather than the backward shift operator alone. As well as introducing a slow acting integrator, this also gives some enhanced numerical properties [10]. In the Δ formulation the T-step ahead predictor (8) takes on the form:

$$y(t+T) = y(t)\alpha(\Delta)y(t) + \beta(\Delta)u(t), \tag{18}$$

and the EHC (9) becomes:

$$u(t) = u(t-1) + (\Sigma\beta_i)^{-1}(y(t+d+T)^* - \beta(\Delta)m(t) - \alpha(\Delta)y(t)).$$

The coefficients of equations (18) and (19) are not necessarily the same as those of equations (8) and (9). But, as before, the coefficients of (18) can be estimated directly by using the modified VFF algorithm. A block diagram of the EHC using the Δ operator formulation is shown in Figure 1.

3. Experimental Results

Algorithm 1 has been extensively tested in simulation and pilot plant experiments. The theoretical results obtained have been supported by these experiments and the adaptive algorithm shows a remarkable degree of robustness and versatility. Two sets of experiments will be discussed here. The first involves the control of the pressure in a large CO_2 absorption desorption unit located at the Imperial College in London, UK. The second involves the control of a steam-water heat exchanger at the University of Massachusetts. The two processes are very different. The first has fast time constants, and is well described by a set of finite dimensional difference equations. The second process is infinite dimensional, but it has longer time constants. Both processes are nonlinear so the linearized dynamics change significantly from one operating point to another.

A simplified diagram of the gas absorption-desorption process is shown in Figure 2. The Extended Horizon Controller with Variable Forgetting Factor estimator has been tested on numerous flow, pressure, temperature and level control loops. Here we report on the pressure control experiments. Other experiments are reported elsewhere [13,15].

The pressure control loop is nearly deterministic (standard dev $\sim 0.5\%$) so a fixed forgetting $\lambda < 1$ leads to covariance windup and subsequent instabilities during quiescent periods. An experiment with $\lambda = 0.99$ is shown in Figure 3. It is clear that $P(t)$ grows exponentially until the diagonal elements are very large. At this point the adaptive controllers show quasi-stable behavior. However, the estimator gain is very high and the parameters are very sensitive to small upsets and disturbances. The estimator is also susceptible to numerical errors. These accumulate rapidly beyond time 150 and some diagonal elements of $P(t)$ become negative. In our experience we have found that numerically stable algorithm like a UDU' or $P^{1/2}$ algorithm solve the stability problem.

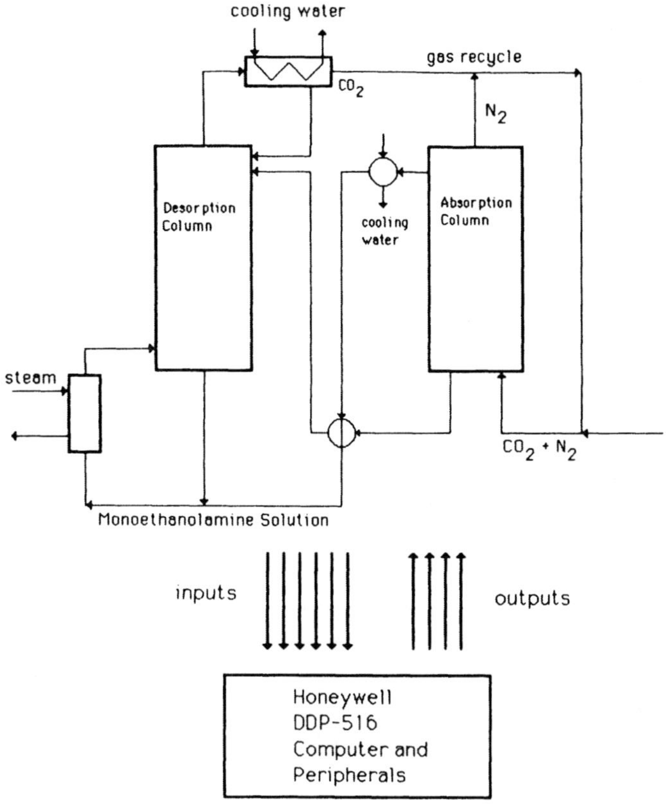

Figure 2: CO_2 absorption-desorption process.

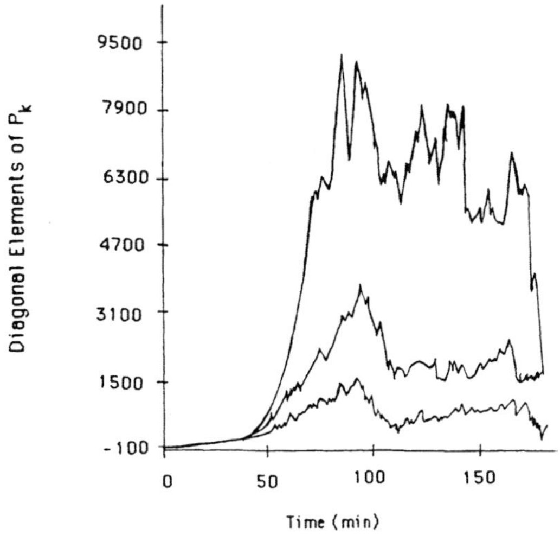

Figure 3: Estimator windup for $\lambda = 0.99$.

The variable forgetting factor algorithm is designed to keep a measure of the information content constant, thereby preventing estimator windup. An experiment with the VFF algorithm (Figure 4) shows that the covariance matrix settles at a steady state level as shown in reference [16]. Further experiments have shown that this level can be tuned by choice of N_o, the asymptotic memory length. The VFF algorithm has also been applied to industrial control systems with success.

Figure 5 shows experiments using the extended horizon controller for $T = 0, 1$ and 2. ($T = 0$ makes the algorithm coincide with the self tuning regulator). The sampling time for the controller is 6 sec. A separate experiment (Figure 6) shows that better set point tracking can be achieved by increasing sampling rate (1 sec) and time horizon ($T = 5$). This could not be done with the self tuning regulator since a sampling time of 1 sec with $d = 6$ yielded an unstable system.

The steam-water heat exchanger (Figure 7) has been subject to simulation and pilot-plant experiments. It has time constants, gains and dead times that vary drastically with the operating conditions. Because of the unknown, varying dead time, the Extended Horizon Controller (EHC) is used. The Variable Forgetting Factor algorithm is once again included to keep a measure of the information content constant. Controllers with a fixed or no forgetting factor did not perform as well.

The analysis of the heat exchanger begins with a steady state heat balance on the water

$$Q = V \rho C_p (T_F - T_o)$$

where Q is the heat transferred, V is the volumetric flow rate, C_p is the heat capacity of the water, ρ is the density of the water, T_F is the exit temperature of water, and T_o is the inlet temperature of water.

The steady state design equation is given by

$$Q = U\, A\, (\Delta T)_{LM}$$

where U is the overall heat transfer coefficient, A is the area of the exchanger, and $(\Delta T)_{LM}$ is the log-mean temperature difference.

In this study, the outlet water temperature is controlled by manipulating the water flow rate. If we combine the two equations to eliminate Q, the steady state nonlinearity for the system can be expressed as

$$\frac{\partial T_F}{\partial v} = N(T_o, v, T_s, \tau) = -\frac{(T_s - T_o) L\, exp(-L/\tau v)}{\tau v^2}$$

where v is the velocity of the water, τ is a parameter containing the overall heat transfer coefficient and fluid properties, L is the length of the exchanger, and T_s is the temperature of the steam.

Figures 8 and 9 show moderate improvement in performance by adding the nonlinear compensator to the Extended Horizon Controller. From the extensive experimentation, we have also seen that the EHC with the nonlinear compensator is less sensitive to the choice of the time horizon in the control algorithm, and less sensitive to the choice of the memory length in the variable forgetting algorithm. Because of this robustness, the nonlinear EHC works well even when the linear EHC and linear STR do not work at all.

4. Conclusions and Discussion

Adaptive controllers have been applied to a wide range of chemical process control problems with encouraging results. Many different algorithms have been successfully implemented and it can be said that many more algorithms can be made to work well, provided that the accompanying tuning parameters are properly chosen. Unfortunately it is often a nontrivial task to choose these tuning parameters and considerable skill and experience are sometimes needed to make the algorithms work at all. Despite considerable enthusiasm for adaptive

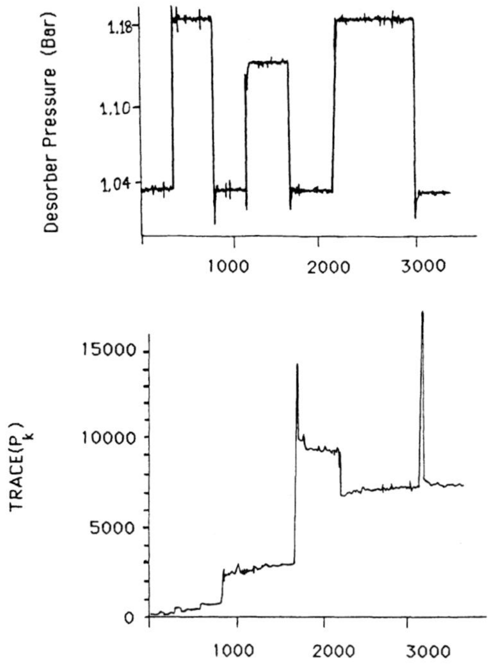

Figure 4: Trace of the covariance matrix using VFF.

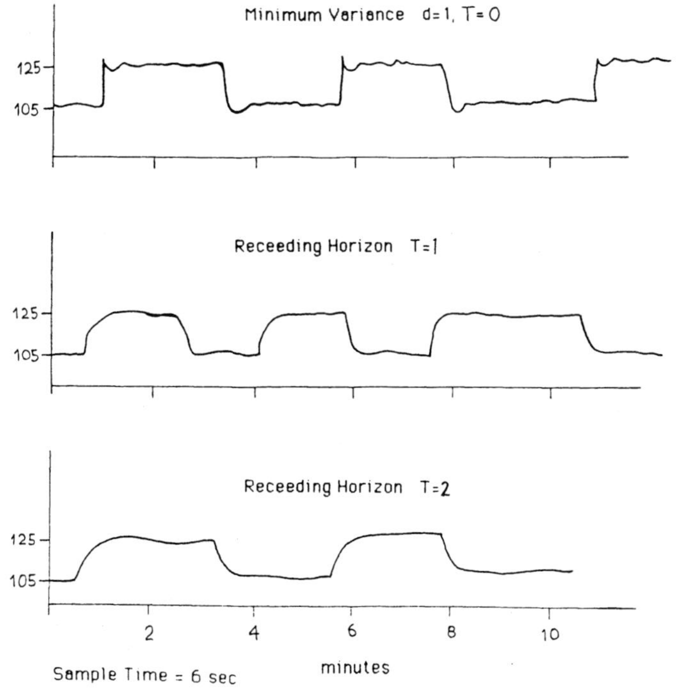

Figure 5: Extended horizon control.

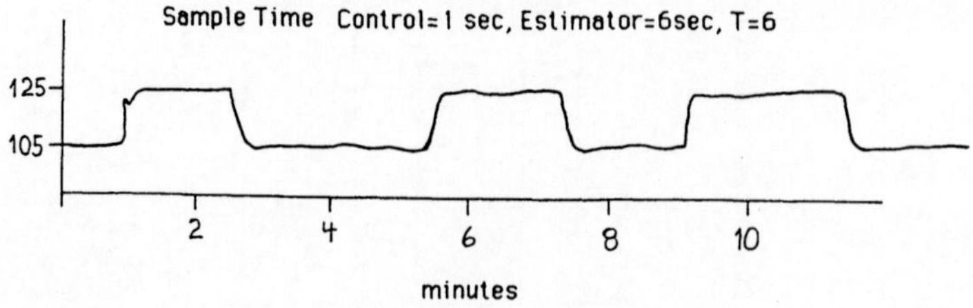

Figure 6: Faster sampling rates ·

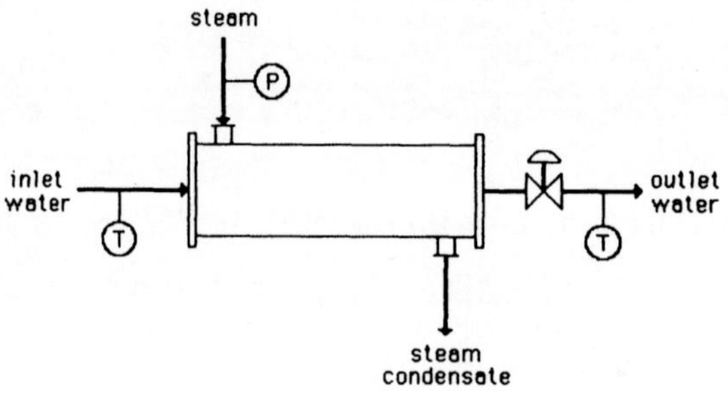

Figure 7: Block diagram of the heat-exchanger.

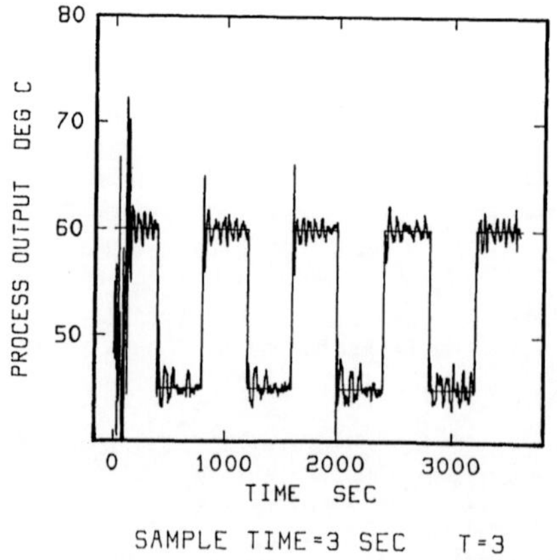

Figure 8: Linear adaptive control (EHC) .

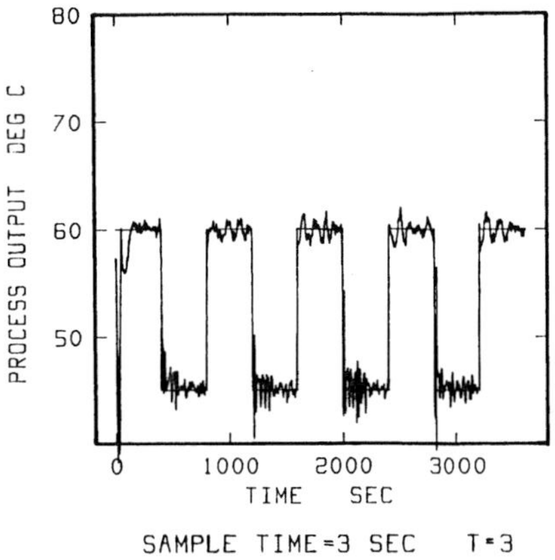

Figure 9: Nonlinear adaptive control (EHC) .

technology among process control engineers, management expectations are low in many chemical engineering companies. This is mainly due to the fact that considerable investment of time and expertise is needed to implement and safeguard the adaptive theory. It is also unclear in many cases as to exactly what returns can be expected if the adaptive theory can be made to work as is hoped for, namely that it adapts and tunes itself to process changes.

In view of the problems outlined above, we have concentrated our efforts in our research on the development of simple algorithms that are intuitively easy to understand (like the STR), but also easy to implement. Some simple modifications have led to algorithms that we have found to be robust and easy to tune. The most important modifications are the Extended Horizon Controller, the Variable Forgetting Factor and the use of delta operators. Recently we have also been experimenting with nonlinear model compensation techniques. These appear to make the algorithm more complex. However, in chemical process applications, steady state models of the processes are often available and they can be easily included into this theory. Even standard PID regulators usually include facilities for nonlinear compensation of valves and measurements as well as low pass filters. It is important that such compensation techniques are included in the adaptive algorithms as well. It is also important that numerically stable algorithms are used for estimation.

A great number of chemical process control problems are multivariable. One major challenge in the area of adaptive control is thus a generalization of the adaptive algorithm to the multivariable case. Though in principle this may appear to be a trivial problem (the one step ahead controller is one example of a simple extension), in practice it turns out to be harder. The potential benefits of multivariable adaptive control are substantial. We have in our research found that it is easy to show that multivariable theory usually works far better than the single input single output PI regulators for a wide range of industrial control problems. However, current algorithms are cumbersome to implement.

A A Matrix Result

Given a matrix $A > 0$ and a matrix $B \geq 0$ then $C = AB$ has non-negative eigenvalues.

Proof: Factor $A = LL'$ where L is an upper triangular square not with positive eigenvalues. The eigenvalues of C then satisfy. $LL'\ B\ v_i = \lambda_i v_i$, where λ_i is the eigenvalue belonging to the eigenvector v_i. This can be rearranged so that $L'BLL^{-1}v_i = \lambda_i L^{-1}v_i$. It follows that λ_i also is an eigenvalue of $L'BL$ (and $L^{-1}v_i$ are eigenvectors). But $L'BL \geq 0$, and the result follows.

References

[1] D.P. Atherton, Nonlinear Control Engineering, Van Nostrand Reinhold, London, 1975.

[2] K.J. Åström and B. Wittenmark, "On Self-Tuning Regulators," *Automatica*, 9, pp. 195-199, 1973.

[3] K.J. Åström and B. Wittenmark, "Self-Tuning Controllers Based on Pole-Zero Placement," *IEE Proc. 127 Pt.D.*, pp. 120-130, 1980.

[4] K.J. Åström and B. Wittenmark, "The Self-Tuning Regulators Revisited," Technical Report, Dept. of Automatic Control, University of Lund, Lund, Sweden, 1985.

[5] D.W. Clark and P.J. Gawthrop, "A Self-Tuning Controller," *IEEE Proc.*, 122, p. 929.

[6] C.R. Cutler, J.J. Haydel and A.M. Moshedi, "An Industrial Perspective on Advanced Control," paper presented at the AIChE Diamond Jubilee Meeting, Washington, D.C., Nov. 1983.

[7] A. Gelb and W.E. Vander Velde, Multiple-Input Describing Functions and Nonlinear System Design, McGraw-Hill, New York, 1968.

[8] M.P. Golden and B.E. Ydstie, "Nonlinear Model Control Revisited," Proceedings of the American Control Conference, Boston, June 1985.

[9] H. Elliott, W.A. Wolovich and M. Das, "Arbitrary Adaptive Pole-Placement for Linear Multivariable Systems," *IEEE Trans.*, AC-29, No. 3, p. 221, 1984.

[10] G. Orlandi and G. Martinelli, "Low Sensitivity Recursive Digital Filters Obtained via the Delay Replacement," *IEEE Trans. Circuits and Systems*, CAS-31, 7, July 1984.

[11] R. Rouhani and R.K. Mehra, "Model Algorithmic Control (MAC); Basic Properties," *Automatica*, 18, pp. 401-414.

[12] D.E. Seborg, S.L. Shah and T.F. Edgar, "Adaptive Control Strategies for Process Control - A Survey," Paper presented at the AIChE Diamond Jubilee Meeting, Washington, D.C., November 1983.

[13] B.E. Ydstie, "Robust Adaptive Control of Chemical Processes," Ph.D. Thesis Imperial College, University of London, June 1982.

[14] B.E. Ydstie and R.W.H. Sargent, "Convergence and Stability Properties of an Adaptive Controller with Variable Forgetting Factor," Proc. of the 6th IFAC Symposium on Identification and System Parameter Estimation, Washington, D.C., July 1982.

[15] B.E. Ydstie, L.S. Kershenbaum and R.W.H. Sargent, "Theory and Application of an Extended Horizon, Adaptive Controller," to appear in the AIChE Journal, 1985.

[16] B.E. Ydstie, "Adaptive Control and Estimation with Forgetting Factors," Proceedings of the IFAC Symposium Identification and System Parameter Estimation, York, July 1985.

[17] B.E. Ydstie, "Extended Horizon Control," Proceedings of the 9th IFAC World Congress, Budapest, July 1984.

Application of Multivariable Adaptive Control Schemes to Distillation Columns

H. Unbehauen and P. Wiemer
Department of Electrical Engineering
Ruhr-University Bochum
Germany

Abstract

Multivariable adaptive control schemes are applied to a binary distillation column. The theoretical background is briefly described. The results are compared with conventional PID-control and state-feedback control. It is shown that the application of adaptive control yields satisfactory disturbance rejection as well as considerable reduction of energy costs. By the high degree of maturity these adaptive techniques represent a powerful practical tool for the control engineer.

1. Introduction

In the last few years adaptive control of distillation columns has evoked much interest in the literature, mainly motivated by the nonlinear and time-varying nature of this process. If linear controllers, e.g. standard PID-controllers, are used, they have to be retuned from time to time in order to provide satisfactory performance under different operating conditions. However, since only a little a priori knowledge about the process dynamics is generally available, the tuning of the controller parameters is a costly and time consuming procedure. With this in mind, it is desirable to have controllers which can compensate for changes in process parameters attaining or holding a defined set of system conditions given an initial uncertainty and/or varying working conditions. This can be attained by adaptive controllers. The first successful application of an adaptive controller, a SISO selftuning controller, to a distillation column was reported by Sastry, Seborg and Wood [18]. Since that time the feasibility of adaptive control for distillation columns has been established [2,3,7,13,15,19].

Nevertheless an overwhelming majority of the regulators used in the chemical industry are conventional PID controllers. This may be due to the fact that the high costs of development and implementation of modern control techniques have to be justified economically. The aim of this paper is to show that adaptive control could satisfy this requirement.

The main control objective is to maintain the top and bottom product composition constant under varying flow rate, composition and temperature at the inlet. Top and bottom composition are controlled via manipulation of the reflux ratio of the distillate at the top and the reboiler heat at the bottom of the column. Classical distillation columns often have large reflux rates and many trays for separating components with similar boiling points. The reflux rate has a major influence on the separation process. Too much reflux flow makes the product over pure, but wastes energy because more reflux liquid has to be vaporized - too little reflux flow causes an impure product.

Fig. 1 shows how product purity is affected by a change in the reflux to feed ratio for a particular column. Changes in the feed flow causes only minor changes in product purity at high reflex rates whereas the sensitivity of the column near the specification point, where the energy consumption is minimal, is increased by a factor of nine. To avoid off-specification products, distillation columns are often operated with higher reflux rates than needed and hence consume more energy than necessary for guaranteeing product specification.

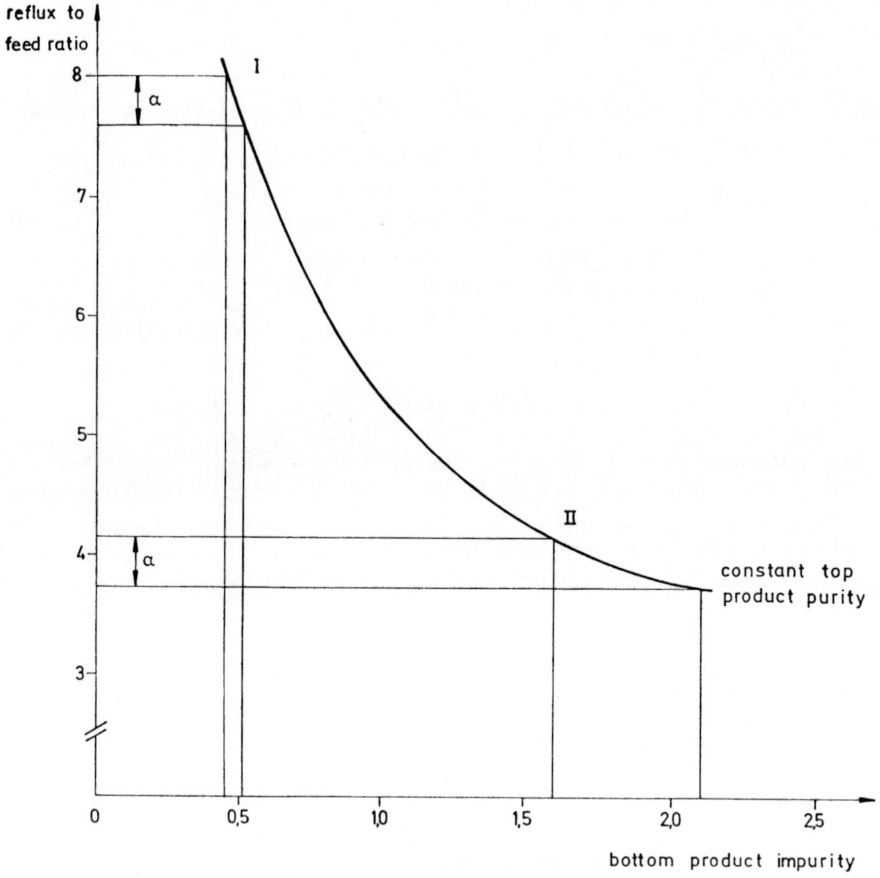

Figure 1: Product impurity as a function of the reflux rate .

Following these guidelines it becomes clear that energy savings can be achieved only by operating the column near the specification point. As the sensitivity of the column with respect to feed flow changes or composition changes is increased at this operating point, the control problem becomes severe and cannot be solved with classical control concepts.

The paper is organized as follows: Section 2 gives a short introduction to the theory of the multivariable model reference adaptive control (MRAC) scheme based on a discrete version of the algorithm proposed by Monopoli and Hsing [14]. Section 3 describes the estimation problem which can be solved by a generalized stochastic approximation method [5]. Though the control scheme is developed as a model reference adaptive controller, it can also be interpreted as an adaptive generalized minimum-variance (AGMVC) controller, because both strategies are mathematically identical [7]. Section 4 gives a brief sketch of this connection. The plant, a pilot scale distillation column, is described in section 5 and in the last section the results of standard linear control (PID, state-feedback) and modern adaptive control (MRAC, AGMVC) are compared.

2. Derivation of the MRAC Control Law

The plant with q inputs $\underline{u}(k)$ and q outputs $\underline{y}(k)$ is described by the transfer function matrix

$$\underline{G}(z^{-1}) = \{\frac{\overline{B}_{ij}(z^{-1})}{\overline{A}_{ij}(z^{-1})}\, z^{-d}\}; i,j = 1,\ldots,q \quad (1)$$

so that

$$\underline{Y}^*(z) = \underline{G}(z^{-1})\underline{U}(z). \quad (2)$$

For the sake of simplicity, let all the control inputs be with the same delay d. An extension to the general case of different arbitrary time delays is given by Hahn [5,6]. $\underline{G}(z^{-1})$ can easily be factorized as

$$\underline{G}(z^{-1}) = \underline{A}^{-1}(z^{-1})\underline{B}(z^{-1})z^{-d} \quad (3)$$

using

$$\underline{A}(z^{-1}) = diag\{A_i(z)\}; \quad i = 1,\ldots,q \quad (4a)$$

and $$A_i(z^{-1}) = \text{least common multiple } \{\overline{A}_{ij}(z^{-1})\}; \quad 1 \le j \le q \quad (4b)$$

$$\underline{B}(z^{-1}) = \{B_{ij}(z^{-1})\}; \quad i,j = 1,\ldots,q \quad (5a)$$

and $$B_{ij}(z^{-1}) = \overline{B}_{ij}(z^{-1})\frac{A_i(z^{-1})}{\overline{A}_{ij}(z^{-1})}. \quad (5b)$$

$\underline{A}(z^{-1})$ contains the least common denominator of each row i of the transfer function matrix $\underline{G}(z^{-1})$ at (i,i). The matrix $\underline{B}(z^{-1})$ contains the extended numerator polynomials.

Usually the output signal vector $\underline{Y}^*(z)$ of the plant contains in addition a $(q \times 1)$ disturbance vector

$$\underline{V}^*(z) = \underline{A}^{-1}(z^{-1})\underline{V}(z) = \underline{A}^{-1}(z^{-1})(\underline{C}(z)\epsilon(z) + \underline{Z}(z)), \quad (6)$$

where $\epsilon(z)$ is a $(q \times 1)$ vector of independent white noise signals and the $(q \times 1)$ vector $\underline{Z}(z)$ describes the unmeasurable deterministic disturbances.

Thus, the total behaviour of the plant is given in the general form

$$\underline{A}(z^{-1})\underline{Y}(z) = \underline{B}(z^{-1})z^{-d}\,\underline{U}(z) + \underline{V}(z). \quad (7)$$

The plant output $\underline{Y}(z)$ is augmented by the signal

$$\underline{Y}_c(z) = \underline{G}_c(z^{-1})\underline{U}(z) \quad (8)$$

with

$$\underline{G}_c(z^{-1}) = \underline{A}_c^{-1}(z^1)\underline{B}_c\,(z^{-1})z^{-d} = diag\{G_{ci}(z^{-1})\}, \quad (9)$$

wherein the filters $G_{ci}(z^{-1}), i = 1,\ldots,q$ must be stable. Such filters were originally introduced in order to make nonminimum phase control possible [8] and are called *correction networks.* The augmented plant output

$$\underline{Y}_a(z) = \underline{Y}(z) + \underline{Y}_c(z) \quad (10)$$

is compared with the output signal

$$\underline{Y}_m(z) = \underline{G}_m(z^{-1})\underline{W}(z), \quad (11)$$

$$\underline{G}_m(z^{-1}) = \underline{A}_m^{-1}(z^{-1})\underline{B}_m\,(z^{-1})z^{-d} = diag\{G_{mi}(z^{-1})\} \quad (12)$$

of the stable reference model. $\underline{W}(z)$ is the reference input signal. In order to reject the influence of deterministic disturbances which have the property

$$\lim_{k\to\infty} \underline{z}(k) = const, \quad (13)$$

it is convenient to force the controller to have integral action. This can easily be done by multiplying eq. (7) with the factor $(1 - z^{-1})$ on both sides. Therefore, eq. (7) is modified to

$$\underline{A}_I(z^{-1})\underline{Y}(z) = \underline{B}(z^{-1})z^{-d}\,\underline{U}_I(z) + \underline{V}_I(z). \quad (14)$$

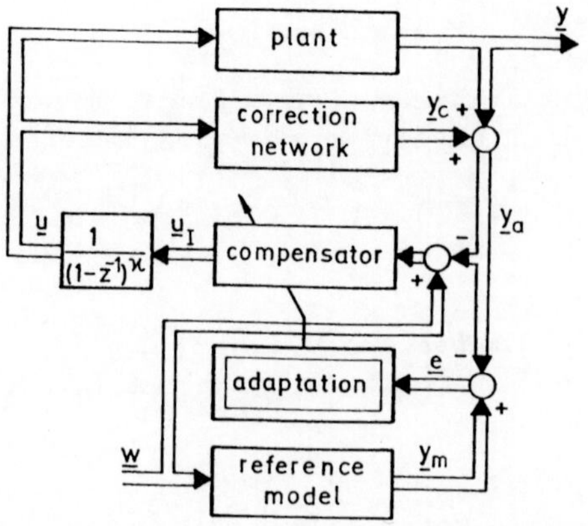

Figure 2: Basic structure of the MRAC scheme.

with

$$\begin{aligned}
\underline{A}_I(z^{-1}) &= (1-z^{-1})^{\mathcal{K}}\underline{A}(z^{-1}), && (15a)\\
\underline{U}_I(z) &= (1-z^{-1})^{\mathcal{K}}\underline{U}(z), && (15b)\\
\underline{V}_I(z) &= (1-z^{-1})^{\mathcal{K}}\underline{V}(z), && (15c)\\
\mathcal{K} &\in \{0,1\}. && (15d)
\end{aligned}$$

Note that under the assumption of eq. (13) and $\underline{\epsilon}(z) = 0$

$$\lim_{k\to\infty} \underline{v}_I(k) = \underline{0} \text{ for } \mathcal{K} = 1. \quad (16)$$

The basic structure of the control scheme is shown in Fig. 2. Using the identities [1,5,10]

$$\begin{aligned}
\underline{A}_m(z^{-1}) &= \underline{S}(z^{-1})z^{-d} + \underline{P}(z^{-1})\underline{A}_I(z^{-1}) && (17a)\\
\underline{A}_m(z^{-1}) &= \underline{S}_c(z^{-1})z^{-d} + \underline{P}_c(z^{-1})\underline{A}_c(z^{-1}) && (17b)
\end{aligned}$$

where $\underline{S}(z^{-1}), \underline{S}_c(z^{-1}), \underline{P}(z^{-1})$, and $\underline{P}_c(z^{-1})$ are diagonal matrices, the error signal

$$\underline{E}(z) = \underline{Y}_m(z) - \underline{Y}_a(z) = \underline{Y}_m(z) - \underline{Y}(z) - \underline{Y}_c(z) \quad (18)$$

can be expressed by

$$\begin{aligned}
\underline{E}_m(z) &= \underline{A}_m(z^{-1})\underline{E}(z)\\
&= \underline{B}_m(z^{-1})z^{-d}\underline{W}(z) - \underline{A}_m(z^{-1})\left(\underline{Y}(z) + \underline{Y}_c(z)\right)
\end{aligned} \quad (19)$$

$$\begin{aligned}
\underline{E}_m(z) = z^{-d}[&\underline{B}_m(z^{-1})\underline{W}(z) - \underline{S}(z^{-1})\underline{Y}(z) - \underline{P}(z^{-1})\underline{B}(z^{-1})\underline{U}_I(z)\\
&- \underline{S}_c(z^{-1})\underline{Y}_c(z) - \underline{P}_c(z^{-1})\underline{B}_c^*(z^{-1})\underline{U}_I(z)] - \underline{V}_{PI} && (20a)
\end{aligned}$$

with
$$\begin{aligned}
\underline{V}_{PI}(z) &= \underline{P}(z^{-1})\underline{V}_I(z)\\
\underline{B}_c(z^{-1}) &= (1-z^{-1})^{\mathcal{K}}\underline{B}_c^*(z^{-1}). && (20b)
\end{aligned}$$

The polynomial matrix products $\underline{P}(z^{-1})\underline{B}(z^{-1})$ and $\underline{P}_c(z^{-1})\underline{B}_c^*(z^{-1})$ can be separated into

$$\begin{aligned}
\underline{P}(z^{-1})\underline{B}(z^{-1}) &= \underline{B}_o + \underline{D}(z^{-1})z^{-1} && (21a)\\
\underline{P}_c(z^{-1})\underline{B}_c(z^{-1}) &= \underline{B}_{co} + \underline{D}_c(z^{-1})z^{-1} && (21b)
\end{aligned}$$

This leads to the error signal

$$\begin{aligned}
\underline{E}_m(z) = \; &\underline{\psi}(z)z^{-d} - (\underline{B}_o + \underline{B}_{co})\underline{U}_I(z) \; z^{-d} - \underline{D}(z^{-1})\, z^{-d-1}\underline{U}_I(z)\\
&- \underline{S}(z)z^{-d}\underline{Y}(z) - \underline{V}_{PI}(z),
\end{aligned} \quad (22)$$

wherein each element of the signal vector $\underline{\psi}(z)$ is defined by

$$\begin{aligned}\psi_i(z) = \;& B_{mi}(z^{-1})W_i(z) - D_{ci}\,(z^{-1}) \quad z^{-1}U_{Ii}(z) \\ & -S_{ci}(z^{-1})Y_{ci}(z); \qquad\qquad i = 1,\ldots,q.\end{aligned} \tag{23}$$

The signals $\psi_i(z)$ are generated only with known parameters and measurable plant input signals.

If the control signal is computed by

$$\underline{U}_I(z) = (\underline{B}_o + \underline{B}_{co})^{-1}[\underline{\psi}\,(z) - \underline{S}(z^{-1})\underline{Y}(z) - \underline{D}(z^{-1})z^{-1}\,\underline{U}_I(z)] \tag{24}$$

for $\underline{V}(z) = 0$ the augmented plant $\underline{Y}_a(z)$ follows the output of the reference model $\underline{Y}_m(z)$ i.e.

$$\underline{E}_m(z) = \underline{0}. \tag{25}$$

Note that in eq. (20b) each of the polynomials $B_{ci}(z^{-1})$ is forced to have a zero at $z = 1$ if integral action is required.

Then, under the assumption

$$\lim_{k\to\infty} u_i(k) = const, \qquad\qquad i = 1,\ldots,q \tag{26}$$

the output $\underline{Y}_c(z)$ of the correction network vanishes for $t \to \infty$. Eq. (26) can only be fulfilled for constant reference signals $W_i(z)$ and step disturbances $z_i(z)$. In this case we have

$$\lim_{k\to\infty} \underline{y}(k) = \lim_{k\to\infty} \underline{y}_m(k) \tag{27}$$

i.e. asymptotic model matching of the plant output. In eq. (24) the coefficients of the polynomial matrices $\underline{S}(z^{-1})$ and $\underline{D}(z^{-1})$ depend on the unknown plant parameters. For the computation of the adaptive control law they have to be replaced by their estimates

$$\underline{U}_I(z) = (\hat{\underline{B}}_o + \underline{B}_{co})^{-1}[\underline{\psi}\,(z) - \hat{\underline{S}}(z^{-1})\underline{Y}(z) - \hat{\underline{D}}(z^{-1})z^{-1}\,\underline{U}_I(z)]. \tag{28}$$

In eq. (28) this is indicated by the sign " ^ ". Eq. (28) can easily be solved provided the matrix $(\hat{\underline{B}}_o + \underline{B}_{co})$ is nonsingular. A necessary condition for this is that the corresponding matrix $(\underline{B}_o + \underline{B}_{co})$ built up by the true parameters is nonsingular. This can always be achieved by choosing an appropriate matrix $\underline{B}_{co}$ for the correction network. With eq. (15b) the output $\underline{U}(z)$ of the adaptive controller with integrator is

$$\underline{U}(z) = \frac{1}{(1 - z^{-1})^K}\,\underline{U}_I(z). \tag{29}$$

3. Derivation of the Adaptive Law for the Parameters

By substituting $\underline{\psi}(z)$ from eq. (28) in eq. (22) we obtain

$$\begin{aligned}\underline{E}_m(z) = \;& -z^{-d}[(\underline{S}(z^{-1}) - \hat{\underline{S}}(z^{-1}))\underline{Y}(z) + (\underline{D}(z^{-1}) - \hat{\underline{D}}(z^{-1}))z^{-1}\underline{U}_I(z) \\ & +(\underline{B}_o - \hat{\underline{B}}_o)\,\underline{U}_I(z)] - \underline{V}_{PI}(z).\end{aligned} \tag{30}$$

Considering now time varying controller parameters, eq. (30) for each element of the error signal $\underline{e}_m(k)$ can be rewritten as

$$e_{mi}(k) = [\underline{p}_i - \hat{\underline{p}}_i(k-d)]^T\underline{x}_i\,(k-d) - v_{PIi}(k), \quad i = 1,\ldots,q \tag{31}$$

where all signals are contained in the *signal vector* $\underline{x}_i(k)$ and all parameters are included into the *parameter vectors* $\underline{p}_i$ and $\hat{\underline{p}}_i(k)$. For the on-line computation of eq. (28), it is necessary to have a short delay between measurement of the plant output and output of the new control signal. Therefore in eq. (28) the parameters estimated at the previous step $(k-1)$ should be used. Then eq. (31) has to be modified as

$$e_{mi}(k) = [\underline{p}_i - \hat{\underline{p}}_i(k-d-1)]^T \underline{x}_i\ (k-d) - v_{PIi}(k). \tag{32}$$

We define an *augmented error signal*

$$\bar{e}_i(k) = e_{mi}(k) + h_i(k), \qquad i = 1, \ldots, q \tag{33}$$

where

$$h_i(k) = [\hat{\underline{p}}_i(k-d-1) - \hat{\underline{p}}_i(k-1)]^T\ \underline{x}_i(k-d). \tag{34}$$

This yields

$$\bar{e}_i(k) = [\underline{p}_i - \hat{\underline{p}}_i(k-1)]^T \underline{x}_i\ (k-d) - v_{PIi}(k). \tag{35}$$

Eq. (35) describes an estimation problem, which has been very well studied in both the disturbance free case as well as the case when the disturbance is white noise (e.g. [12]). This estimation problem can be formulated as follows: Find a law for the adaptation of the controller parameters $\hat{\underline{p}}_i[i = 1, \ldots, q]$ such that

$$\lim_{k\to\infty} \bar{e}_i(k) = 0, \qquad i = 1, \ldots, q. \tag{36}$$

There are different solutions available for this estimation problem. As the least squares estimation takes more time for computing, in our approach a *generalized stochastic approximation method* [5], similar to that proposed by Goodwin et al. [4], was applied for the estimation of the controller parameters:

$$\hat{\underline{p}}_i = \hat{\underline{p}}_i(k-1) + r_i(k)\underline{G}_i\ \underline{x}_i(k-d)\bar{e}_i(k) \tag{37}$$

where

$$r_i(k) = \begin{cases} r_i^*(k) \text{if} r_i^*(k) \geq \frac{1}{\rho_i \underline{x}_i^T(k-d)\ \underline{G}_i \underline{x}_i(k-d)+\gamma_i} \\ \frac{1}{\rho_i \underline{x}_i^T(k-d)\underline{G}_i \underline{x}_i\ (k-d)+\gamma_i} \end{cases} \tag{38}$$

$$\frac{1}{r_i^*(k)} = \frac{\lambda_{1i}(k)}{r_i(k-1)} + \lambda_{2i}(k)\ \underline{x}_i^T(k-d)\underline{G}_i \underline{x}_i(k-d) + \lambda_{3i}(k) \tag{39}$$

and the values of

$$\underline{G}_i = \underline{G}_i^T > 0;\ r_i(-1) > 0 \tag{40a}$$

$$0 \leq \lambda_{1i}(k) \leq \bar{\lambda}_{1i}; \frac{1}{2} \leq \lambda_{2i}(k)\ \leq \bar{\lambda}_{3i} \tag{40b}$$

$$0 \leq \lambda_{3i}(k) \leq \bar{\lambda}_{3i}; \rho_i > \frac{1}{2}; \gamma_i\ \geq 0 \tag{40c}$$

can be freely chosen. It is advisable to choose $\rho_i >> \bar{\lambda}_{2i}$ and large values for γ_i for $i = 1, \ldots, q$. This algorithm (eqs. (37-40)) contains the "classical" adaptation algorithm with non-decreasing (fixed) gain estimation (e.g. [9]) as a special case, with

$$\lambda_{1i} = 0; \lambda_{2i} > 0.5; \lambda_{3i} = 1. \tag{41}$$

The stability of the total algorithm in the disturbance free case, i.e. $v_i(k) \equiv 0$ for all $i = 1, \ldots, q$, is guaranteed if $\underline{G}_c(z^{-1})$ and $\underline{G}_m(z^{-1})$ are stable and $\underline{G}(z^{-1}) + \underline{G}_c(z^{-1})$ has no zeros for $|z| \geq 1$. Then all signals are bounded. The algorithm can be extended to the disturbed case by introducing a "dead zone" into the adaptation law [16]. For the stability proof see [5].

As the reference model is stable, the convergence of the original errors $E_i(z)$ directly follows. The influence of step disturbances on the estimation procedure is at least asymptotically rejected if an explicit integrator ($\mathcal{K} = \infty$) is used because then eq. (16) holds.

In order to overcome the possible numerical difficulties in the inversion of $(\underline{\hat{B}}_o(k) + \underline{B}_{co})$ during start-up, a lower bound d_{min} of $det(\underline{\hat{B}}_o(k) + \underline{B}_{co})$ has been set up and the calculation of the inverse has been modified as:

$$(\underline{\hat{B}}_o(k) + \underline{B}_{co})^{-1} = \begin{cases} (\underline{\hat{B}}_o(k) + \underline{B}_{co})^{-1} \text{ if } det(\underline{\hat{B}}_o(k) + \underline{B}_{co})^{-1} > d_{min} \\ (\underline{\hat{B}}_o(k-1) + \underline{B}_{co})^{-1} \text{ otherwise.} \end{cases}$$

With $det(\hat{B}_o(0) + \underline{B}_{co}) > d_{min}$ this setting up of the lower bound ensures the successful generation of the control input.

4. Comparison with the AGMVC Scheme

Although model reference adaptive control and the adaptive generalized minimum variance control are very different from their origins, they seem to be very similar [7,11]. As shown by Hahn et al. [7] the proposed MRAC scheme can be used for minimum variance control with slight modifications. Following the ideas of Koivo [10] a generalized output signal

$$\underline{E}_v(z) = \underline{R}(z^{-1})z^{-d}\underline{W}(z) \quad - \underline{Q}(z^{-1})(\underline{Y}(z) + \underline{Y}_c(z)) \tag{42}$$

is introduced. With

$$\underline{R}(z^{-1}) = \underline{B}_m(z^{-1}) \tag{43a}$$
$$\underline{Q}(z^{-1}) = \underline{A}_m(z^{-1}) \tag{43b}$$

this generalized output signal is identical with the error signal in eq. (19). Therefore the derivation of the control law is the same as for the MRAC scheme, if the matrix $\underline{C}(z^{-1})$ of eq. (6) instead of $\underline{A}_m(z^{-1})$ is used in eqs. (17a,b). As shown by Hahn et al. [7] the control law minimizes the variance of the signal

$$\underline{E}_q(z) = (1 - z^{-1})^{-\mathcal{K}}\underline{Q}^{-1}(z^{-1})\ \underline{E}_v(z). \tag{44}$$

Eq. (44) indicates that no explicit integrator ($\mathcal{K} = 0$) should be used for stochastic disturbances. Depending on the stochastic environment for the estimation of the controller parameters, a recursive least square estimation method is preferred.

5. The Plant

In a typical distillation column (Fig. 3) the feed stream enters near the center of the column and flows down. Vapors that are released by heating rise to the top, are condensed and can be removed as the overhead product or distillate. Any liquid that is returned to the column is called reflux. The reflux flows down the column and joins the feed stream.

Conventional distillation column control mostly uses temperatures as controlled variables using a fixed reflux rate. It is generally only one temperature near the feed, which is controlled via the reboiler heat. Additional control of a temperature near the top of the column by the reflux rate hardly produces an improvement since both controllers work independently and do not take into account the interactions of the two loops. This can result in both controllers working against each other producing oscillations or even destabilizing the control.

An improvement in distillation control, therefore, can only be achieved by using multivariable control, modelling the column as a two-input-two-output system with the inputs, reboiler heat and reflux rate, and the top and bottom product compositions forming the outputs. This is still complicated by the fact that measurements on product specification can

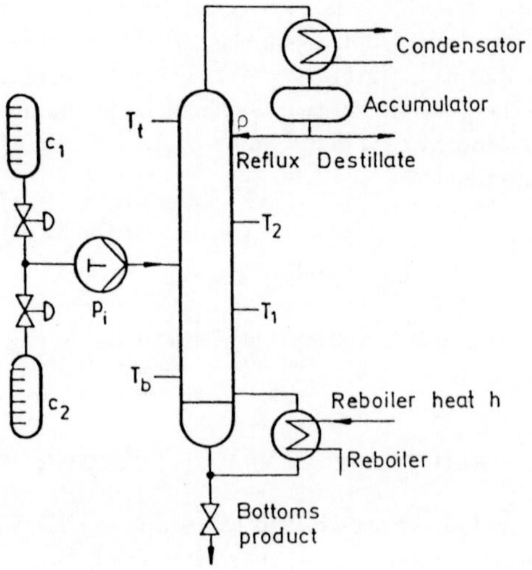

Figure 3: Schematic diagram of the distillation column.

$c_1; c_2$: containers with different concentrations of the feed product, p_i: inlet pump, h: heating, ρ: reflux flow ratio, T_t: top temperature, T_b: bottom temperature

only be done by on-line chromatographs, which will add delays and maintenance problems. As in binary distillation under isobaric conditions the concentrations are directly correlated to the temperatures, the temperatures near the top T_t and the bottom T_b of the column are selected as controlled variables. Deterministic disturbances at the inlet are generated by changing the speed of the inlet pump (feed disturbance) and switching the feed between two containers filled with different concentrations of the components of the feed stream (concentration disturbance).

Fig. 4 shows the measured step responses of the top and bottom temperatures for a change of the reboiler heat (Fig. 4a), the reflux rate (Fig. 4b), a concentration disturbance (Fig. 4c) and a feed disturbance (Fig. 4d).

The dynamic behavior of the plant may be expressed by a 2×2 transfer matrix. Three elements of this matrix are estimated as transfer functions of 2nd order and one transfer function is assumed to be only of first order. An a priori model of the distillation column has been obtained as

$$\underline{G}(s) = \begin{bmatrix} \frac{3.31\cdot10^{-6}+9.75\cdot10^{-4}s+1.41\cdot10^{-4}s^2}{5.07\cdot10^{-6}+8.08\cdot10^{-3}s+s^2} & \frac{-8.54\cdot10^{-6}-3.0\cdot10^{-4}s+9.03\cdot10^{-4}s^2}{3.36\cdot10^{-6}+4.19\cdot10^{-3}s+s^2} \\ \frac{1.04\cdot10^{-5}+1.22\cdot10^{-3}s-5.71\cdot10^{-5}s^2}{6.41\cdot10^{-6}+5.58\cdot10^{-3}s+s^2} & \frac{-1.15\cdot10^{-2}}{1.3\cdot10^{-3}+s} \end{bmatrix}$$

For the MRAC scheme the discrete time delay d of each transfer function is set to 1, and 21 parameters were estimated on-line. For set point changes the two measured temperatures T_t and T_b have rise-time from 45 to 70 minutes. However, according to the dynamics of some fast disturbances a sampling time of 30 sec is chosen. The poles and transmission zeros of the corresponding discrete time transfer function matrix are listed in Table 1.

Table 1. Poles and zeros of the discrete a priori model

No.	poles		No.	zeros	
	real part	imaginary part		real part	imaginary part
1	0.801	0.0	1	0.857	$3.43 \cdot 10^{-2}$
2	0.979	0.0	2	0.857	$-3.43 \cdot 10^{-2}$
3	0.887	0.0	3	0.990	$2.15 \cdot 10^{-2}$
4	0.952	0.0	4	0.990	$-2.15 \cdot 10^{-2}$
5	0.911	0.0	5	0.962	0.0
6	0.968	0.0			
7	0.961	0.0			

All zeros of the a priori model are inside the unit disk centered at the origin. Therefore theoretically a correction network is not necessary. However to make the adaptive controller robust against changes in the dynamic behavior of the plant, a simple correction network was used in all experiments.

6. Experimental Results

The experimental results are shown in Figures 5, 6 and 7. The ordinate-axes are scaled in relative values related to a certain set point. Because of the plotter used, the time bases of the different plots are slightly shifted with respect to each other.

Four control strategies were tested at the column:

a) conventional PID-controllers
b) multivariable linear state feedback controller with PI-action (LSFC)
c) generalized multivariable self-tuning minimum variance controller (AGMVC) with least squares parameter estimation
d) multivariable model reference adaptive controller (MRAC) with fixed gain parameter estimation

Both adaptive control schemes are designed for

$$\underline{A}_m(z^{-1}) = \underline{B}_m(z^{-1}) = \underline{A}_c(z^{-1}) = \underline{C}(z^{-1}) = \underline{I}, \underline{K} = 0 \tag{45a}$$
$$\underline{B}_c(z^{-1}) = diag\{\mu_i\} \quad \text{for AGMVC} \tag{45b}$$
$$\underline{B}_c(z^{-1}) = diag\{(1 - z^{-1})\mu_i\} \quad \text{for MRAC.} \tag{45c}$$

For a stable plant the sign of the parameters μ_i depends on the gain of each transfer function [5,6], if $B_{ci}(z^{-1})$ has a zero at $z = 1$ (eq. (45c))

$$sgn(\mu_i) = sgn\{ \frac{det\ \underline{G}_i(1)}{det\ \underline{G}_{i-1}(1)} \} ; \qquad i = 1, \ldots, q \tag{46}$$

wherein $\underline{G}_i(1)$ denotes the north-west sub gain matrix of dimensions $i \times i$ and $det\ \underline{G}_o(1) = 1$.

With the simple correction networks eqs. (45a-c) smooth control actions by weighting the control cost or the control increments can be achieved.

The correction network eq. (45c) may be interpreted as a parallel high pass filter, which was introduced by Rohrs and Shortelle [17] to make the adaptive controller robust against unmodelled parasitic dynamics.

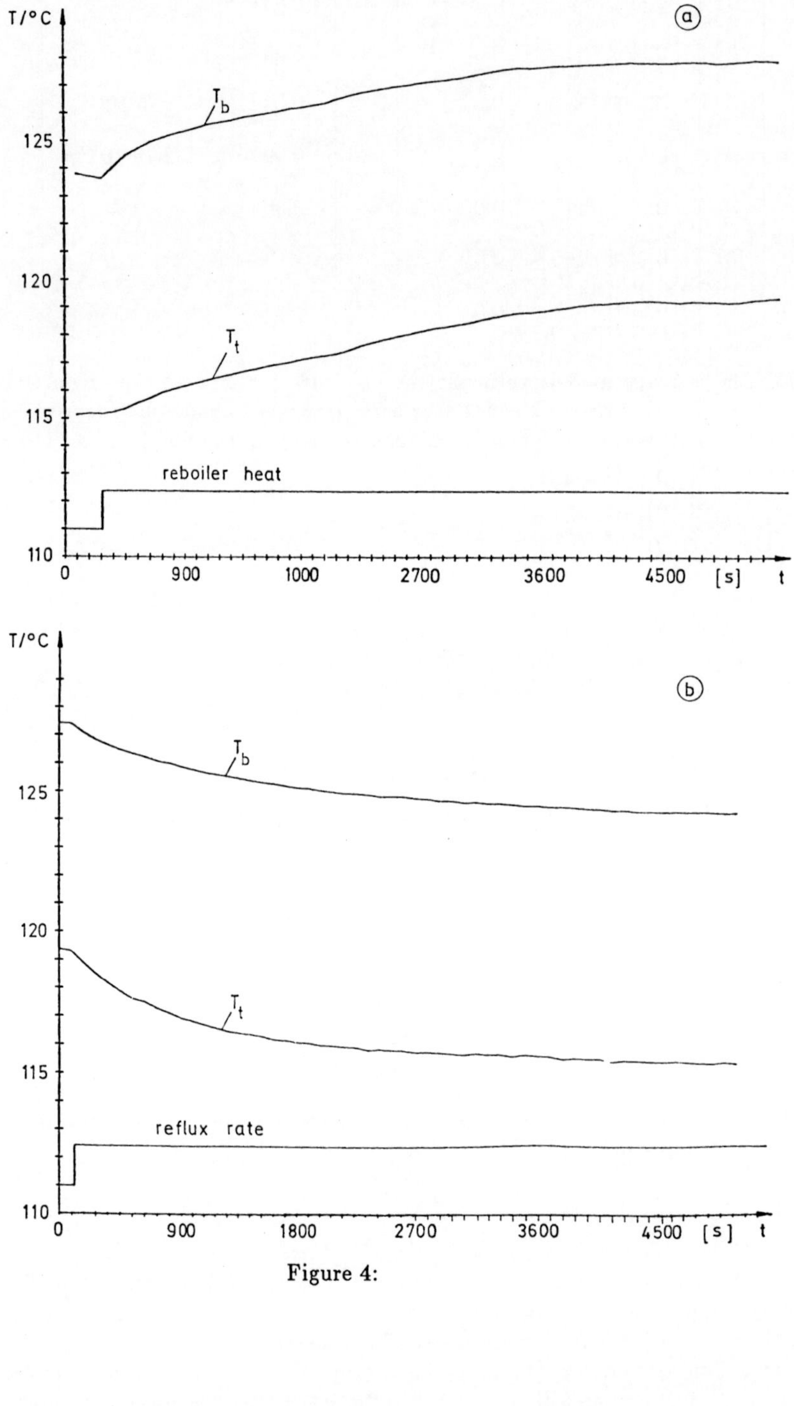

T/°C
(a)
T_b
125
120
T_t
115
reboiler heat
110
0
900
1000
2700
3600
4500
[s]
t
T/°C
(b)
T_b
125
120
T_t
115
reflux rate
110
0
900
1800
2700
3600
4500
[s]
t

Figure 4:

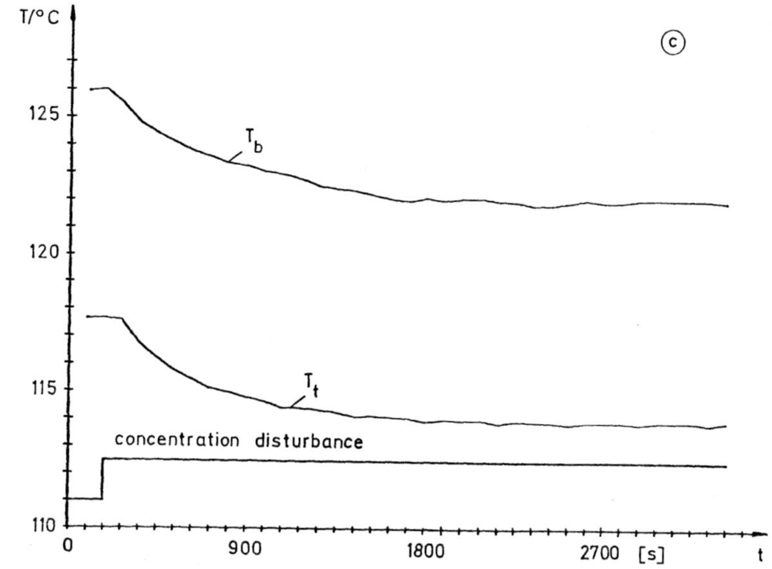

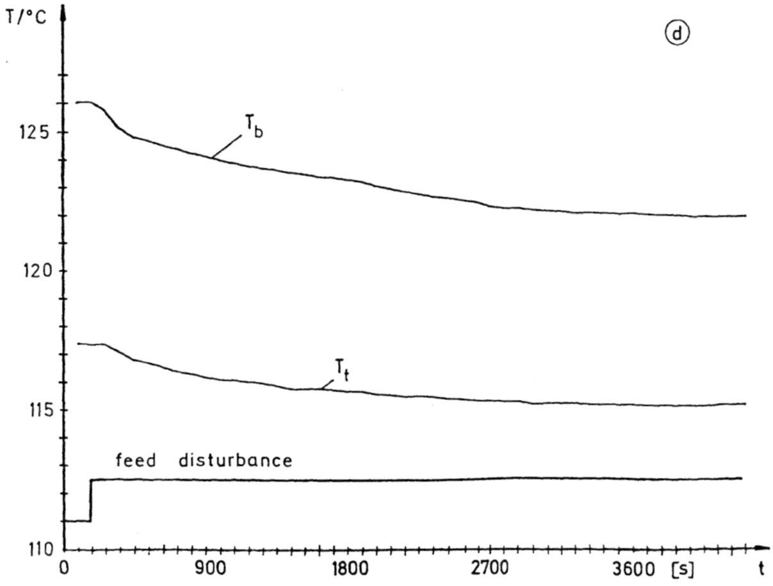

Figure 4: Measured step responses of the top and bottom temperatures, a) reboiler heat, b) reflux rate, c) concentration disturbance, d) feed disturbance.

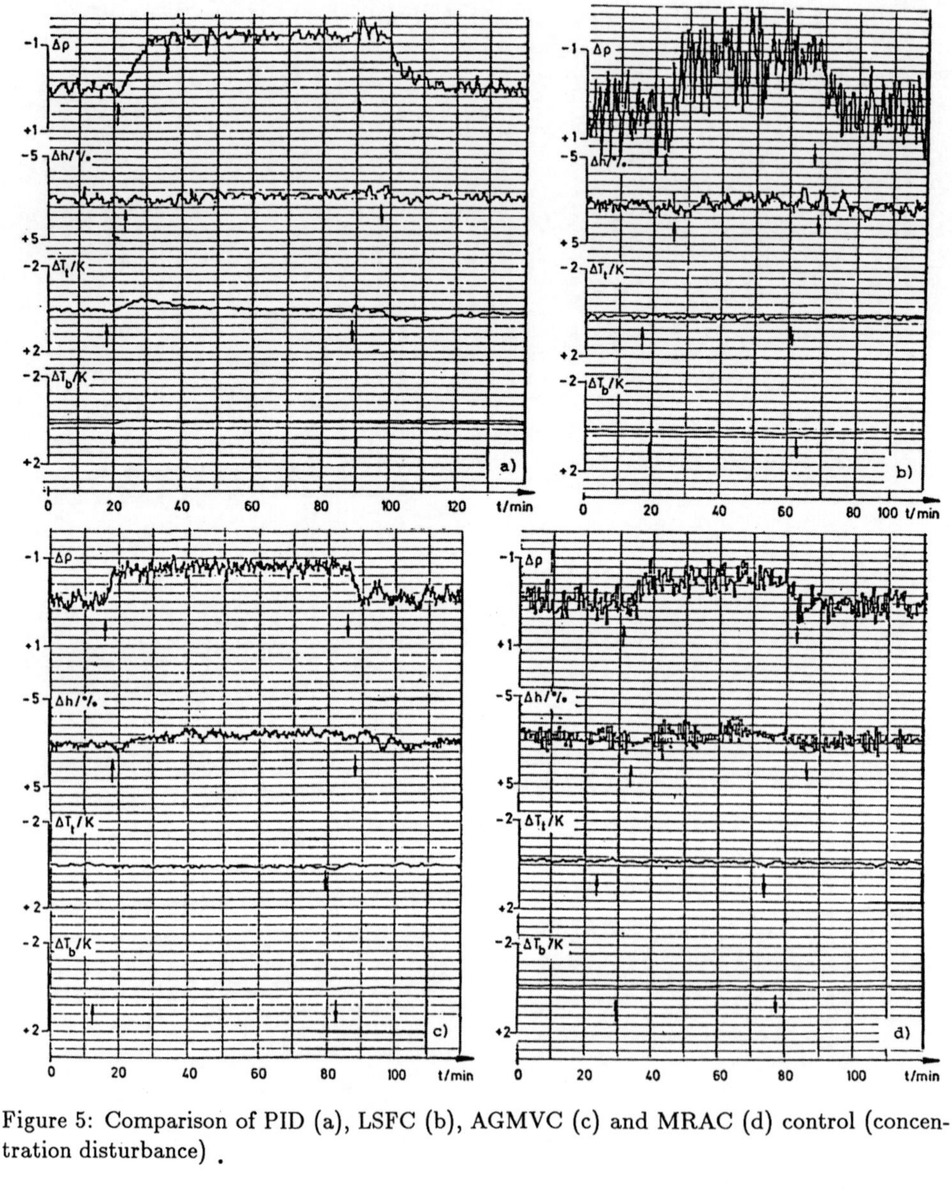

Figure 5: Comparison of PID (a), LSFC (b), AGMVC (c) and MRAC (d) control (concentration disturbance) .

Fig. 5 shows the dynamic behavior of the four control strategies under a 20% concentration disturbance. In all cases the influence of the disturbance is well rejected by decreasing the reflux flow ratio ρ. All controller parameters are well adjusted using an a priori model of the plant and on-line tuning of the linear controllers. For the PID controllers which don't take into account the dynamic coupling of the plant, the top and bottom temperatures vary about $0.5°C$ and $0.3°C$ about the set point. The nonlinear behavior of the plant is indicated by the different behavior for positive (↑) and negative (↓) disturbances. In all multivariable control schemes the deviation from the set points is negligibly small. The linear state feedback controller is very sensitive to stochastic disturbances caused by the measurement of the top temperature. This results in a large variance of the reflux flow ratio. It should be noted that the linear state feedback controller is also very sensitive to changes in the operation point so that its parameters have to be retuned from time to time.

In Fig. 6 the results of both adaptive control schemes under a 25% feed disturbance and a 20% concentration disturbance are compared. While the MRAC controller has no temperature deviation from the set points, a small deviation of about $0.1 - 0.3°C$ can be observed with the AGMVC-scheme. The AGMVC-scheme has some advantages under stochastic disturbances. This is indicated by the variance of the control signals in Fig. 6, which is about 50% smaller.

It should, however, be mentioned that the full least squares estimation combined with AGMVC control leads to larger storage requirements and a computation time which is about five times greater than that for the MRAC scheme. In order to keep the top and bottom temperatures constant, the temperature profile in the column must change due to changes at the inlet. In Fig. 6 this is indicated by two other temperatures T_1 and T_2 measured above and below the inlet (Fig. 3).

The results of the comparison of these 4 different controllers are listed in Table 2.

Table 2. Settling time and overshoot for all 4 types of controllers

Type	Settling time[min]		Overshoot [°C]		conc.	feed
	top	bottom	top	bottom	dist.	dist.
PID	25	10	0.5	0.3	X	
LSFC	0	0	0	0	X	
AGMVC	8	0	0.2	0	X	
MRAC	8	0	0.1	0	X	
AGMVC	10	5	0.3	0.1	X	X
MRAC	-	. 0	-	0		

From Fig. 7 the performance of adaptive transients can be seen for the MRAC controller. At time $t = 0$ the adaptation is started with all initial values of the controller set to zero. Fig. 7a shows the dynamic behavior of the inputs and outputs of the plant under concentration and feed disturbances. Already the first disturbance step (indicated by ↓ and ↑) is rejected in a satisfactory manner. Because of the adaption of the controller parameters the performance of the control system is considerably improved after further disturbance changes.

So far only the disturbance behavior of the adaptive control scheme has been described because this is very important in chemical industry. If the controller is implemented in a distillation column with great uncertainty about its dynamic behavior, it is convenient to fasten the adaptation by exciting the control system with small set point changes. As shown in Fig. 7b, small set point changes about $1°C$ are sufficient for a fast estimation of the controller parameters. In order to avoid heavy control actions, the set point changes are

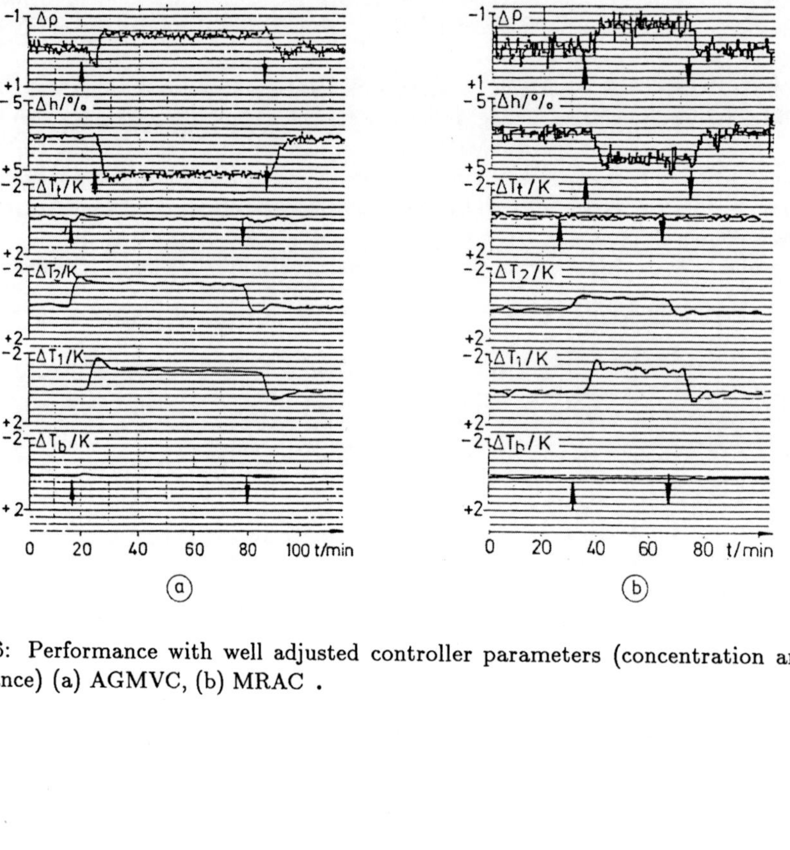

Figure 6: Performance with well adjusted controller parameters (concentration and feed disturbance) (a) AGMVC, (b) MRAC .

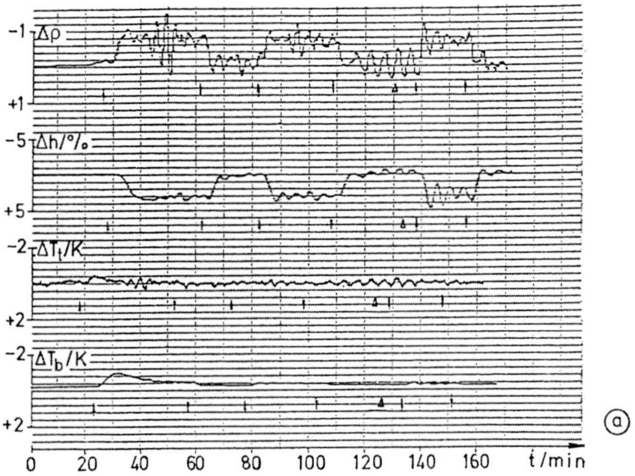

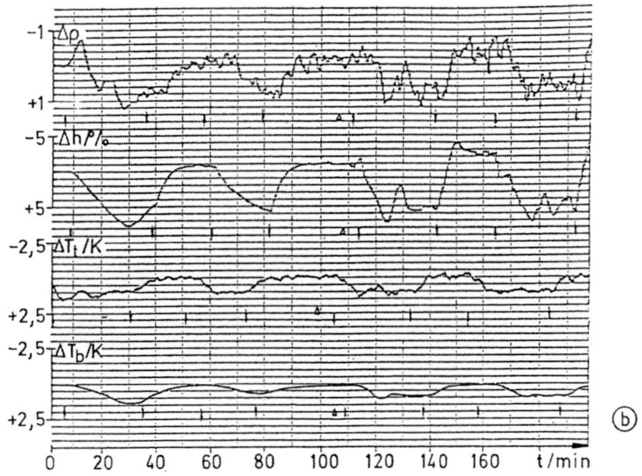

Figure 7: Performance of adaptive transients: (a) concentration and feed disturbance, (b) set point changes $+/-1°(\uparrow/\downarrow)$.

filtered by low pass filters of first order with time constants about 270 sec for top and 190 sec for bottom temperature.

The adaptive controller discussed here has been successfully applied since 1981 to large technical recycling columns. As a result, energy has been saved in the range of 10% compared to conventional control. At the same time the product continuity has been increased and the bottom's product purity has been improved from 200 ppm with conventional PID control to 20 ppm.

7. Conclusions

Multivariable adaptive control has been successfully applied to a binary distillation column. Though model reference schemes have originally been designed for servo problems, they can also be used for regulation in a deterministic and with slight modifications in a stochastic environment. The comparison between MRAC and AGMVC schemes shows nearly identical results. However, parameter adaptation with fixed gain or generalized stochastic approximation methods seem to be much simpler than the full least squares estimation.

At the present time adaptive control theory has reached a high degree of maturity, making it a very powerful tool for the experienced control engineer, capable of exploiting all the inherent degrees of freedom offered by the adaptive control scheme. The high costs of development, implementation and operation of such control systems are justified economically when compared with the classical PID control. The adaptive control system is capable of replacing the time consuming process of plant experiments and tuning of the parameters. In addition a remarkable progress in energy saving could be achieved.

Further research has to be undertaken, for making it accessible to many process engineers, i.e. reducing the degrees of freedom by providing appropriate tuning elements to simplify the handling of the adaptive controller.

References

[1] K.J. Åström , Introduction to Stochastic Control, Academic Press, London, 1970.

[2] S.A. Dahlqvist, "Application of Selftuning Regulators to the Control of Distillation Columns," Prepr. 6th IFAC/IFIP Int. Conf. on Digital Computer Appl. to Process Control, Düsseldorf, 1980.

[3] J.P. Gerry, E.F. Vogel and T.F. Edgar, "Adaptive Control of a Pilot Scale Distillation Column," Proc. of the 1983 American Control Conference, San Francisco, 1983.

[4] G.C. Goodwin, P.J. Ramadge and P.E. Cains, "Discrete Stochastic Control," *SIAM Journ. of Control and Opt. 19*, pp. 829-853, 1981.

[5] V. Hahn, "Direct Adaptive Control Schemes for Discrete Time Control of Multivariable Systems," (in German), Ph.D. Thesis, Ruhr-Universität Bochum, 1983.

[6] V. Hahn, "A Direct Adaptive Controller for Non-minimum Phase Multivariable Systems with Arbitrary Time Delays," Proc. of the 7th IFAC/IFIP/IMACS Conference on Digital Computer Applications to Process Control, Wien, Sept. 1985.

[7] V. Hahn, H. Röck, Chr. Schmid and P. Wiemer, "Some Experiences with the Application of Multivariable Adaptive Control in Chemical and Electromechanical Plants," *Optimal Control Application & Methods*, vol. 6, pp. 225-248, 1985.

[8] V. Hahn and H. Unbehauen, "Direct Adaptive Control of Non-minimum Phase Systems," Prepr. IEEE Conf. on Applications of Adaptive and Multivariable Control, Hull, pp. 170-175, 1982.

[9] T. Ionescu and R.V. Monopoli, "Discrete Model Reference Adaptive Control with an Augmented Error Signal," *Automatica 13*, No. 5, pp. 507-517, 1977.

[10] H.N. Koivo, "A Multivariable Self-tuning Controller," *Automatica 16*, No. 4, pp. 351-366, 1980.

[11] I.D. Landau, "Combining Model Reference Adaptive Controllers and Stochastic Self-tuning Regulators," *Automatica 18*, No. 1, pp. 77-82, 1982.

[12] R. Lozano and I.D. Landau, "Redesign of Adaptive Control Schemes," *Intern. J. Contr. 33*, No. 2, pp. 247-268, 1981.

[13] J.M. Martin-Sánchez and S.L. Shah, "Multivariable Adaptive Predictive Control of a Binary Distillation Column," *Automatica 20*, No. 5, pp. 607-620, 1984.

[14] R.V. Monopoli and C.C. Hsing, "Parameter Adaptive Control of Multivariable Systems," *Intern. J. Contr. 22*, No. 3, pp. 313-327, 1975.

[15] A.J. Morris, Y. Nazer, R.K. Wood and H. Lieuson, "Evaluation of Self-tuning Controllers for Distillation Column Control," Prepr. 6th IFAC/IFIP Int. Conf. on Digital Computer Appl. to Process Control, Düsseldorf, 1980.

[16] B.B. Peterson and K.S. Narendra, "Bounded Error Adaptive Control," *IEEE Trans. on Aut. Contr. 27*, pp. 1161-1168, 1982.

[17] C.E. Rohrs and K. Shortelle, "Conditioning a Plant for Frequency Selective Adaptive Control with Improved Robustness," Proc. of the 1984 American Control Conference, San Diego, pp. 1579-1583, 1984.

[18] V.A. Sastry, D.E. Seborg and R.K. Wood, "Self-tuning Regulator Applied to a Binary Distillation Column," *Automatica 13*, pp. 417-424, 1977.

[19] D.R. Yang and W. Lee, "Experimental Investigation of Adaptive Control of a Distillation Column," Proc. of the 1984 American Control Conference, San Diego, pp. 893-898, 1984.

The EXACT Pattern Recognition Adaptive Controller, a User-Oriented Commercial Success

E. H. Bristol
The Foxboro Company
Foxboro, Massachusetts

Abstract

A pattern recognition-based adaptive control concept (termed EXACT [1]) has been commercialized successfully as a universal successor to The Foxboro Company's current electronic controllers. The delay in market entry was a consequence of a design, incredible (for different reasons) to both academic and practitioner. Yet it evolved not as a regression from the now standard, theory motivated, model-based designs, but as an artificial intelligence related advance over them, suggested by studies on the effects of mismodeling. In this regard, it offers superior adaptation through direct performance feedback. In the absence of a general adaptation theory, an essential element contributing to the development of the EXACT concept was an experimental analysis technique which permitted rigorous demonstration of the general applicability of the method.

Its current success is also derived from a clear understanding of the fundamental role of the PID controller in process control, a role that has allowed the PID controller to prevail and outperform practically its many proposed competitors, each of which, by itself, exhibits apparently superior theoretical performance. The distinctive character makes the design an interesting case study subject of the interplay between market, technology, and innovation. The EXACT Controller has been recognized by three commercial awards.

1. Introduction

The EXACT pattern recognition adaptive controller [1] is a recently commercialized, general-purpose, PID-based adaptive controller for process control. Unlike most current model-based adaptive designs, it responds to directly observed deviations from desired control performance. It feeds back a response shape error using refined tuning rules not basically different from those used by experts in the field. In this sense, it is an early example of an Expert System. The design differs from related model reference designs in that the pattern features used to measure shape are normalized so that they always force a feasible control.

Its success is derived from both technical and market factors: the nature of process control, its processes, disturbances, and operational needs. The design emphasizes predictable, trouble-free, basic control as the major benefit of adaptation. It supports the normal role of the PID controller as the control of first choice and a module extendable to suit more complex needs. In this respect, it sidesteps the more usual, but overstated, view of adaptation as a tool to improve control of economically critical processes. Economically critical processes can motivate better engineering understanding and will benefit more from careful, fixed, nonlinear, multivariable design.

Instead, EXACT is oriented to the larger benefits coming from the systematic predictable trouble-free system maintenance of a fully adapted control performance for the entire plant. The requisite trouble-free adaptation has always been the basic design goal of the different pattern recognition designs. This notwithstanding, some EXACT applications have shown startling payback.

[1] EXACT is a trademark of The Foxboro Company.

Technically feasible adaptive process control designs date back to the early 1950's, [2] but their realization has been subject to a number of dilemmas. The full implications of the supporting digital system technology, in all its roles, has taken many years to master. The absence of accepted relevant theory has prevented the practical experience which would have led to deeper application. It is the continual academic promoting and probing which has kept interest alive and created the competitive environment which brought the EXACT design to the surface. But the emphasis on a purely theoretical resolution has been misguided.

The absence of a relevant accepted adaptive theory has not prevented the necessary practical understanding, but it has made communication of that understanding quite difficult and bred unnecessary and confusing rediscoveries. On the other hand, most experimental work has been completely without formal structure. A more integrated combination of structured experiment is needed to address a diverse reality beyond possible current theory by itself. In fact, a systematic experimental analytic technique, general and robust enough to take the place of a formal theory, ultimately made the current EXACT control design comprehensible and possible.

The pattern recognition adaptive control concept was initially developed in the mid-1960's but faced its own commercialization dilemmas. It overcame experimentally observed deficiencies in model-based designs [3,4,5]. The direct feedback measurement of performance guaranteed a convergence to intended performance without regard to errors in model structure.

Unfortunately, most adaptive work has avoided rigorous examination of performance; hence the practical consequences of mismodeling are not generally understood. Because the EXACT Controller solves these problems, as what is now called an "Expert System," in terms of tuning experience not normally presented in theoretical terms, this concept was unnatural to those with a theoretical background. On the other hand, the practical people were generally not used to seeing their tuning practices embedded in the complex and interacting form necessary for robust adaptation.

2. The Process Control Environment and the PI or PID Controller

The process control environment has a number of basic characteristics that PI and PID controllers respond to:

1. One-of-a-kind process designs for which an elaborate modeling and debugging history is not usually available. Such designs require simple, tunable, general-purpose controllers.
2. Processes are high order and nonlinear but generally self-stabilizing with smooth, monotonic, S-shaped step responses. These arise naturally because of the nonlinear partial differential equations underlying most processes. Under practical circumstances, unstable or difficult-to-control processes tend to breed themselves out of the technology. For the more usual examples of single-loop processes, the PID is near optimal.
3. The economics and technology of control actuators require that they be sized without excess range. The nonlinearities of the process are on the same scale as the disturbances; the assumption of local linearity is only partially valid. Since the actuator limits are on the same scale as the disturbance, the control benefits little from high-order dynamic correction.
4. Major disturbances are the transient consequences of human or natural events (Poisson, not Gaussian!).
5. Low-level Gaussian measurement noise disturbances are on the same scale as actuator hysteresis and unlikely to give reresentative model identification. For both these reasons, adaptation favors the use of transient responses.

The PI or PID controller (in practice, the PI is four or five times more likely to be used) has general-purpose characteristics convenient to this need:

1. The PID is a simple approximation to all standard control forms - a minimum state controller. (However, this fact makes rigorous PID adaptation with analytical methods difficult.)
2. Most industrial single-loop processes, not optimally controlled by the PID, are, nonetheless, robustly and stably controlled by it.
3. Eighty percent of the control loops in the process are so much faster than the process time constants that their precise tuning would have no economic effect. In this case, predictable responsive control is enough.
4. The PID combines well with other nonlinear and multivariable control elements to achieve any form of complex control. The PID is the "AND gate" of control, combinable by its own "De Morgan's rules" to accomplish any objective, with any desired level of performance.
5. The tuning of the PID to fit arbitrary processes and disturbances is well-understood. This expertise is the basis not only of our "Expert System," but of our user acceptance.

The qualitative character of our process and control understanding leads to effective control designs because the design principles are essentially tautologous. Consider Ziegler, Nichols: If one drives a process loop to the lowest unstable gain and divides the result by two, then, of course, the control is stable and acceptable. If the Integral time is set wider than the resonant period, then, of course, it introduces little destabilizing phase shift. And if the Derivative time is set in proportion to the Integral time, about the resonant time, then, of course, it will offset any undesirable Integral phase shift.

Practical experience has made the tautologies well-known to working engineers. They fully define the behavior of the usual processes, and can be refined to tune these processes to any desired performance. In the pattern recognition designs, these tautologies are refined and automated. They provide its basic robustness in the face of the high process order and nonlinearity.

3. The Mismodeling Experiments

The experiments which led to the pattern recognition concept were carried out on a successful on-line adaptive [3,4] design, shown in Figure 1. Identification was based on a direct and continuous solution of the convolution sum model used more recently for the off-line tuned IDCOM and DMC systems. The control of parameter memory and forgetting was carried out by an averaging of the parameter value with a time constant derived from a calculated sensitivity of the solution. The design was too successful! It worked in the face of extreme modeling error! Modeling error in this case arose from the process nonlinearity and incorrectly assumed worst case process settling time.

Two consequences were observed:

1. Faulty initial identification.
2. A convergence to satisfactory control performance sometimes accompanied by continuously varying tunings as the system traded the control parameters against the implied desired control performance of the complete design. Of course, some adopted process loops diverged rather than converged.

The convergence to suitable control was possible because the all-zero model was capable of mimicking the higher control frequency behavior closely under the closed-loop conditions. Accordingly, as the control system's operation excited these frequencies, the adaptor returned an identification consistent with good control under proper settings. But in the end a good

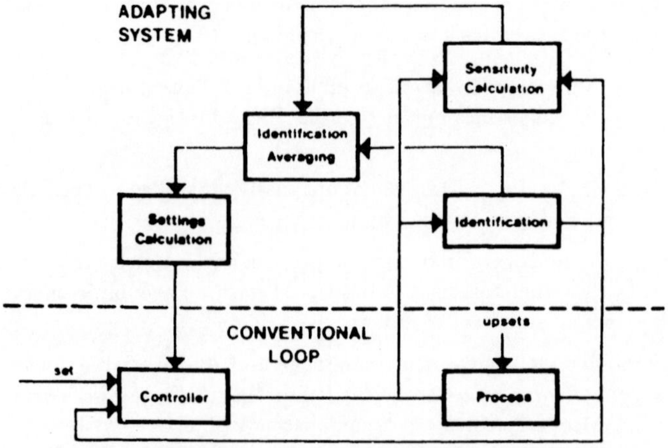

Figure 1: Adaptive control system

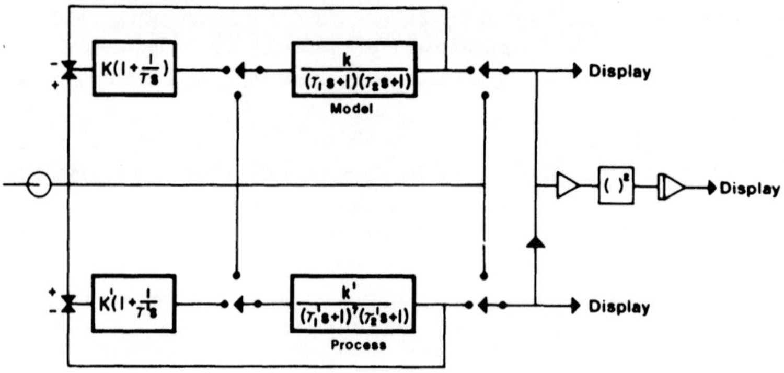

Figure 2: Test setup comparing two loop with different process dynamic structures .

commercial design must be understood in success as well as in failure. And complexities not understood may not be needed.

Later experiments [3,4] designed to demonstrate the earlier mismodeling issues showed similar behavior (Items 1 and 2 above) when adaptation was abstracted as least squares fitting of a model in a loop against the identically controlled process loop in Figure 2. Here, successful convergence was a consequence of the fact that the system forced a qualitative match of the closed loop, but the resulting identification equations were nonlinear and not suitable to adaptive design.

The earlier experiments justified pattern recognition as a method which bypassed the now invalidated analytic complexity in favor of the more easily implemented and studied pattern convergence process. Each transient was summarized by several pattern features. Properly chosen features could clearly summarize all significant response data characteristics.

Given the qualitative basis of accessible analysis, it was essential to reduce the dimensionality of the problem to allow effective experimental analysis. The convergence could then be studied experimentally in terms of the much simpler finite dimensional function space between pattern features and controller tunings. Since the features adequately summarized the data, their functions adequately summarized the adaptive problem. Practical experience predicted that these functions would be monotonic and easily solved. In the serious commercialization, we began to study these functions more formally in order to refine the

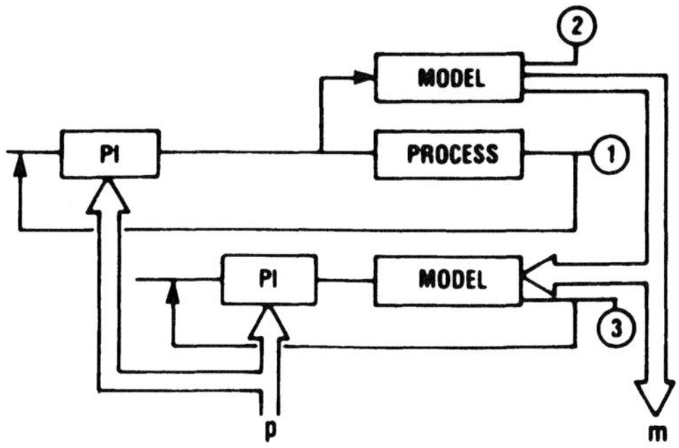

Figure 3: Idealized self-tuning regulator.

design.

When these same ideas were applied to study convergence of the self-tuning regulator, a new convergence difficulty was observed [5]:

3. Convergence to unacceptable control performance could arise as a consequence of mismodeling.

In Figure 3, the self-tuning regulator structure matches a model excited by the process loop against the process (Signals 1 and 2). The resulting fit emphasizes a low-frequency model match which may poorly predict the behavior of the same model in its own closed loop. The resulting poor adaptive convergence is believed to be a more important factor in the well-known nonminimum phase adaptation difficulties, than is the absence of simple relations between nonminimum phase models and their required controls. Current literature is beginning to show analytic examples of the possible difficulties [6].

4. The Exact Controller

In the evolution of the pattern recognition concept, there were two designs: the initial conservative design of the author [2]; and the more refined commercial design of the development group defined by Kraus et al [1] which evolved after the techniques for experimental analytic validation became available. The techniques differed in several respects:

- The treatment of noisy feature signals
- The treatment of resonant and damped responses
- The treatment of unusual response shapes

The earlier design emphasized noise insensitivity, the ability to specify highly damped and undamped responses, and the ability to experimentally tune the adaptor. The later design emphasized far-reaching convergence and minimal user intervention.

Both designs track the development of the disturbance response in terms of a state diagram, "parsing" the response in the same manner as a compiler parses a character string. In the state diagram of the earlier design, shown in Figure 4, the system first seeks a disturbance, then its initial error peak, and then its overshoot and so on [3]. The early design was based simply on a well-known tendency ("tautology"): undamped loops stabilize on decreasing gain. Thus the magnitude of the signal in the part of the recovery corresponding to the normal overshoot peak of a transient (Figure 5) tends to vary monotonically with the controller gain. Similarly, the overall recovery tends to vary monotonically with the integral time. This allowed closure of two simple adaptive loops about the two control parameters, which updated control every time a transient settled.

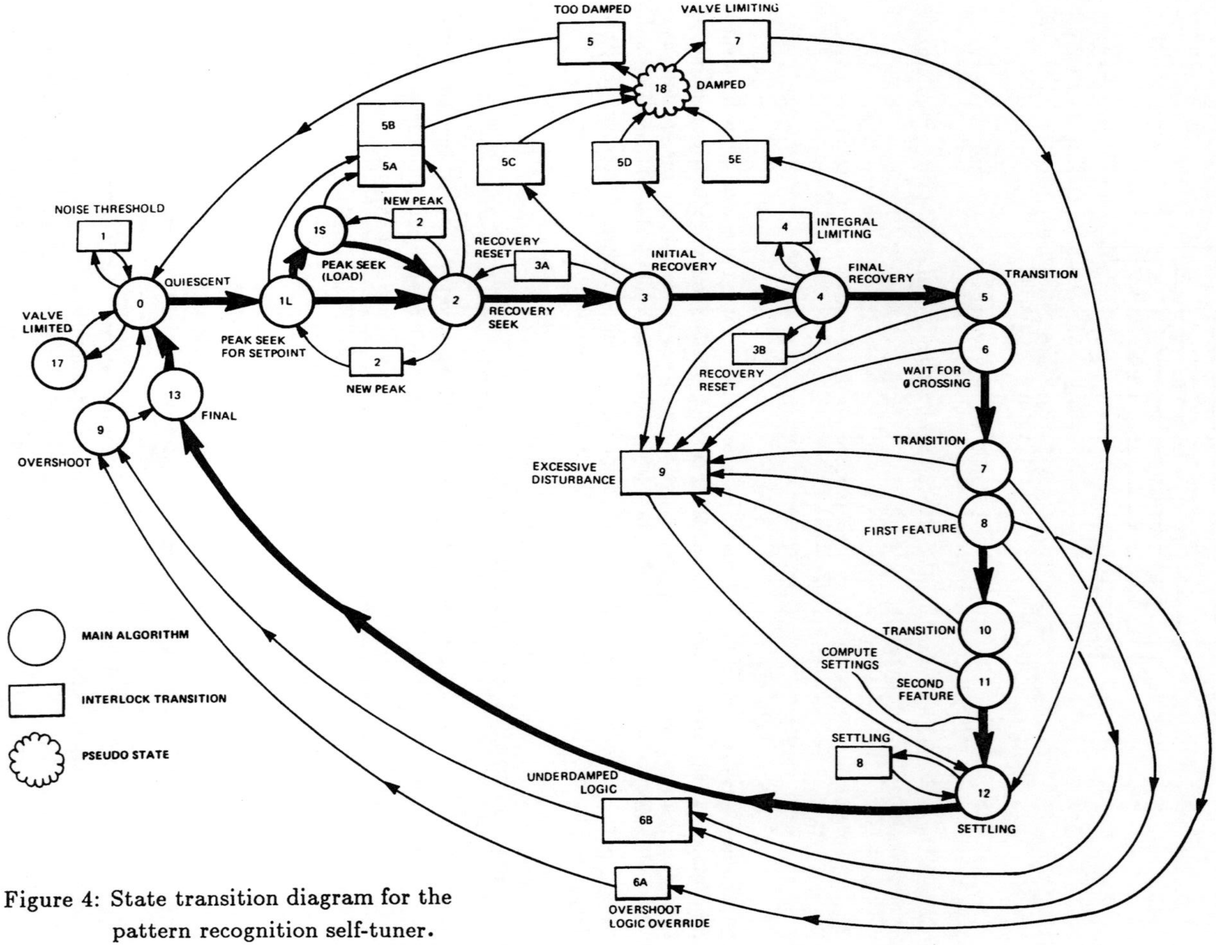

Figure 4: State transition diagram for the pattern recognition self-tuner.

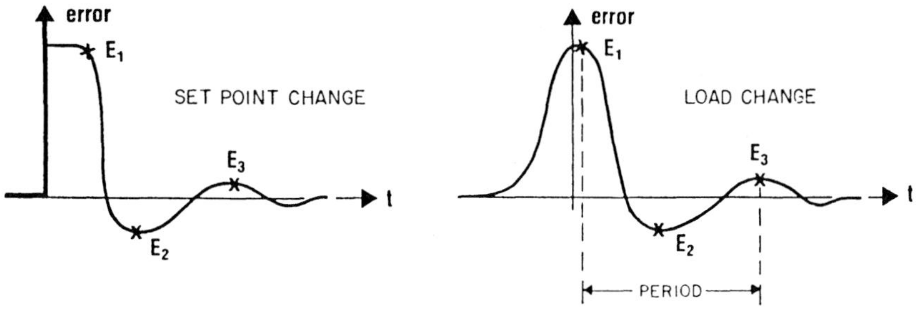

$$\text{OVERSHOOT} = -\frac{E_2}{E_1} \qquad \text{DAMPING} = \frac{E_3 - E_2}{E_1 - E_2}$$

Figure 5: Transients

As experiments justified other process independent strategies, these adapting loops were evolved to include decoupling and nonlinear compensation. Similarly, derivative tuning was included. It was obvious that the behavior of the adaptive loops would be nonlinear and occasionally nonmonotonic outside the normal, nearly optimal, parameter range. Initially, these situations were addressed by identifying them with additional feature tests within the state diagram.

Interlock actions then directly altered parameters to bring the behavior within normal boundaries. For example, interlocks recognizing excessive overshoot or persistent instability or offset all forced immediate, large corrective parameter changes. Eventually, there were a dozen distinct situations (Figure 6), each addressed by rule in the state diagram (Figure 4). The rules generalized standard tuning attitudes such as Ziegler-Nichols tuning.

When the complex interactions of the adaptive feedbacks and interlocks became difficult to explain, we began to develop systematic plots of the mappings between control parameters and the pattern features for differing processes and disturbance conditions, as in Figure 7. Similarly, we generated plots for different processes and load conditions, with and without derivative control, showing the regions of activation of the different interlocks (Figures 8-11).

The plots are included for illustrative purposes only, since the complete design definition requires many plots. Figure 7 examines two features and two control parameters under two kinds of disturbances applied to a 5-lag plus dead-time process. The axes of the plot are the Proportional and Integral control parameters. The contour lines represent lines of constant feature. The left-hand mapping is under load disturbances; the right-hand mapping, under reference changes. The point of the plots is that with careful design, they are easily backsolved to derive tuning parameters from the features. The design problem is to identify strategies that apply equally well over the entire range of such plots.

Figures 8-11 show the high variability of the regions defined by interlock tests. Since these were the basis of qualitative actions only, this was satisfactory.

These plots allowed systematic study of the convergence character of the system, and finally convinced the members of the development team of the practicality of pattern-based adaptation. Because the plots were simple, they could be analyzed intuitively. Modifications based on the "tautologies" of experience could be proposed and tested for their process generality over a wide range of process/disturbance situations. The proposals could be motivated by theoretical insights, but did not depend on the theory for validation. In this way the necessarily greater generality was achieved.

At this point, the current EXACT concept began to evolve largely under Kraus [1]. Early measures, based on overshoot and recovery, were replaced by more intuitive measures

PATTERN ADAPTATION INTERLOCKS

0. MAIN PEAK SUPERSEDED
1. TOP OF PEAK TOO BROAD
2. RECOVERY TOO SLOW
3. INCONSISTENT WITH ZIEGLER NICHOLS TIMES
4. PEAK TOO BROAD
5. RECOVERY FLATTENING TOO SOON
6. RECOVERY TOO UNDERSHOT
7. EXCESSIVE OVERSHOOT

8, 9. TOO UNSTABLE WITHOUT OVERSHOOT

ALSO ABORTS ADAPTATION IF:

OVERLAPPING DISTURBANCES

OUTPUT LIMITING

Figure 6: Table

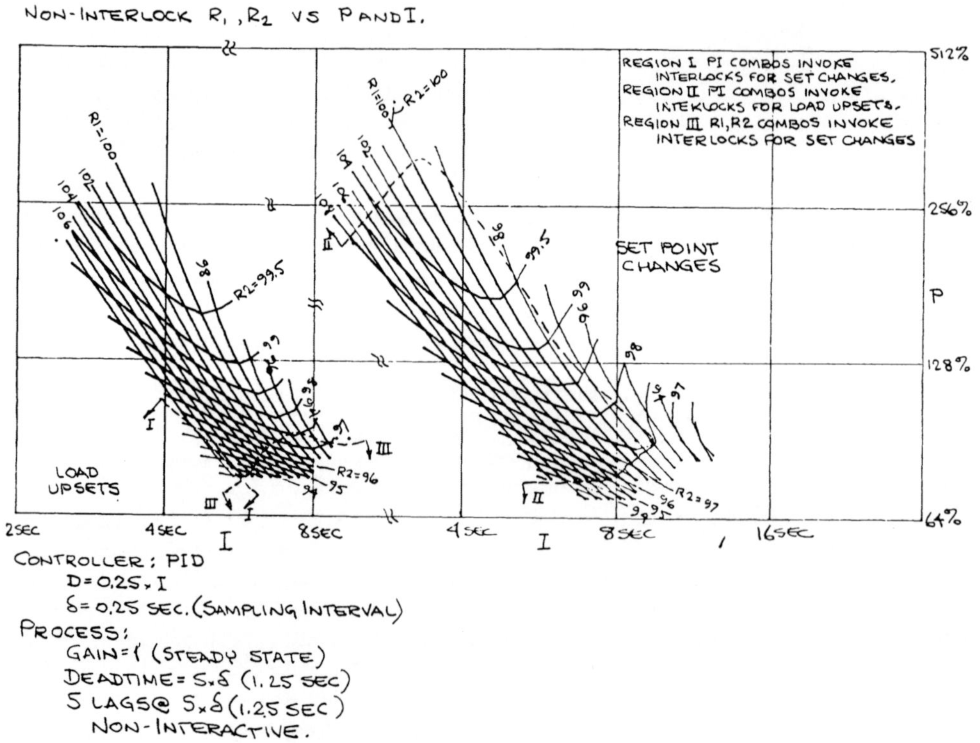

Figure 7: Tuning/Features Map

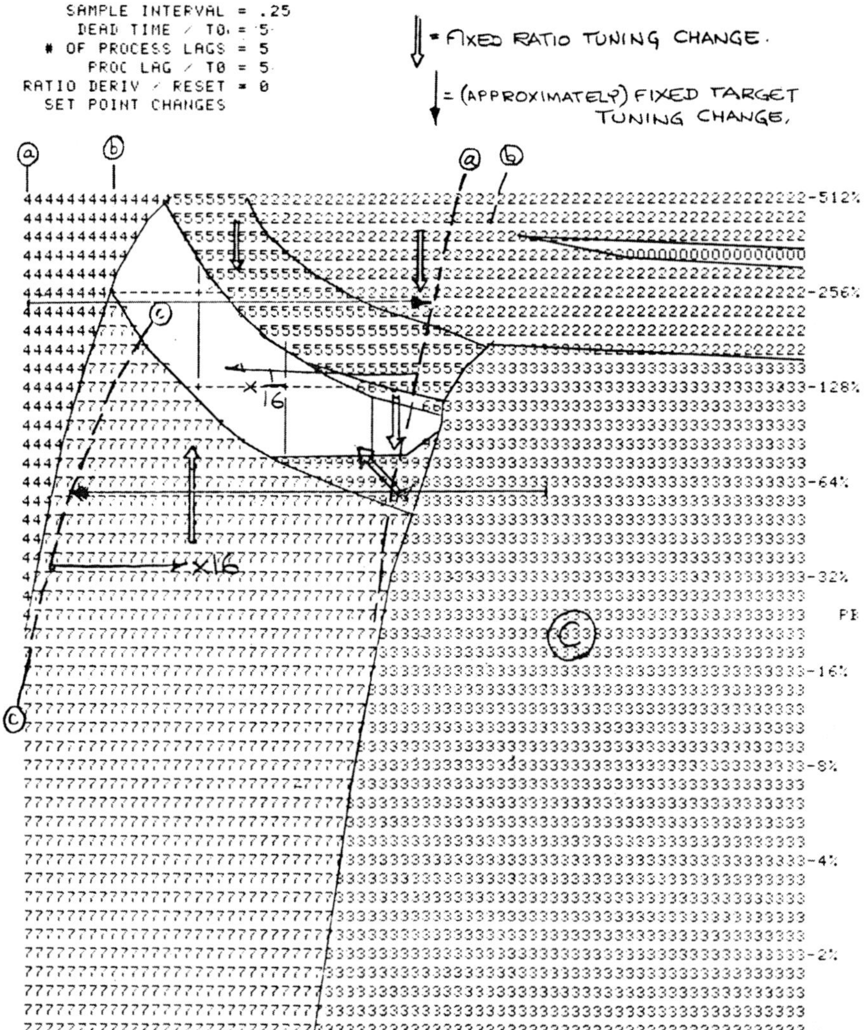

Figure 8: Interlock Map

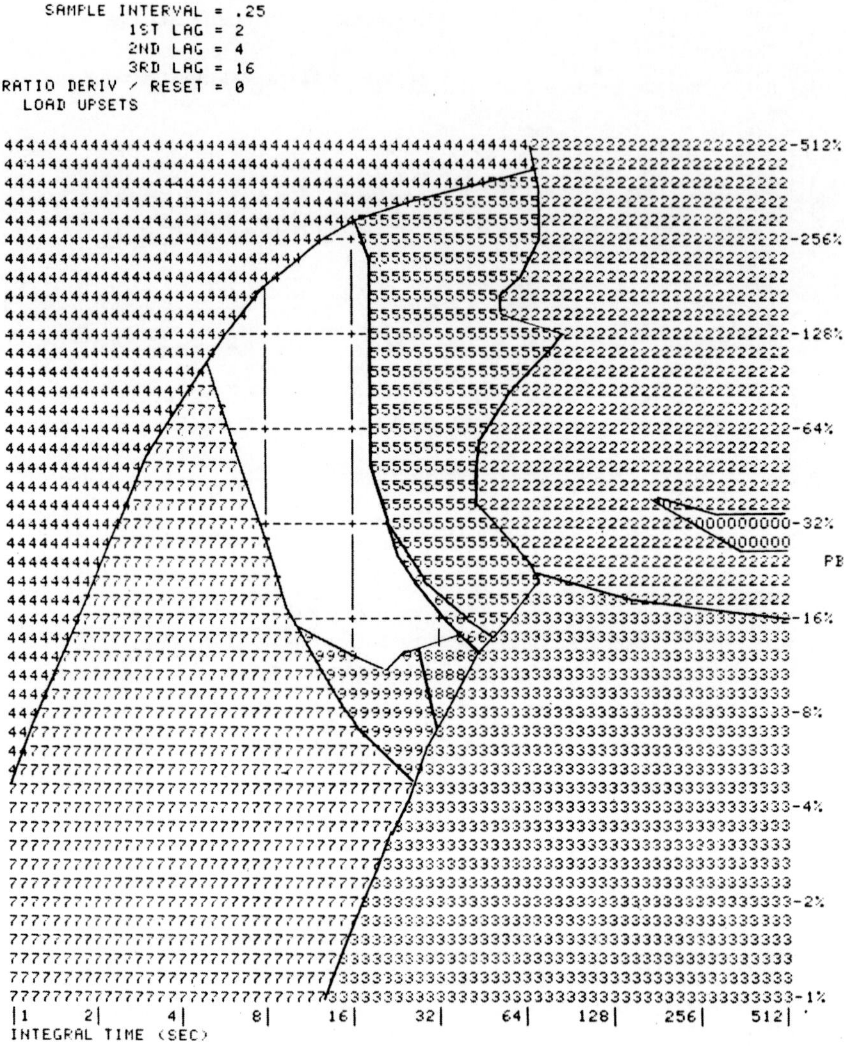

Figure 9: Interlock Map

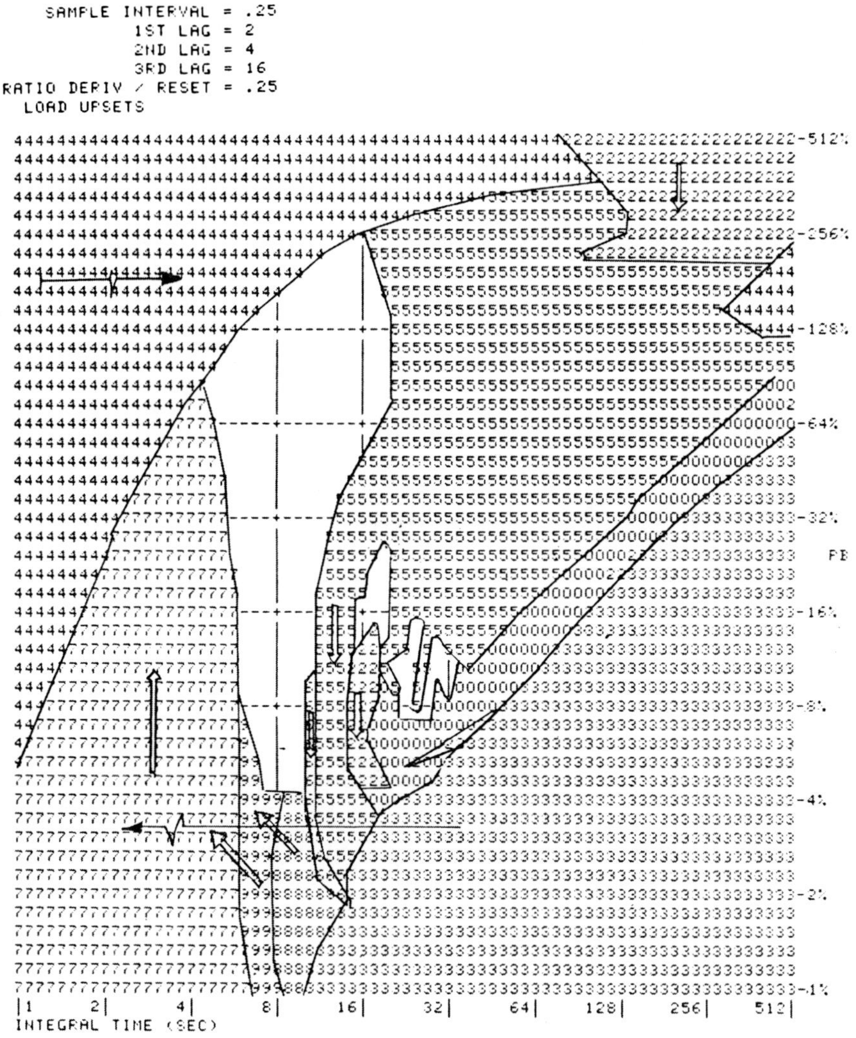

Figure 10: Interlock Map

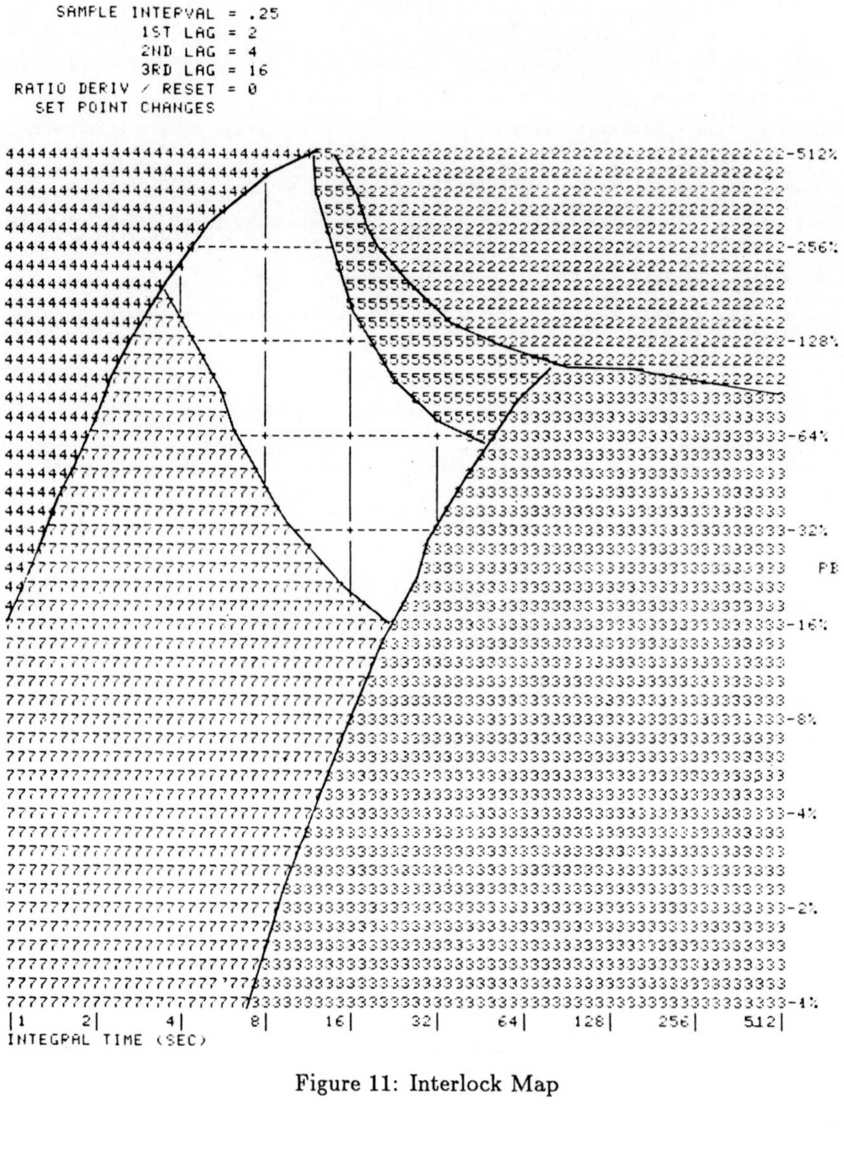

Figure 11: Interlock Map

of overshoot and damping, which were used to define two performance constraints that the user was able to impose, at levels he chose, on the running algorithm. A series of rules for setting integral and derivative parameters was developed, based on phase and period arguments. The interlocks were minimized in favor of improving the basic adaptive loops.

The results were validated by generating hundreds of plots such as those in Figure 7, and computing for every feature state the corresponding convergence trajectory under the proposed algorithm. This process was repeated for an essentially continuous range of processes, starting with a pure integrating process and ending with a pure dead-time process, to show the practical invariance of the key design properties.

A dozen or more field trials (now extended by hundreds of commercial installations) further validated the design, including special situations not previously covered: finite right-half plane zeros, cascaded loops, etc. Some of the benefits of the earlier design were set aside, but in general, the effort represented an order of magnitude improvement.

The present design is parameterized in a number of dimensions: initial settings, noise threshold, maximum settling times, maximum allowed parameter changes. These values can be defaulted, automatically set by the user, or set by the system in an automated pretuning procedure. None of these choices is usually crucial to successful tuning. But they allow greater user control and application refinement.

The system recognizes and accommodates overriding disturbances and output constraints. The system is regularly in the hands of unsophisticated users and may be applied by people without formal control engineering backgrounds. There is a natural fear of something not understood, and confidence requires the knowledgeable user. But, in this case, success doesn't. The EXACT Controller replaces and appears to the operator like any PID controller. The standard controller faceplate is unchanged. The unit is provided with a hand-held configuring device which allows the interested user to follow the EXACT Controller's "thinking." In the future, more centralized designs and more elaborate operational uses of this information are likely.

5. Theoretical Outlook

One might argue that the pattern recognition design is an eccentricity not requiring any permanent theoretical or academic response. But the comparative robustness of the design and the kinds of systematic evolution possible from its associated methodologies argue otherwise. Obviously a quite different kind of theory would be needed [4]. The success of the design suggests that this theory would help adaptive thinking in general.

Part of the problem is that there are really three different kinds of theory required. There is a need for a foundational theory which studies why adaptive control should work in the high-order nonlinear reality. There is a need for teaching theory which makes it possible to provide some mnemonic basis for discussing adaptation generally to the future user and specialist alike. And there is need for a design theory which allows the knowledgeable designer to advance the practical art in the face of experimental reality and competitive refinements. It is natural for the teacher to confuse these in developing the subject with students, but the requirements are quite different.

A foundation theory for adaptive process control and the EXACT design would address two different topics. On one hand, there is a need for a functional analytic theory which once and for all explains the success of linear theory in qualitatively explaining feedback control of our high-order nonlinear processes. I believe that it will turn out to be possible to usefully generalize most of the results applicable to our monotonic processes in linear theory, to their natural, smooth, nonlinear, infinite order counterparts. The necessary results will have to take a topological rather than computational flavor. Their value will be philosophical. Why do the tautologies work so well?

The second foundational issue is the information theoretic study of the adaptive mappings. Let it be given that experiment or theory derives mappings between control parameters and meaningful response features, varying somewhat with process. What are useful ways of cataloging these mappings and their consequences? [4]

For the pattern recognition design, a teaching theory is likely to be simple. It might consist of a body of examples which illustrate the natural tendency of processes to behave in certain broad ways under tuning. It would include examples of exceptional behavior covered by a practical design.

The role of a design theory is filled by the already discussed design mappings. The role of series approximations in traditional nonlinear theory is not appropriate for a pattern recognition design theory. Computer simulation gives adequate data in easier and more general terms. And the design requires generality. For now, the strategy of testing and generating mappings is effective. It naturally lends itself to computerized automation based on a large data base of simulations, mappings, and potential feature calculations. Effective automatable tests for useful mapping generality would expand this art. A design theory of this sort is necessarily specialized to the particular adaptive design that it is refining. The earlier paper [5] showed the potential value of such a methodology even for analyzing model-based designs.

6. The Future

All of us have our list of required commercial advances in the field:

- Difficult single-loop processes such as dead time
- Multivariable processes and feedforward
- Nonlinear processes

Dificult single-loop processes turn out to be a red herring. The PID is nearly as good for most of these situations as any nonconventional form. Some of the reasons for this were explained earlier. Experiments with obvious candidates (i.e., the Smith Predictor Controller for dead-time control) to improve on the PID typically improve the control criterion by 10 percent or less and then under very special disturbance circumstances. Even so, pattern recognition has already been used to adapt for dead time [7]. And the pattern recognition concept was originally successfully tested on non-PID control structures, including Bang Bang control, and is easily extended to any others.

Nonlinear processes can be explicitly adapted and controlled in terms of their nonlinearity. Or the linear control settings can be stored as a function of operating point. In either case, the result is better than having to relearn optimum settings for each operating point, particularly for multivariable processes.

Adaptation of more complex processes is difficult in part because of complex human requirements that an adaptive system cannot be expected to anticipate. These requirements may become so varied as to be beyond the capability of any "canned" technique, requiring human design rather than adaptation. Thus, effective adaptive designs must allow the configuration of complex systems out of simple self-adaptive modules.

The pattern recognition concept lends itself to modular combinations and does not depend on the analytic properties of the elements to which it is combined. Thus, it is as flexibly combined into multiloop and feedforward combinations as the PID itself. And it is robust enough to operate under the resulting interaction, without special accommodation.

The multivariable distinctions that the practical control makes between single-loop feedback, feedforward, and decoupling beneficially affect practical adaptation because these functions have quite different mismodeling effects. Feedforward and decoupling are generally insensitive to high frequency effects, whereas feedback is insensitive to low frequency effects. Thus, the characteristics of pattern feedback and model-based feedforward adaptation can be made to complement each other.

We would hope that this commercial success with adaptation will have a positive effect on the forms of theory and pedagogy brought to bear on the field. Its underlying knowledge is real and manageable, and eventually reducible to theory. Such a theory would be derived from the study of the mappings and from the underlying information set theoretic relations indicated in the references. [4,5]

From a larger perspective, the academic world could benefit in improved control and adaptive teaching by greater theoretical and pedagogical attention to the practice of manual tuning. Practical feedback control understanding is largely based on this practice. It is an important source of readily accessible and extremely general knowledge that the academic community has ignored or rejected.

Apart from the valid forced experience of seeing the eigenfunctions actually occurring live, tuning is the control equivalent of computer program debugging, the one being as essential to its area of practice as the other. A control model is as unavoidably error-prone as any first draft FORTRAN program. But, whereas debugging may perfect the FORTRAN program, model errors are a permanent consequence of our theoretical approximations. No calculations based on our models will derive the best performance, and manual tuning and its automatic equivalent as embodied in EXACT control will remain a central practice.

7. Concluding Comments

Adaptation is here. But the opportunities go beyond analytic adaptations. Some of the richest possibilities will come from automating practical human common sense. In this, the field of Artificial Intelligence comes closer than generally recognized to defining the necessary flexible models.

In process control, most users will prefer flexible systems put together out of small, independent, "smart" modules, rather than inflexible monolithic designs, however effective. In this case, successful commercial adaptation will avoid a dependence on techniques which depend on global models and strict interdependencies. Instead, it will develop techniques which allow related adaptive controls to be set up independently.

Commercial efforts will develop other kinds of control intelligence and will be designed to accommodate other operational issues: intervention by human beings and higher level systems. Adaptation by pattern recognition illustrates the possibilities.

References

[1] T.W. Kraus and T.J. Myron, "Self-Tuning PID Controller Uses Pattern Recognition Approach," *Control Engineering*, vol. 31, No. 6, pp. 106-111, June 1984.

[2] E.H. Bristol and T.W. Kraus, "Life with Pattern Recognition," 1984 ACC, San Diego, Ca., June 6-8, Session TP3, pp. 888-892.

[3] E.H. Bristol, "Adaptive Control Odyssey," ISA Conference, Philadelphia, October 1970, Paper 561-70.

[4] E.H. Bristol, "Pattern Recognition: An Alternative to Parameter Identification in Adaptive Control," *Automatica*, vol. 13, pp.197-202, March 1977.

[5] E.H. Bristol, "Experimental Analysis of Adaptive Controllers," Third Yale Workshop on Applications of Adaptive Systems Theory, Center for Systems Science, Yale University, New Haven, Ct., June 15-17, 1983, pp. 249-253.

[6] C.E. Rohrs, G. Stein and K. Åström , "Uncertainty in Sampled Systems," 1985 ACC, Boston, Ma., June 19-21, 1985, pp. 95-97.

[7] H.J. Chizeck, G.I. Voss and P.G. Katona, "A Self Tuning Controller for Nonminimum Phase Plants," Internal Memo, Case Western Reserve University, Cleveland, Oh. 44106, November 1984.

Self-Cohering a Large, Distorted Antenna for Microwave Imaging

Bernard D. Steinberg
Valley Forge Research Center
Moore School of Electrical Engineering
University of Pennsylvania
Philadelphia, Pennsylvania

Abstract

A large phased array designed for microwave imaging consists of many independent antenna elements each weighted appropriately to provide the desired radiation pattern, and the weighted outputs are summed to form the system output. The arguments of the weights determine beam direction. The argument is a sensitive function of element location in the array and phase shifts in the circuits of each element receiver channel. Uncompensated position and/or phase errors result in severe image distortion.

The Valley Forge Radio Camera is a large imaging microwave radar. Its 83 m antenna is very flimsy, highly distorted and nonrigid. The distortions are severe, bordering on 1 wavelength rms in the three position coordinates at the operating wavelength of 3 cm. An adaptive subsystem dynamically calibrates the array based upon measurements made of the radiation field at the array due to reflections from targets. Described is one of the algorithms used, the nature of the experiments conducted and the evidence of the validity of the adaptive beamforming method. High resolution microwave images of a town in the neighborhood of the laboratory are shown.

1. Introduction

A data adaptive system modifies the algorithm or its coefficients in accordance with the data upon which the system is operating. In an open-loop data adaptive system, control is exercised by the input data. In a closed-loop system, control is exercised by the output data, that is, control is performance driven.

At the Valley Forge Research Center of the University of Pennsylvania, data adaptive techniques have been applied for the purpose of achieving unusually high resolution microwave imagery. Rather than the usual time-domain control, in this problem signal processing and adaptivity are performed in the spatial domain. The high resolution imaging system is called a Radio Camera.

The system input has many ports or antenna elements. The signals that drive the adaptive process are samples of a complex radiation field; hence, the system may be called radiation-field-adaptive.

Both open- and closed-loop algorithms have been designed to accomplish the objective. This paper concentrates upon a single, successful open-loop algorithm. It is based upon a minimum-variance data search and a coefficient adjustment that results in making the radiation pattern of the antenna array the angular response of an approximate matched filter to the spatial reflection properties of the external source, called the beamformer, whose radiation drives the system.

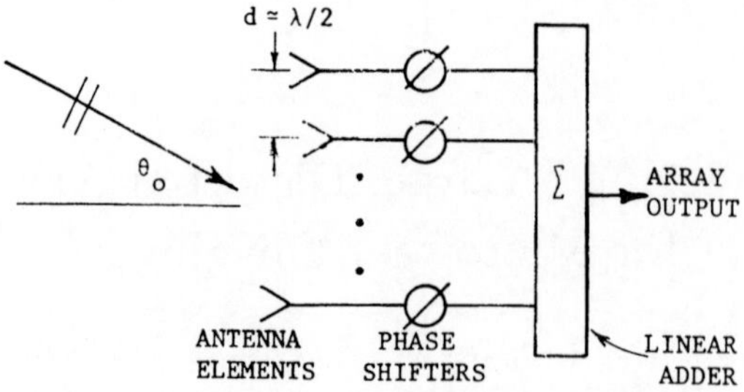

Figure 1: A plane wave arrives from direction θ_o relative to the array normal. The phase shifters cophase the signals at the adders. The differential phase shift between channels is $\Delta\phi = 2\pi d sin\theta_o/\lambda$.

2. Background

Diffraction theory teaches that the available angular resolution from a lens or mirror or microwave dish or any other transducer is approximately fixed by the number of wavelengths of the radiation across the aperture. The familiar expression $\Delta\theta(rad) \cong \lambda/D$ gives the beam cross section $\Delta\theta$, radians, of the radiation pattern of an aperture of size D when operating at wavelength λ. The human eye is a few thousand wavelengths in size. A microwave antenna as part of an imaging system designed to achieve comparable resolution would have to be hundreds of meters in size. The lens of a small camera stopped down for picture taking in bright sunlight is ten thousand wavelengths in diameter; that of a 2-in optical telescope is one hundred thousand wavelengths. The microwave aperture having equivalent resolution at a typical wavelength such as 10 cm is 10 km, the size of a small city.

Such antennas cannot be microwave dishes; they must be phased antenna arrays instead. A phased array is a linear system with many input ports and a single output port. Each channel consists of an antenna element or transducer which converts a sample of the radiation field to an electrical current, and a phase shifter. The collection feeds a linear summer, the output of which is the array output. Figure 1 is an illustration of such an array. The spacing between antenna elements is customarily fixed, usually approximately one-half wavelength. The elements are customarily deployed on a straight line or in a plane. The purpose of the phase shifters is to alter the phase of signals arriving from some arbitrary direction θ_o, or from some arbitrary point in space $\underline{r}$, so that these signals arrive cophased at the adder. If the signal waveform is wideband, time delays must be used instead of phase shifters.

This cophasing procedure (or achievement of simultaneity when time delays are used instead of phase shifters) is the electrical equivalent of the geometric modification to the phasefront of the radiation field as accomplished by a parabolic mirror or a lens. The curvature of the mirror causes rays arising from a source at a great distance to be reflected to a point called the focus and to arrive there simultaneously. The lens, by introducing differential time delays through the differing thicknesses of the lens, causes rays emerging from its backside to achieve spatial focus and temporal simultaneity.

3. The Problem

Let $\underline{r}'$ be the location of an antenna element in the array and $\underline{r}$ the location of a field point to which the array is to be focused. The phase delay of a wave in transit between these two points is

$$\phi = \frac{2\pi}{\lambda}|\underline{r} - \underline{r}| \tag{1}$$

A phase advance of exactly this much (plus any arbitrary phase that is constant across the aperture) corrects for the varying phase delay from field point to elements in the aperture so that the signals at the adder are cophased. Equation (1), however, neglects the realities of large antenna array design. When the aperture is hundreds of meters to several kilometers in size, the locations of the antenna elements are not accurately known, the refractive index of the medium is not necessarily constant in the region between the array and its focal point, and the elctromagnetic coupling from antenna element to its local environment may also vary from element to element. The result is a phase shift better represented by

$$\begin{aligned} \phi + \delta\phi &= \frac{\omega(r_o+\delta r)}{C_o+\delta C} + \xi \\ &\simeq kr_o + k\delta r - \frac{kr_o\delta C}{C_o} + \xi, k = \frac{2\pi}{\lambda} = \frac{\omega}{C_o} \end{aligned} \tag{2}$$

where the lead term is the expected value of the phase delay given by (1). The second term represents the distance error. The third term shows the effect of random perturbations of the propagation speed due to the refractive index variation. The fourth term is the random effect upon the driving point impedance of the antenna element due to the variation in local electromagnetic coupling. The proper phase shift given by the negative of (2) is only partially introduced when (1) is assumed to be the correct phase delay. The rms tolerance on the error is between a few hundredths and a tenth of a wavelength, which is very difficult to maintain in a microwave aperture large enough for high resolution imaging. The designer, therefore, is intrinsically denied the necessary detailed information to properly design the imaging system. In short, some adaptivity based upon measurements of the radiation field is required to correct the error.

4. The Self-Cohering Procedure

A procedure that has been found to work satisfactorily repeatedly in experimental on-the-air microwave tests uses a minimum echo amplitude variance test to search for a beamforming source and a phase conjugating procedure to correct the errors. The beam pattern that is formed is approximately the complex conjugate of the reflectivity profile of the beamforming source; the adaptive processor is an approximate spatial matched filter [1].

Figure 2 illustrates the concept. An array is shown distorted in the center of the figure: the elements are not colinear and are not equally spaced. A point source of radiating energy, which may be reflecting energy from a radar transmitter, is assumed to radiate via free-space propagation (no multipath) to the array. Assuming that the distance is large compared to the size of the array, the radiation is constant in amplitude in the array and its phasefront is spherical. Measurement of the phase at each element discloses (2) modulo 2π. Differential phase measurements with some reference element in the array, denoted by the black dot in the horn, exactly provide the relative phase shifts needed to cophase the signals arriving from the beamforming source and thereby focus the array on the source. The comparator feedback circuit of Figure 2 provides this function. The second bank of phase shifters is used for open-loop scanning of the beam after the initial self-cohering process. In practice, the two phase shifters in each channel are combined.

Figure 3 schematizes the algorithm and Figure 4 describes its steps.

5. The Algorithm

In the upper left of Figure 3 is a radar transmitter which emits periodic short pulses in the direction of the target area to be imaged, which is shown in the upper right. In the vicinity of the target area is a prominent reflector of large microwave cross section but small physical size. It is modeled as a corner reflector in the figure. The RF echo waveform contains clutter, the reference target echo, and echoes from the target area plus clutter. These waves arrive at the distorted antenna array and are received by antenna elements that drive coherent receivers. That is, the receivers measure both the amplitude and phase, each as a function of time, of the received waves. Other data are sampled and stored in the format shown in the lower right. The abscissa is sample time which is proportional to scatterer range. Each

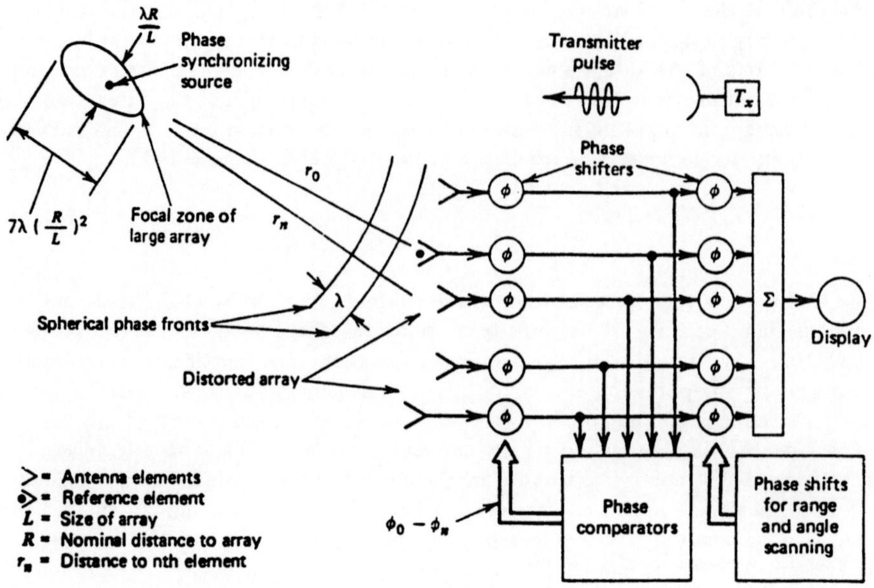

Figure 2: Distorted array self-coheres on a phase-synchronizing source: $>$ = antenna elements; $\cdot >$ = reference element; L = size of array; R = nominal distance to array; r_n = distance to nth element (from [2]) .

row is the complex data sequence from a single antenna element. The ordinate is antenna number which is monotonic with element position in the array. Each data entry is a complex number representing the complex field strength of the radiation field at a particular antenna element at a particular time.

If the phase synchronizing source were truly a point source and propagation were truly in free space, the amplitude of the signal at each receiver would be the same. The phases of course would all differ because the distance from each receiver to the reference reflector is a random variable. The system searches through the data for a spatial echo sequence (at constant range) in which the echo strength is nearly constant. If such a sequence is found, its echo range R_o is identified as the reference range and it is assumed that the source of energy at that distance is approximately a point source radiating in free space. Based upon this assumption, the phases of the signals received at each antenna element may be identified with (2). Hence the negative of these phases are the proper phase shifts to correct for all the distortion sources.

In addition, these measured phases also contain phasefront curvature information. That is, if the reference source were at a greater or a lesser distance than R_o, the measured phases would contain substantially the same error contributions given in (2) but the lead term in that equation would be altered by the exact geometry of the source in relation to the array. This focal-distance effect is easily calculated from the diffraction integral and corrected open-loop, as is seen in Figure 4.

The first step in the procedure following the selection of the reference range R_o is to introduce a complex weight vector into the array equal to the complex conjugate of the column vector at R_o. Doing so cophases all signals arising from the reference source at the adder and thereby focuses the array at that source. The beam is focused without error at the source even in the absence of any knowledge regarding the locations of the antenna elements. This is purely a retrodirective process of adaptive beamforming. It can be shown that the spatial linear filter so formed is a finite sample approximation to the spatial matched filter, meaning that the radiation pattern is approximately the complex conjugate of the complex reflection profile of the beamforming source.

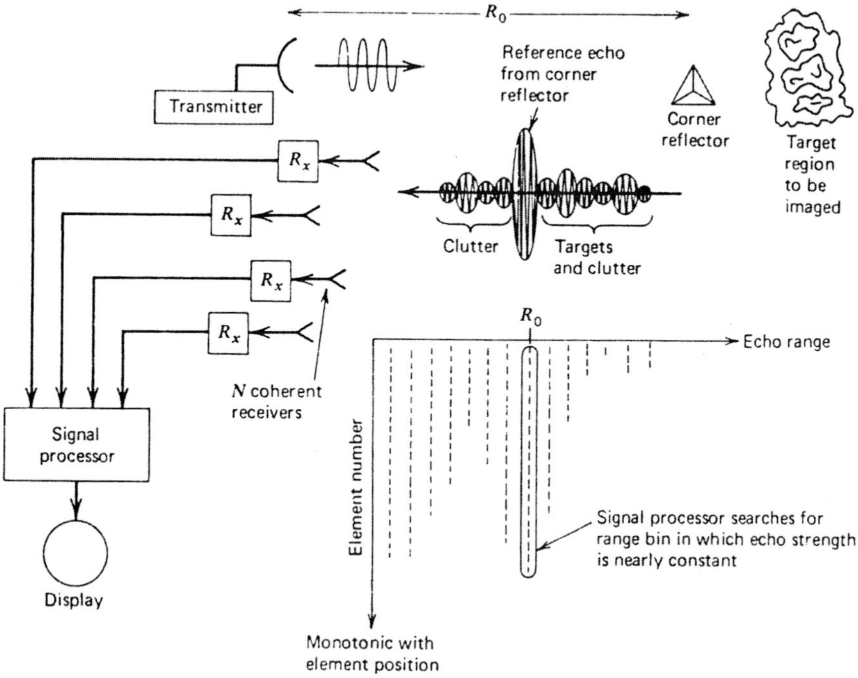

Figure 3: Radio camera data format and procedures

STEP		
1	Measure and store complex envelopes of echo samples	$V_{in} e^{j\psi_{in}}$ range bin ↗ ↖ element number
2	Correct amplitudes by dividing by element pattern estimate $\hat{f}_n$	$A_{in} e^{j\psi_{in}}$
3	Find R_0 such that $A_{0n} \approx A$, all n	$A e^{j\psi_{0n}}$
4	Phase rotate at R_0 by phase conjugate in relation to reference element, $\exp j(\psi_{00} - \psi_{0n})$	$A e^{j\psi_{00}}$
5	Phase rotate at all range elements	$A_{in} e^{j(\psi_{in} - \psi_{0n} + \psi_{00})}$
6	Focus at each range R_i	$A_{in} e^{j[\psi_{in} - \psi_{0n} - \psi_{00} + (kx_n^2/2)(1/R_i - 1/R_0)]} \triangleq B_{in}$
7	Phase shift linearly with angle	$B_{in} e^{-jkx_n u}$
8	Sum at each range element	$\hat{s}_i(u) = \sum_{n=1}^{N} B_{in} e^{-jkx_n u}$

Figure 4: The radio camera algorithm, experiments and techniques. Steps in radio camera imaging (from [2]).

The retrodirective beamforming process is adequate as described for point-to-point communications. It is not adequate for imaging, however. Imaging requires that the beam be scanned in both range and cross range (azimuth) so that echoes from a two-dimensional sector can be used to form a two-dimensional image. The first step toward this objective is to refocus the antenna array simultaneously in every range cell. Because the phasefront curvature built into the weights is easily calculable from the knowledge of the distance R_o to the beamforming source and from estimates of the element positions, the phase variation due to the curvature associated with R_o can be removed, the correct phase variation associated with the curvature from every other range can be calculated and the weight vector so modified for each range. In this way, the system becomes a focused aperture at all ranges simultaneously. It is the equivalent of a large aperture with an infinite depth of field. Following range-focus correction, which is an open-loop process, the array is scanned in angle by conventional electronic phased array scanning. In this way, the beam, focused simultaneously at all ranges, sweeps to the left and to the right of the reference reflector.

The detailed steps are given in Figure 4. An amplitude correction is made on line 2 to account for the nonuniform radiation pattern of the antenna element. The range-focus correction introduced in line 6 is shown as a quadratic function, which is the Fresnel approximation to the spherical phasefront. The quadratic approximation is illustrated here because it is a very simple function to write down; in computer implementation of the algorithm, there is no advantage to using the approximate phase variation. The linear phase scan for moving the beam in azimuth is introduced in line 7.

6. Experiments

Figure 5 shows an experimental antenna array 83 m long. The wavelength used is 3 cm. This array, when measured in units of wavelength, is approximately as large as the human eye. Its structure is a nonrigid cable strung between the towers. A time-shared experiment is conducted by pulling a microwave receiver along a trolley from one end of the array to the other. A radar transmitter, located on the ground under the middle tower, launches short pulses from the antenna mounted on the middle tower. The size of the range cell is 3 m. The reflected echo sequence from a town several km away is received by the microwave receiver as it slowly transits the cable. The microprocessor store on the ground is turned on approximately 300 times during the experiment as the receiver moves from the east to the west tower. The receiver is pulled by clothesline, which is also a nonrigid body, bobbles up and down and sways in the wind. The rms uncertainty in antenna position at any instant of time is the order of 1 wavelength in all directions.

Figure 6 shows an image of a housing development in the town at a distance of 4.5 km. The cross range dimension is 1.8 m, which is about 1/45th of the aperture length. This implies that the distance to this housing scene is about 1/45th of the distance to the transition from the near field to the far field.

The effectiveness of the adaptive beamforming process is demonstrated in the next two figures. Figure 7 shows two streets of houses in the same town. The image was formed based upon the algorithm described above. In Figure 8, the same microwave data were used with the conventional, nonadaptive beamforming algorithm. Whereas the adaptive beamforming process provided a diffraction-limited image of the scene (that is, the best that could be obtained in principle from an aperture of the same size), the nonadaptive imaging procedure provided a useless image in which the image energy was quasi-randomly distributed. The fact that the image of Figure 7 was diffraction-limited was established by careful measurement of the output angle scans (at constant range) of the adaptive processor.

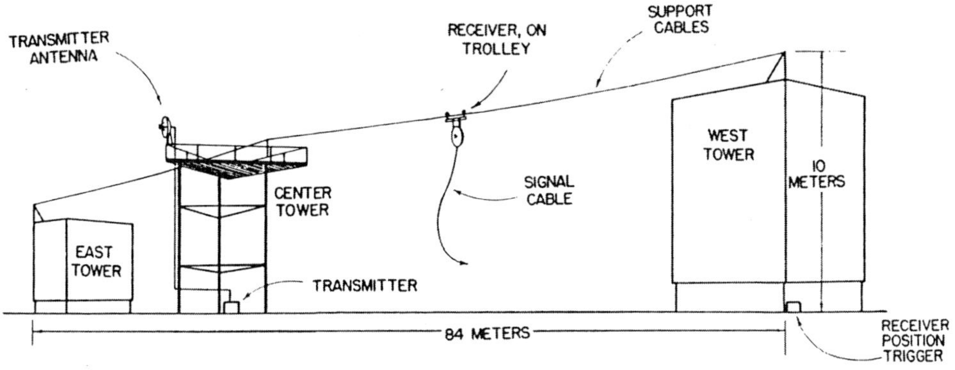

Figure 5: Sketch of 83 m cable-supported x-band array .

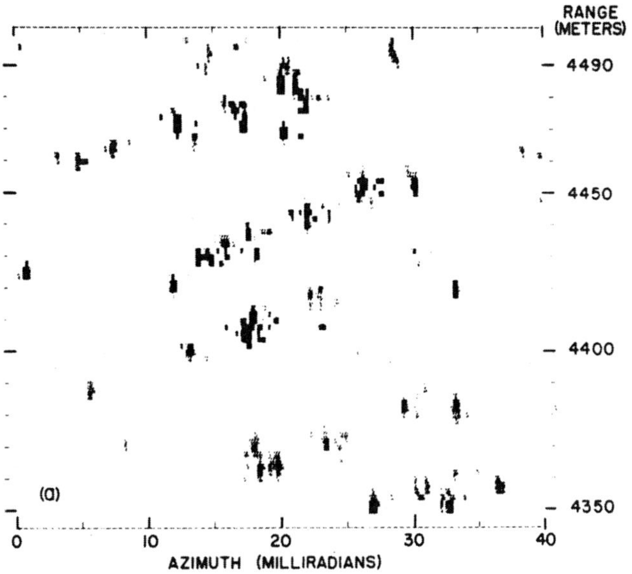

Figure 6: Two-dimensional high resolution microwave images of a housing development of single homes in Phoenixville, Pa. $\lambda = 3cm$, range resolution = 3 m. Cross-range resolution = 1.8 m .

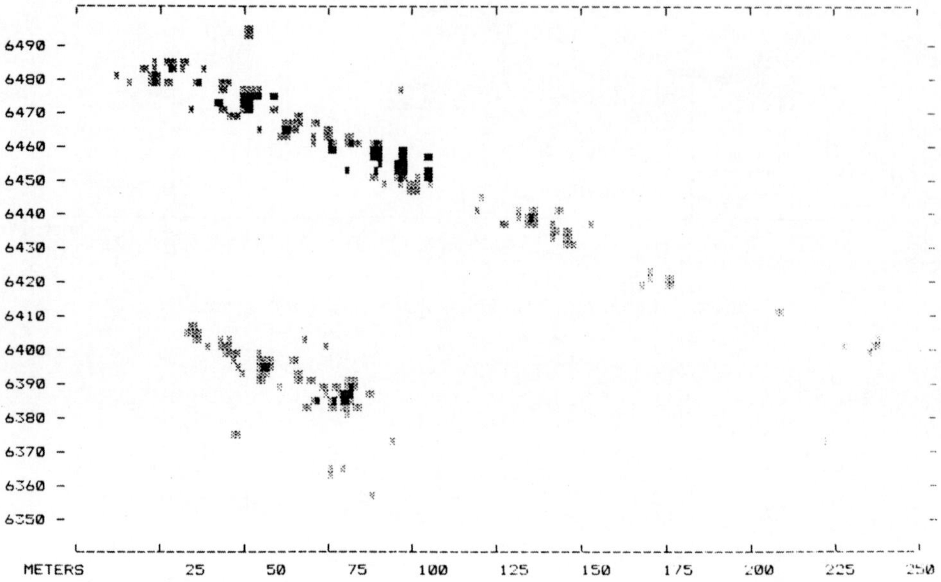

Figure 7: Radio camera images of row houses on streets in Phoenixville, Pa.

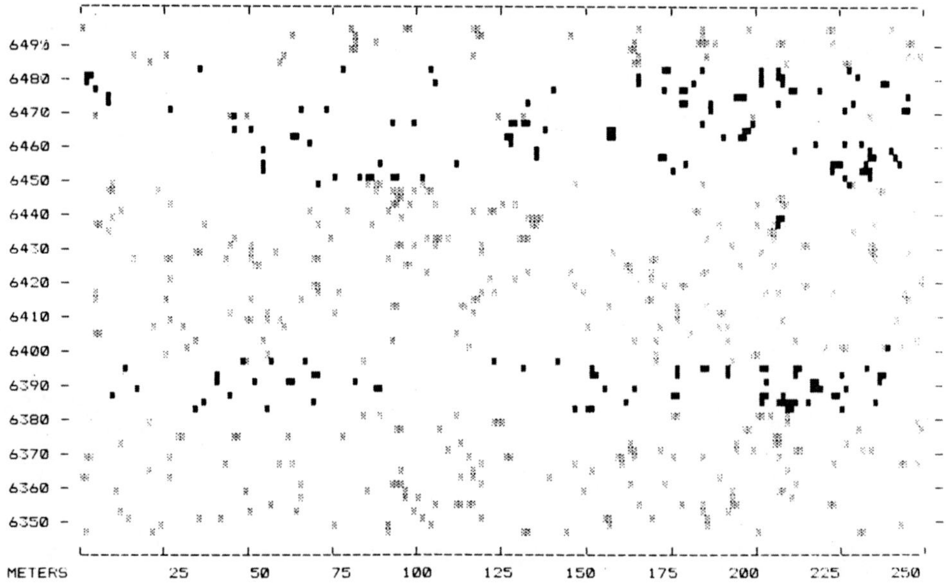

Figure 8: Same echo-trace data as for Figure 7; adaptive beamforming not used .

7. Summary

Data-adaptive techniques can be used to control the complex weights that distort a phased antenna array so as to compensate for the position errors of the elements as well as for electrical sources of distortion due to medium turbulence and system errors. Spatial data consisting of simultaneous measurements of the complex radiation field are obtained from the antenna elements in the array. An algorithm is described that searches through the radar echoes received by an array for echoes from a target that approximates a point source radiating in free space. The complex conjugates of the measured signals are applied as weights in the antenna channels. The beam pattern that results from this process is approximately the complex conjugate of the reflectivity profile of the target. Thus, a physically small beamforming source results in a narrow beam cross section. Following adaptivity, the beam is scanned in range and angle (cross range) to obtain the microwave image of the target area.

The method was developed to overcome intrinsic errors in the very large antenna systems required to achieve high angular resolution at microwaves. Demonstration of microwave imagery obtained with an 83 m radio camera shows the validity of the technique.

Acknowledgment

This work has been principally supported by the Office of Naval Research under Contract No. N00014-79-C-0505, and by the Army Research Office under Contract No. DAAD07-84-R-0081.

References

[1] B.D. Steinberg, Microwave Imaging with Large Antenna Arrays, John Wiley & Sons, New York, 1983.

[2] B.D. Steinberg, "Radar Imaging from a Distorted Array: The Radio Camera Algorithm and Experiments," *IEEE Trans. Antennas and Prop.*, vol. AP-20, No. 5, September 1981.

Identification of Robot Dynamics: An Application of Recursive Estimation

Charles P. Neuman*and Pradeep K. Khosla†
Department of Electrical and Computer Engineering
The Robotics Institute
Carnegie-Mellon University
Pittsburgh, Pa. 15213

Abstract

To synthesize robust robot parameter identification algorithms, we outline the fundamental properties of the Newton-Euler (N-E) and Lagrange-Euler (L-E) formulations of robot dynamics. We transform the nonlinear (in dynamic parameters) N-E dynamic robot model into the equivalent linear (in dynamic parameters) L-E dynamic robot model. We cast the L-E torque/force error model into the series and parallel identifier structures for on-line and off-line robot parameter estimation. To illustrate our approach, we identify (in simulation) the dynamic parameters of the cylindrical prototype robot and the three degree-of-freedom positioning system of the Stanford manipulator. Our identification algorithm is directly amenable to the real-time identification of the pay-load inertial characteristics and the dynamic frictional coefficients for precise trajectory control.

1. Introduction

Robots are chains of rigid bodies (links) which are connected serially by joints. Each joint is actuated independently to produce the relative motion between the links, thus positioning the end-effector. Precise positioning of the end-effector demands accurate modeling of robot dynamics and a robust control algorithm.

Robot dynamics are characterized by systems of coupled nonlinear second-order differential equations [1,2]. Dynamic robot models describe the temporal interactions of the joint motions in response to the inertial, centrifugal and Coriolis, gravitational and actuating torques/forces. Formulations of complete dynamic robot models are documented in the literature. The structured, closed-form Lagrange-Euler (L-E) formulation [1,2] leads to physical insight into the coupled nonlinear robot dynamics and is attractive from both the dynamic modeling and control engineering points-of-view. The computationally efficient, recursive Newton-Euler (N-E) formulation [3] sacrifices the compact structure of the Lagrange-Euler formulation and is more amenable to simulation and real-time control applications [4]. Both closed-form and recursive formulations are essential ingredients of the control engineering repertoire [5].

The robot control problem centers around the computation of the actuating torques/forces to produce the desired motion of the end-effector [6]. To achieve this objective, a plethora of control algorithms has been proposed in the literature [2]. The mathematically sophisticated algorithms are difficult to implement. Practically-oriented algorithms incorporate physically-motivated assumptions to reduce the computational requirements for real-time implementation. Examples include the computed-torque [7] and resolved-acceleration [8]

*Professor of Electrical and Computer Engineering.
†Graduate Student, Department of Electrical and Computer Engineering.

Table 1: Denavit-Hartenberg Link Parameters (Scalars)

θ_i is the angle of rotation about the z_{i-1} axis
a_i is the length of translation along the rotated x_i axis
d_i is the length of translation along the z_{i-1} axis
α_i is the angle of rotation about the x_i axis

controllers. The fundamental assumption is that the robot dynamics are modeled accurately. To make these controllers become a practical reality requires precise knowledge of the kinematic (Denavit-Hartenberg) parameters and the dynamic parameters (link masses and inertias, and centers-of-mass) of the robot.

In this paper, we introduce a recursive algorithm to estimate the kinematic and dynamic parameters of a robot from input (actuating torques/forces)—output (joint positions, velocities and accelerations) measurements [9]. The parameter identification problem can be approached from both the closed-form and recursive dynamic robot model points-of-view. We focus on both the recursive and closed-form dynamic robot models. We cast the identification problem into the series-identifier (input-error) structure for on-line applications and the parallel-identifier (output-error) structure for simulation studies [10-12]. We demonstrate our identification algorithm through simulation experiments on the three degree-of-freedom cylindrical robot [6] and the positioning system (i.e., the first three degrees-of-freedom) of the Stanford manipulator [1]. Throughout the paper, we use standard robotic terminology [1,2].

This paper is organized as follows. In Section 2, we review the Newton-Euler and Lagrange-Euler formulations and identify the applicable properties of these formulations for robotic parameter identification. In Section 3, we apply these properties to the cylindrical robot. (We address the Stanford manipulator in Appendix A.) We then derive (in Section 4) our identification algorithm for a general-purpose N degree-of-freedom robot and generate the torque/force error model. We cast (in Section 5) this model in the series and parallel identifier structures [10-12] and highlight the properties of our identification algorithm. In Section 6, we evaluate the performance (in simulation) of our algorithm on the two case study robots, and in Section 7, we outline the practical implementation issues. Finally, in Section 8, we draw our conclusions.

2. Dynamic Robot Modeling

In this section, we outline the properties of the $O(N)$ recursive Newton-Euler and $O(N^4)$ closed-form Lagrange-Euler formulations for parameter identification in robot dynamics.

2.1 Newton-Euler Formulation

We begin by summarizing the Newton-Euler formulation [3,13-15]. To formulate the dynamic equations-of-motion for an N degree-of-freedom (DOF) manipulator, the Denavit-Hartenberg convention [16] defines $N+1$ coordinate frames (the i-th coordinate frame being attached to the i-th link, beginning with link 0 - the base, and ending with link N - the end-effector). The four Denavit-Hartenberg parameters, which define homogeneous transformations between successive coordinate frames, are listed in Table 1. For each joint, one of the four Denavit-Hartenberg parameters is a variable, and the remaining three are constants. For a revolute joint i, the joint position variable q_i is the angle θ_i; for a translational joint i, the joint position variable is the distance d_i.

The kinematic and dynamic parameters and variables, which appear in the Newton-Euler algorithm, are compiled in Table 2. We outline (in Table 3) the forward recursions (from the base to the end-effector) for the kinematic variables. The initial conditions (for i = 0) assume that the manipulator is at rest in the gravitational field. This practical assumption for industrial manipulators is not required for our identification algorithm (in Section 4) which processes a complete set of measured initial conditions.

Table 4 displays the backward recursions (from the end-effector to the base) of the forces and moments, and culminates in (9) with the calculation of the joint torques/forces. The terminal conditions (for $i = N + 1$) on the forces and moments at the end-effector are included in the table. The $O(N)$ recursions in Tables 3 and 4 compute the inverse dynamics (i.e., the joint torques/forces are calculated from the joint positions, velocities and accelerations). The authors [13] have detailed the computational requirements of the minimal general-purpose and customized recursive Newton-Euler algorithms for kinematically and dynamically structured manipulators (e.g., manipulators with parallel/perpendicular joint axes, spherical wrists, or sparse center-of-mass vectors and inertia tensors).

The identification problem is to estimate all of the kinematic and dynamic parameters that affect the link torques/forces. The Denavit-Hartenberg parameters (defined in Table 1) constitute the kinematic parameters and the link masses, classical link inertia tensors and center-of-mass vectors (enumerated in Table 2) are the dynamic parameters. The number of independent kinematic and dynamic parameters is determined by the robot configuration. Since the $L - E$ dynamic robot model is nonlinear in the kinematic parameters, and linear in the classical link inertia tensors and center-of-mass vectors [5,18] (if the link masses are known), we assume initially that the kinematic parameters and link masses are known. For expository convenience, we henceforth use the term dynamic parameters to refer only to the classical link inertia tensors and center-of-mass vectors. We apply standard identification algorithms [12,19,20] to estimate the dynamic parameters. In Section 4, we relax this assumption and include the kinematic parameters and link masses in our identification algorithm.

From Tables 3 and 4, we note the three properties of the Newton-Euler dynamic robot model for robot parameter identification [9]:

1. The Newton-Euler dynamic robot model is linear in the classical link inertia tensors I_i.

   This property follows directly from the backward recursions in (5)-(9). The joint torque/force $\tau_i(t)$ in (9) is linear in the moment n_i/force f_i exerted on link i by link $i-1$. In the linear recursion for the moment n_i in (8), the force f_i, and the net moment and force N_i and F_i exerted on link i appear additively. In the linear recursion for the force f_i in (7), the net force F_i appears additively. Finally, the moment N_i in (6) is linear in the classical link inertia tensor I_i and the force F_i in (5) is linear in the link mass m_i. The $N - E$ dynamic robot model is thus linear in the classical link inertia tensors I_i.

2. For rotational joints, the Newton-Euler dynamic robot model is nonlinear in the center-of-mass vectors s_i.

   From (4) and (5), the link force F_i is linear in the center-of-mass vector s_i. The vector cross product $s_i \times F_i$ in (8) is thereby nonlinear in s_i. Hence, the torque $\tau_i(t)$, for a rotational joint, in (9) is nonlinear in the center-of-mass vector s_i.

3. The dynamic equations of links $i + 1$ through N are independent of the mass m_i and the classical inertia tensor I_i of link i.

   This physically intuitive property of the Newton-Euler dynamic robot model is an immediate consequence of the backward recursions in Table 4.

In the $N - E$ dynamic robot model, the classical link inertia tensors I_i and the link masses m_i appear linearly. The link masses are multiplied by linear or quadratic functions of the center-of-mass vectors s_i, and nonlinear functions of the joint position variables. In the Appendix, we illustrate that the three degree-of-freedom Stanford manipulator is an

Table 2: Kinematic and Dynamic Parameters

m_i	Total mass of link i
$\tau_i(t)$	Joint torque/force at joint i
ω_i and $\dot{\omega}_i$	Angular velocity and acceleration of the ith coordinate frame
v_i and $\dot{v}_i$	Linear velocity and acceleration of the ith coordinate frame
v_i^* and $\dot{v}_i^*$	Linear velocity and acceleration of the center-of-mass of link i
F_i and N_i	Net force and moment exerted on link i
f_i and n_i	Force and moment exerted on link i by link $(i-1)$
p_i	Position of the ith coordinate frame with respect to the $(i-1)^{st}$ coordinate frame: $p_i = [a_i \; d_i sin(\alpha_i) \; d_i cos(\alpha_i)]^T$
s_i	Position of the center-of-mass of link i : $s_i = [s_{ix} \; s_{iy} \; s_{iz}]^T$
z_o	$= [0\;0\;1]^T$
A_i	Orthogonal rotation matrix which transforms a vector in the ith coordinate frame to a coordinate frame which is parallel to the $(i-1)^{st}$ coordinate frame: $A_i = \begin{bmatrix} cos(\theta_i) & -cos(\alpha_i)sin(\theta_i) & sin(\alpha_i)sin(\theta_i) \\ sin(\theta_i) & cos(\alpha_i)cos(\theta_i) & -sin(\alpha_i)cos(\theta_i) \\ 0 & sin(\alpha_i) & cos(\alpha_i) \end{bmatrix}$ for $i = 1, 2, \ldots, N$, where $A_{N+1} \triangleq 1$.
I_i	Classical inertia tensor [1,17] of link i about the center-of-mass of link i (and parallel to the i-th coordinate frame); with principal inertias I_{ixx}, I_{iyy} and I_{izz}; and cross-inertias I_{ixy}, I_{ixz} and I_{iyz}.

Table 3: Forward Newton-Euler Recursions

Iterate for $i = 0, 1, \ldots, N-1$

$$\omega_{i+1} = \begin{cases} A_{i+1}^T[\omega_i + z_o\dot{\theta}_{i+1}] & \text{if joint } i+1 \text{ is rotational} \\ A_{i+1}^T[\omega_i] & \text{if joint } i+1 \text{ is translational} \end{cases} \quad (1)$$

$$\dot{\omega}_{i+1} = \begin{cases} A_{i+1}^T[\dot{\omega}_i + z_o\ddot{\theta}_{i+1}] + \omega_i \times (z_o\dot{\theta}_{i+1})] & \text{if joint } i+1 \text{ is rotational} \\ A_{i+1}^T[\dot{\omega}_i] & \text{if joint } i+1 \text{ is translational} \end{cases} \quad (2)$$

$$\dot{v}_{i+1} = \begin{cases} A_{i+1}^T[\dot{v}_i] + \dot{\omega}_{i+1} \times p_{i+1} + \omega_{i+1} \times (\omega_{i+1} \times p_{i+1}) & \text{if joint } i+1 \text{ is rotational} \\ A_{i+1}^T[\dot{v}_i + z_o\ddot{d}_{i+1} + 2\omega_i \times (z_o\dot{d}_{i+1})] + \dot{\omega}_{i+1} \times p_{i+1} + \omega_{i+1} \times (\omega_{i+1} \times p_{i+1}) & \text{if joint } i+1 \text{ is translational} \end{cases} \quad (3)$$

Initial Conditions

$\omega_0 = \dot{\omega}_0 = v_0 = 0$

$\dot{v}_0 = [g_x g_y g_z]^T$ (gravitational acceleration of the manipulator base)

Table 4: Backward Newton-Euler Recursions

Iterate for $i = N, N-1, \ldots, 1$

$$\dot{v}_i^* = \dot{\omega}_i \times s_i + \omega_i \times (\omega_i \times s_i) + \dot{v}_i \quad (4)$$

$$F_i = m_i \dot{v}_i^* \quad (5)$$

$$N_i = I_i \dot{\omega}_i + \omega_i \times (I_i \omega_i) \quad (6)$$

$$f_i = A_{i+1}[f_{i+1}] + F_i \quad (7)$$

$$n_i = A_{i+1}[n_{i+1}] + p_i \times f_i + N_i + s_i \times F_i \quad (8)$$

$$\tau_i(t) = \begin{cases} n_i^T (A_i^T z_o) & \text{if joint } i \text{ is rotational} \\ f_i^T (A_i^T z_o) & \text{if joint } i \text{ is translational} \end{cases} \quad (9)$$

Terminal Conditions

f_{N+1} is a known externally applied force at the end-effector.
n_{N+1} is a known externally applied moment at the end-effector.

exception because the link masses are multiplied by linear functions of the center-of-mass vector and not by nonlinear functions of the joint position variables. Properties 1 and 2 lay the foundation for our identification algorithm and Property 3 (as we show in Section 4) reduces the complexity of our identification algorithm.

2.2 Lagrange-Euler Formulation

We proceed to outline the properties of the closed-form Lagrange-Euler dynamic robot model. The Lagrange-Euler formulation is obtained by differentiating the kinetic and potential energies of the manipulator with respect to the joint positions and velocities to generate the torques/forces acting on the manipulator [1,2,21,22]. Paul [1] utilizes this formulation for open kinematic linkages, represented by the Denavit-Hartenberg convention (in Table 1), to derive the closed-form dynamic equations-of-motion for an N degree-of-freedom manipulator. Upon neglecting mechanical dissipation, the torque/force at link i is:

$$\sum_{j=1}^{N} d_{ij}\ddot{q}_j + \sum_{j=1}^{N}\sum_{k=1}^{N} \dot{q}_j c_{jk}(i)\dot{q}_k + G_i = \tau_i(t) \qquad \text{for } i = 1, 2, \ldots, N \quad (1)$$

where the inertial, centrifugal and Coriolis, and gravitational coefficients $d_{ij}, c_{jk}(i)$ and G_i are defined according to:

$$d_{ij} = \sum_{k=max(i,j)}^{N} Tr\left\{{}^{p}U_{kj}\ J_k\ {}^{p}U_{ki}^T\right\} \qquad \text{where } p \triangleq min(i,j) - 1 \quad (2)$$

$$c_{jk}(i) = \sum_{m=max(i,j,k)}^{N} Tr\left\{{}^{p}U_{mjk}\ J_m\ {}^{p}U_{mi}^T\right\} \qquad \text{where } p \triangleq min(i,j,k) - 1 \quad (3)$$

and

$$G_i = -\sum_{k=i}^{N} m_k g^T\, {}^{0}U_{ki}\bar{r}_k. \quad (4)$$

We compile in Table 5 [21], the kinematic and dynamic parameters and variables which appear in the Lagrange-Euler dynamic robot model in (1).

Whereas the Newton-Euler formulation (in Tables 2-4) utilizes geometrical 3-vectors, the Lagrange-Euler dynamic robot model in (1) is founded upon position 4-vectors. The exogenous actuating torque/force $\tau_i(t)$ in (1) is balanced by three torque/force components.

Table 5: Lagrange-Euler Nomenclature

N	Number of degrees of freedom (DOF)
r_i	Position 4-vector of a point on link i, in the ith coordinate frame: $r_i = [r_{ix}\ r_{iy}\ r_{iz}\ 1]^T$
T_i	Homogeneous (4×4) coordinate transformation matrix, *from* the ith coordinate frame *to* the $(i-1)^{st}$ coordinate frame; T_i, which is defined completely by the four kinematic link parameters $(\theta_i, d_i, a_i, \alpha_i)$, is: $T_i = \begin{bmatrix} cos(\theta_i) & -cos(\alpha_i)sin(\theta_i) & sin(\alpha_i)sin(\theta_i) & a_i\ cos(\theta_i) \\ sin(\theta_i) & cos(\alpha_i)cos(\theta_i) & -sin(\alpha_i)cos(\theta_i) & a_i\ sin(\theta_i) \\ 0 & sin(\alpha_i) & cos(\alpha_i) & d_i \\ 0 & 0 & 0 & 1 \end{bmatrix}.$
${}^pT_k$	Homogeneous (4×40 coordinate transformation matrix from the kth coordinate frame to the pth coordinate frame; ${}^pT_k$ is calculated according to: ${}^pT_k = T_{p+1}\ T_{p+2} \ldots T_{k+1}\ T_k \quad \text{for } 0 \le p < k \le N \text{ and } {}^pT_p \triangleq 1$
${}^0T_N$	Forward solution (i.e., homogeneous (4×4) coordinate transformation matrix) from the end-effect (N^{th}) coordinate frame to the base (zeroth) coordinate frame
J_i	Pseudo-inertia (4×4) matrix of link i : $\int_{link\ i}[r_i\ r_i^T]dm_i$
$\bar{s}_i$	Center-of-mass 4-vector of link i, in the ith coordinate frame: $\bar{s}_i = [\bar{s}_{ix}\ \bar{s}_{iy}\ \bar{s}_{iz}\ 1]^T$
m_i	Mass of link i
g	Gravity 4-vector in the base coordinate frame:$g = [g_x\ g_y\ g_z 0]^T$
${}^pU_{ki}$	First partial-derivative (4×4) matrix of ${}^pT_k$ with respect to q_i
${}^pU_{mjk}$	Second partial-derivative (4×4) matrix of ${}^pT_k$ with respect to q_j and q_m
$q_i(\dot{q}_i,\ \ddot{q}_i)$	Joint coordinate (velocity, acceleration) of joint i : $q_i = \theta_i$ if joint i is revolute; and $q_i = d_i$ if joint i is prismatic.
$q(\dot{q},\ \ddot{q})$	Joint coordinate (velocity, acceleration) N-vector

The first models the inertial torque/force coupling between the joints, the second includes the centrifugal and Coriolis torques/forces, and the third is the gravitational torque/force.

From (1)-(4), we observe the three companion properties of the Lagrange-Euler dynamic robot model for robot parameter identification [9]:

1. The Lagrange-Euler dynamic robot model is linear in the pseudo-inertia matrices J_i [5,18].

   The homogeneous transformation T_i matrices, and hence the forward solution ${}^pT_k$ and partial-derivative ${}^pU_{mjk}$ matrices (in Table 5) are nonlinear functions of the kinematic parameters. The dynamic inertia parameters appear linearly in the pseudo-inertia J_i matrices. In (2) and (3), the pseudo-inertia matrices are premultiplied and postmultiplied by the partial-derivative ${}^pU_{mjk}$ matrices. Since matrix multiplication, trace and summation are linear operations, the inertial and centrifugal and Coriolis coefficients d_{ij} and $c_{jk}(i)$ are linear functions of the pseudo-inertia matrices. We conclude from (1) that the link i torque/force $\tau_i(t)$ is linear in the pseudo-inertia matrices (and hence the dynamic parameters) and nonlinear in the kinematic parameters.

2. The Lagrange-Euler dynamic robot model is linear in the center-of-mass vectors (if the link masses are known).

   The link torque/force is linear in the pseudo-inertia matrix and the components of the center-of-mass vector appear linearly in the pseudo-inertia matrix and the link masses m_i are multiplied by linear functions of the center-of-mass vectors s_i.

3. The dynamic equations of links $i+1$ through N are independent of the pseudo-inertia matrix of link i.

   In (2) and (3), the inertial coefficients d_{ij} and the centrifugal and Coriolis coefficients $c_{jk}(i)$ are functions of J_k and J_m, respectively. The indices k and m range from $max(i,j)$ to N and $max(i,j,k)$ to N, respectively, where N is the number of degrees-of-freedom of the manipulator. The link i torque/force, which is a function of the d_{ij} and $c_{jk}(i)$ coefficients, is thus independent of the inertias of links $i+1$ through N.

From the viewpoint of parameter identification, the Lagrange-Euler dynamic robot model in (1) is thus linear in the dynamic parameters (the pseudo-inertia matrices and center-of-mass vectors), if the link masses are known, and nonlinear in the kinematic (Denavit-Hartenberg) parameters [5,18].

2.3 Relationships between the Newton-Euler and Lagrange-Euler Formulations

We have presented three properties of the Newton-Euler and Lagrange-Euler formulations. While Properties 1 and 3 are identical, we notice an anomaly between the second property of the two formulations. We elaborate upon this anomaly to formulate our identification algorithm. In Section 3, we illustrate these properties for the cylindrical robot.

The Lagrange-Euler formulation utilizes the pseudo-inertia matrices. These J_i matrices can be obtained from the classical link inertia tensors (of the Newton-Euler formulation) by a nonlinear transformation from the center-of-mass of the link to the link coordinate system, and creating the pseudo-inertia matrix from the three-by-three transformed classical link inertia tensor [1]. Because of the different representations of the link inertia, the Lagrange-Euler formulation is linear and the Newton-Euler formulation is nonlinear in the center-of-mass vectors.

Silver [23] has demonstrated the equivalence of the Newton-Euler and Lagrange-Euler formulations. Furthermore, these dynamic robot models are exact for robots with rigid links [3]. To facilitate the development of our identification algorithm, we apply the Newton-Euler formulation. Our task is to transform the link torques/forces which are nonlinear in the center-of-mass vectors (in the Lagrange-Euler formulation). We thus transform the classical link inertia tensor about the center-of-mass to an equivalent inertia tensor about the link coordinate frame to create the pseudo-inertia matrix about the link coordinate frame. This transformation leads to the Lagrange-Euler dynamic robot model which (according to the equivalence of robot dynamics formulations [23]) is equivalent to the Newton-Euler dynamic robot model. In Section 3, we illustrate our approach for the cylindrical robot.

3. Identification Algorithm for the Cylindrical Robot

3.1 Introduction

In this section, we review the Newton-Euler and Lagrange-Euler models of the three degree-of-freedom positioning system of the cylindrical robot [2,6]. We then illustrate the properties outlined in Section 2. Finally, we transform the nonlinear center-of-mass vectors of the Newton-Euler model into the linear center-of-mass vectors of the Lagrange-Euler model to introduce our identification algorithm.

The prototype cylindrical robot, depicted schematically in the literature [2,4,6], consists of three degrees-of-freedom: a rotation θ_1, a vertical translation d_2 and a radial translation d_3. The Denavit-Hartenberg parameters of the cylindrical robot are listed in Table 6.

The coordinate vector of the cylindrical robot is thus $q = [\theta_1 \ d_2 \ d_3]^T$ and the joint coordinates are the cylindrical world coordinates. Even though the cylindrical robot exhibits

Table 6: Denavit-Hartenberg Parameters of the Cylindrical Robot

Link	θ	α	a	d
1	θ_1	0°	0	0
2	0	-90°	0	d_2
3	0	0°	0	d_3

a relatively simple dynamic model, it preserves all of the inherent coupling and nonlinear characteristics of robot dynamics [6].

3.2 Newton-Euler Model

We derive the Newton-Euler model of the cylindrical robot by applying the Denavit-Hartenberg parameters (in Table 6) and expanding the Newton-Euler recursions (in Tables 3 and 4). We assume that the classical link inertias of the three links are diagonal and that only the s_{1z}, s_{2y} and s_{3z} components of the center-of-mass vectors are non-zero. These assumptions (without any loss of generality) simplify the dynamic model. The complete $N - E$ dynamic model of the cylindrical robot is:

$$\tau_1(t) = [(I_{1zz} + I_{2yy} + I_{3yy}) + m_3(s_{3z} + d_3)^2]\ddot{\theta}_1 + [2m_3(s_{3z} + d_3)]\dot{d}_3\dot{\theta}_1 \tag{5}$$

$$\tau_2(t) = (m_2 + m_3)\ddot{d}_2 + (m_2 + m_3)g \tag{6}$$

$$\tau_3(t) = m_3\ddot{d}_3 - m_3(s_{3z} + d_3)\dot{\theta}_1^2. \tag{7}$$

We observe that the Newton-Euler model in (5)-(7) is indeed linear in the classical link inertia tensors, nonlinear in the link masses and center-of-mass vector (due to the presence of s_{3z}^2 in (5)), and that the dynamic parameters of link i do not affect the torques/forces of links $i + 1$ through N (for the cylindrical robot, $N = 3$). The Newton-Euler model of the cylindrical robot thus satisfies the three properties outlined in Section 2.1.

3.3 Lagrange-Euler Model

Upon applying the Denavit-Hartenberg parameters (in Table 6), the dynamic characteristics (outlined in Section 3.2), and the Lagrange-Euler formulation in (1)-(4), the complete $L - E$ dynamic model of the cylindrical robot is:

$$\begin{aligned} \tau_1(t) = \ & [(J_{1xx} + J_{1yy} + J_{2xx} + J_{2zz} + J_{3xx} + J_{3zz}) + m_3d_3(2s_{3z} + d_3)]\ddot{\theta}_1 \\ & +[2m_3(s_{3z} + d_3)]\dot{d}_3\dot{\theta}_1 \end{aligned} \tag{8}$$

$$\tau_2(t) = (m_2 + m_3)\ddot{d}_2 + (m_2 + m_3)g \tag{9}$$

$$\tau_3(t) = (m_3\ddot{d}_3 - m_3(s_{3z} + d_3)\ddot{\theta}_1^2. \tag{10}$$

The dynamic robot model in (8)-(10) shows that the vertical d_2 motion is decoupled from the coupled and nonlinear rotational θ_1 and radial d_3 motions. The Lagrange-Euler model in (8)-(10) is indeed linear in the pseudo-inertia matrices and center-of-mass vectors, and the torques/forces of links $i + 1$ through N are independent of the dynamic parameters of link i. We observe that the link mass multiplies the z-component s_{3z} of the center-of-mass vector throughout the torque/force equations of the links. The model thus satisfies the three properties of the Lagrange-Euler formulation outlined in Section 2.1.

Upon interpreting (in Table 7) the physical parameters of the cylindrical robot, the $L - E$ model in (8)-(10) is equivalent to the standard closed-form dynamic model of the cylindrical robot in Table 8 [6].

In the next section, we transform the nonlinear dynamic parameters of the Newton-Euler model into equivalent linear parameters for the Lagrange-Euler model.

Table 7: Parameter Relationships for Closed-Form Dynamic Models of the Cylindrical Robot

L-E Model in (17)-(19)	Standard Model in Table 8
$J_{1zz} + J_{1yy} + J_{2zz} + J_{2zz} + J_{3zz} + J_{3zz}$	J
$m_3 d_3 [2 s_{3z} + d_3]$	$j(d_3)$
$2 s_{3z}$	R
$m_2 + m_3$	M
m_3	$m_R + m_L$

Legend

J	is the *constant inertia* of the vertical column.
$j(d_3) = (m_R + m_L) d_3^2 + m_R R d_3$	is the *coordinate-dependent* inertia of the radial link and is a *quadratic* function of the radial displacement.
m_R	is the *mass* of the radial link.
m_L	is the *mass* of the payload which is concentrated at the tip of the radial link; i.e. at $d_3 = R$.
m_V	is the *mass* of the vertical column.
$M = m_V + m_R + m_L$	is the vertically translated mass.
R	is the length of the radial link.
$\tau_1(t), \tau_2(t), \tau_3(t)$	are the external joint torques/forces that actuate the θ_1, d_2, and d_3 DOF, respectively.

Table 8: Closed-Form Dynamic Model of the 3 DOF Cylindrical Robot

$$
\begin{aligned}
[J + j(d_3)]\ddot{\theta}_1 + \frac{\partial j(d_3)}{\partial d_3}\dot{\theta}_1 \dot{d}_3 &= \tau_1(t) \\
M\ddot{d}_2 + Mg &= \tau_2(t) \\
(m_R + m_L)\ddot{d}_3 - \frac{1}{2}\frac{\partial j(d_3)}{\partial d_3}\dot{\theta}_1^2 &= \tau_3(t)
\end{aligned}
$$

3.4 The Nonlinear Transformation

In Section 2.3, we observed an anomaly between the second property of the Newton-Euler and Lagrange-Euler dynamic robot models. This anomaly arises from a nonlinear transformation implicit in the Lagrange-Euler formulation. In this section, we derive the nonlinear transformation to convert the Newton-Euler dynamic robot model (which is nonlinear in the dynamic parameters) into the equivalent Lagrange-Euler dynamic robot model (which is linear in the dynamic parameters).

Let (X_0, Y_0, Z_0) be a coordinate frame fixed at the center-of-mass of link i and let (X_i, Y_i, Z_i) be a Denavit-Hartenberg coordinate frame for the $i-th$ link of the manipulator. Let ${}^0A_i$ be the orthogonal (3×3) rotation matrix which transforms the $i-th$ coordinate frame into the $0-th$ coordinate frame, and let s_i be the translational vector from the origin of the $i-th$ coordinate frame to the origin of the $0-th$ coordinate frame. If I_i is the classical link inertia tensor about the center-of-mass of link i (and because the coordinate frames of the manipulator are described by the Denavit-Hartenberg convention), the corresponding classical inertia tensor I_i' about the $i-th$ link coordinate frame is computed according to the nonlinear transformation of the parallel-axis theorem or Steiner's law [17]:

$$I_i' = I_i + m_i(s_i^T s_i E - s_i s_i^T) \tag{11}$$

where E is the three-by-three identity matrix.

We compute the diagonal elements of the pseudo-inertia matrix J_i from the classical moments-of-inertia (about the $i-th$ link coordinate frame) according to [1]:

$$J_{ixx} = \frac{-I'_{ixx} + I'_{iyy} + I'_{izz}}{2} \tag{12}$$

$$J_{iyy} = \frac{I'_{ixx} - I'_{iyy} + I'_{izz}}{2} \tag{13}$$

$$J_{izz} = \frac{I'_{ixx} - I'_{iyy} - I'_{izz}}{2}. \tag{14}$$

The off-diagonal elements of J_i equal the classical cross-products of inertia.

Conversely, we can express the classical moments-of-inertia in terms of the diagonal pseudo-inertia elements according to:

$$I'_{ixx} = J_{iyy} + J_{izz} \tag{15}$$

$$I'_{iyy} = J_{ixx} + J_{izz} \tag{16}$$

$$I'_{izz} = J_{ixx} + J_{iyy}. \tag{17}$$

Equations (12)-(17) place in perspective the relationship between the classical moments-of-inertia and the diagonal pseudo-inertia elements. We proceed to transform the Newton-Euler dynamic robot model in (5)-(7) into the equivalent Lagrange-Euler dynamic robot model in (8)-(10) which is more appealing from the identification point-of-view.

We apply (11) to transform the classical link inertias of the Newton-Euler model in (5)-(7) into equivalent inertias about the link coordinate frames of the cylindrical robot according to:

$$I'_{3yy} = I_{3yy} + m_3 s_{3z}^2 \tag{18}$$

$$I'_{2yy} = I_{2yy} \tag{19}$$

$$I'_{1zz} = I_{1zz}. \tag{20}$$

Upon substituting (18)-(20) into (15)-(17) and the results into the N-E model in (5)-(7), we obtain the L-E model in (8)-(10) of the cylindrical robot.

We note that the Newton-Euler equation-of-motion in (5) contains the nonlinearity $m_3 s_{3z}^2$, which is absorbed by the transformation in (18), and that I_{2yy} and I_{1zz} are retained in (19) and (20). From this observation, we conclude that once the nonlinearities in an equation-of-motion are identified to be associated with link i, we may proceed to transform the inertias of link i only. This fact reduces the computation involved in transforming linearly, the nonlinearity-free link inertias. In the next section, we derive the torque/force error model of the complete Lagrange-Euler dynamic robot model of the cylindrical robot in (8)-(10). In Section 4, we outline the procedure for developing the identification model for an N degree-of-freedom robot.

3.5 The Torque/Force Error Model

If we assume that we have nominal values of the dynamic parameters (from engineering drawings), and that we measure the link masses and kinematic parameters, then identification of the dynamic parameters is a problem of linear estimation [12,20]. If the link masses or the kinematic parameters are not known precisely, robot parameter identification is a problem of nonlinear estimation [19].

We linearize the torque/force model in (8)-(10) about the nominal values of the kinematic and dynamic parameters to obtain the torque/force error model [9]:

$$\epsilon_i(t) = \tau_i(t) - \tau_i^0(t) = \varphi_i^T(t)\theta_i \qquad \text{for } i = 1, 2, \ldots, N \tag{21}$$

where $\tau_i(t)$ is the applied torque/force to link i, $\tau_i^0(t)$ is the nominal value of the torque/force, as computed by an inverse dynamics model [3], $\epsilon_i(t)$ is the input torque/force error of link i, θ_i is the vector of unknown dynamic parameters that affect the torque/force of link i and $\varphi_i(t)$ is the nonlinear vector function of the kinematic parameters and the output measurements: joint positions, velocities and accelerations. We call (21) the torque/force error model of link i. The torque/force error model relates the error torque/force of link i to corresponding modeling inaccuracies in the kinematic and dynamic parameters of the closed-form dynamic robot model.

For dynamic parameter estimation, (21) is a linear algebraic equation in μ_i unknowns, where μ_i is the number of dynamic parameters in θ_i that affect the torque/force of link i and can be estimated. We step sequentially through the manipulator, from the tip to the base, and identify the dynamic parameters that affect the link torques/forces of each link.

For the cylindrical robot, $\mu_3 = 1$ (the unknown dynamic parameter is s_{3z} in (10)), $\mu_2 = 0$ (since we measure the link masses m_2 and m_3 in (9)), and $\mu_1 = 1$ (because we can only identify the sum of the six pseudo-inertias in (8)). We measure the input torque/force to link i and the position, velocity and acceleration of link i at N sampling instants (where $N > \mu_i$), compute the nominal values of $\tau_i^0(t)$ (according to the inverse dynamics model in (8)-(10)), and apply a recursive linear least-squares algorithm [12,20] to estimate the dynamic parameters of link i. We formulate (in Section 5) our identification algorithms and highlight (in Section 6) our simulation results for the cylindrical robot.

4. Identification Model for an N-DOF Robot

In Section 3, we compared the Newton-Euler and Lagrange-Euler models of the cylindrical prototype robot and introduced a procedure to transform the Newton-Euler model (which is nonlinear in the dynamic parameters). In this section, we generalize our approach to formulate the identification model for an N degree-of-freedom manipulator. We adopt a hybrid approach to utilize the aforementioned properties of the recursive Newton-Euler and closed-form Lagrange-Euler formulations. We expand the Newton-Euler recursions to derive the closed-form torque/force equations-of-motion for the manipulator links, and solve the parameter estimation problem for the Lagrange-Euler model. We transform the Newton-Euler model into the Lagrange-Euler model according to (11), and the development parallels our formulation (in Section 3) for the cylindrical robot.

We apply the third property of the Newton-Euler formulation (in Section 2.1) to simplify the derivation of the torque/force equations-of-motion for the N links of the manipulator. In Section 2.1, we observed that the torque/force equations-of-motion of links $i + 1$ through

N are independent of the dynamic parameters of link i. We thus treat the parameter identification problem sequentially (starting from link N, the tip, and proceeding to link 0, the base) and estimate the dynamic parameters of each link. The identified dynamic parameters of link i become known parameters in the torque/force equation of link $i-1$. This sequential procedure reduces the number of dynamic parameters which must be estimated for each link in the identification algorithm. Furthermore, since the dynamic parameters are constant, we can estimate these parameters off-line [24]. If we lock the first $i-1$ joints mechanically (to set the velocities and accelerations of joints one through $i-1$ to zero), we reduce dramatically the complexity of the torque/force error model of link i (for parameter estimation). This simplification can only be achieved in off-line estimation. In Appendix A, we apply this procedure to derive the torque/force error model of the positioning system (i.e., the first three degrees-of-freedom) of the Stanford manipulator for on-line estimation. When the constant offset is set equal to zero, the positioning system of the Stanford manipulator reduces to the spherical prototype robot [6].

We summarize our off-line identification procedure for an N degree-of-freedom robot in the following steps [9]:

1. Expand the recursions of the Newton-Euler formulation (in Tables 2-4) to obtain the closed-form link torques/forces of the manipulator. To derive the torque/force equation for link i, set the velocities and accelerations of links one through $i-1$ to zero.

   Since the dynamic parameters affecting the torques/forces of links $i+1$ through N have already been identified, these dynamic parameters become known numerical quantities when the Newton-Euler recursions are expanded to obtain the torque/force equation for link i.

2. Apply (11) to convert the Newton-Euler model into the equivalent Lagrange-Euler model. This is achieved by transforming the classical inertias about the origin of link i to equivalent inertias about the $i-th$ link frame and using (12)-(14) to compute the diagonal pseudo-inertias.

   Simplify the transformation by associating each nonlinearity in the dynamic equation-of-motion of link i with link $j (i \leq j \leq N)$ by transforming the classical inertias of link i only.

3. Incorporate the nominal values of the dynamic parameters to be estimated in the torque/force equation (in Step 1) and generate the torque/force error model in (21).

4. Identify all of the $L-E$ dynamic parameters in the torque/force error model in (21) that affect the torque/force of link i.

5. Apply (15)-(17) to recover the $N-E$ parameters of link i.

The on-line identification procedure for an N degree-of-freedom robot parallels Steps 1 through 5, with the exception that in Step 1 we do not lock mechanically the first $i-1$ joints and derive the torque/force error model by allowing the positions, velocities and accelerations of links $i-1$ through one to respond to the actuating torques/forces.

The closed-form dynamic model of a six degree-of-freedom robot (such as the Stanford manipulator or Puma robot) is complex [25] and the corresponding torque/force error model for on-line estimation of the robot parameters is of comparable complexity. To facilitate the estimation process, a viable approach is to generate the torque/force error model in (21) for off-line estimation and identify, on-line, the inertial characteristics of the pay-load. This approach requires the real-time identification of the parameters of only the last link of the robot. Our identification algorithm is directly amenable to the real-time identification of the pay-load inertial characteristics for dynamic feedback control. Frictional torques/forces in robots can be modeled as linear functions of the joint velocities. To achieve accurate trajectory control, our identification algorithm can also estimate on-line the dynamic frictional coefficients for dynamic compensation.

Each link of a robot is characterized by a maximum of ten dynamic parameters: the link mass, six classical inertias and the three elements of the center-of-mass vector. In practice, only a fraction of these ten dynamic parameters affect the torque/force $\tau_i(t)$ of link i. Identification of all of the parameters that affect the torque/force of link i involves estimating a fraction of the ten parameters of link i and a fraction of the $10(N-i)$ dynamic parameters of links $i+1$ through N that affect the torque/force of link i. We emphasize that we can only estimate the dynamic parameters that affect the joint torques/forces. An example is the I_{3xx} classical moment-of-inertia parameter of the cylindrical robot (in Section 3.2) that does not affect the dynamic robot model in (5)-(7) and hence cannot be identified.

The recovery (in Step 5) of the classical inertias from the pseudo-inertias may not be unique. This happens, for example, when a linear combination of the link inertias appears in an equation-of-motion. An example is the dynamic equation of the first link of the cylindrical robot in (8) for which we can estimate only the sum $(J_{1xx} + J_{1yy} + J_{2xx} + J_{2zz} + J_{3xx} + J_{3zz})$ of the link pseudo-inertias. Such an estimate is sufficient for computing the closed-form inverse dynamics in (8)-(10). If, however, the inverse dynamics are implemented by the recursive $N-E$ formulation (in Tables 2-4), the numerical values of all of the classical link inertias are required. To obtain numerical values for the classical link inertias, we constrain the identification problem by setting the values of five of the six classical link inertias to their nominal values and estimate the sixth accordingly. Even though the numerical values of these parameters may not have physical significance, they do lead to the arithmetically-correct closed-form expansion of the recursions of the Newton-Euler formulation (in Tables 2-4).

5. Identification Algorithms

In Section 4, we formulated the torque/force error model of an N degree-of-freedom manipulator. In this section, we cast the model into the series and parallel identifier structures and highlight the applicable parameter estimation algorithms [9].

5.1 Introduction

A parameter identifier consists of three components: the system to be identified, a postulated model and an adaptation algorithm which updates the model based upon an error criterion. If the adaptation algorithm updates the model based upon the error between the inputs to the model and the system, the structure is called the series or input error identifier. If the adaptation algorithm processes the difference between the outputs of the model and the system, the structure is called the parallel or output-error identifier [10,12]. The series (or input error) structure is depicted schematically in Figure 1 and the parallel (or output error) structure is depicted schematically in Figure 2.

In the robotics literature [2], the system is called the Forward Arm and the model is called the Inverse Arm. The Forward Arm (which is the physical robot) is actuated by input torques/forces and the outputs are the link positions, velocities and accelerations. The inputs to the Inverse Arm are the link positions, velocities and accelerations and the outputs are the link torques/forces. The Inverse Arm, which is the inverse model of the Forward Arm, plays the role of the inverse system in control engineering [12].

In practice, the inputs to the robot are the joint torques/forces and the outputs are the joint positions, velocities and accelerations. In the series structure, the outputs of the robot are used to compute the inverse dynamics torques/forces and the error signal, which is the difference between the applied and the computed input torques, drives the model identification algorithm. This input-error structure has direct applicability to on-line identification. In design and development, the engineer seeks to evaluate the robustness of estimation algorithms through simulation studies. Although the series identifier structure can be used for these practical studies, its implementation suffers from the intensive computational requirements of the forward arm simulation. The reduced computational requirements [13] of

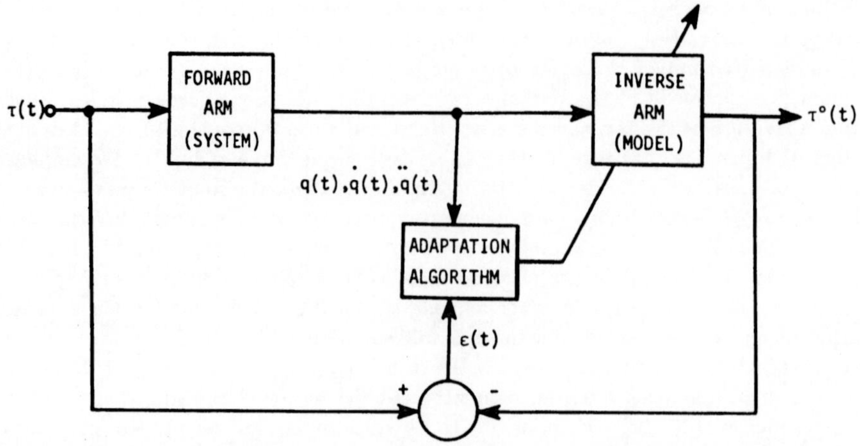

Figure 1: Series robot identifier structure.

the inverse dynamics formulation has direct applicability to this engineering evaluation. For simulation studies, we thus introduce the parallel (or output error) structure (in Figure 2). The difference between the output torques/forces of the system and the model is the error signal which drives the identification algorithm. Since the error signal (which is the difference between the system and the model torques/forces) is identical in both the series and parallel identifiers, the two structures can accommodate identical estimation algorithms. This is a practical advantage because the applicability and performance of an on-line estimation algorithm can be evaluated through off-line simulation studies.

5.2 Linear Estimation Algorithms

If the link masses and the kinematic (Denavit-Hartenberg) parameters are assumed to be known, then identifying the dynamic parameters is a problem of linear estimation. For dynamic parameter identification, there is thus a plethora of established estimation algorithms [11,12,20,24]. In our simulation experiments (in Section 6), we apply both least-squares and recursive least-squares algorithms.

5.3 Nonlinear Estimation Algorithms

If we relax the assumption that the link masses or the kinematic parameters are known, robot parameter identification becomes a problem of nonlinear estimation. This problem is more complicated than its linear counterpart and an identification algorithm must be customized for each application.

For a robot (which can be modeled by exact dynamic equations-of-motion) initial estimates of the dynamic parameters are available from engineering drawings. We utilize practical knowledge about the range of these dynamic parameters to solve the nonlinear estimation problem by constrained optimization of the least-squares algorithm [19]. In this technique, the parameters are constrained to lie in a convex region and a solution outside the region is penalized.

6. Simulation Results

In this section, we highlight our preliminary simulation experiments for identifying the dynamic parameters of the cylindrical robot and the first three degrees-of-freedom of the Stanford manipulator. In our simulation experiments, we generated two models. The first models the physical arm and the second is the model (or inverse arm). To reduce the computational

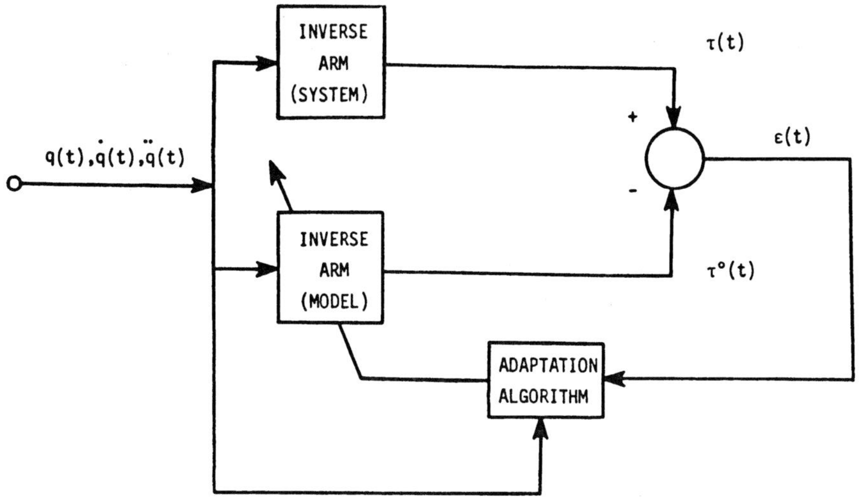

Figure 2: Parallel robot identifier structure.

effort, we conducted the experiments with the parallel (output error) identifier. We commanded each joint to move in a sinusoidal trajectory, beginning from the rest position of zero degrees to ninety degrees and returning in one second. We sampled the trajectory at 20 ms intervals and used the first-half of the trajectory (or 25 data points) in our identification experiments. We applied both the least-squares and the recursive least-squares algorithms under the assumption that the input-output measurements were noise-free. This assumption is justified practically by the current availability of high resolution resolvers (16 bits/revolution) and tachometers [26].

We call the measurement of the applied torque/force and the position, velocity and acceleration of a link (at one point along the trajectory) one set of data. In Table 9, we summarize our simulation results for the cylindrical robot.

We first implemented the least-squares algorithms for the torque/force error model in (21) of the third link using the 25 data sets and identified the z-component of the center-of-mass vector of the third link s_{3z}. We then identified the sum of the link pseudo-inertias from the torque/force error model of the first link by again using the 25 data sets along the trajectory. In practice, we initialize the least-squares algorithms with parameter values from engineering drawings of the manipulator. Our simulation results, listed in the estimated value column, match exactly the entries in the true value column.

Our simulation experiments for the Stanford manipulator are highlighted in Table 10. We used the 25 data sets to identify successfully the dynamic parameters which affect the torques/forces of each of the three links. The recursive least-squares algorithm computed identical parameter values (from the first 5 data sets for the first link, the first 10 data sets for the second link and the first 3 data sets for the third link) when we used the stopping criterion that the difference between two successive parameter estimates be less than 10^{-5}.

The recursive least-squares algorithm updates its estimates of the dynamic parameters in response to each new data set and stops the estimation process when the user-specified convergence criterion is satisfied. The recursive least-squares algorithm thus uses the minimum number of data sets required for convergence. In contrast, the least-squares algorithm processes simultaneously all of the data sets to estimate the dynamic parameters. From our experience, we need about twice the number of data sets as the number of the parameters being estimated.

Table 9: Simulation results for the Cylindrical Robot

Link	Dynamic Parameter (Dimensions)	Initial Value	True Value	Estimated Value
3	$s_{3z}(m)$	0.3	0.5	0.5
1	$J_{1xx} + J_{1yy} + J_{2xx} + J_{2zz} + J_{3xx} + J_{3zz}$ $(kg - m^2)$	2.0	2.5	2.5

Table 10: Simulation results for the Stanford Manipulator

Link	Dynamic Parameter (Dimensions)	Initial Value	True Value	Estimated Value
3	$s_{3x}(m)$	0.0	0.1	0.1
	$s_{3y}(m)$	0.0	0.1	0.1
	$s_{3z}(m)$	-0.56	-0.7	-0.7
2	$J_{3xx} + J_{2xx}(kg - m^2)$	0.3	0.1	0.0999
	$J_{3xy} + J_{2xy}(kg - m^2)$	0.05	0.1	0.1001
	$J_{3xz} + J_{2xz}(kg - m^2)$	0.3	0.1	0.0999
	$J_{3yz} + J_{2yz}(kg - m^2)$	0.1	0.2	0.2
	$J_{3zz} + J_{2zz}(kg - m^2)$	0.4	0.6	0.5999
	$s_{2x}(m)$	0.1	0.2	0.2
	$s_{2z}(m)$	0.1	0.15	0.15
1	$J_{2yy} + J_{1xx} + J_{1zz}(kg - m^2)$	2.0	2.3	2.2999
	$s_{2y}(m)$	0.05	0.1	0.1001

7. Practical Implementation Issues

Our preliminary simulation results are encouraging and show that the dynamic parameters can be estimated accurately. Achieving such accuracy may be impractical because of the finite dynamic ranges of the computer, the analog-to-digital and digital-to-analog interfaces, and the sensors. For example, in our six degree-of-freedom articulated direct drive DD Arm II [13,26], we use a 16-bit absolute resolver to measure the angular position of the joints and a 12-bit digital-to-analog converter (DAC) to transmit the computed torque values to the AC servomotors. Even if our estimates of the dynamic parameters were perfect, we would not be able to apply the correct torque values because of the finite 12-bit dynamic range of the DAC.

The computational requirements of the identification algorithms are intimately related to the desired estimation accuracy. We can reduce these computational requirements by specifying the estimation accuracy of the robot parameters in terms of the dynamic ranges of the devices used for implementing the robot control algorithms. To reduce the effects of these quantization errors, enhanced analog and digital feedback control algorithms, which are robust in the presence of modeling and parameter errors, and unmodeled dynamics, have been introduced recently in the computed-torque framework [27,28].

8. Conclusions

In this paper, we have addressed robot parameter identification. We have compared the Newton-Euler and Lagrange-Euler formulations and identified three fundamental properties which enable us to formulate the robot parameter estimation problem. We then transformed the nonlinear (in dynamic parameters) dynamic robot model based upon the Newton-Euler formulation into the equivalent linear (in dynamic parameters) dynamic robot model based upon the Lagrange-Euler formulation. The torque/force error model was then formulated from the linear (in dynamic parameters) dynamic robot model, and cast into the series and parallel identifier structures for on-line and off-line estimation, respectively. Our identification algorithm is directly amenable to the real-time identification of the pay-load inertial characteristics and can estimate on-line the dynamic frictional coefficients for accurate trajectory control. We are currently directing our research efforts towards implementing our identification scheme to estimate the dynamic parameters of the DD Arm II.

A Torque/Force Error Model of the Stanford Manipulator

In this appendix, we present the torque/force error model of the first three degrees-of-freedom of the Stanford manipulator [1]. We applied ARM [25] to develop the dynamic equations-of-motion of the Stanford manipulator and followed the approach introduced in Section 4 to generate the torque/force error model in (21) for on-line estimation. Throughout this appendix, we assume that the Denavit-Hartenberg parameters and the link masses of the Stanford manipulator are known.

The vectors of unknown dynamic parameters θ_i and the nonlinear vector functions of the kinematic parameters, link masses and output measurements $\varphi_i(t)$, for the 3 DOF positioning system of the Stanford manipulator, are:

$$\begin{aligned}
\theta_3^T &= [s_{3x}\ s_{3y}\ s_{3z}] \\
\varphi_3^T(t) &= [-m_3(\ddot{\theta}_2 + C_2 S_2 \dot{\theta}_1^2)\ \ -m_3 S_2 \ddot{\theta}_1\ \ -m_3(\dot{\theta}_2^2 - S_2^2 \dot{\theta}_1^2)] \\
\theta_2^T &= [J_{3xx} + J_{2xx}\ \ J_{3xy} + J_{2xy}\ \ J_{3xz} + J_{2xz}\ \ J_{3yz} + J_{2yz}\ \ J_{3zz} + J_{2zz}\ \ s_{2x}\ \ s_{2z}] \\
\varphi_2^T(t) &= [\ddot{\theta}_2 + C_2 S_2 \dot{\theta}_1^2\ \ S_2 \ddot{\theta}_1\ \ \dot{\theta}_1^2(S_2^2 - C_2^2)\ \ -C_2 \ddot{\theta}_1, \\
&\quad \ddot{\theta}_2 - C_2 S_2 \dot{\theta}_1^2\ \ m_2(d_2 S_2 \ddot{\theta}_1 - g C_2)\ \ m_2(-d_2 C_2 \ddot{\theta}_1 - g S_2)] \\
\theta_1^T &= [J_{2yy} + J_{1xx} + J_{1zz}\ \ s_{2y}] \\
\varphi_1^T(t) &= [\ddot{\theta}_1\ 2 m_2 d_2 \ddot{\theta}_1]
\end{aligned}$$

where $C_2 = cos(\theta_2)$ and $S_2 = sin(\theta_2)$.

In Section 4, we noted that:

- Each link of a robot is characterized by a maximum of ten dynamic parameters: the link mass, six classical inertias and the three elements of the center-of-mass vector. Since we assume that the link masses are known, we need to identify a maximum of nine parameters for each link; and
- In practice, we can identify only a fraction of the nine dynamic parameters that affect the torque/force $\tau_i(t)$ of link i.

From the torque/force error model of the positioning system of the Stanford manipulator, we observe that:

- The torque of the third link is affected only by the three elements of the center-of-mass vector s_3 which can be identified.
- The torque of the second link is affected by five pseudo-inertias of the third link, five pseudo-inertias of the second link and two elements of the center-of-mass vector of the second link. Only seven of these twelve parameters affecting the torque of the second link can be identified because the pseudo-inertia elements appear in linear combination in the vector θ_2 of unknown dynamic parameters; and
- The torque of the first link is affected by two independent dynamic parameters (the sum of one pseudo-inertia of the second link and two pseudo-inertias of the first link, and the y component of the center-of-mass vector of the second link) which appear in the parameter vector θ_1 and can be identified.

In summary, for the three degree-of-freedom positioning system of the Stanford manipulator, we can identify nineteen (twelve of which are independent) of the twenty-seven unknown dynamic parameters.

Acknowledgments

This research is supported by the Office of Naval Research through Contract Number N00014-81-0503, and the Robotics Institute and the Department of Electrical and Computer Engineering, Carnegie-Mellon University. The authors express their appreciation to Professor Takeo Kanade, Head of the Vision Laboratory (Robotics Institute, Carnegie-Mellon University), for his support and cooperation throughout the course of their research.

References

[1] R.P. Paul, Robot Manipulators: Mathematics, Programming and Control, MIT Press, Cambridge, Ma., 1981.

[2] M. Brady, et al. (editors), Robot Motion: Planning and Control, MIT Press, Cambridge, Ma., 1982.

[3] J.Y.S. Luh, M.W. Walker and R.P. Paul, "On-Line Computational Scheme for Mechanical Manipulators," *Journal of Dynamic Systems, Measurement, and Control*, vol. 102, No. 2, pp. 69-76, June 1980.

[4] C.S.G. Lee, "Robot Arm Kinematics, Dynamics and Control," *Computer*, vol. 15, No. 12, pp. 62-80, December 1982.

[5] C.P. Neuman and J.J. Murray, "Computational Robot Dynamics: Foundations and Applications," *Journal of Robotic Systems*, vol. 2, No. 4, 1985.

[6] C.P. Neuman and V.D. Tourassis, "Robot Control: Issues and Insight," Proceedings of the Third Yale Workshop on Applications of Adaptive Systems Theory, K.S. Narendra, ed., Yale University, New Haven, Ct., June 15-17, 1983, pp. 179-189.

[7] B.R. Markiewicz, "Analysis of the Computed-Torque Drive Method and Comparison with the Conventional Position Servo for a Computer-Controlled Manipulator," Technical Memorandum 33-601, Jet Propulsion Laboratory, Pasadena, Ca. March 1973.

[8] J.Y.S. Luh, M.W. Walker and R.P. Paul, "Resolved-Acceleration Control of Mechanical Manipulators," *IEEE Transactions on Automatic Control*, vol. 25, No. 3, pp. 468-474, June 1980.

[9] C.P. Neuman and P.K. Khosla, "Identification of Robot Dynamics: An Application of Recursive Estimation," Proceedings of the Fourth Yale Workshop on Applications of Adaptive Systems Theory, K.S. Narendra, ed., Yale University, New Haven, Ct., May 29-31, 1985, pp. 42-49.

[10] Y.D. Landau, Adaptive Control: The Model Reference Approach, Marcel Dekker, New York, 1979.

[11] K.S. Narendra and R.V. Monopoli, (editors), Applications of Adaptive Control, Academic Press, New York, 1980.

[12] G.C. Goodwin and K.S. Sin, Adaptive Filtering, Prediction and Control, Prentice-Hall, Englewood-Cliffs, N.J., 1984.

[13] P.K. Khosla and C.P. Neuman, "Computational Requirements of Customized Newton-Euler Algorithms," *Journal of Robotic Systems*, vol. 2, No. 3, pp. 309-327, 1985.

[14] N.M. Swartz, "Arm Dynamics Simulation," *Journal of Robotic Systems*, vol. 1, No. 1, pp. 83-100, 1984.

[15] J.J. Murray and C.P. Neuman, "Symbolic Linearization of the Newton-Euler Dynamic Robot Model," Technical Report, Department of Electrical and Computer Engineering, Carnegie-Mellon University, 1984.

[16] J. Denavit and R.S. Hartenberg, "A Kinematic Notation for Lower Pair Mechanisms Based on Matrices," *Journal of Applied Mechanics*, vol. 77, No. 2, pp. 215-221, June 1955.

[17] L.D. Landau and E.M. Lifshitz, Mechanics, Pergamon Press, Oxford, 1976.

[18] S.V. Gusev and V.A. Yakubovich, "Adaptive Control Algorithm for a Manipulator," *Automation and Remote Control*, vol. 41, No. 9, pp. 1268-1277, September 1980.

[19] Y. Bard, Nonlinear Parameter Estimation, Academic Press, New York, 1974.

[20] L. Ljung and T. Soderstrom, Theory and Practice of Recursive Identification, MIT Press, Cambridge, Ma., 1983.

[21] C.P. Neuman and J.J. Murray, "Linearization and Sensitivity Functions of Dynamic Robot Models," *IEEE Transactions on Systems, Man, and Cybernetics*, vol. 14, No. 6, pp. 805-818, November/December 1984.

[22] M.S. Pfeifer and C.P. Neuman, "VAST: A Versatile Robot Arm Dynamic Simulation Tool," *Computers in Mechanical Engineering*, vol. 3, No. 3, pp. 57-64, November 1984.

[23] W.M. Silver, "On the Equivalence of Lagrangian and Newton-Euler Dynamics for Manipulators," *The International Journal of Robotics Research*, vol. 1, No. 2, pp. 60-70, Summer 1982.

[24] G. Solbrand, A. Ahlen and L. Ljung, "Recursive Methods for Off-Line Identification," *International Journal of Control*, vol. 41, No. 1, pp. 177-191, January 1985.

[25] J.J. Murray and C.P. Neuman, "ARM: An Algebraic Robot Dynamic Modeling Program," Proceedings of the First International IEEE Conference on Robotics, R.P. Paul, ed., Atlanta, Ga., March 13-15, 1984, pp. 103-114.

[26] T.K. Kanade, P.K. Khosla and N. Tanaka, "Real-Time Control of the CMU Direct Drive Arm II Using Customized Inverse Dynamics," Proceedings of the 23rd IEEE Conference on Decision and Control, M.P. Polis, ed., Las Vegas, Nv., December 12-14, 1984.

[27] V.D. Tourassis and C.P. Neuman, "Robust Nonlinear Feedback Control for Robotic Manipulators," *IEE Proceedings - D: Control Theory and Applications, Special Issue on Robotics*, vol. 132, No. 4, pp. 134-143, July 1985.

[28] C.P. Neuman and V.D. Tourassis, "Robust Discrete Nonlinear Feedback Control for Robotic Manipulators," Technical Report, Department of Electrical and Computer Engineering, Carnegie-Mellon University, 1985.

LEARNING SYSTEMS

The trouble with people is not that they don't know but that they know so much that ain't so.

Josh Billings
(Henry Wheeler Shaw)

A cursory review of the papers in this section reveals striking differences in comparison with the remainder of the book, both in the general approach used as well as in specific details. The adaptive algorithms used in the latter are deeply rooted in the theory of dynamical systems and differential equations. In contrast to this the first three papers in this section have their origins in mathematical learning theory, while the last three are closer in spirit to the rule-based expert systems of artificial intelligence. Their inclusion in this volume shows our deep conviction that, even though stated differently, these learning paradigms are closely related in many ways to adaptive control paradigms. The questions of decentralization, convergence and robustness which arise in the control of most complex systems with large uncertainties can, in our opinion, be effectively attempted only by combining the powerful tools developed in these diverse fields.

The first paper by Narendra and Wheeler surveys the major developments in learning automata theory in the last two decades. They argue that the automaton approach has the greatest potential in problems involving many decision makers interacting in a decentralized manner and discuss a recent result in the decentralized control of Markov chains. Based on the evidence of extensive applications, they also contend that the problem must possess certain specific characteristics if the control algorithms are to be most effective.

The problem of routing in telephone networks represents one of the successful applications of learning algorithms. In their paper, Mason and Gu extend the application of such algorithms to packet switching networks and compare the power performance, transient response and stability characteristics of various control architectures. A novel approach to image data compression is proposed by Hashim, Amir and Mars. Based more on identification than control, the method uses automata at the transmitting and receiving ends of a communication channel to adaptively code and decode the data being sent. While the initial experiments are promising, decisions as to whether this is a viable area of application of learning automata theory can be made only after the approach has been compared with well known methods for data compression.

The last three papers present intriguing models of behavior under uncertainty. Barto, Anandan and Anderson combine aspects of learning automata theory to supervised pattern classification tasks. By properly coding the states, they claim that experience based on some environmental states can be generalized to other states as well, reducing storage requirements and increasing learning rate. They also suggest the possibility of solving difficult nonlinear associative learning problems using networks of cooperating interconnected learning automata. Goldberg discusses genetic algorithms involving reproduction, crossover, and mutation and points out the differences between such algorithms and normal search methods encountered in engineering optimization. The Animat model of Wilson performs actions which alter the signals it receives from the environment. How it acts both externally and internally to optimize the occurrence of specific signals is described. These papers, while interesting in their own right, achieve a broader significance by introducing novel problem domains which may provide challenging environments for the study of many other adaptive and learning techniques.

Recent Developments in Learning Automata

Kumpati S. Narendra and Richard M. Wheeler, Jr.
Center for Systems Science
Yale University

Abstract

The paper surveys the major developments in learning automata theory and applications in the last two decades. Since a survey article on the subject appeared in 1977, the emphasis is on subsequent developments. Of particular importance is some recent work on the use of many automata interacting in a decentralized manner. This framework provides a conceptual focus and an analytical basis for future research on modeling and control of complex systems. Applications of the theory are also reviewed, with special attention devoted to the problem of traffic routing in telecommunication networks.

1. Introduction

Adaptation and learning are terms used to describe behavior modification in natural organisms as well as machines. Biologists and mathematical psychologists have been primarily concerned with the modeling questions associated with these phenomena while systems theorists have addressed the problem of synthesizing machines which exhibit these properties. Simple quantitative models have been studied in various disciplines which deal with the sequential choice of actions from a finite set to optimize responses from a random environment. It is now generally realized that the design of experiments in statistics, learning automata in Systems Theory, the k-armed bandit problem in Operations Research and stimulus sampling models in Mathematical Psychology have a common statistical basis.

The subject of this paper is the intuitive yet analytically tractable paradigm of the learning automaton. An attempt is made to survey the major developments in learning automata theory and applications in the last two decades and is written with three main objectives. The first is to summarize the theory that has been developed over this period and to convey the spirit of the approach. Since two other survey articles have appeared in 1974 and 1977 [1,2], the emphasis is on developments since the latter article. The second objective is to place special emphasis on current work dealing with many automata interacting in a decentralized manner and to suggest that this area forms a compelling rationale for the future study of learning automata. The third objective is to survey recent and current applications of the theory. The aim here is to identify some characteristics of systems which make the use of learning automata attractive for control purposes.

The simple model of an organism choosing actions sequentially in a random environment has been extensively studied over the past 25 years. Early work is well documented in [3] and [4]. A rigorous mathematical framework for the study of learning models is found in [5] and [6]. Tsetlin [7] introduced the use of deterministic automata operating in random environments as models of learning. This was followed by the concept of variable structure stochastic automata, in which the probabilities of the actions of an automaton are updated on the basis of the response of the environment to one of its actions [8]. Since then, two principal avenues have been taken, primarily concerned with the variable structure formulation. The first has been to develop new algorithms that exhibit various types of behavior and to clarify the subtle stochastic convergence questions arising in such models. The second has been to extend the basic model to address such important topics as nonstationary environments, multiple environments and decentralized automata interacting in various ways. A special volume on learning automata [9] and a recent book [10] provide many of the details associated with both avenues up to about 1980.

Of all the extensions considered to date, the most significant is the study of decentralized automata interacting through competition and cooperation. The potential for local learning schemes producing desirable global behavior has long been recognized (e.g.[11]). However, theoretical results of the last few years, discussed in section 3, have illuminated in detail the power as well as the limitations of decentralized automata. Recently, networks of automaton-like elements have been suggested for use in conjunction with supervised learning schemes to solve difficult pattern classification tasks [12]. Such an approach appears to have application to the study of neural nets in biology and artificial intelligence. It appears that the theoretical results of learning automata form a tractable method for analyzing such systems. Of particular importance is the result for a class of automata games in which the whole group, even though decentralized, achieves global optimality [37]. Another learning approach relevant to the automaton formulation is the genetic algorithm [13]. Such an algorithm can be described as a probabilistic "survival of the fittest" search procedure and is discussed elsewhere in this collection of papers [14,15].

Numerous applications of learning automata theory have been suggested in the past (e.g. see [1]). However, the most successful one to date is that of routing in telecommunication networks. This application was first proposed in [16] and further developed in [17-19]. Methods for modeling the network as a nonstationary environment were proposed in [20-22]. The theoretical studies, combined with extensive computer simulations, have led to the stage where the use of automata for network routing is a viable method. A comprehensive discussion of this approach is contained in [23] and [24]. This problem, as well as the problem of flow control in networks (e.g.[25]), underscores the importance of understanding the group behavior of decentralized automata. A new application in the area of data compression has also been proposed [26] and the preliminary results appear promising.

2. Theoretical Basis

As mentioned in the introduction, we view the principal motivation for the study of learning schemes to be situations involving many decentralized automata. This problem has a greater richness for dealing with complex systems than do the topics of nontationary or multiple environments, also mentioned in the introduction. However, the general formulation of the problem, for example as an abstract game, yields little insight into overall learning behavior without an understanding of what can be expected of a single automaton operating alone. Hence, results for the single automaton case are useful in the many-automata case for two reasons. First, in some cases the results can be extended to multi-automata problems and the analysis in the latter is much simplified. Second, knowledge of a single automaton's behavior gives a clue as to how difficulties peculiar to the game problem arise, not just that they do arise. Therefore, the basic notation, problem statement and convergence properties of the single automaton are presented below.

Since most of the content of this section has appeared in the two earlier surveys [1,2], the aim is only to provide a context on which subsequent discussion in sections 3 and 4 can draw. Also, comments will be made where appropriate to indicate how experience with the automaton approach has influenced current thinking about the fundamental topics.

a. The Basic Model

A learning automaton is a feedback system connecting an automaton, which chooses an action at each time instant, and an environment, which produces responses to those actions. In a modeling context, where a simple environment can be imposed, the complexity is in the description of a learning mechanism which can explain observed behavior. In the control context, of principal interest here, the complexity of the system is reflected by the complexity of the environment. Different applications require different environment models and it is important that the performance of automata remain attractive over as wide a range of environments as possible. The term 'basic model' is useful to describe the simplest variable structure stochastic automaton and refers to the following environment and automaton pair.

(i) Environment. A general environment model has as its input an element $\alpha(n) \in \underline{\alpha} = \{\alpha_1, \alpha_2, ..., \alpha_r\}$, where $\underline{\alpha}$ is the finite action set of the automaton and as its output (or response) a random variable $\beta(n) \epsilon [0, 1]$, where 0 and 1 correspond to the maximum degree of success and failure, respectively. This normalization of the output makes easy the task of conserving probability measure in the learning schemes described below. The environment is characterized by the set $\underline{c} = \{c_1, c_2, ..., c_r\}$ where c_i is the expected failure when the input is α_i.

$$c_i = E[\beta(n)|\alpha(n) = \alpha_i]. \tag{1}$$

The values c_i are unknown and it is assumed that $\{c_i\}$ has a unique minimum.

Virtually all of the early convergence results considered only the case where $\beta(n)$ is binary, taking on the values 0 or 1. The definition (1) allows $\beta(n)$ to assume either an arbitrary finite number of values or any value in [0,1]. It has been found that the general case is typically more useful as an environment model than the binary case and examples are given in sections 3 and 4. An important fact, however, is that the same convergence analysis used for the binary model also applies to the general model. Hence, theoretical results are often given for the binary model for ease of exposition even when the natural application is for the general model.

(ii) Automaton. The automaton chooses an action $\alpha(n)$ from its action set according to its current action probability distribution $p(n) = [p_1(n), p_2(n), ..., p_r(n)]$ and subsequently receives the environmental response $\beta(n)$, defined above. The vector $p(n)$ is then updated using the learning (or reinforcement) scheme

$$p(n+1) = T[p(n), \alpha(n), \beta(n)] \tag{2}$$

where T is a mapping which determines the action probabilities at stage $n+1$ on the basis of the action probabilities at stage n, the action $\alpha(n)$ chosen and the response $\beta(n)$ received.

The basic design problem is to specify T in such a manner that the automaton improves its performance in some sense. Typically, this means decreasing over time the performance index

$$M(n) \triangleq E[\beta(n)|p(n)] = \sum_{i=1}^{r} p_i(n) c_i \tag{3}$$

where $M(n)$ is the conditional average failure at stage n. Since $p(n)$ is also random, the performance of the automaton is judged by the asymptotic behavior of $E[M(n)]$. An automaton is said to be optimal if $\lim_{n\to\infty} E[M(n)] = \min_i\{c_i\} = c_\ell$. A slightly weaker form of behavior, $\epsilon - optimality$, occurs if $\lim_{n\to\infty} E[M(n)] < c_\ell + \epsilon$ can be realized by the proper choice of the learning scheme parameters, for an arbitrary $\epsilon > 0$.

(iii) Learning Schemes. The basic idea behind a learning scheme is rather simple. If the automaton chooses an action α_i which results in a favorable response (low value of $\beta(n)$), the probability $p_i(n)$ is increased and all other components of $p(n)$ are decreased. Similarly $p_i(n)$ is decreased if the response is unfavorable. The precise manner in which $p_i(n)$ is to be increased or decreased is given by the learning scheme and determines the asymptotic behavior of the automaton. Since most of the schemes currently in use were known at the time of the survey paper [1], the reader is referred to that article for further details on specific algorithms.

A general learning scheme can be defined as follows:

$$\begin{aligned} p_i(n+1) &= p_i(n) - [1-\beta(n)] g_i(p(n)) + \beta(n) h_i(p(n)), && \text{if} \alpha(n) \neq \alpha_i \\ p_i(n+1) &= p_i(n) + [1-\beta(n)] \textstyle\sum_{j\neq i} g_j(p(n)) - \beta(n) \textstyle\sum_{j\neq i} h_j(p(n)), && \text{if} \alpha(n) = \alpha_i \end{aligned} \tag{4}$$

where g_j and h_j are non-negative continuous functions satisfying the conditions $0 < g_j(p(n)) < p_j$ and $0 < \sum_{i \neq j}[p_i + h_i(p(n))] < 1$ for all j = 1,2,...r. In the binary model, where $\beta(n) \in \{0, 1\}$, note that there are four possibilities for changes in probabilities. $p_i(n)$ can increase if $\alpha(n) = \alpha_i$ and the outcome is a success ($\beta(n) = 0$) or $\alpha(n) \neq \alpha_i$ and the outcome is a failure ($\beta(n) = 1$). Similarly there are two ways in which $p_i(n)$ can decrease. The problem is to determine the 2r functions g_j and $h_j (j = 1, 2, ..., r)$ so that the automaton exhibits $\epsilon - optimal$ behavior.

The scheme (4) becomes linear if g_j and h_j are chosen to be $g_j(p(n)) = ap_j(n)$ and $h_j(p(n)) = b[\frac{1}{r-1} - p_j(n)]$ with $0 < a < 1$ and $0 \leq b < 1$. a is called the reward parameter and b the penalty parameter. Three linear schemes have been extensively analyzed and are found to exhibit interesting and significantly different behavior. In the L_{R-I} (Linear Reward-Inaction) scheme, $b = 0$ and all probabilities remain unchanged if $\beta(n) = 1$. In an L_{R-P} (Linear Reward-Penalty) scheme, $b = a$. Finally, in an $L_{R-\epsilon P}$ scheme, the penalty parameter b is much smaller than the reward parameter. While many nonlinear schemes have also been proposed, these three linear schemes are sufficient to produce the behavior modes observed in all learning automata.

(iv) Convergence. Detailed convergence analysis of the above learning schemes has attracted much attention over the past decade and a unified approach based on the theory of compact Markov processes is now well understood. The approach has been outlined in [2] and the details are presented comprehensively in [10]. For the present discussion, it is sufficient to note that schemes of the form (4) are distance-diminishing [6] and hence asymptotic analysis is possible. As mentioned above, the L_{R-I}, L_{R-P} and $L_{R-\epsilon P}$ schemes show markedly different behavior and in each case we are interested in the evolution of the mean and variance of $p(n)$ as well as the sample path behavior. In the L_{R-I} case, every sample path converges to choosing only one action with probability one. The problem of interest lies in determining the probability of converging to the optimal action. A bound on this probability can be obtained [6] which depends upon the reward parameter a and can be made arbitrarily close to one if a is sufficiently small. Hence, the L_{R-I} scheme is $\epsilon - optimal$.

In contrast, the L_{R-P} and $L_{R-\epsilon P}$ schemes do not exhibit sample path convergence to a single action. These schemes are ergodic and hence converge in distribution to a limiting distribution (on every sample path) which is independent of the initial distribution p(0). It is found that the mean of the limiting distribution depends upon the size of b relative to a while the variance depends upon a (a smaller a gives a smaller variance). These moments are adequate since the limiting distribution is asymptotically Gaussian in the vicinity of its mean value. Like the L_{R-I} scheme, the $L_{R-\epsilon P}$ scheme is $\epsilon - optimal$, but the L_{R-P} is not.

From a practical viewpoint, the behavior of the L_{R-I} and $L_{R-\epsilon P}$ schemes is similar. Although analysis of L_{R-I} schemes is often easier in modified environments or collective situations, in practice the $L_{R-\epsilon P}$ scheme would be used to avoid the possibility of becoming forever locked in on a nonoptimal action. This is important when the environment is non-stationary, a problem frequently encountered in adaptive and learning systems.

(v) Recent Developments. While most of the recent efforts in learning automata have been in applications and extensions to the basic model, some attempts have been directed at the basic model itself. The idea of making explicit use of estimates of the c_i has been around for a long time (e.g.[27]). However, only recently has it been incorporated directly into the scheme (4) [28]. This has the effect of improving the rate of convergence without sacrificing accuracy. Even so, the convergence rate remains a serious problem if the number of actions is more than three or four.

It is now recognized that updating need not occur at every time step. An averaging method, which uses the same action repeatedly over an iterval and is updated with a large step size, will also work [29]. In this scheme the updating is based on the total degree of performance normalized by the length of the interval. This idea also appears in the multiple environment extension (see below) as well as in the decentralized control of Markov chains (see section 3).

Also related to the convergence rate issue is the problem of finite time behavior. While the known theoretical results are all asymptotic, an important question still needing detailed analysis is the trade-off between the number of steps available for decision making and the confidence in the best finite-time decision.

b. Extensions to the Basic Model

While the detailed analysis of the basic model has been essential to understanding the behavior of various automata, the real motivation for the use of learning comes from more complex situations. The three principal extensions which continue to be of interest are non-stationary environments, one automaton in multiple environments, and many automata in one environment. Of these, the most general, from both modeling and control viewpoints, is the problem of many automata interacting in a decentralized manner, the subject of section 3. Here, we merely summarize the general results recently available in these three areas.

(i) Nonstationary Environments. In the basic model, the c_i are assumed to be constant. However, this description of the environment in many situations is inadequate. For example, when an automaton is used to route traffic in a network, as described in section 4, the c_i are affected by the actions of the automaton. This led to the introduction of abstract nonstationary models [20,21] as well as dynamic models of the environment [22], needed to explain observed transient behavior. Under certain assumptions on the nature of the nonstationarity, namely that

(i) $c_i(p)$ are continuous functions of $p (i = 1, 2, ... r)$

(ii) $\frac{\partial c_i}{\partial p_i} > 0 \quad \forall i$

(iii) $\frac{\partial c_i}{\partial p_i} >> \frac{\partial c_i}{\partial p_j} \qquad \forall i, j \qquad i \neq j,$

it is shown [21] that the L_{R-P} scheme converges in distribution to a limiting distribution whose mean value is close to a value p^* (differs by 0(a)), where p^* is given by $p_i^* c_i(p^*) = p_j^* c_j(p^*) \forall i, j$. In the $L_{R-\epsilon P}$ case, the same analysis results in a different equalizing behavior, with the mean of the limiting distribution close to p^{**} where $c_i(p^{**}) = c_j(p^{**}) \forall i, j$. The dynamic nonstationary environment in [22] exhibits identical steady-state behavior, but more reasonable transient behavior, and the result appears to carry over to environments described by difference equations of arbitrary order.

In a different direction, the use of an intermediate parameter vector $\theta(n)$, on which the action probabilities depend, has been suggested as a way of coping with a simple nonstationary environment [12]. Here the environment can be in any of a finite number of states, known to the automaton, and the object is to update $\theta(n)$ so as to asymptotically learn to choose the best action in each environment. The environment changes state randomly, but independently of the actions of the automaton. The learning scheme is similar in spirit to the L_{R-P} scheme and the result is given in [12] as an alternative approach to two-class pattern classification problems.

(ii) Multiple Environments. The problem of a single automaton operating in many environments simultaneously has been studied by several authors [30-32] and is essentially equivalent to the problem of multi-objective decision making. Optimality must have a meaning and therefore regularity conditions are assumed about the environments in [30-32]. For example, in [32] it is shown that the scheme (4) results in $\epsilon-$*optimality* if there are N binary environments provided that

$$\sum_{j=1}^{N} c_{\ell j} < \sum_{j=1}^{N} c_{ij} \qquad \forall i \neq \ell \tag{5}$$

where c_{ij} represents the failure probability of the jth environment if the ith action is used. Here α_ℓ is the optimal action and $\beta(n) = \frac{m}{N}$, where m is the number of failures obtained at time n. This type of weighting of responses to form $\beta(n)$ has also been used to study the behavior of interconnected automaton-environment pairs [33], discussed further in section 3.

(iii) Games of Automata. The possibility of using collectives of learning automata to model the complex behavior of large self-organizing systems led to the formulation of automata game problems [34]. Recent interest in the design of large networks of decentralized yet interconnected components continues to motivate the study of automata behavior in abstract games. Much of the early work on games is contained in the surveys [1] and [2]. More recently, rigorous results have been obtained for the two-automaton zero-sum game [35,36] and the N-automata identical payoff game [37-39], also referred to as a 'cooperative' game. In the former it is shown that if the two automata use L_{R-I} or $L_{R-\epsilon P}$ schemes, they can converge arbitrarily closely to the optimal min-max strategies by the proper choice of the learning parameters. In the latter, it is shown that the use of L_{R-I} schemes by all players results in convergence to the group optimum, provided a unique Nash equilibrium point exists.

The power of these results lies in the fact that no information about the game is assumed by the players a priori and no communication among players occurs. The players operate as though they were each in the environment of the basic model and yet can still perform optimally in the game. This intriguing property is relevant to more general decentralized decision making problems, the subject of the next section. In the much more complex domain of N player nonzero-sum games, very little is known about automata behavior and its relationship to game theoretic concepts of group rationality such as the core, the various value solutions or even Pareto optimality. Some preliminary recent work on automata behavior in nonzero-sum games includes [40] and [41]. In general, only when the game has very special structure, such as dominant strategies or a unique equilibrium, can known automata results be applied directly to describe the group behavior.

3. Decentralized Decision Making Using Automata

The model of a single automaton in a stationary random environment, discussed in section 2, has limited scope for modeling and control of complex systems. For example, if the number of actions is large or if a dynamical model is available, then the basic automaton model is not appropriate. The use of a nonstationary environment model extends the domain of applicability of the single automaton somewhat. However, it is in problems involving many decision makers interacting in a decentralized manner that the automaton approach has the greatest potential. The automata game is only an abstract representation of such a problem, but nonetheless provides a theoretical framework for the study of complex decision-making systems. The basic question is: how do agents who are interacting via some unknown mechanism behave when facing extreme uncertainty about the world in which they operate? Conceptually, automata have long been considered in such a framework [11,42]. However, it is only recently that rigorous analysis has begun to lend insight into this rich domain.

A significant practical problem which renewed interest in decentralized automata was that of routing traffic through a telephone network. This is discussed in section 4 as a major application, but is of interest here as a motivation for decentralized control problems in general. Consider the following situation. Let one automaton be used at each node of a network to route traffic from various sources to destinations. They do not communicate with each other and each changes its routing strategy based only on information concerning completion or blocking of past calls. Note that the completion rate depends upon the actions taken by all automata, so in effect they are participating in a game. In general, the structure of the game may be complex (e.g. many equilibria may exist). However, to the extent that the models of game theory are accurate models of decentralized decision making under uncertainty, a strong motivation exists for studying the behavior of automata in abstract games.

As stated in section 2, a number of powerful results on learning in games now exists.

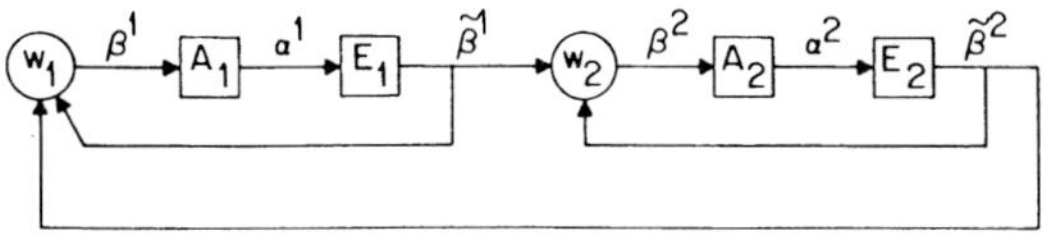

Figure 1: A simple feedback interconnection of automata.

However, a vast array of open problems remains, both in modeling decentralized systems as games as well as in studying the behavior of automata in games with general payoff structures. In the former area, a recent result on the decentralized control of unknown Markov chains, discussed in section 3b below, demonstrates the power of the automaton approach. It now seems apparent that this example is only one of many important problems that can be modeled as a decentralized automata problem.

a. Some Simple Decentralized Models

Decentralized models using learning automata have been recently introduced which retain tractability in that they can be represented as games with particular structures [33]. The basic idea is to consider interconnections of individual automaton-environment pairs, where the behavior of each pair is well understood. Two principal questions naturally arise: 1) Can such interconnections be found which model more complex systems adequately? and 2) What convergence properties does the collective have, given the known convergence properties of the individual pairs? In [33], a number of interconnections are presented including both synchronous and sequential models.

(i) Synchronous Models. In this class of models, regardless of the manner in which the automata are interconnected, each chooses an action at the same time and subsequently receives a response. In general, the responses to the automata are different. Further, each automaton updates its action probabilities at the same time. Fig. 1 shows a simple feedback interconnection and indicates the manner in which the output $\tilde{\beta}^1$ of automaton A_1's local environment E_1 can serve as the input, perhaps weighted with other signals, to A_2.

Since the mechanism of the interconnection is identical to that of a game, the interconnection can be represented by a game payoff structure. For two automata, as in Fig. 1, this is simply a $(r_1 \times r_2)$ matrix, where r_i is the number of actions of automaton A_i. The payoff to A_i is simply $w_{i1}\tilde{\beta}^1 + w_{i2}\tilde{\beta}^2$, where $w_i = (w_{i1}, w_{i2})$ and the local environments are assumed to be binary, $\tilde{\beta}^i \epsilon \{0, 1\}$. For this structure, each automaton has a dominant strategy, corresponding to its best action locally, provided the weight on its own response is not zero. As such, from each automaton's viewpoint the environment is nonstationary but very regular – the expected failure of the actions may change, but their ordering remains the same. It is known that an automaton facing this type of environment is $\epsilon -$ *optimal* [43].

Many other interconnections have been given and each one is amenable to an automata game formulation. For certain interconnections, the corresponding game has a structure for which automata behavior is known (e.g. dominance or identical payoff with a unique equilibrium). However, in some cases, the structure is of a general nonzero-sum type and little is known about convergence of standard algorithms such as L_{R-I}.

(ii) Sequential Models. In contrast to the above models, in sequential models only one automaton acts at a time. The action chosen determines who acts next and in addition generates a random response (output of a local environment). These models can be viewed as networks of controllers in which control passes from node to node. Clearly, the telephone traffic routing application is a particular example of a sequential model. We note that a

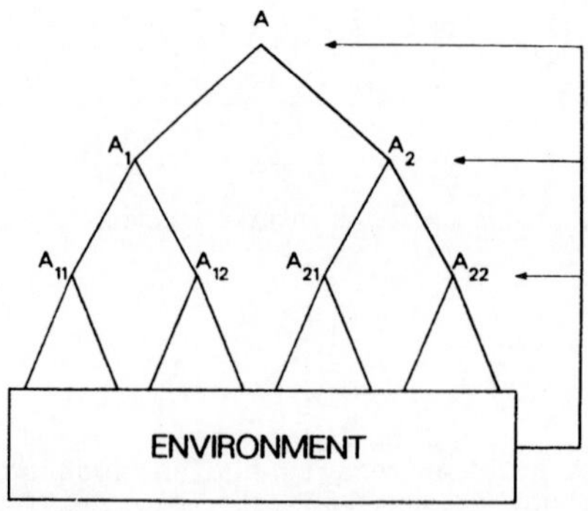

Figure 2: Hierarchy of automata–a simple sequential model.

node in a sequential model can represent a synchronous model, so it is possible to combine the two types in a more general model.

One example of a sequential model is the hierarchy shown in Fig. 2. This structure was proposed in [44] as a means of improving convergence rates, but the task was identical to that of the basic model. The difference is that instead of a single automaton with many actions, a group of several automata, each with a few actions was used. In Fig. 2 each has two actions, with those of the higher level automata corresponding to the selection of an automaton at the next lower level. An $\epsilon-$ *optimal* algorithm was given, but it required knowledge at each level about the action probabilities used at the next higher level. A heuristic approach to reorganizing the hierarchical structure on line to gain faster convergence in an arbitrary environment has also been suggested [45].

While in the scheme given in [44] all automata who participate in any path through the hierarchy update their probabilities simultaneously, this need not be so in general. Updating of an automaton might occur, for example, whenever control is passed to that automaton. A very general sequential model, the decentralized control of Markov chains using automata, was posed in [33] and subsequently analyzed in [39]. The method used is similar to that for the synchronous models and views the decentralized control problem as a game. This problem is discussed below.

b. Decentralized Control of Markov Chains: A Learning Approach

The sequential models described above provide the context for the application of decentralized automata to the control of unknown Markov chains. In particular, if each node of a sequential model corresponds to a state of a Markov chain, then the chain proceeds according to transition probabilities given by the actions chosen in the various states. A learning automaton is associated with each state and acts only when the chain is in that state. The automaton updating occurs only after the chain returns to that state and is based on the performance of the system during the intervening period.

While the Markov chain control problem has been studied before (e.g.[46,47]), the decentralized learning approach is new and attractive due to its simplicity. No explicit parameter estimation is required and, as in the automata game, there is no direct communication between the automata. By showing that the control problem has a special property, under an ergodicity assumption it is concluded that the automata, operating in nearly total ignorance of their environment, implicitly coordinate themselves to lead to $\epsilon-$ *optimal* behavior [39]. The problem statement and solution are summarized below.

(i) Statement of the Problem. The state space of a finite Markov chain $\{x(n)\}_{n\geq 0}$ is denoted by $\Phi = \{\phi_1, \phi_2, ..., \phi_N\}$. At every state ϕ_i, a finite action set $\alpha^i = \{\alpha^i_1, \alpha^i_2, ..., \alpha^i_{r_i}\}$

is defined. The one-step transition probability $t_j^i(k)$ and one-step reward $r_j^i(k)$ depend on the starting state ϕ_i, the ending state ϕ_j and the action α_k^i chosen in ϕ_i. The action $\alpha(n)$ is the action taken at stage n when the chain is in state $x(n)$. The performance criterion J is defined as the expected reward per transition and is given by

$$J \triangleq \lim_{n\to\infty} \frac{1}{n} E[\sum_{t=0}^{n-1} r(x(t), x(t+1),\ \alpha(t))] \tag{6}$$

where $r(x(t), x(t+1), \alpha(t))$ is the reward generated by a transition from $x(t)$ to $x(t+1)$ using action $\alpha(t)$. The control problem is to determine the policy α, consisting of one action at every state, which optimizes the performance index J. While centralized methods are known to precompute the solution to this problem if the parameters $t_j^i(k)$ and $r_j^i(k)$ are known [47], it is the fact that they are unknown that creates the adaptive or learning problem. Centralized indirect adaptive approaches to this problem have been suggested [48,49].

The decentralized approach discussed here to solve the above problem is to associate one agent, modeled as a learning automaton, with every state ϕ_i. When the process is in ϕ_i at stage n, automaton A_i chooses an action α_k^i using a probability distribution $p^i(n)$ on the set of allowable actions at state ϕ_i. On the basis of the response of the system, as described below, the learning scheme A_i updates $p^i(n)$. The principal features of the decentralized approach are the following:

1. The automaton A_i at state ϕ_i is unaware of the existence of automata at the other states, the elements of their action sets or the particular actions chosen by them at any instant. Even the number of states N is not needed by A_i.

2. Every automaton operates in the Markov chain environment exactly in the same fashion as it would if it were to be $\epsilon -$ *optimal* in a stationary environment. Specifically, we assume that each A_i uses an identical L_{R-I} algorithm.

3. When the chain is in state ϕ_i, automaton A_i uses the most recently available measurement of the system performance to update its action probabilities. This results in the implementation of a nonstationary randomized policy at every state. Optimality implies that the policy used by each A_i and hence the overall policy converges to a stationary nonrandomized policy.

(ii) Choice of Action at State ϕ_i. The action set of the automaton A_i was defined as $\{\alpha_1^i, \alpha_2^i, \ldots, \alpha_{r_i}^i\}$. It is assumed that the first time the Markov chain enters state ϕ_i, automaton A_i chooses its actions with equal probabilities. If the Markov chain is at state ϕ_i at time $t = 0$, an action is chosen by automaton A_i. This results in a transition from the state ϕ_i to a new state ϕ_j. Now A_j chooses an action from its action set. As the Markov chain moves from one state to another, information concerning the cumulative reward generated by the process is passed to the new state. If $x(n) \neq \phi_i$, automaton A_i does not operate. However, when the chain with the cumulative reward returns to the state ϕ_i, A_i can compute the total gain since the previous visit. Since the total elapsed time is also available[1], A_i can estimate the performance criterion (obtained while the action chosen at the previous visit was in effect) as the $\beta(n)$ ratio of the total reward to the total time. This in turn is used to update the probabilities of all the actions according to algorithm (4).

From the above discussion it is clear that the structure of the model itself is such that information required by an automaton at one of the states to improve its performance is obtained even as the process evolves. Hence inter-automata exchange of information is minimal; even the source of the information (i.e. the previous state) is not needed.

(iii) Convergence to the Optimal Policy. The principal steps involved in the proof of global $\epsilon -$ *optimality* are presented briefly in this section.
Step 1: Since the action set at state ϕ_i contains r_i elements, there are $\prod_{i=1}^{N} r_i$ stationary policies for the control of the Markov chain. Each policy corresponds to the choice of a pure

[1]Either A_i has a clock running at the rate of the chain or the current global time is passed to ϕ_i along with the cumulative reward.

strategy at every state. It is assumed that the Markov chain corresponding to each policy α is ergodic. This assumption assures that the process is not absorbed in any state or subset of states. An equivalent statement is that the chain visits every state an infinite number of times.

Step 2: It is assumed that L_{R-I} algorithms are used by the automata $A_i(i = 1, 2, \ldots N)$. Hence there are $\prod_{i=1}^{N} r_i$ absorbing states for the group (one for each possible policy). However, by choosing the step size of the learning algorithms sufficiently small, any initial set of action probability distributions is made to evolve to the optimal policy.

Step 3: An automata game result, given in section 2, states that if an identical payoff game has a unique equilibrium L_{R-I} schemes converge to the global optimum. If the Markov chain control problem using automata is represented as a finite identical payoff game Γ, this result applies if Γ has a unique equilibrium. The intriguing result that this is indeed the case is shown in [39] using the following property. For any policy α leading to an ergodic finite Markov chain, the performance criterion J defined in (6) can be written equivalently as

$$J = \sum_{i=1}^{N} \pi_i(\alpha) \sum_{j=1}^{N} t_j^i(\alpha) r_j^i(\alpha) \tag{7}$$

where $\pi_i(\alpha)$ is the stationary probability of being in state ϕ_i under policy α and $t_j^i(\alpha)$ and $r_j^i(\alpha)$ are the one step transition probabilities and rewards. The values of J in (7), one for each policy, are taken as the payoffs in an identical payoff game Γ. The proof that Γ has a unique equilibrium is first given for the case of a two state Markov chain in which the automaton at each state has two actions. This result is then extended to an N state Markov chain with N automata $A_i(i = 1, 2, \ldots N)$ where A_i has r_i actions. It is the uniqueness of an equilibrium in the game representation of the controlled Markov chain that enables global $\epsilon - optimality$ to be achieved by the decentralized automata.

4. Applications

a. Introduction

Although the basic learning automaton model was posed originally in the context of modeling complex biological systems [11], the recent application of the theory has been primarily to control problems in engineering systems. In both cases, it is the use of a number of interconnected automata that makes the application plausible. In the modeling area, examples exploring the automata game formulation have been given in the context of market price formation and general resource allocation [50]. Some ideas on the use of collective automata models in adaptive networks, relevant to both animal and machine learning, have also been given [51]. These represent only preliminary results and much remains to be done, both theoretically and empirically, if these areas are to become very successful applications.

By contrast, the application of automata theory to control problems has received much more attention over the past ten years. This research has dealt almost exclusively with problems arising in telecommunication networks, in particular the problem of routing traffic through a network. Extensive experience with this application has led to the view that the problem should have the following characteristics if learning schemes are to prove truly effective for control purposes:

(i) The system must be sufficiently complex and involve large operational uncertainties so that an adequate dynamic model does not exist.

(ii) The system must be amenable to decentralized control. Also, at each location where control can be exercised, the number of actions must be small.

(iii) Feedback to every decentralized controller must be provided by some random realization of a global performance criterion.

The problem of network routing to a large extent possesses these characteristics. Before devoting the remainder of this section to this one successful application, we mention some other applications which have been considered.

b. Some Potential Applications

For the most part, applications other than network routing are also drawn from problems in telecommunications or computers. The important problem of flow control (limiting the number or distribution of messages in the network) has been formulated as a learning problem [25]. Here, all messages entering the network must have a permit and each source node has a limited number of permits. The task of each automaton, one located at each destination node, is to allocate free permits to the source nodes. The performance feedback (or environment response) can assume a variety of forms, for example delay, power or permit population. A second communication network application, which has an automata game flavor, is the multi-access problem. A number of users of a computer network must share a common channel. In efficient operation conflicts over the use of the channel inevitably arise. Rules must be adopted for retransmitting messages in such cases. Some preliminary research has been reported in which simple automata make this decision [52].

A second, and somewhat related application area, is the control of queues in general and the scheduling of tasks in a multi-processor computer system in particular. The use of learning automata for these purposes has been discussed in [53] and [54], respectively. It is not yet clear that this is in fact a viable application.

A final potentially fruitful application of automata to data compression was suggested recently in [26]. A pair of identical automata, using the same random number generator, are used at the transmitting and receiving ends of a communication channel to adaptively code and decode the data being sent. In this case, the task is identification rather than control. Since many other approaches exist for data compression, it is premature to predict the success of this application. However, initial experiments have been promising [26].

c. Routing in Telecommunication Networks

A communication network can be modeled as a set of terminals generating messages, a connecting network providing the physical paths over which communication takes place and a control system which provides the supervisory signals. The traffic behavior of the incoming messages is modeled by a Poisson process with point to point loads given in some convenient units as λ_{ij} which can be collected in the form of a matrix Λ. The capacity of the communication channel from node i to node j is denoted by l_{ij} and is denoted by the ij^{th} element of a matrix L. Changes in load and network operating conditions are efficiently described by changes in Λ and L respectively.

In large systems, such as communication networks, investment in the operating facilities is very high and hence a small improvement in efficiency will result in considerable savings. Further, such systems are generally designed for average conditions which rarely prevail. Hence, existing facilities can be used more effectively by using on-line information. This is the logic generally used for suggesting adaptive or learning algorithms in real systems. If such schemes are to prove truly effective, they must converge rapidly and accurately to better strategies so that the above mentioned advantages are realized.

The routing problem in its simplest form may be stated as follows. At every node, a message has to be routed to a destination along one of a finite number of paths. A performance measure such as delay in transmission, probability of call completion or maximization of throughput has to be optimized. Since generally there is considerable ignorance regarding the state of the overall system, a learning automaton can be a candidate for effecting the routing. For the automaton to update itself it is essential that information fed back from the destination node be available at the node at which the automaton is located. Finally, since in any complex system additional information is invariably available, there must be some means of incorporating it in the operation of the automaton or any other decision maker that is used.

(i) Telephone Networks. The application of learning automata to telephone traffic routing has been described in [16-24]. Extensive simulation studies have been carried out and there is general awareness among workers in the field regarding the capabilities of the various learning schemes. At any node i an automaton A_{ij} routes all calls destined for node j. Any sequence of trunk groups along which routing of a call can be attempted corresponds to an action of the automaton. Call completion and call rejection are used as outputs and hence a binary environment model is ideally suited for describing the problem. Simulation results indicate that when the load is low, any scheme performs well. When the network is congested, very little appears possible to improve the system. However, when loads are high but additional capacity is available in the network, the automata can use it efficiently. In the actual telephone network, with years of experience and data available, a fair amount of information is available about traffic patterns and hence sophisticated control action may be possible. Automata schemes, as used at present, are effective only when the level of ignorance concerning the system is high.

(ii) Message Switched Networks. The telephone network is an example of a circuit switched network. In contrast to this, in a message switched network, the information is transmitted from node to node till it reaches its destination. The time taken to reach the destination may be used as a performance criterion and since this is a continuous random variable it entails the use of a continuous environment model. Further, when a message reaches a node, if the control system selects a specific link over which transmission is in progress, the message is placed in a queue. Hence, in this case the number of actions is equal to the number of links along which the message can be transmitted at that node.

(iii) Comments. Extensive simulation studies of telephone and communication networks have clearly demonstrated that automata schemes are generally better than fixed rule routing schemes. The equalization of blocking probabilities when $L_{R-\epsilon P}$ schemes are used and of blocking rates when L_{R-P} schemes are used has been verified in numerous cases. Hence there is very little doubt that automata schemes behave in the manner predicted by the theory presented in section 2. Since such schemes are simple, easy to implement and require only changes in software they are also attractive in many situations. Robustness in the presence of noisy information, delay in the feedback of data, the manner in which such data is to be collected and transmitted, and engineering and administrative feasibility are all important considerations.

From a theoretical point of view, the greatest drawback of the automaton approach is its inability to use additional information. If the method is to compare favorably with other schemes, procedures for incorporating such information in the selection of routes are essential. Methods for anticipating where extra capacity may be in the network, suppressing problematic sources and including additional choices for the automaton on the basis of prevailing conditions will also be needed if the automaton is to be a truly viable contender for routing in networks. The versatility of the automaton approach allows specialized automata assigned to these other tasks to be used conjunctively with routing automata in a hierarchical or a decentralized fashion. The results given in section 3 provide a theoretical justification for such an approach, but naturally much empirical work is required to demonstrate the performance of the more intelligent yet more complex automata system.

5. Conclusion

The last two decades have witnessed significant developments in both the theory and application of learning automata. Many theoretical issues have been clarified and resolved and features of a system required for a successful application have been identified. Recent work on the behavior of decentralized automata, as players in a game, as controllers of a finite

Markov chain and as traffic routers in communications networks, provides a new impetus for the use of learning automata in complex modeling and control problems.

References

[1] K.S. Narendra and M.A.L. Thathachar, "Learning Automata - A Survey," *IEEE Transactions on Systems, Man and Cybernetics*, vol. SMC-4, pp. 323-334, 1974.

[2] K.S. Narendra and S. Lakshmivarahan, "Learning Automata - A Critique," *Journal of Cybernetics and Information Science*, vol. 1, pp. 53-65, 1977.

[3] R.R. Bush and F. Mosteller, Stochastic Models for Learning, Wiley, New York, 1958.

[4] R.C. Atkinson, G.H. Bower and E.J. Crothers, An Introduction to Mathematical Learning Theory, Wiley, New York, 1965.

[5] M. Iosifescu and R. Theodorescu, Random Processes and Learning, Springer, New York, 1969.

[6] M.F. Norman, Markov Processes and Learning Models, Academic, New York, 1972.

[7] M.L. Tsetlin, "On the Behavior of Finite Automata in Random Media," *Automation and Remote Control,* vol. 22, pp. 1345-1354, 1961.

[8] V.I. Varshavskii and I.P. Vorontsova, "On the Behavior of Stochastic Automata with Variable Structure," *Automation and Remote Control,* vol. 24, pp. 327-333, 1963.

[9] "Special Issue on Learning Automata," *J. of Cybernetics and Information Science,* vol. 1, No. 2-4, 1977.

[10] S. Lakshmivarahan, *Learning Algorithms Theory and Applications,* New York: Springer-Verlag, 1981.

[11] M.L. Tsetlin, *Automaton Theory and Modeling of Biological Systems,* New York: Academic, 1973.

[12] A.G. Barto, P. Anandan and C.W. Anderson, "Cooperativity in Networks of Pattern Recognizing Stochastic Learning Automata," (this volume).

[13] J.H. Holland, Adaptation in Natural and Artificial Systems, University of Michigan Press, Ann Arbor, 1975.

[14] S.W. Wilson, "Knowledge Growth in an Artificial Animal," (this volume).

[15] D.E. Goldberg, "Controlling Dynamic Systems with Genetic Algorithms and Rule Learning," (this volume).

[16] L.G. Mason, "Self-Optimizing Allocation Systems," Ph. D. dissertation, University of Saskatchewan, Saskatoon, Sask., Canada, 1972.

[17] K.S. Narendra, L.G. Mason and S.S. Tripathi, "Application of Learning Automata to Telephone Traffic Routing Problems," Becton Center Tech. Report CT-69, Dept. of Engg. and Applied Science, Yale University, New Haven, Ct., Jan. 1974.

[18] K.S. Narendra, A.E. Wright and L.G. Mason, "Application of Learning Automata to Telephone Traffic Routing and Control," *IEEE Transactions on Systems, Man and Cybernetics*, vol. SMC-7, pp. 785-792, 1977.

[19] P. Mars and M. S. Chrystall, "Real-Time Telephone Traffic Simulation using Learning Automata Routing," S&IS Report No. 7909, Dept. of Engg. and Applied Science, Yale University, New Haven, CT, Nov. 1979.

[20] K.S. Narendra and M. A. L. Thathachar, "On the Behavior of a Learning Automaton in a Changing Environment with Application to Telephone Traffic Routing," *IEEE Trans. on Systems, Man and Cybernetics,* vol. SMC-10, pp. 262-269, 1980.

[21] P. R. Srikantakumar and K.S. Narendra, "A Learning Model for Routing in Telephone Networks," *SIAM J. Control and Optimization,* vol. 20, pp. 34-57, 1982.

[22] O.V. Nedzelnitsky, Jr. and K.S. Narendra, "A Learning Approach to Routing in Data Communication Networks," Tech. Report No. 8212, Center for Systems Science, Yale University, New Haven, Ct.

[23] K.S. Narendra and P. Mars, "The Use of Learning Algorithms in Telephone Traffic Routing - A Methodology," *Automatica*, vol. 19, pp. 495-502, 1983.

[24] K.S. Narendra and R.M. Wheeler, Jr, "Routing in Communication Networks - A Case Study of Learning in Large Scale Systems," *Journal of Large Scale Systems*, vol. 8, pp. 211-222, 1985.

[25] L.G. Mason and X.D. Gu, "Learning Automata Models for Adaptive Flow Control in Packet-Switching Networks," (this volume).

[26] A.A. Hashim, S. Amir and P. Mars, "Application of Learning Automata to Image Data Compression," (this volume).

[27] L.P. Devroye, "A Class of Optimal Performance Directed Probabilistic Automata," *IEEE Trans. on Systems, Man and Cybernetics,* vol. SMC-6, pp. 777-784, 1976.

[28] M.A.L. Thathachar and P.S. Sastry, "A New Approach to the Design of Reinforcement Schemes for Learning Automata," Tech. Report EE/60, Dept. Elec. Engg., Indian Inst. Sci., Bangalore, India, Dec. 1983.

[29] O.V. Nedzelnitsky, Jr., "The Application of Learning Methodology to Message Routing in Data Communication Networks," Ph.D. Thesis, Dept. Elec. Engg., Yale University, Dec. 1983.

[30] D.E. Koditschek and K.S. Narendra, "Fixed Structure Automata in a Multi-Teacher Environment," *IEEE Transactions on Systems, Man and Cybernetics*, vol. SMC-7, pp. 616-624, 1977.

[31] M. A. L. Thathachar and R. Bhakthavathsalam, "Learning Automaton Operating in Parallel Environments," *J. of Cybernetics and Informatin Science - Special Issue on Learning Automata,* vol. 1, pp. 121-127, 1977.

[32] N. Baba, "On the Learning Behaviors of Variable-Structure Stochastic Automaton in the General N-Teacher Environment," *IEEE Trans. Syst., Man and Cybernetics*, vol. SMC-13, pp. 224-231, 1983.

[33] R.M. Wheeler, Jr. and K.S. Narendra, "Learning Models for Decentralized Decision Making," *Automatica*, vol. 21, pp. 479-484, 1985.

[34] V.Y. Krylov and M.L. Tsetlin, "Games Between Automata," *Autom. and Remote Cont.*, vol. 24, pp. 975-987, 1963.

[35] S. Lakshmivarahan and K.S. Narendra, "Learning Algorithms for Two-Person Zero-Sum Stochastic Games with Incomplete Information," *Mathematics of Operations Research,* vol. 6, pp. 379-386, 1981.

[36] S. Lakshmivarahan and K.S. Narendra, "Learning Algorithms for Two-Person Zero-Sum Stochastic Games with Incomplete Information: A Unified Approach," *SIAM J. Control and Optimization,* vol. 20, pp. 541-552, 1982.

[37] K.S. Narendra and R.M. Wheeler, Jr., "An N-Player Sequential Stochastic Game with Identical Payoffs," *IEEE Trans. on Syst., Man and Cybernetics,* vol. 13, pp. 1154-1158, 1983.

[38] K.R. Ramakrishnan, "Hierarchical Systems and Cooperative Games of Learning Automata," Ph.D. Thesis, Dept. Elec. Engg., Indian Inst. Sci., Bangalore, India, July 1982.

[39] R.M. Wheeler, Jr. and K.S. Narendra, "Decentralized Learning in Finite Markov Chains," Tech. Report No. 8410, Center for Systems Science, Yale University, New Haven, Ct., Dec. 1984; To appear in *IEEE Transactions on Automatic Control.*

[40] Y.M. El-Fattah, "Multi-Automaton Games: A Rationale for Expedient Collective Behavior," *Systems and Control Letters,* vol. 1, pp. 332-339, 1982.

[41] Y.M. El-Fattah, "Fairness and Mutual Profitability in Collective Behavior of Automata," *IEEE Trans. Syst. Man and Cybern.*, vol. SMC-13, pp. 236-241, 1983.

[42] V.I. Varshavskii, "Collective Behavior and Control Problems," in Machine Intelligence, vol. 3, D. Michie (ed), Edinburgh University Press, Edinburgh, 1968.

[43] N. Baba and Y. Sawaragi, "On the Learning Behavior of Stochastic Automata Under a Nonstationary Random Environment," *IEEE Trans. Syst., Man and Cybern.*, vol. SMC-5, pp. 273-275, 1975.

[44] M.A.L. Thathachar and K.R. Ramakrishnan, "A Hierarchical System of Learning Automata," *IEEE Trans. Syst., Man and Cybern.*, vol. SMC-11, pp. 236-242, 1981.

[45] B.T. Mitchell and D.I. Kountanis, "A Reorganization Scheme for a Hierarchical System of Learning Automata," *IEEE Trans. Syst., Man and Cybern.*, vol. SMC-14, pp. 328-334, 1984.

[46] R.E. Bellman, "A Markovian Decision Process," *J. Math. Mech.*, vol. 6, pp. 679-684, 1957.

[47] R.A. Howard, Dynamic Programming and Markov Processes, MIT Press, Cambridge, Ma., 1960.

[48] V. Borkar and P. Varaiya, "Adaptive Control of Markov Chains, I: Finite Parameter Set," *IEEE Trans. Autom. Cont.*, vol. AC-24, pp. 953-958, 1979.

[49] P.R. Kumar and W. Lin, "Optimal Adaptive Controllers for Unknown Markov Chains," *IEEE Trans. Autom. Cont.*, vol. AC-27, pp. 765-774, 1982.

[50] Y.M. El-Fattah, "Stochastic Automata Modeling of Certain Problems of Collective Behavior," *IEEE Trans. Syst., Man and Cybern.*, vol. SMC-10, pp. 304-314, 1980.

[51] A.G. Barto and P. Anandan, "Pattern Recognizing Stochastic Learning Automata," *IEEE Trans. Syst., Man and Cybern.*, vol. SMC-15, pp. 360-375, 1985.

[52] P.R. Srikanta Kumar, "Application of Learning Theory to Communication Networks Control," Proc. of the Third Yale Workshop on Applications of Adaptive Systems Theory, New Haven, Ct., pp. 135-141, 1983.

[53] M.R. Meybodi and S. Lakshmivarahan, "A Learning Approach to Priority Assignment in a Two Class M/M/1 Queuing System with Unknown Parameters." Proc. of the Third Yale Workshop on Applications of Adaptive Systems Theory, New Haven, Ct., pp. 106-109, 1983.

[54] R.M. Glorioso, Engineering Intelligent Systems, Digital Press, Bedford, Ma., 1980.

Learning Automata Models for Adaptive Flow Control in Packet-Switching Networks

L. G. Mason and XueDuo Gu
INRS-Telecommunications

Abstract

Performance and Stability results for three adaptive isarithmic flow control systems for packet-switching networks are described. The performance models, which are based on the BCMP theory for closed networks of queues, are exact under stationary conditions. The control architectures considered include one centralized and two decentralized schemes. The decentralized architectures include single-chain and multiple-chain cases. The controllers are modeled by L_{R-I} learning automata. Four types of network feedback responses were considered. These are loop permit delay, loop population, loop power and path delay, where a loop includes the controller, the source queue, the network path to the message destination node and the path back to the controller. The model has been verified by Monte Carlo event simulation, thus demonstrating the feasibility of the proposed control systems and the accuracy of the analytic performance model. The various control architectures and algorithms are compared in regard to their power performance, transient response and stability characteristics. Several areas for further research are then identified.

1. Introduction

Packet-switching networks are typically dimensioned to achieve message transport delay objectives under normal busy hour traffic conditions. From time to time circumstances arise which place more load on the network than it was designed to handle. These overloads may be general in nature, or they may be focused, where certin nodes or links become congested and the resulting delays become excessive. Under such conditions it is desirable to control the rate at which messages enter the network to avoid performance degradation and unfair allocation of capacity.

In the *isarithmic* flow control scheme [1] each message arrival must secure a permit to gain entry to the network. When the message has been delivered to its destination the permit is released and becomes available for other message arrivals. By limiting the total number of permits cycling in the network the rate of message influx is controlled. In the original scheme [1] the permits are carried from node to node by the exogenous traffic. While this approach results in little overhead, as permits are associated with message traffic, it can lead to an undesirable distribution of permits under asymmetric traffic conditions. To mitigate this condition a limit is placed on the number of permits which may collect at a node. Excess permits are "piggy backed" on other message-permit combinations. The blocking feature renders the scheme intractable to an exact performance analysis and as a result the isarithmic scheme has only been studied through simulation.

An alternative approach to avoid permit collection at slow nodes involves employing an adaptive controller for permit disbursal. In this arrangement a permit which has just been released by a delivered message is allocated to the queues associated with source nodes by an adaptive controller. Here the permit does not wait at the destination node for an exogenous message arrival to carry it to another location. While this approach will increase the control overhead somewhat, in that permits do not carry messages on the return path, the advantage afforded by the flexibility it allows in permit distribution can more than offset the overhead

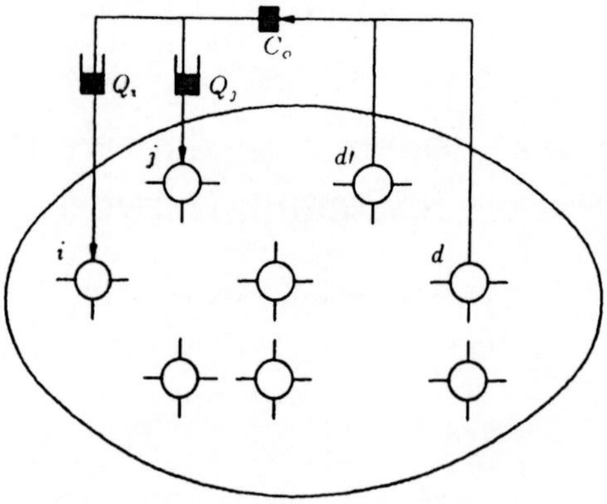

Figure 1: Centralized Adaptive Flow Control Architecture.

factor. In particular the distribution of permits among the source nodes can be optimized and made adaptive to network conditions.

The objective of this paper is to evaluate the feasibility and effectiveness of a number of learning automata control schemes for parameter optimization. In section 2 three different flow control architectures are described which are based on the isarithmic principle, where a fixed number of permits regulate the rate of message acceptance to the distributed packet-switching network. Section 3 summarizes the approach taken to performance modeling. A detailed description of the performance models is given in [13]. In section 4 a recursion is given for parameter adaptation where these parameters are permit routing probabilities. The updates are functions of the network response or feedback to a prior control action, where a control action corresponds to the permit destination selected. The types of feedback considered include round trip delay, loop power, number of permits per loop and path delay. A loop is a closed path consisting of the controller, the source node to which the permit is sent and the path through the network to the message destination node and the path from the destination back to the controller. For some cases a loop may be bifurcated. In section 5 numerical results are described and discussed. Section 6 concludes with a discussion of on-going research.

2. Isarithmic Flow Control Architectures

Several isarithmic flow control architectures admit an exact performance analysis under stationary conditions by the BCMP theory for closed networks of queues. In reference [13] several adaptive and non-adaptive control schemes are described. Three of the adaptive schemes will now be considered.

2.1 Centralized Adaptive Flow Control

In this model, shown in Figure 1, the permits are disbursed by a centralized controller to the message sources. Real-time co-ordination between source arrivals and permit allocation is not attempted. The controller measures the network response, (delay, power, permit population etc.) taken for a permit to traverse the loop. To do this it is necessary to tag a permit at the time of disbursal with the node identity to which it is routed as well as with a permit sequence number so it will be recognized when it returns and the appropriate loop will be identified. The single controller, C_0, updates its action probabilities each time a permit passes through it.

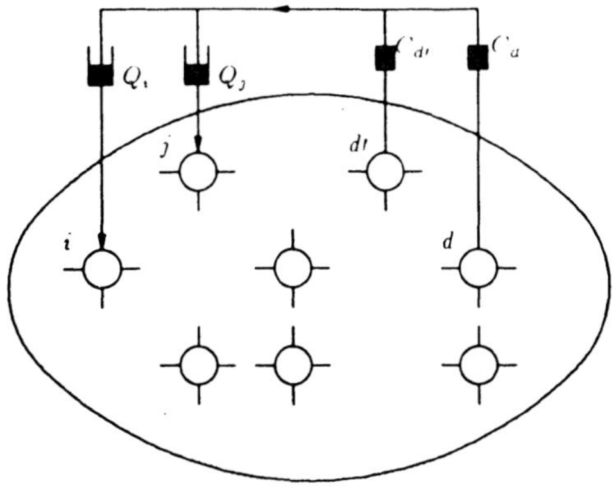

Figure 2: Decentralized Adaptive Flow Control (Single-Chain).

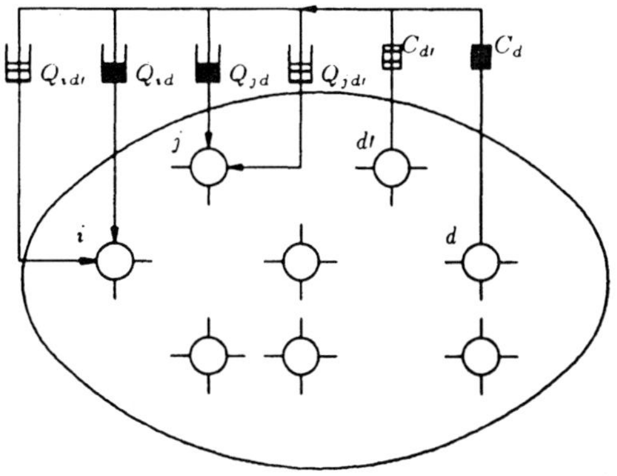

Figure 3: Decentralized Adaptive Flow Control (Multiple-Chain).

2.2 Decentralized Adaptive Flow Control (Single-Chain)

In this case a single class of permits is distributed by N controllers, $C_d, d = 1,..N$, with a controller associated with each destination node as shown in Figure 2. The controllers are L_{R-I} learning automata with N actions, one for each source node. The remarks made in connection with permit tagging apply here as well. In addition the permit must also be tagged with the destination identity of the message which secures the permit. For the single-chain implementation the automaton at the destination node, C_d, only updates its action probabilities when a permit marked with message destination d is received.

2.3 Decentralized Adaptive Flow Control (Multi-Chain)

In this variation, shown in Figure 3, there are N permit classes, with a permit class and an $N-1$ action L_{R-I} controller, C_d, associated with each destination node, d. The action probabilities of an automaton are updated each time a permit passes through it. Each source node maintains $N-1$ source queues where free permits are held, with a queue associated with each of the distinct permit classes.

3. Performance Models

Reiser et al [2] have described a technique for modeling window controlled virtual circuit packet-switching networks by a closed network of queues. We have employed the same technique for modeling the isarithmic global flow control schemes. In this approach a fixed number of permits circulate in a closed network of queues. In addition to the link queues there are queues associated with message sources. Free permits reside in the source queues until they are "served" by a message arrival. Accordingly the service rate of the source queues is given by the message arrival rate to these sources. Messages which arrive when its source queue does not contain permits are lost. When a message obtains a free permit at the source queue, the message and the permit are routed through the transmission network to the message destination, where the permit is released for subsequent use by other messages. The various isarithmic flow control schemes considered here differ in the manner in which free permits are disstributed.

Schemes 2.1 and 2.2 are models with a single permit class, and accordingly the general BCMP model reduces to the class of closed queueing networks originally analyzed by Gordon and Newell [3]. Case 2.3 is on the other hand a multi-chain model and requires the more general BCMP theory for analysis [4].

For adaptive control schemes it is assumed that the network dynamics are fast in comparison to the speed of adaptation, in other words we are considering the case of quasi-static flow control. For a higher rate of adaptation, which correspond to a large adaptive loop gain, the quasi-static performance models apply to the equilibrium conditions only.

4. Adaptive Control Algorithms

We have modeled the adaptive controllers by L_{R-I} learning automata [7]. The expected action probabilities for the centralized adaptive control scheme are given by

$$g_i(t+1) = g_i(t)(1 + G(F_i(t) - \sum_{j=1}^{N} F_j(t)g_j(t))) \quad i = 1, \dots N \tag{1}$$

where t is the epoch at which the allocation is performed and $F_i(t)$ is the normalized network response strength associated with that action. G is an adaptive loop gain parameter. For the decentralized control schemes there are N controllers competing in a game situation with a controller associated with each destination node. The action probabilities are updated according to the recursion

$$\begin{aligned} g_{di}(t+1) = \; & g_{di}(t)(1 + G(F_{id}(t) - \textstyle\sum_{j=1}^{N} F_{jd}(t)g_{dj}(t))) \\ & i = 1, \dots N \;\; d = 1, \dots N \;\; \neq d \end{aligned} \tag{2}$$

Four types of network feedback have been considered, namely loop delay, loop population, loop power and path delay. We refer to these algorithms as A_{TL}, A_K, A_P, A_T respectively. All algorithms except A_T can be implemented without any additional state measurements on the basis of the permit flows through the controllers. The A_T algorithm requires supplementary information concerning the expected delays in the source queues and for this reason is less practical. It has been studied to see if such state information can improve performance. It turns out that it cannot.

4.1 Algorithm A_{TL}

In this case the controller measures the round trip delay for the permit to cycle through the source node, the network links and back to the controller.

It should be pointed out that the formulae used in the numerical study were for the expected behavior of the system under slow-learning conditions. In an on-line implementation the delays would of course be measured rather than calculated. To verify the correctness

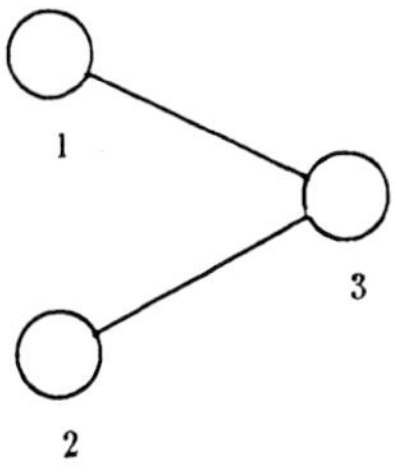

Figure 4: Simple Network Topology.

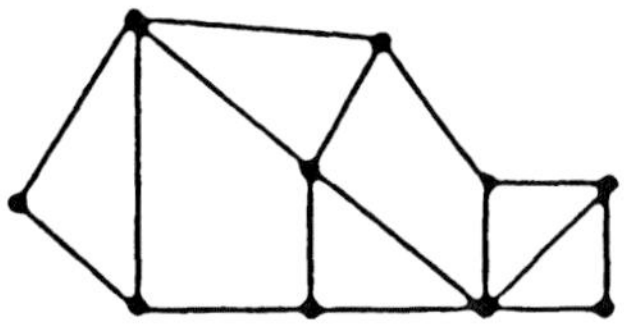

Figure 5: 10-Node Network Topology.

of the deterministic model a Monte Carlo event simulation was carried out where measurements of the network random delays were taken and L_{R-I} learning automata were used to implement the controllers. Excellent agreement with our deterministic model was found.

4.2 Algorithm A_K

The reward strengths are defined as the 1's complement of the normalized loop permit population. This can be "measured" on-line by updating the exact loop population with each permit arrival and disbursal at the controller(s). In our deterministic model we calculate the expected permit population from the BCMP theory.

4.3 Algorithm A_P

The reward strength is the normalized loop power, which is defined as the loop throughput-delay ratio. This can be implemented by measuring the loop throughput and delay associated with the various control actions.

4.4 Algorithm A_T

The reward strength is defined as the 1's complement of the normalized path delay associated with each loop. In these schemes the delay of the source queue is not included. Accordingly, the path delay, which is defined as the loop delay less the source queue delay, is not directly measurable from the permit flows.

5. Numerical Results

The centralized adaptive control architecture for the four control algorithms were tested on a very simple 3-node network, (Figure 4), and on a 10-node 32-link network, (Figure 5), which has been previously used for studying adaptive routing schemes [8,9].

The simple network was considered to facilitate insight into the algorithmic behavior. For the simple network, the optimal permit size and allocation can be found by exhaustive search. This serves as a benchmark when evaluating algorithmic performance. The larger

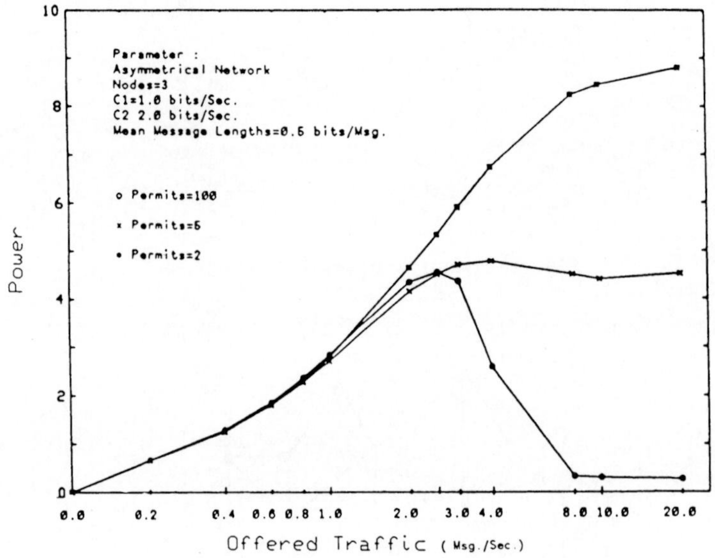

Figure 6: Network Power vs. Load (Simple Network).

network was considered to test the feasibility of the proposed control systems in a realistic network environment.

For the *10-node* network, three traffic demand models were considered. For the *uniform* case the arrival rate between origin-destination pairs was 0.317 messages/sec. for the light traffic case. For the nominal traffic case, all values were increased by a factor of 10. For the *non-uniform* case, node 5 has an arrival rate of 1.217 messages/sec to all destination nodes, while the arrival rate for all other origin destination pairs was 0.217 messages/sec. For the *2-chain* traffic model the only non-zero traffic demands are λ_{19} and λ_{59}. For this case three traffic distributions were considered, where the total arrival rate was 28.48 messages/sec. For this case, the mean message length was 256 bits, while the simplex link capacities were 9600 bits/sec.

5.1 Equilibrium Performance for the Simple Network

In Figure 6, the network power is plotted as a function of offered traffic for three different permit populations, ($W = 2, 5, 50$). The 50 permit case approximates the open or uncontrolled network. While the uncontrolled network gives marginally better power performance for light traffic conditions, the rapid decline in performance under heavy traffic conditions demonstrates the need for controlling the rate of traffic influx to the network.

In Figure 7, the optimal permit population W^*, and allocation, g_1^*, is shown as a function of traffic distribution, where the portion of traffic is defined as the percentage of the total traffic which arrives at node 1. The optimal permit population is a function of traffic distribution, suggesting that permit population should also be adaptively controlled. The optimal permit size increases as the traffic demand becomes more uniform and decreases to a single permit for highly skewed traffic distributions. If we define fairness to be the condition under which permits are distributed in proportion to node message arrival rates we see that the optimal power criterion is fair or nearly fair for a wide range of traffic distributions. It becomes unfair under highly skewed traffic distributions.

Figures 8 and 9 plot network power performance vs. permit population for various permit allocation algorithms for different traffic distributions and link capacities. The equilibrium performance is shown for the four adaptive control algorithms A_{TL}, A_K, A_P, A_T as well as

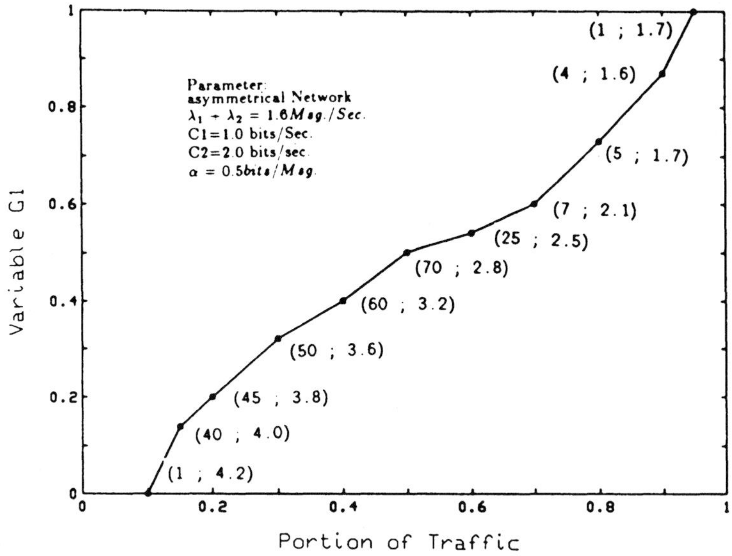

Figure 7: Optimal Population and Allocation vs. Traffic Distribution.

for the optimal allocation algorithm. These figures demonstrate the need for optimizing both the permit population and its allocation. The A_{TL} algorithm performs optimally in terms of network power for highly skewed traffic distributions where the optimal permit size is unity. The power performance degrades as the permit population increases. Optimal performance for $W = 1$ is to be expected since minimizing loop delay will also maximize loop throughput and consequently loop power. When the allocation is deterministic the loop power of the selected loop equals the network power and consequently network power will be optimized.

The algorithm A_K attempts to equalize the loop permit populations. For small permit populations the performance is significantly below the optimal level. The performance tends to improve as the population size increases. In this sense the algorithms A_{TL} and A_K are complementary. The A_K algorithm has a built in fairness property in that the permit allocation cannot be deterministic even under highly skewed traffic conditions. Thus while A_{TL} gives better power performance for small permit populations, A_K provides a fairer allocation of permits.

The A_P algorithm performs reasonably well for small permit populations and heavy traffic in balanced networks, (i.e. networks where traffic and capacity are nearly uniform). The performance degrades with network asymmetry and with increasing permit population. This is to be expected since power is not decentralizable for unbalanced networks. For large permit populations the loop power gives a poor estimate of path power due to large delays in the source queues. For $W = 1$, A_P converges to A_{TL}, and hence gives optimal power but is unfair.

The A_T algorithm performs poorly for large W whether the network is balanced or not. For small W, A_T converges to A_{TL} since the source queueing delay becomes negligible.

5.2 Equilibrium Performance for the 10-node Network

Figures 10 and 11 display the power performance of the four algorithms for the centralized adaptive control architecture, under uniform and non-uniform traffic respectively, as a function of the permit population. For the low traffic levels considered the optimal permit size is quite large.

It turns out that for the uniform case the A_{TL}, and A_K algorithms yield the same performance over the entire range of permit populations considered, while the A_P algorithm gives

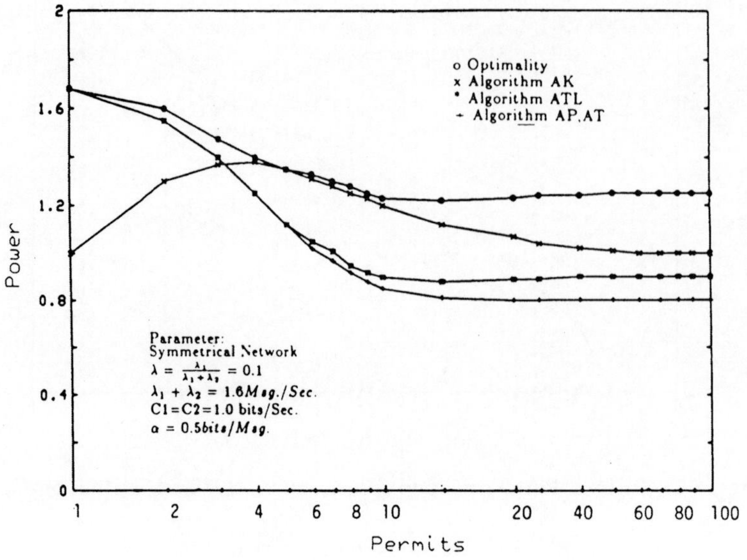

Figure 8: Skewed Traffic Symmetric Network.

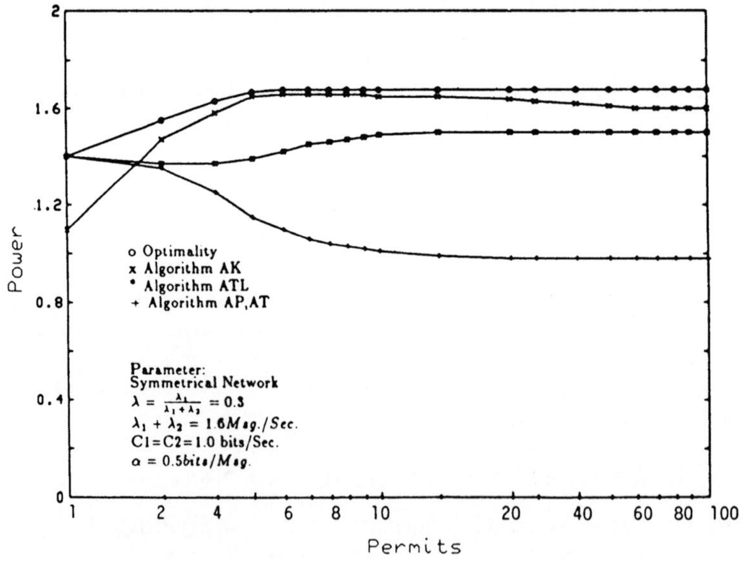

Figure 9: Nominal Traffic Symmetric Network.

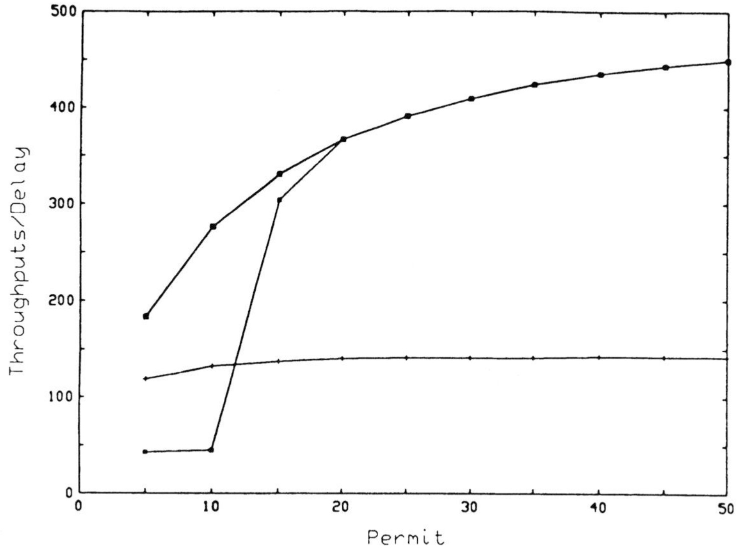

Figure 10: Uniform Traffic, Centralized Control.

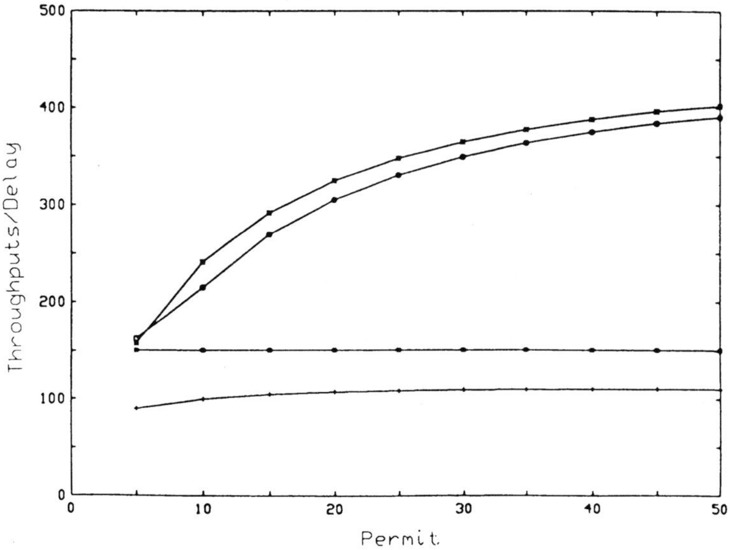

Figure 11: Non-uniform Traffic, Centralized Control.

identical performance when W exceeds 20. For smaller values of permit size the performance of the A_P algorithm is inferior. The A_T algorithm gives lower performance than A_{TL} and A_K over the entire range of W considered. For non-uniform traffic conditions, with one dominant traffic source, the performance of all algorithms differ. The ranking in performance from best to worst is A_K, A_{TL}, A_P, A_T. In this case both A_P and A_T are significantly worse than A_{TL} and A_K.

Figures 12 and 13 display the results for the same network and traffic models for the decentralized single chain architecture. Comparison with Figures 10 and 11 reveals that the A_{TL}, A_K and A_P algorithms for the decentralized single chain architecture yield almost the same equilibrium performance as in the centralized architecture. There is a significant difference in the case of the A_T algorithm, however, with the decentralized case performing significantly better.

Finally the decentralized multiple-chain case has been compared with the centralized case for a two-chain traffic model for the A_{TL} and A_K algorithms. Figure 14 illustrates that for the same total permit population, the centralized and single chain decentralized models outperform the decentralized multiple chain model. The performances of A_K and A_{TL} are very similar for the range of permit populations considered with A_{TL} marginally better. Whether the algorithms A_K and A_{TL} are optimal or near optimal for the network is at present unknown as we do not, at present, have an effective means of determining the optimal flow control.

5.3 Transient Response

The transient response of network delay, throughput, power and permit allocation probabilities have been studied for the various control architectures and adaptive control algorithms for the simple and larger network topologies. The transient response depends of course on the adaptive loop gain parameter G as well as upon the control architecture and adaptive permit allocation algorithm. For conciseness we only display typical cases in Figures 15 and 16 for the 10-node network.

For a fixed value of adaptive loop gain the response of the A_K and A_{TL} algorithms are comparable and faster than those for the A_P and A_T. As far as the different control architectures are concerned, the multiple-chain decentralized model is fastest followed by the centralized scheme while the decentralized single chain case is the slowest for the network examples studied. Whether this ranking holds true in general is unclear; however, for the various traffic demand models considered, the time constant associated with the transient response does not change significantly. The transient response time is roughly inversely proportional to the adaptive loop gain parameter, G. Accordingly one minimizes the time taken to compute the equilibrium performance by choosing the gain parameter as large as possible consistent with system stability. When the gain parameter is larger than a critical value the system becomes unstable (i.e. sustained oscillation occurs and no steady state equilibrium value is reached).

5.4 Stability

We have investigated the system stability of the 10-node network using the centralized control architecture and the A_{TL} permit allocation algorithm for nominal and low levels of non-uniform traffic demand. For the system considered, the adaptive routing algorithm described in [9] was employed, where the adaptive gain associated with routing is G_r and the adaptive flow control gain is G. The stable region in the $G - G_r$ plane is shown in Figure 17, for light and nominal traffic levels. These curves show the effect of interaction of the two adaptive loops in regard to system stability. These curves reveal that, for light and nominal traffic levels, the critical gain, associated with the inner loop (adaptive routing), is an order of magnitude larger than that for adaptive flow control. The curves also show that, as the traffic level increases, for a fixed permit population, the flow control critical gain increases while the adaptive routing critical gain decreases. For saturated networks, not shown here,

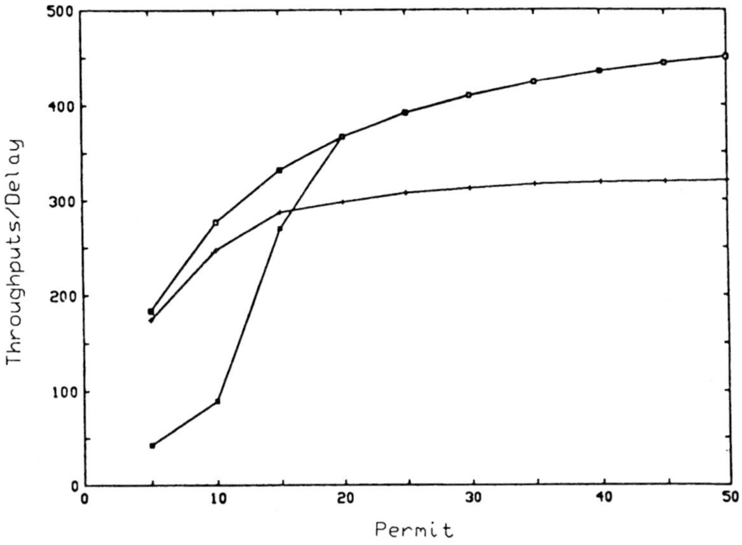

Figure 12: Uniform Traffic, Decentralized, Single-Chain .

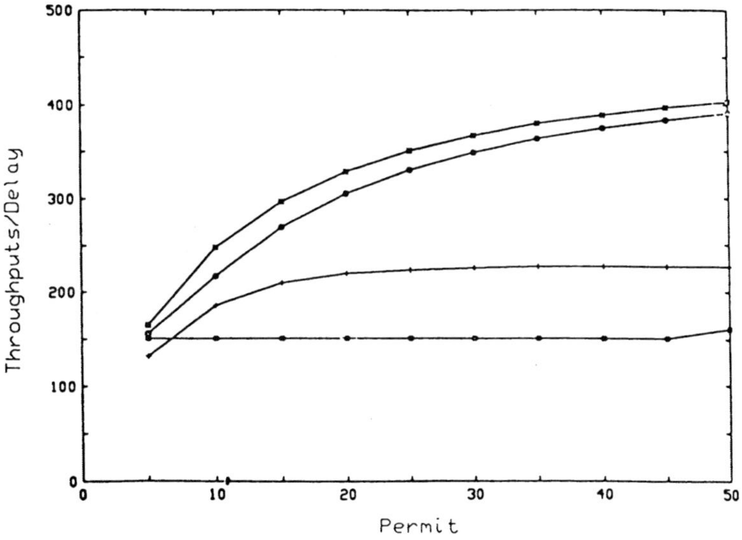

Figure 13: Non-uniform Traffic, Decentralized, Single-Chain .

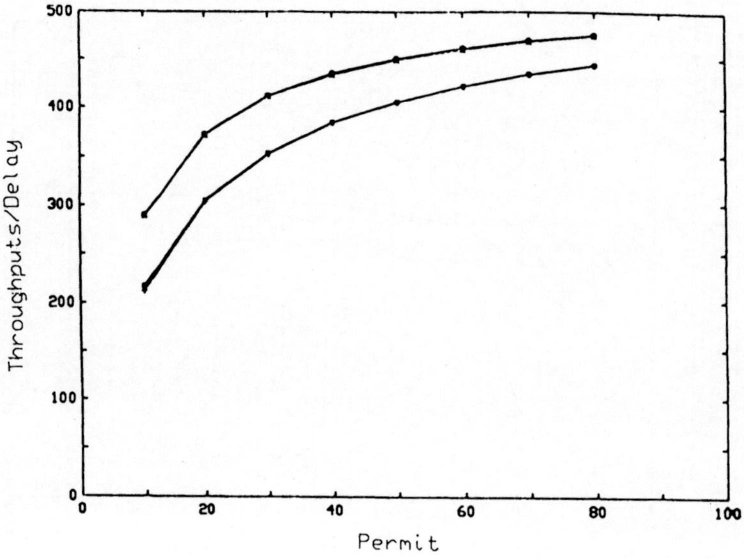

Figure 14: Performance Comparison Single-Chain and Multi-Chain .

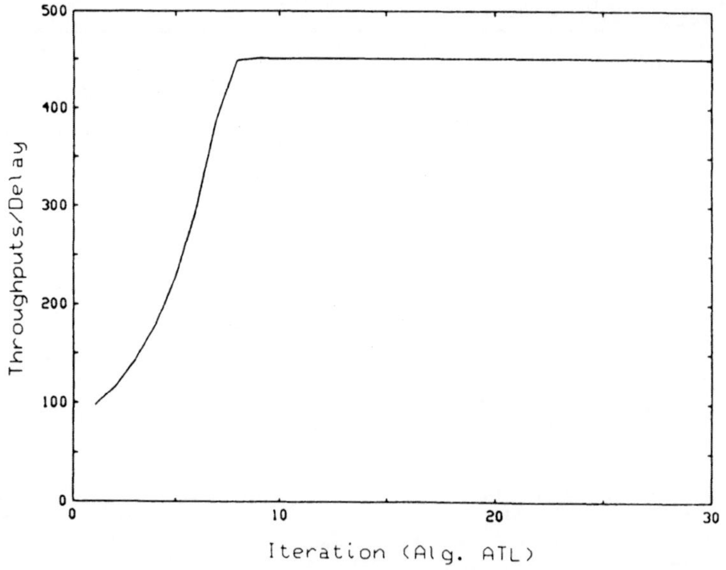

Figure 15: Centralized Control, A_{TL}.

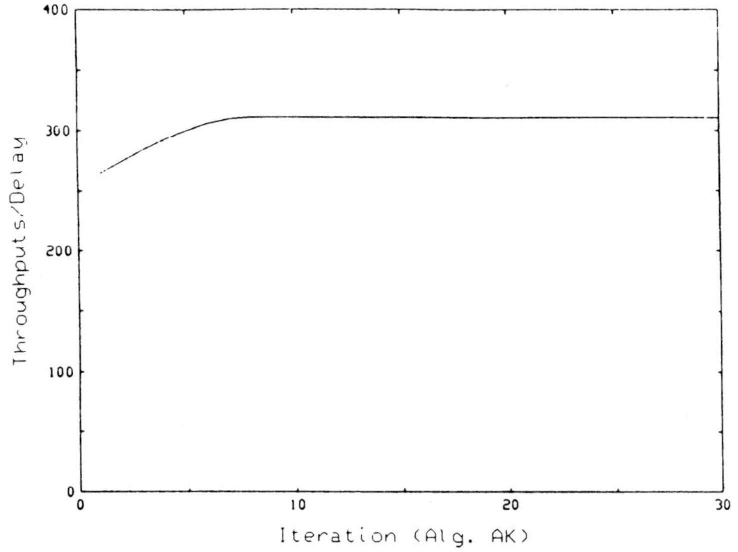

Figure 16: Decentralized Multiple-Chain, A_K.

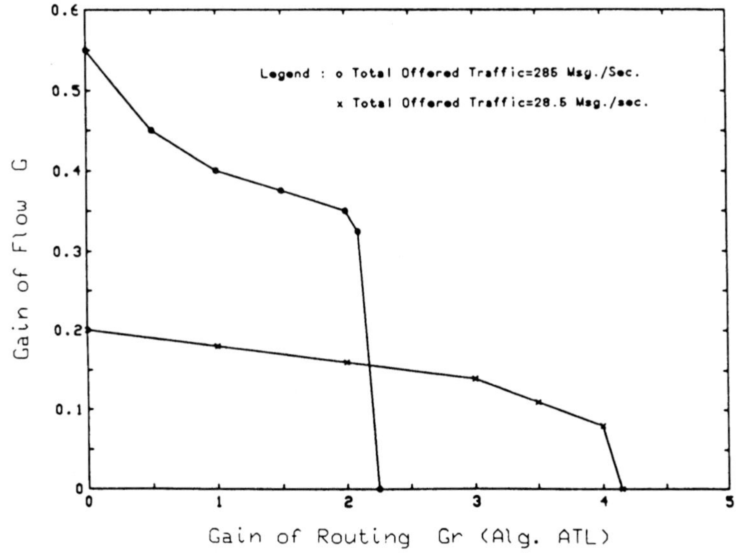

Figure 17: Stability Region, Centralized Control, A_{TL}.

the adaptive routing gain parameter becomes critical before the adaptive flow control gain parameter. This behavior can be explained on the basis that network delay is more sensitive to changes in routing under heavy traffic, while the network power performance is most sensitive to changes in permit allocation for large permit populations under light traffic.

6. Conclusions

Exact analytic performance models have been developed for a number of isarithmic flow control schemes for distributed packet-switching networks. The performance models for the centralized adaptive flow control architecture and two decentralized architectures have been implemented in software. A Monte Carlo event simulation of centralized adaptive control in the simple network has been carried out and excellent agreement with the analytic model was obtained. The L_{R-I} automata were implemented in the simulation along with the random network response to the automata actions, thereby demonstrating the feasibility of the proposed control systems.

A numerical study of performance and stability, based on the analytic performance models, was carried out for both simple and complex networks under various traffic conditions. If such an extensive study had been done by simulation, exorbitant amounts of CPU time would have been required. The study revealed that network performance and stability is a very complex function of the system parameters. The study also indicated that two of the network feedback mechanisms, A_K and A_{TL} show promise for the three control architectures considered. On the other hand the A_P and A_T algorithms do not perform well and will not be considered further. Among the architectures studied, the single chain models yield the best equilibrium power performance and stability characteristics for the same total permit population. As the permit population was not optimized we cannot conclude definitely that the single-chain case will be absolutely superior to the multi-chain case.

The results reported herein are preliminary and further study is required to determine whether they are robust under all network conditions. Several performance criteria have been reported in the literature [10-12], which account for fairness. As the power measure considered herein is inherently unfair in some instances, the behavior of the adaptive isarithmic schemes with respect to these alternate criteria should be investigated. Such a study is now in progress and will be reported in due course.

References

[1] D.W. Davies, "The Control of Congestion in Packet Switching Networks," *IEEE Trans. on Comm.*, vol. COM-20, June 1972.

[2] M. Reiser, "A Queueing Network Analysis of Computer Communication Networks with Window Flow Control," *IEEE Trans. on Comm.*, vol. COM-27, No. 8, August 1979.

[3] W.J. Gordon and G.F. Newell, "Closed Queueing Systems with Exponential Servers," *Operations Research*, vol. 15, pp. 254-265, 1967.

[4] F. Baskett, K.M. Chandy, R.R. Muntz and F. Palacios, "Open, Closed and Mixed Networks of Queues with Different Classes of Customers," *Journal of the ACM*, vol. 22, No. 2, April 1975.

[5] S.S. Lam, "Dynamic Scaling and Growth Behavior of Queueing Network Normalization Constants," *Journal of the ACM*, vol. 29, No. 2, pp. 492-513, April 1982.

[6] K.M. Chandy and C.H. Sauer, "Computational Algorithms for Product Form Queueing Networks," *Communications of the ACM*, vol. 23, No. 10, Oct. 1980.

[7] L.G. Mason, "An Optimal Learning Algorithm for S-Model Environments," *IEEE Trans. on Automatic Control*, Oct. 1973.

[8] P. Mars, K.S. Narendra and M. Chrystal, "Learning Automata Control of Computer Communication Networks," Proceedings of the Third Yale Workshop on Applications of Adaptive Systems Theory, New Haven, June 1983.

[9] L.G. Mason, "Equilibrium Flows, Routing Patterns and Algorithms for Store-and-Forward Networks," *Journal of Large Scale Systems*, LSS 285, 1985.

[10] M. Gerla and M. Staskauskas, "Fairness in Flow Controlled Networks," *Journal of Telecommunication Networks*, 1982.

[11] R.G. Gallager and S.J. Golestanni, "FlowControl and Routing Algorithms for Data Networks," Proceedings of Intl. Conf. on Computer Communications, Atlanta, Georgia, Oct. 1980, pp. 779-784.

[12] J.M. Jaffe, "A Decentralized 'Optimal' Multiple-User Flow Control Algorithm," Proceedings of Intl. Conf. on Computer Communications, Atlanta, Georgia, Oct. 1980, pp. 839-844.

[13] L.G. Mason and X. Gu, "Adaptive Isarithmic Control," INRS-Tech. Rep. No. 85-18, May 1985.

Application of Learning Automata to Image Data Compression

A. A. Hashim, S. Amir and P. Mars
Leicester Polytechnic
U.K.

Abstract

A novel approach to image data compression is proposed which uses a stochastic learning automaton to predict the conditional probability distribution of the adjacent pixels. These conditional probabilities are used to code the gray level values using a Huffman coder. The system achieves a 4/1.7 compression ratio. This performance is achieved without any degradation to the received image.

1. Introduction

With the continuing growth of modern communication technology, the demand for image transmission and storage is increasing rapidly. Advances in computer technology for mass storage and digital processing have paved the way for implementing advanced data compression techniques to improve the efficiency of the transmission and storage of images.

Image data compression is concerned with the minimization of the number of information-carrying units used to represent an image. For digital image transmission and storage, the conventional method is the Pulse Code Modulation (PCM) technique. The continuous image is first sampled at Nyquist rate in the spatial domain to produce an NxN array of discrete samples. Sampling of a band-limited image signal is the simplest and most dramatic form of data compression. The samples thus obtained may have an infinite number of amplitude levels and hence may require infinite bandwidth for transmission. Therefore each image sample, also called pel or pixel, must be represented by a finite number of levels (2^k, where k is the number of bits per sample) in order to transmit them over a digital channel. Normally the number of quantization levels in PCM is 64 or 128 corresponding to 6 or 7 bits respectively [1,2]. Thus the PCM technique requires KxNxN bits per image. This needs a large bandwidth for image transmission. Also there is degradation in the picture quality where K is reduced to six or fewer bits per pixel.

Fortunately there is an alternative approach for solving this problem, by exploiting the correlation between the adjacent pixels. Normally an image source is very highly correlated both spatially and temporally, there is a strong dependence among the value of individual picture elements (pixels). The dependency can be regarded as statistical redundancy. Taking the pixels correlation into consideration, will reduce the bits needed for representing each pixel. Many techniques have been suggested in the literature to remove the redundancy from the image without much or any degradation in the picture quality. Picture quality does not depend on the compression method only, but also on the quantization strategy employed with the method.

Two important classes of compression schemes that make use of the statistical redundancy in the image are linear transformation coding and linear predictive coding, where the psychovisual redundancy can be exploited by using an appropriate quantization technique.

Many linear transforms have been used in image data compression such as Fourier transform [3,4], Hadamard transform [5], Karhnuner-Love transform [6]. Slant transform was specially developed for image coding [7]. The optimum transformation would be one that results in statistically independent coefficients and minimum mean square error, but this requires knowledge of higher order statistics of the image. Although Shreiber [8] measured a few third order statistics, only first and second moment can be measured in detail. Furthermore, even if higher order statistics are known, the problem of determining a reversible transformation that results in independent coefficients remains unsolved.

The second approach uses a classical prediction theory; the Differential Pulse Code Modulation (DPCM) is an example of this technique. DPCM system was first introduced by Cutler [9], and due to its simplicity is widely used. However it requires a prior statistical knowledge of the image source, and the system suffers from the inherent problem of high quantization errors.

The important point is that since both of the above schemes involve some image degradation they are not directly applicable to data compression.

In this paper a novel approach to image data compression is proposed which uses a stochastic learning automaton to predict the conditional probability distribution of the adjacent pixels. These conditional probabilities are used to code the gray level values using a Huffman coder. The system uses two learning automatons with identical pseudo-random generators. One automaton is within the transmitter and the second in the receiver.

It is important to distinguish between the learning automata approach and previous work involving finite state machines [13]. Essentially two different interpretations of optimality are involved. In the context of learning automata we are concerned with the learning time for the estimation of a distribution. Optimality in the finite state machine approach relates to the problem of encoding procedures which give an average word length approaching the entropy of the string.

2. Compression Strategy

The objective is to compress a digital image of NxN pixels, each pixel quantized to L gray levels. The gray level value of the $n-th$ pixel (called pixel value) is denoted by $g(n)$, such that, $g(n) = \{1, 2, \ldots, L\}$. The probability $P(i) = P(g(n) : g(n) = i)$, is called the pixel probability.

The compression strategy is based on the classical technique of variable length coding, which assigns codeword lengths on the basis of pixel probability. Thus frequent pixel values are given shorter codewords than less common pixel values. The variable length code is said to be optimum, if the average codeword length is equal to the entropy of the image. Huffman [10] suggested an algorithm which constructs an optimum code. Huffman's algorithm generates a code which is optimal in the sense that no other prefix code will achieve a lower average codeword length. As the code is based on knowledge of the pixel probabilities, the performance of the code is dependent on the accuracy with which the probabilities have been estimated, and on the adaption of the pixel probabilities with time or space. A learning automaton was used to provide an estimate for $P(i) : i = 1, 2, \ldots, L$.

The data compression system Figure 1 consists of a transmitter and a receiver. At the $n-th$ stage both learning automata in the transmitter and receiver contain the state probability vector $P(i-1)$. The coder generates the appropriate binary codeword $C(n)$ corresponding to the value $g(n)$ and the estimated probability vector $P(i)$. $\{$Assumed to be equal to $P(i-1)\}$. The decoder at the receiver will decode the received codeword $C(n)$ to $g(n)$, according to the Huffman algorithm and the estimated probability vector $P(i)$ at the receiver learning automaton. Both automatons then update their vectors according to reinforcement scheme U.

3. Learning Automata

The concept and theory of learning automata has been the subject of a large volume of publications in the form of books and papers. References [11,12] give an excellent survey of

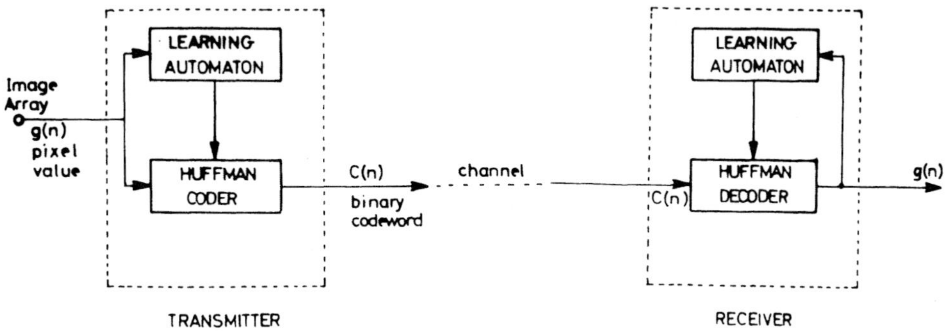

Figure 1: Data compression system .

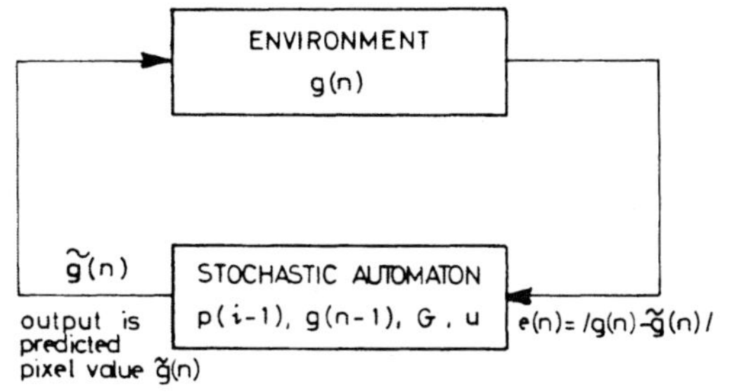

Figure 2: Automaton - environment configuration .

the available results in the area of learning automata, and provide a good introduction to its fundamentals.

The automaton in both the transmitter and receiver employs an identical pseudo-random generator. Figure 2 shows the configuration of the automaton-environment used in both transmitter and receiver. At the $n - th$ stage the environment and automaton contains the $n - th$ pixel value $g(n)$, and the estimated state probability vector $P(i-1)$ respectively. The output of the automaton $g(n)$ is the predicted value of $g(n)$ given $g(n-1)$, and its probability vector $P(i-1)$. $g(n)$ is determined by a stochastic function G which employs a pseudo-random generator. Generators in the transmitter and receiver are synchronized. The environment response is $e(n) : e(n) = g(n) - \tilde{g}(n)$.

4. Reinforcement Scheme

Both automatons use identical updating schemes U for changing $P(i) : i = 1, 2, \ldots, L$. Although various updating schemes have been studied the most satisfactory performance was obtained using a Q-model L_{RP} scheme. This scheme is described by:

$$P_i(n+1) = P_i(n) + \tfrac{B}{C}[C - /E_R/][P_i(n)]$$
$$P_{j\neq i}(n+1) = P_j(n) - \tfrac{B}{C}[C - E_R/][1 - P_i(n)]$$

where $/E_R/ = /g_n - \tilde{g}_n/$

$\tilde{g}_n =$ the predicted value of g_n

C and $B =$ *constants*

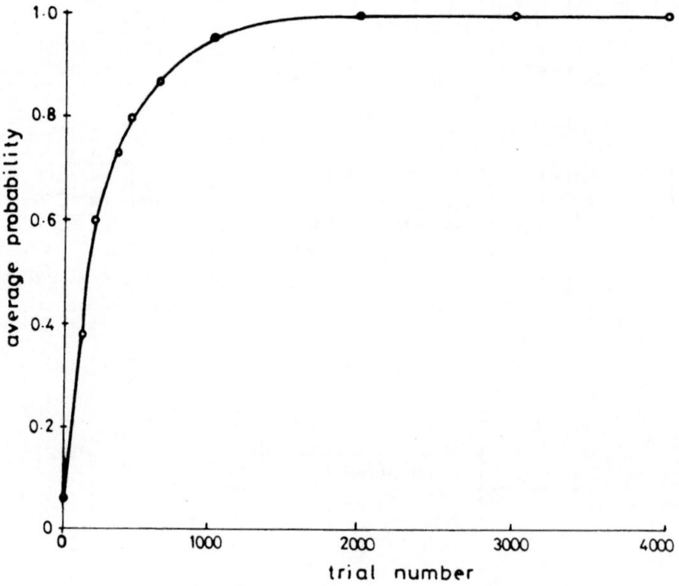

Figure 3: Average probability as a function of trial number.

5. Simulation Studies

Several types of image compression system were simulated on a PDP11/60 minicomputer with a real time image capture and storage system. The objective of the simulations was to study various updating schemes and to compare the performance of the proposed scheme with that of the best known data compression systems. Various linear and non-linear reinforcement schemes have been simulated which result in very poor compression ratio due to either the automaton failing to converge, or the convergence process was too slow for efficient compression. It should be noted that the compression ratio improves significantly with the increase of the speed of convergence. The problem of speeding up the convergence process was closely examined by using the known S- and Q-model schemes. A heuristic approach was adopted, making use of the hybrid scheme concepts. An updating function was proposed using this principle to achieve the best compression ratio.

The scheme presented in the previous section was tested using a 256x256 image with 16 gray levels. The first entropy of the image was computed and found to be 3.3 bit per pixel. Initially the probabilities of the gray levels were set to 1/16, the constant C set to equal 3. B is calculated so that the initial probability reaches the maximum probability after 100-150 successive maximum reward iterations. B was found to be 0.005. With these parameters the probability $P(n)$ reaches 0.95 within 1000 trials. Figure 3 shows a plot of the average probability versus trial number.

Huffman codes have the disadvantages that the source statistics must be known a priori and that only stationary sources can be used. The probability distribution of the learning automaton provides a source of information which allows the Huffman algorithm to be used in a readily adaptive manner to overcome both of these difficulties. Using the learning automaton probability distributions, the average Huffman wordlength was found to be 1.6925 bit per pixel. This compression was achieved without degradation of the image. Although image degradation is a subjective concept, nevertheless the mean squares error (MSE) between the actual image and the resulted image after compression is used as a measure of degradation. Obviously higher degradation results in a large MSE value. Table 1 lists the performance of various compression schemes. The results are generally favorable to the DPCM system.

Table 1: Data Compression System Performance

System	No. of Quantized Levels (bits)	Entropy Compressed Image	MSE	Comment
L.A.	Use Huffman	1.695	0	No degradation
DPCM	3	1.55	12.55	Very noisy background deblared edges
DPCM	4	1.95	10.4	Noticeable noise deblared edges
DPCM	5	2.3	8.4	No noticeable noise

Note: (i) Image 256×256 pixel, 16 gray level/pixel image 1st order entropy 3.3.

(ii) Quantizer includes feedback loop and quantizer output entropy encoded.

6. Conclusions

It has been shown that a learning automaton may be used successfully in an image data compression system. The system uses a Q-model L_{RP} updating scheme. Computer simulations have been used to demonstrate the nature of convergence and to compute the compression ratio.

A large number of algorithms have been developed for compression of data, but the lack of maturity is manifested in the data dependence of the algorithms. A select group works well on speech, another set works well on images, yet another on data. There has been no unified theory to define the optimal compression technique. The proposed procedure inherently seems to offer a unified data compression technique whatever the nature and the statistics of the data.

This paper has reported initial promising results based on using learning automata for the adaptive estimation of probability distributions coupled with Huffman coding. In contrast to previous work [13] the method involves coding on a character by character basis instead of a string. Instead of defining efficiency as the average length of a string the learning automata approach relates efficiency as a function of the estimation of a probability distribution of characters or gray levels taking into account the spatial position of the character.

In future it is hoped to study closely the optimality of the updating scheme and its relationship with the compression ratio. Application of the system to other types of data such as ASCII and speech data is under investigation. The object of the research is to develop an integrated data compression system.

Acknowledgments

The authors wish to gratefully acknowledge the support of other members of the signal processing group at Leicester Polytechnic.

References

[1] R.L. Cabrey, "Video Transmission over Telephone Cable Pairs by Pulse Code Modulation," *Proc. IRE*, vol. 48, pp. 1546-1561, Sept. 1960.

[2] L.H. Harper, "PCM Picture Transmissions," *IEEE Spectrum*, vol. 3, pp. 146, June 1966.

[3] H.C. Andrew and W.K. Pratt, "Fourier Transform Coding of Images," Proc. Hawaii Inter. Confer. System Sci., Western, Jan. 1968, pp. 677-679.

[4] C.B. Anderson and T.S. Huang, "Piecewise Fourier Transformation for Picture Bandwidth Compression," *IEEE Trans. Commun. Tech.*, vol. COM-19, No. 2, pp. 133-140, April 1971.

[5] W.K. Pratt, J. Kane and Andrews, "Hadamard Transform Image Coding," *Proc. IEEE*, vol. 57, No. 1, pp. 58-68, 1969.

[6] M. Tasb and P.A. Winta, "Image Coding by Adaptive Block Quantization," *IEEE Trans. Commun. Tech.*, vol. COM-19, pp. 957-972, 1971.

[7] W.K. Pratt, W.H. Chen and L.R. Welch, "Slant Transform Image Coding," *IEEE Trans. Commun. Tech.*, vol. COM-22, pp. 1075-1093, 1974.

[8] W.F. Schreiber, "The Measurement of Third Order Probability Distribution of Television Signals," *IRE Trans. Inform. Theory*, vol. IT-2, pp. 45-105, Sept. 1956.

[9] C.C. Cutler, "Differential Quantization of Communication Signals," Patent No. 2, 605, 361, July 29, 1952.

[10] D.A. Huffman, "A Method for the Construction of Minimum Redundancy Codes," *Proc. IRE*, 40, pp. 1098-1111, 1952.

[11] K.S. Narendra and M.A.L. Thathachar, "Learning Automata - A Survey," *IEEE Trans. on Systems, Man and Cyb.*, vol. SMC-4, pp. 323-334, July 1974.

[12] K.S. Narendra, "Special Volume on Learning Automata," *J. Cybern. & Inf. Sci.*, 1, 2, 1977.

[13] J. Ziv and A. Lempel, "Compression of Individual Sequences via Variable-Rate Coding," *IEEE Trans. on Information Theory*, IT-24, pp. 530-536, Sept. 1978.

Cooperativity in Networks of Pattern Recognizing Stochastic Learning Automata

Andrew. G. Barto, P. Anandan, and Charles W. Anderson
Department of Computer and Information Science
University of Massachusetts, Amherst MA 01003

Abstract

A class of learning tasks is described that combines aspects of learning automaton tasks and supervised learning pattern-classification tasks. We call these *associative reinforcement learning* tasks. An algorithm is presented, called the *associative reward-penalty*, or A_{R-P}, algorithm, for which a form of optimal performance has been proved. This algorithm simultaneously generalizes a class of stochastic learning automata and a class of supervised learning pattern-classification methods. Simulation results are presented that illustrate the associative reinforcement learning task and the performance of the the A_{R-P} algorithm. Additional simulation results are presented showing how cooperative activity in networks of interconnected A_{R-P} automata can solve difficult nonlinear associative learning problems.

1. Introduction

The ability of stochastic learning automata [1,2] to produce effective collective behavior in the absence of direct inter-automaton communication is an important capability that can be exploited in applications to control problems in distributed systems [3–5]. In this paper, however, we focus on collective behavior produced by networks of *mutually communicating* stochastic learning automata. If a suitable communication medium can be utilized effectively, component automata can condition their actions on both the state of the collection's environment and on the actions of other automata in the collection—yielding cooperative phenomena beyond the range of conventional formulations.

Our approach to providing a communication medium is to give the component learning automata sensitivity to input signals other than reward/penalty signals. One way to do this is to maintain a separate action probability vector for each possible state of an automaton's environment (which includes the actions being chosen by other automata) and then use one of the many conventional stochastic learning automaton algorithms to update the probability vector corresponding to the appropriate state of the environment. This approach amounts to the use of a lookup table for acquiring and accessing action probabilities and has been applied to several types of control problems [5–8] Although this method permits immediate extension of existing learning automaton results to allow dependency on environmental state, it offers no possibility for generalizing among states in order to reduce storage requirements and increase learning rate.

In a previous paper, Barto and Anandan [9] present an algorithm that is a combination of a stochastic learning automaton algorithm and a pattern-classification algorithm. This algorithm, called the *associative reward-penalty*, or A_{R-P}, algorithm, adjusts the parameters of a function that maps pattern input to action probabilities. Unlike a conventional learning automaton algorithm, which converges to a single optimal action, Barto and Anandan [1] prove that under certain conditions the A_{R-P} algorithm asymptotically forms an optimal mapping from input vectors (or patterns) to actions. The result of the learning process is the same mapping that would be achieved by a conventional supervised learning pattern-classification method (see, for example, [10]), but it is formed under the influence of probabilistic reward/penalty feedback that is less informative than the training information required by conventional learning algorithms for pattern classification.

Using this scheme, instead of being maintained in separate slots in a lookup table, the action probabilities corresponding to each environmental state are computed from state information in such a way that similarly coded states yield similar action probabilities. The measure of similarity depends upon the nature of this parameterized computation. If this measure is appropriate for a given task, then action probabilities adjusted on the basis of experience with some environmental states will tend to extrapolate, or generalize, correctly to other states, thereby reducing the need for repeated experience with each state. Consequently, a mapping from the set of possible environmental states to action probabilities may be learned on the basis of a relatively small number of trials.

The well-known deficiencies of this type of procedure arise when the measure of similarity implicit in the computation is not appropriate to the problem, or when the required input/action mapping is not representable by any selection of parameter values. In the case of the A_{R-P} algorithm, since the mapping is restricted to being a linear threshold function, the class of input/action mappings implementable by a single A_{R-P} automaton is limited. It is precisely this form of limitation, however, that is addressed by the *collective behavior* of A_{R-P} automata. As we illustrate by means of computer simulations, interacting A_{R-P} automata (which we also call A_{R-P} *elements*) are able to form complex nonlinear mappings if required to do so to maximize reward probability. They utilize communication links between them to coordinate their actions so as to achieve reward with higher probability than would be possible if the component automata acted independently. As these nonlinear mappings develop, the measure of similarity underlying generalization effectively changes. Networks of this kind differ from the hierarchical networks of learning automata previously studied [11] since here all the component automata participate in parallel at each step to construct a mapping from input vectors to actions. Although there is not as yet a theorem covering this form of collective behavior, our major interest in the A_{R-P} algorithm is its role in this technique for learning nonlinear mappings that are required for complex pattern classification and control problems.

The A_{R-P} algorithm is a result of our study of the learning and computational abilities of networks of relatively simple neuron-like processing elements [12–17]. There is renewed interest in this approach in the allied fields of Cognitive Science and Artificial Intelligence, where the term "connectionism" has been revived to refer to it (see, for example, [18, 19] for collections of relevant papers). Not only have advances in microelectronics made the physical realization of brain-like hardware more of a possibility, but advances in our understanding of some of the problems involved in vision, motor control, and knowledge representation suggest that such hardware offers advantages over conventional computational architectures and may be necessary for real-time performance. For example, much attention has been devoted to "associative memory networks" that store information in such a way that it is distributed across the network and is accessed by an association process instead of the lookup-table addressing scheme used in computer memories [19–20] These memory networks typically consist of single layers of linear threshold elements, and information storage is accomplished by using some form of supervised learning pattern-classification algorithm. Numerous properties make associative memory networks attractive both as computational devices and as models of biological memory: their operation tends not to be seriously disrupted by noise, and information can be retrieved based only on partial matches of storage and retrieval keys. Part of our motivation for studying A_{R-P} elements (and for using the term "associative") is that the mappings they form are of this kind. Other aspects of A_{R-P} elements were suggested by the ideas put forward by Klopf in [21].

One of the difficulties with associative memory networks, however, has been the lack of effective algorithms extending single-element learning results to multilayered networks that can implement nonlinear associative mappings. A key requirement for such algorithms is that they cannot depend on *a priori* knowledge in the environment about the complete structural detail of the desired associative mappings. Straightforward extensions of single-element gradient-descent learning procedures to the task of adjusting all a network's parameters fail due to the difficulty of determining the gradient locally and the multimodality of the error functional. Recent research efforts have resulted in several novel approaches to these problems. Ackley, Hinton and Sejnowski [22] utilize principles from statistical thermodynamics to derive a learning algorithm for symmetrically connected networks; Rumelhart, Hinton, and Williams [23] have developed an algorithm that uses recursive back-propagation of error information for layered networks. Our approach relies on the synthesis of algorithms from pattern classification and the theory of learning automata that will be described here. Comparison of these various methods with regard to robustness and computational efficiency are currently underway.

2. Associative Reinforcement Learning and the A_{R-P} Algorithm

The A_{R-P} algorithm is designed to solve what we call *associative reinforcement learning* tasks. In these tasks the learning system and its environment interact in a closed loop. At each discrete time step, or trial, k, the environment provides the learning system with a pattern vector, x_k, selected from a finite set $X = \{x^1, \ldots, x^m\}$, $x^i \in \Re^n$; the learning system emits an action, a_k, chosen from the finite set $A = \{a^1, \ldots, a^r\}$; the environment receives a_k as input and sends to the learning system a reward/penalty signal $b_k \in \{-1, +1\}$ that evaluates the action a_k, where -1 and $+1$ respectively indicate *penalty* and *reward.* The environment determines the evaluation according to a map $d : X \times A \to [0,1]$, where $d(x,a) = Pr\{b_k = +1 \mid x_k = x, a_k = a\}$. Ideally, one wants the learning system eventually to respond to each input vector $x \in X$ with action a_x with probability 1, where a_x is such that $d(x, a_x) = \max_{a \in A}\{d(x,a)\}$.

As pointed out in [9], in the case of a single input vector, this task reduces to the task usually studied by learning automaton theorists [1, 2] (which, according to the terminology used here, is a *nonassociative* reinforcement learning task). On the other hand, in the case of two actions ($|A| = 2$) the task reduces to a conventional formulation of supervised learning pattern classification (see [10]) if for each $x \in X$, $d(x,a^1) + d(x,a^2) = 1$. This restriction (assuming it is known to hold) implies that feedback received from performing one action provides information about the other action. This makes the task much easier and allows conventional supervised learning pattern-classification algorithms (slightly modified) to succeed (see [9] for details).

The A_{R-P} algorithm's decision rule is parameterized at step k by a vector $\theta_k \in \Re^n$:

$$a_k = \begin{cases} +1, & \text{if } \theta_k^{\mathrm{T}} x_k + \eta_k > 0; \\ -1, & \text{otherwise;} \end{cases} \tag{1}$$

where $\theta_k^{\mathrm{T}} x_k$ is the inner product of θ_k and x_k, and the η_k are independent identically distributed random variables, each having distribution function Ψ. The parameter vector is updated according to the following equation:

$$\theta_{k+1} - \theta_k = \begin{cases} \rho_k[a_k - E\{a_k|\theta_k, x_k\}]x_k, & \text{if } b_k = +1 \text{ (reward)}; \\ \lambda\rho_k[-a_k - E\{a_k|\theta_k, x_k\}]x_k, & \text{if } b_k = -1 \text{ (penalty)}; \end{cases} \tag{2}$$

where $0 \le \lambda \le 1$ and $\rho_k > 0$. The expected value in (2) is a known function of θ_k and x_k that depends on Ψ (see below).

According to (1), the action probabilities at step k are conditional on the input vector in a manner determined by the parameter vector θ_k. In particular

$$p_k^{-1x} = Pr\{a_k = -1|x_k = x\} = Pr\{\theta_k^{\mathrm{T}} x + \eta_k \le 0\} = \Psi(-\theta_k^{\mathrm{T}} x),$$

and

$$p_k^{+1x} = Pr\{a_k = +1|x_k = x\} = 1 - p_k^{-1x}.$$

If, for example, each random variable η_k has zero mean, then when $\theta_k^T x = 0$, the probability that each action is emitted given input vector x is .5, and the action expectation, $E\{a_k|\theta_k, x_k\}$, in (2) is 0; when $\theta_k^T x$ is positive, action $a_k = +1$ is the more likely action, and the action expectation is positive; when $\theta_k^T x$ is negative, action $a_k = -1$ is the more likely, and the action expectation is negative. As $|\theta_k^T x|$ increases for all $x \in X$, the mapping (1) approaches a deterministic linear discriminate function. In the case of reward, according to (2), θ changes so as to reduce the discrepancy between the action expectation and the action actually chosen, a_k. In the case of penalty, θ changes so as to reduce the discrepancy between this expectation and the action *not* chosen, $-a_k$. Note that the parameter λ in (2) determines the degee of asymmetry in the magnitude of the parameter change for these two cases.

It is shown in [9] that the A_{R-P} algorithm reduces under various restrictions to more conventional algorithms. It reduces to the two-action (nonassociative) linear reward-ϵ-penalty ($L_{R-\epsilon P}$) learning automaton algorithm [1] when each η_k in (1) is uniform in the interval $[-1, 1]$, the input pattern is constant and nonzero over time steps ($x_k \equiv \hat{x}$), and the initial parameter vector θ_1 is such that $\theta_1^T \hat{x} \in [-1, 1]$. If additionally $\lambda = 0$, then the A_{R-P} algorithm reduces to the linear reward-inaction (L_{R-I}) algorithm [1]. On the other hand, when the A_{R-P} algorithm is made deterministic by letting $\eta_k = 0$ for all k (i.e., the distribution fuction Ψ is the step function), it becomes the perceptron algorithm if one uses the product $b_k a_k$ as the training signal giving the desired response or correct classification. With a slight modification, the A_{R-P} algorithm can be reduced to the pattern-classification method introduced by Widrow and Hoff [24] (the adaline algorithm), which can be viewed as an application of the Robbins-Monro stochastic approximation algorithm [25]. Finally, note that if the set X of input vectors consists of the standard unit basis vectors $x^i = (0, \ldots, 1, \ldots, 0)^T$, then the A_{R-P} algorithm reduces to the (two-action) lookup-table method mentioned in the introduction. Consequently, the A_{R-P} algorithm not only extends learning automata capabilities but also occupies the intersection of important classes of learning algorithms. The A_{R-P} algorithm is most closely related to the "selective bootstrap adaptation" algorithm of Widrow, Gupta, and Maitra [26], to which it is compared in [9].

A convergence theorem is proven by Barto and Anandan [91] by extending to the associative case results proven by Lakshmivarahan [7, 27]. It holds under the following conditions: (C1) the set of input vectors $X = \{x^1, x^2, \ldots, x^m\}$, $x^i \in \Re^n$, is a linearly independent set; (C2) for each $x^i \in X$ and $k \geq 1$, $Pr\{x_k = x^i\} = \xi^i > 0$; (C3) the independent, identically distributed random variables η_k in (1) have a continuous and strictly monotonic distribution function Ψ; and (C4) the sequence ρ_k in (2) is such that $\rho_k \geq 0$, $\sum_k \rho_k = \infty$, $\sum_k \rho_k^2 < \infty$. We can prove the following theorem:

Theorem. *Under conditions (C1)-(C4), for each $\lambda \in (0, 1]$, there exists a $\theta_\lambda^\circ \in \Re^n$ such that the random process $\{\theta_k\}_{k \geq 1}$ generated by the A_{R-P} algorithm in an associative reinforcement learning task converges to θ_λ° with probability 1 (that is, $Pr\{\lim_{k \to \infty} \theta_k = \theta_\lambda^\circ\} = 1$), where for all $x \in X$,*

$$Pr\{a = 1|\theta_\lambda^\circ, x\} \quad \begin{matrix} > 1/2, & \text{if} \quad d(x, +1) > d(x, -1) \\ < 1/2, & \text{if} \quad d(x, +1) < d(x, -1). \end{matrix}$$

In addition, for all $x \in X$,

$$\lim_{\lambda \to 0} Pr\{a = 1|\theta_\lambda^\circ, x\} = \begin{cases} 1, & \text{if } d(x, +1) > d(x, -1); \\ 0, & \text{if } d(x, +1) < d(x, -1). \end{cases}$$

According to the usual performance criteria for learning automata [1], this result implies that for each $x \in X$, the A_{R-P} algorithm is ϵ-optimal. In fact, it implies a strong form of ϵ-optimality for each $x \in X$ [7]. It is highly unlikely that this result is the most general that can be proved about this class of algorithms (see [9]). As is often done when using similar pattern-classification algorithms, in the simulations presented here we hold ρ_k constant in order to increase learning speed even though a weaker form of convergence in this case has not yet been proven. We have not yet investigated elaborations of the A_{R-P} algorithm that reduce to recursive least squares methods based on the Newton's algorithm, but these have the possibility for showing improved convergence rates. We view condition (C1) that the set of input vectors is linearly independent as the most serious restriction required for the present theorem. It is likely that this restriction can be removed and a result proved that involves some form of operator pseudoinverse.

3. Simulation of a Single A_{R-P} Automaton

In order to illustrate the A_{R-P} algorithm's performance, we describe the results of simulating a single A_{R-P} element in a simple associative reinforcement learning task that requires discrimination between two linearly independent, but non-orthogonal, input vectors. We use as a measure of performance the probability that the element will receive reward on the average time step given its current parameter vector. We denote this M_k when computed based on the parameter vector θ_k:

$$\begin{aligned} M_k &= \sum_{x \in X} \xi^x [Pr\{b_k = +1 | x_k = x\}] \\ &= \sum_{x \in X} \xi^x [d(x,+1) p_k^{+1x} + d(x,-1) p_k^{-1x}]. \end{aligned}$$

This measure is maximized when the optimal action for each input pattern occurs with probability 1, in which case it is

$$M_{\max} = \sum_{x \in X} \xi^x \max\{d(x,+1), d(x,-1)\}.$$

The distribution function of the random variables used in all the simulations described here is the logistic distribution given by $\Psi(r) = 1/(1+e^{-r/T})$, where T is a parameter. This is a sigmoidal function that is similar to a normal distribution function but is easier to evaluate. It is also used in the studies of statistical cooperativity (e.g., [22]), where T is the "computational temperature" of the system. As T approaches zero, it approaches a step function, which means that the algorithm more closely approximates a deterministic system. Given this distribution function, the term $E\{a_k|\theta_k, x_k\}$ in (2) becomes a specific function of $\theta_k^{\mathrm{T}} x_k$:

$$\begin{aligned} E\{a_k|\theta_k, x\} &= -1 \cdot p_k^{-1x} + 1 \cdot p_k^{+1x} \\ &= 1 - 2\Psi(-\theta_k^{\mathrm{T}} x) \\ &= \frac{e^{\theta_k^{\mathrm{T}} x/T} - 1}{e^{\theta_k^{\mathrm{T}} x/T} + 1}. \end{aligned}$$

This is an odd function of $\theta_k^{\mathrm{T}} x$ with limits of -1 and $+1$ for $\theta_k^{\mathrm{T}} x$ respectively approaching $-\infty$ and $+\infty$. In all simulations presented here, we set $T = .5$, so that this function has a slope of 1 at the origin (except where otherwise noted).

In the first simulation the input vectors are: $x^1 = (1,0)^{\mathrm{T}}$ and $x^2 = (1,1)^{\mathrm{T}}$, which are linearly independent but not orthogonal. These vectors are equally likely to occur on each trial ($\xi^1 = \xi^2 = .5$). The parameter vector θ is zero at the start of each sequence of trials, which makes the actions initially equiprobable for both input vectors. The reward probabilities implemented by the element's environment are given by the following table:

x	$d(x,-1)$	$d(x,+1)$
x^1	.6	.9
x^2	.4	.2

Thus it is optimal for the learning system to respond to $(0,1)^{\mathrm{T}}$ with action $+1$ to obtain reward with probability .9, and to respond to $(1,1)^{\mathrm{T}}$ with action -1 to obtain reward with probability .4. Therefore, in this task $M_{\max} = (.9+.4)/2 = .65$, and the initial overall reward probability is $(.6+.9+.4+.2)/4 = .525$. Note that any nonassociative learning automaton

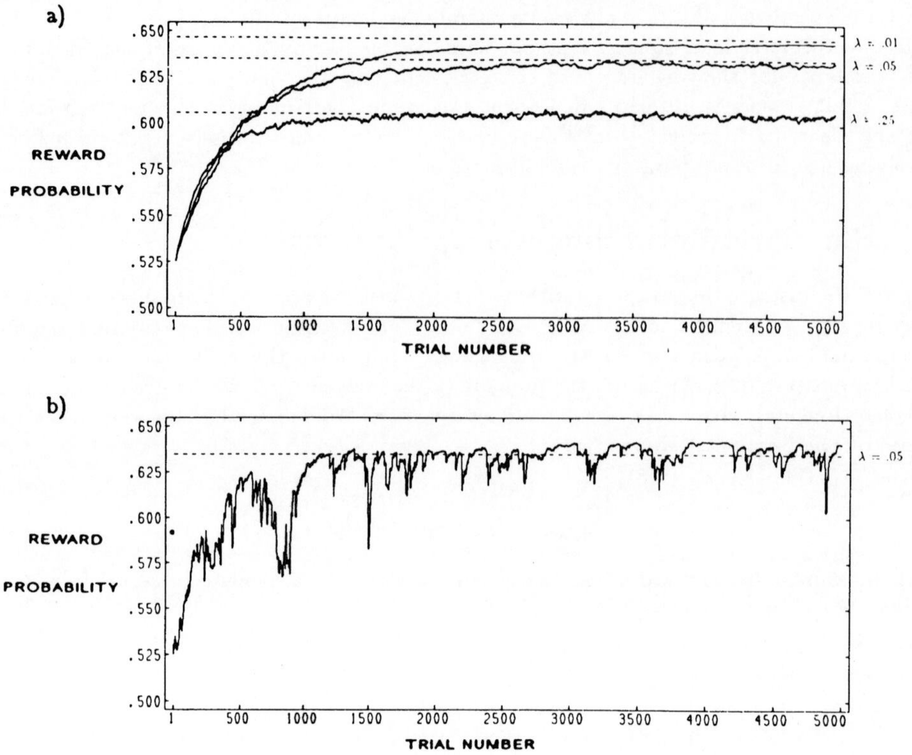

Figure 1:

algorithm will be able to achieve a reward probability of at most $(.9+.2)/2 = .55$ by learning to perform action $+1$ at all times. Also note that for each input x, the reward probablities are either both greater than .5 or both less than .5, making this task considerably more difficult than one with the reward probabilties placed above and below .5 for each x.

Fig. 1a shows results of simulating an A_{R-P} element in this task with three different values of λ: .01, .05, and .25. We held the parameter ρ_k at the value .5 for all k. Plotted for each trial k is the average of M_k over 100 runs, where a run is a sequence of 5000 trials. The dashed lines show theoretical asymptotic performance levels for the three values of λ (if ρ_k were decreasing according to (C4)). Note that this asymptote approaches the optimal performance level .65 as λ decreases and that the learning rate decreases as λ decreases. The average final parameter vectors for $\lambda = .01$, .05, and .25 are respectively $(2.99, -4.04)^T$, $(2.73, -3.08)^T$, and $(1.91, -1.71)^T$. Fig. 1b shows a plot of M_k for one of the runs contributing to the average shown in Fig. 1a for $\lambda = .05$. Although this task involves only two-dimensional pattern vectors, it illustrates the essential difficulties of learning to discriminate between patterns that are similar by virtue of sharing a subset of feature values.

4. Nonlinear Problems

In the task described above, parameter vectors exist that produce performance as close to optimal as desired. In fact, this is the only case covered by the convergence theorem since the condition that the set of input vectors, X, is linearly independent implies that any mapping from X to $\Re$ can be implemented by a suitable value of θ. If X is not a linearly independent set, several different cases can arise. One case corresponds to the linearly separable case in pattern classification. In this case it is also possible to find parameter vectors that yield

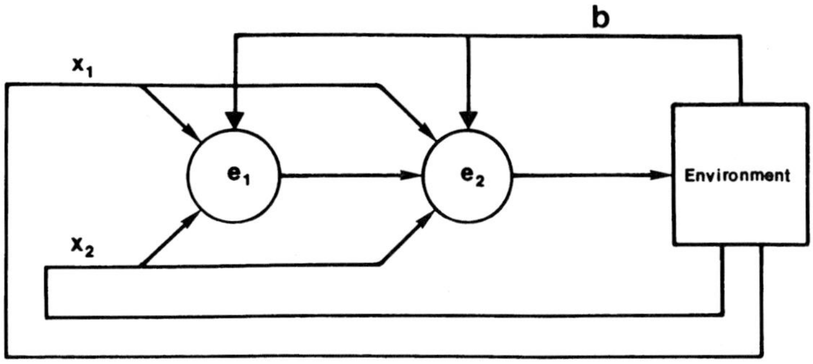

Figure 2:

performance as close to optimal as desired, but we do not yet know if the A_{R-P} algorithm will always converge to one of them. Another case occurs when such a vector does not exist. The requirement of obtaining optimal performance, or even better than chance performance, for one input vector may preclude doing so for another input vector. This is the nonseparable case. We have not yet thoroughly studied the performance of a single A_{R-P} element in these cases. We have instead investigated, by means of computer simulation, the performance of collections of interconnected A_{R-P} elements facing nonseparable associative reinforcement learning tasks. These simulations suggest that through cooperative collective activity, networks of A_{R-P} elements can reliably achieve performance as close to optimal as desired in nonseparable problems. We now describe two examples.

A. The Exclusive-Or Task

A network of two A_{R-P} elements, e_1 and e_2 (Fig. 2), is placed in a task requiring it to form the two-feature exclusive-or mapping The stimulus patterns are all the two-component binary vectors, $x^0 = (0,0)^T$,$x^1 = (0,1)^T$,$x^2 = (1,0)^T$, $x^3 = (1,1)^T$. These vectors are equally likely to occur on any trial. On each trial, both elements receive one of these input vectors, but e_2 additionally receives the action of e_1 as an input component (which we recode to take the values 0 and 1 instead of -1 and $+1$). Only the action of e_2 directly affects the reward/penalty feedback (labeled b in Fig. 2), which is delivered to both elements. Thus, the elements form a "team" in the sense studied by learning automaton theorists except that the elements receive information indicating environmental state, and e_2 receives a signal indicating the action of e_1. Each element in this and the following simulation has an additional scalar input that is equal to 1 for all steps k. By adjusting the component of θ corresponding to this input via (2), an element can effectively adjust the mean value of its threshold. Since this input component is always present, we do not include it when we list the input vectors, and we do not show it in the figures.

The reward probabilities are given by the following table:

x	$d(x,-1)$	$d(x,+1)$
x^0	.9	.1
x^1	.1	.9
x^2	.1	.9
x^3	.9	.1

Table entry $d(x,a)$ is the reward probability given that e_1 and e_2 receive x as input and e_2 produces action a. The optimal reward probability is $M_{\max} = .9$, which is obtained when the action of e_2 is the exclusive-or of the input vector components. Note that a nonassociative learning automaton can obtain a reward probability of at most .5. A single A_{R-P} element

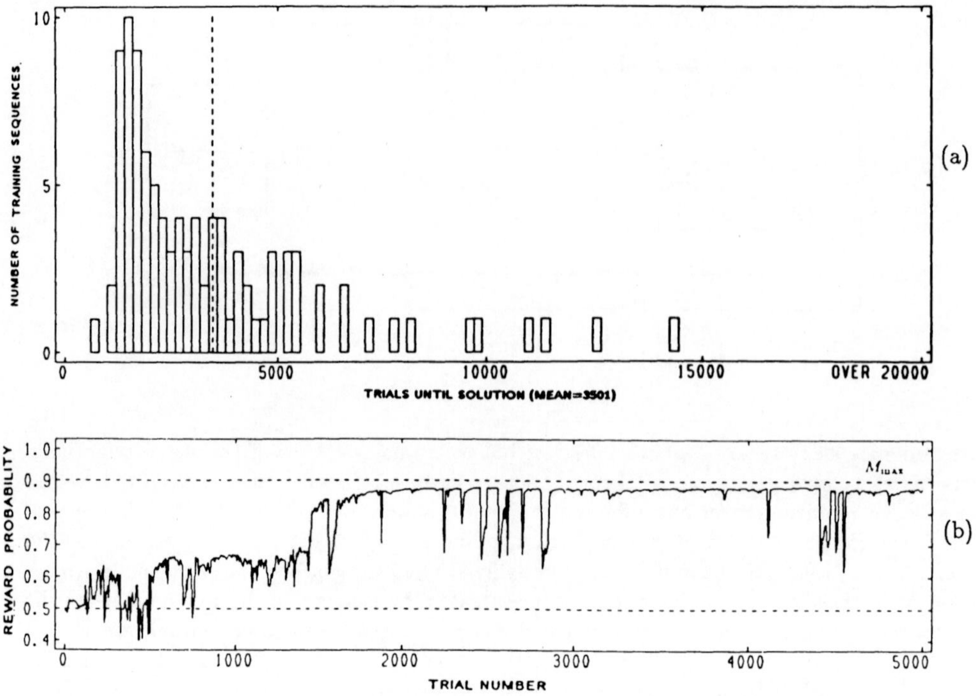

Figure 3:

can be correct for at most three of the four input vectors, yielding a reward probability of $(.9 + .9 + .9 + .1)/4 = .7$, since parameters do not exist that allow a single linear threshold element to respond correctly to all four vectors. However, the performance of the network shown in Fig. 2 can approach $M_{\max}$ if e_1 learns to respond to the fourth case and e_2 takes advantage of this signal to "debug" its mapping. Fig. 3a shows a histogram of the number of trials until a criterion of 95% of $M_{\max}$ is attained for each of 100 runs with $\rho_k \equiv 1.5$ and $\lambda = .08$. The average number of trials until criterion is 3501, or about 875 steps for each input vector. In all of the runs the network reached this criterion before 15 000 trials.

The typical learning process over a sequence of trials has the following character. Performance quickly improves until the network responds correctly to three of the four input vectors with a high probability (yielding reward probability of near .7). For example, the network may form the inclusive-or function and therefore produce the incorrect action (with a high probability) for the input vector $x^3 = (1,1)^{\mathrm{T}}$. This error is the result of incorrect generalization from the cases x^1 and x^2, whose sum is x^3. If the correct function had been inclusive-or, then this generalization would have resulted in faster learning than that of a table-lookup method. For exclusive-or, however, performance levels off at the suboptimal level of .7, but after remaining at this level for a period of time, it quickly jumps to a value near the optimum of .9. This occurs when e_1 becomes a functional part of the network by learning, for example, to respond with action +1 only to x^3 while e_2 simultaneously assigns a negative weight to e_1's action so that it correctly produces action -1 whenever x^3 is present. If necessary, therefore, the network structure develops so as to alter the natural generalizations produced by the initially given representation of the input vectors. Fig. 3b shows an example of how network performance changes for a single run. This profile showing periods of near constant performance separated by rapid increases in performance is characteristic of the behavior of A_{R-P} networks in many tasks.

A significant aspect of this illustration is that the network's environment need not know

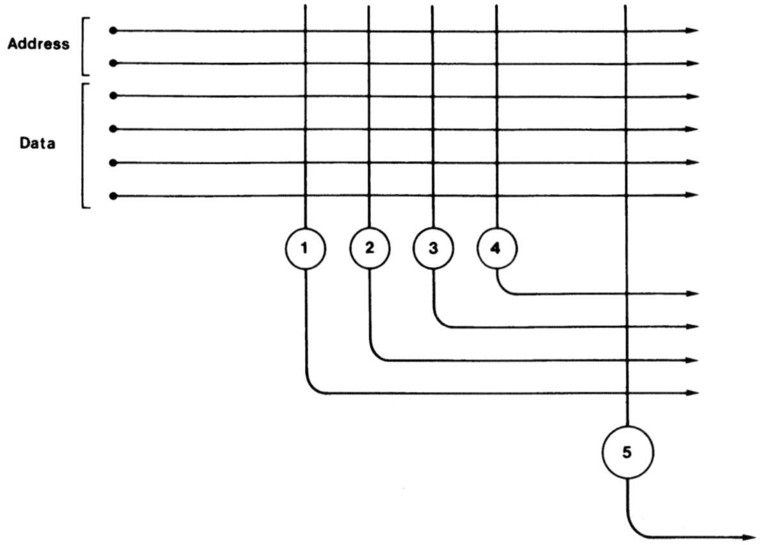

Figure 4:

the structural details about how the exclusive-or mapping can be realized by linear-threshold primitives. We initially provided a sufficient number of elements and potential interconnections to realize the required mapping, but the reward probabilities implemented by the environment relate only to the overall behavior of the network—not to its structure. If there had been more elements than required, then solutions could leave some elements unused in the computation. These extra elements could then become engaged in the computation if the requirements of the task were to change. Some of these features of the learning process are illustrated more vividly in the task we describe next.

B. A More Difficult Nonlinear Task

The network shown in Fig. 4 has six input components and a single main output (from element 5, or e_5). There are 39 parameters to adjust: one associated with each of the pathway intersections and one associated with the constant input to each element (not shown). There is also a pathway that delivers reward/penalty feedback uniformly to each element (also not shown). The reward probabilities implemented by the network's environment force the network to learn to realize a multiplexer circuit in order to maximize reward frequency. A multiplexer is a device with n address input lines and 2^n data input lines, each of which is associated with a distinct n-bit address. Given a pattern over the address lines, i.e., an address, a multiplexer's output is equal to whatever signal (0 or 1) appears on the data line with that address. This is a highly nonlinear mapping and learning it is difficult since some of the natural generalizations produced as a result of similarities among the input vectors are irrelevant or misleading with respect to the required actions of the network. Elements 1–4 must tune to certain constellations of the components of the input vectors in order to disrupt misleading generalizations.

For each of the 64 possible input vectors, we rewarded each element of the network with probability 1 if e_5 produced the correct output, and we penalized each element with probability 1 otherwise. The input vectors were chosen randomly for presentation to the network. All of the elements implement the A_{R-P} algorithm with $T = .5$ except for e_5 which uses $T = 0$. This implies that e_5 is deterministic—it essentially implements the perceptron algorithm [9]. Fig. 5 is a histogram of the number of trials required for the network to respond correctly for 99% of 1000 consecutive trials in each of 30 runs with $\rho_k \equiv 1$ and $\lambda = .01$. The average number of trials required was 133 149, or about 2080

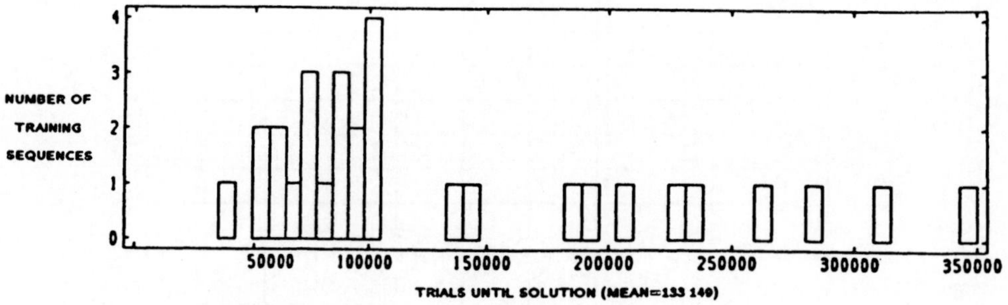

Figure 5:

presentations of each input vector. In every run the network reached the criterion within 350 000 trials. We used this criterion because an exact computation of M_k for each trial k requires a considerable amount of computation for a network of this size.

There are many different ways in which this network can realize the multiplexer function, each of which corresponds to a different functionally equivalent logical expression. Different training sequences result in different structures, some of which are quite complex but nevertheless correct. This reflects the fact that the environment's evaluation process provides functional, but not structural, information to the adaptive elements that make up the network.

5. Discussion

The learning in nonseparable tasks illustrated by the simulations suggests that layered networks of A_{R-P} elements are able to learn to implement mappings that are beyond the capabilities of individual elements. Although we have not yet proved convergence for networks of A_{R-P} elements, all of our simulations suggest extremely reliable performance. However, these results also suggest that the process may take a considerable amount of time. It is difficult to evaluate the learning rate of A_{R-P} networks without comparing their performance with that of other learning algorithms, and we are currently in the process of performing comparative simulation studies using a variety of algorithms.

Despite these questions about learning rate, interacting A_{R-P} elements show potentially useful collective behavior. The ability to learn to form nonlinear mappings in the manner illustrated by the computer simulations described here is significant because the evaluative information provided by the network's environment need not be based on any knowledge of the structural details of the mapping. Reward and penalty signals can be generated based upon knowledge of "what" the network as a whole should be doing. Knowledge of "how" the network can accomplish this is not required. This suggests that the collective bahavior of pattern recognizing stochastic learning automata may provide a basis for extending parametric adaptation techniques toward more powerful structural learning methods.

Acknowledgment

This research was supported by the Air Force Office of Scientific Research and the Avionics Laboratory (Air Force Wright Aeronautical Laboratories) through contract F33615-83-C-1078.

References

[1] K.S. Narendra and M.A.L. Thathachar, "Learning Automata—A Survey," *IEEE Trans. Syst., Man, Cybern.*, vol. 4, pp. 323–334, 1974.

[2] K.S. Narendra and S. Lakshmivarahan, "Learning Automata—A Critique," *J. Cybern. and Inf. Sci.,* vol. 1, pp. 53–65, 1977.

[3] P. Mars, K.S. Narendra, and M. Crystall, "Learning Automata Control of Computer Communication Networks," Proc. Third Yale Workshop on Applications of Adaptive Systems Theory, 1983.

[4] L.G. Mason, "Learning Automata and Telecommunications Switching," Proc. Third Yale Workshop on Applications of Adaptive Systems Theory, 1983.

[5] R.M. Wheeler and K.S. Narendra, "Models for Decentralized Decisionmaking," Report No. 8403, Electrical Engineering, Yale University, 1984.

[6] R.A. Jarvis, "Teaching a Stochastic Automaton to Skillfully Play Hand/Eye Games," *J. of Cybern. and Inf. Sci.*, vol. 1, pp. 161–177, 1977.

[7] S. Lakshmivarahan, Learning Algorithms and Applications, Springer-Verlag, New York, 1981.

[8] I.H. Witten, "An Adaptive Optimal Controller for Discrete-time Markov Environments," *Inf. and Contr.*, vol. 34, pp. 286-295, 1977.

[9] A.G. Barto and P. Anandan, "Pattern Recognizing Stochastic Learning Automata," *IEEE Trans. on Syst., Man, Cybern.*, vol. 15, pp. 360–375, 1985.

[10] R.O. Duda and P.E. Hart, Pattern Classification and Scene Analysis, Wiley, New York, 1973.

[11] M.A.L. Thathachar and K.R. Ramakrishnan, "An Automaton Model of a Hierarchical Learning System," Proc. 8th Triennial World Congress, IFAC Control Science and Technology, Kyoto, Japan, pp. 1065–1070, 1981.

[12] A.G. Barto, Editor. "Simulation Experiments with Goal-seeking Adaptive Elements," Air Force Wright Aeronautical Laboratories/Avionics Laboratory Technical Report AFWAL-TR-84-1022, Wright-Patterson AFB, Ohio, 1984.

[13] A.G. Barto, C.W. Anderson, and R.S. Sutton, "Synthesis of Nonlinear Control Surfaces by a Layered Associative Search Network," *Biol. Cybern.*, vol. 43, pp. 175–185, 1982.

[14] A.G. Barto and R.S. Sutton, "Landmark Learning: An Illustration of Associative Search," *Biol. Cybern.*, vol. 42, pp. 1–8, 1981.

[15] A.G. Barto, R.S. Sutton, and C.W. Anderson, "Neuronlike Elements That Can Solve Difficult Learning Control Problems," *IEEE Trans. on Syst., Man, Cybern.*, vol. SMC-13, pp. 834–846, 1983.

[16] A.G. Barto, R.S. Sutton, and P.S. Brouwer, "Associative Search Network: A Reinforcement Learning Associative Memory," *Biol. Cybern.*, vol. 40, pp 201–211, 1981.

[17] R.S. Sutton and A.G. Barto, "Toward a Modern Theory of Adaptive Networks: Expectation and Prediction," *Psych. Rev.,* vol. 88, pp. 135–171, 1981.

[18] J.A. Feldman (Ed.), Special Issue on Connectionist Models and Their Applications, *Cognitive Science*, vol. 9, 1985.

[19] G. Hinton and J. Anderson, Parallel Models of Associative Memory, Erlbaum, Hilsdale, N. J., 1981.

[20] T. Kohonen, Associative Memory: A System Theoretic Approach, Springer, Berlin, 1977.

[21] A.H. Klopf, The Hedonistic Neuron: A Theory of Memory, Learning, and Intelligence, Hemisphere, Washington, D.C., 1982.

[22] D.H. Ackley, G.E. Hinton, and T.J. Sejnowski, "A Learning Algorithm for Boltzmann Machines," *Cognitive Science*, vol. 9, pp. 147–169, 1985.

[23] D.E. Rumelhart, G.E. Hinton, and R.J. Williams, "Learning Internal Representations by Error Propagation," ICS Report 8506, Institute for Cognitive Science, University of California, San Diego, 1985.

[24] B. Widrow and M.E. Hoff, "Adaptive Switching Circuits," *1960 WESCON Convention Record Part IV*, pp. 96–104, 1960.

[25] R.L. Kasyap, C.C. Blaydon, and K.S. Fu, "Stochastic Approximation," in Adaptation, Learning and Pattern Recognition Systems: Theory and Applications, J.M. Mendel and K.S. Fu, Eds. Academic Press, New York, 1970.

[26] B. Widrow, N.K. Gupta, and S. Maitra, "Punish/Reward: Learning with a Critic in Adaptive Threshold Systems," *IEEE Trans. on Syst., Man, Cybern.*, vol. 5, pp. 455–465, 1973.

[27] S. Lakshmivarahan, "ϵ-optimal Learning Algorithms—Non-absorbing Barrier Type," Technical Report EECS 7901, School of Electrical Engineering and Computer Sciences, University of Oklahoma, Norman, Oklahoma, 1979.

The Genetic Algorithm Approach: Why, How, and What Next?

David E. Goldberg
Dept. of Engineering Mechanics
The University of Alabama
University, Alabama 35486

1. Introduction

When man wanted to fly, he first turned to natural example – the bird – to develop his early notions of how to accomplish this difficult task. Notable failures by Daedulus and numerous bird-like contraptions (ornithopters) at first pointed in the wrong direction, but eventually, persistence and the abstraction of the appropriate knowledge (lift over an airfoil) resulted in successful glider and powered flight. In contrast to this example, isn't it peculiar that when man has tried to build machines to think, learn, and adapt he has ignored and largely continues to ignore one of nature's most powerful examples of adaptation, genetics and natural selection. The primary mechanisms for adaptation in most optimization and learning systems depend upon man's own artificial creations such as calculus and counting. The rich and efficient performance of nature's own adaptation algorithm-of-choice is just starting to receive the attention it deserves in artificial system adaptation and learning.

In this paper, we examine a methodology of adaptation and machine learning rooted in an appropriate abstraction of genetics and natural selection. This methodology, loosely referred to as the genetic algorithm method, combines a Darwinian survival-of-the-fittest among discrete string structures (artificial chromosomes) and a structured, yet randomized, information exchange among these structures to motivate a search heuristic with surprising robustness. In the remainder of this paper, we examine the mechanics of a simple genetic algorithm (GA), discuss its highly leveraged power of effect, sketch one application in a rule learning system, and outline research needs in the theory and application of GA's to both optimization and machine learning systems.

2. A Simple Genetic Algorithm

Genetic algorithms are different from the normal search methods encountered in engineering optimization in the following ways:

1. GA's work with a coding of the parameter set not the parameters themselves.
2. GA's search from a population of points.
3. GA's use probabilistic not deterministic transition rules.

Genetic algorithms require the natural parameter set of the optimization problem to be coded as a finite length string. A variety of coding schemes can and have been used successfully. Because GA's work directly with the underlying code they are difficult to fool because they are not dependent upon continuity of the parameter space and derivative existence.

In many search methods, we move gingerly from a single point in the decision space to the next using some decision rule to tell us how to get to the next point. This point-by-point method is dangerous because it often locates false peaks in multimodal search spaces. GA's work from a database of points simultaneously (a population of strings) climbing many peaks in parallel, thus reducing the probability of finding a false peak.

Unlike many methods, GA's use probabilistic decision rules to guide their search. The use of probability does not suggest that the method is some kind of random walk. Genetic algorithms are quite rapid in locating improved performance.

For our work, we may consider the strings in our population of strings to be expressed in a binary alphabet containing the characters {0,1}. Each string is of length ℓ and the population contains a total of n such strings. Of course, each string may be decoded to a set of physical parameters according to our design. Additionally, we assume that we may evaluate a fitness value for each string (parameter set). Fitness is defined as the non-negative figure of merit we are maximizing. Thus, the fitness in genetic algorithm work corresponds to the objective function in normal optimization work.

A simple genetic algorithm which gives good results is composed of three operators:

1. Reproduction
2. Crossover
3. Mutation

With our simple genetic algorithm we view reproduction as a process by which individual strings are copied according to their fitness. Highly fit strings receive higher numbers of copies in the mating pool. There are many ways to do this; we simply give a proportionately higher probability of reproduction to those strings with higher fitness (objective function value). Reproduction is thus the survival-of-the-fittest or emphasis step of the genetic algorithm. The best strings make more copies for mating than the worst.

After reproduction, simple crossover may proceed in two steps. First, members of the newly reproduced strings in the mating pool are mated at random. Second, each pair of strings undergoes crossing over as follows: an integer position k along the string is selected uniformly at random on the interval $(1, \ell - 1)$. Two new strings are created by swapping all characters between positions 1 and k inclusively.

For example, consider two strings A and B of length 7 mated at random from the mating pool created by previous reproduction:

$$\left.\begin{array}{lccccccc} A = & a1 & a2 & a3 & a4 & a5 & a6 & a7 \\ B = & b1 & b2 & b3 & b4 & b5 & b6 & b7 \end{array}\right\}$$

Suppose the roll of a die turns up a four. The resulting crossover yields two new strings A' and B' following the partial exchange:

$$\left.\begin{array}{lccccccc} A' = & b1 & b2 & b3 & b4 & a5 & a6 & a7 \\ B' = & a1 & a2 & a3 & a4 & b5 & b6 & b7 \end{array}\right\}$$

The mechanics of the reproduction and crossover operators are surprisingly simple, involving nothing more complex than string copies and partial string exchanges; however, together the emphasis step of reproduction and the structured, though randomized, information exchange of crossover give genetic algorithms much of their power. At first this seems surprising. How can such simple (computationally trivial) operators result in anything useful let alone a rapid and robust search mechanism? Furthermore, doesn't it seem a little strange that chance should play such a fundamental role in a directed search process? The answer to the second question was well recognized by the mathematician J. Hadamard [1]:

> We shall see a little later that the possibility of imputing discovery to pure chance is already excluded....On the contrary, that there is an intervention of chance but also a necessary work of unconsciousness, the latter implying and not contradicting the former....Indeed, it is obvious that invention or discovery, be it in mathematics or anywhere else, takes place by combining ideas.

The suggestion here is that while discovery is not a result of pure chance, it is almost certainly guided by directed serendipity. Furthermore, Hadamard hints that a proper role for chance is to cause the juxtaposition of different notions. It is interesting that genetic algorithms adopt Hadamard's mix of direction and chance in a manner which efficiently builds new solutions from the best partial solutions of previous trials.

To see this, consider a population of n strings over some appropriate alphabet coded so that each is a complete *idea* or prescription for performing a particular task. Substrings within each string (*idea*) contain various *notions* of what's important or relevant to the task. Viewed in this way, the population contains not just a sample of n *ideas*, rather it contains a multitude of *notions* and rankings of those *notions* for task performance. Genetic algorithms carefully exploit this wealth of information about important *notions* by 1) reproducing quality *notions* according to their performance and 2) crossing these *notions* with many other high performance *notions* from other strings. Thus, the act of crossover with previous reproduction speculates on new *ideas* constructed from the high performance building blocks (*notions*) of past trials.

If reproduction according to fitness combined with crossover give genetic algorithms the bulk of their processing power, what then is the purpose of the mutation operator? Not surprisingly there is much confusion about the role of mutation in genetics (both natural and artificial). Perhaps it is the result of too many B movies detailing the exploits of mutant eggplants that devour portions of Chicago, but whatever the cause for the confusion, we find that mutation plays a decidedly secondary role in the operation of genetic algorithms. Mutation is needed because, even though reproduction and crossover effectively search and recombine extant *notions*, occasionally they may become overzealous and lose some potentially useful genetic material (1's or 0's at particular locations). The mutation operator protects against such an unrecoverable loss. Mutation is the occasional random alteration of a string position. In a binary code, this simply means changing a 1 to a 0 and vice versa. By itself, mutation is a random walk through the string space. When used sparingly with reproduction and crossover it is an insurance policy against premature loss of important *notions*.

The underlying processing power of genetic algorithms is understood in more rigorous terms by considering the notion of a *notion* more carefully. If two or more strings (*ideas*) contain the same *notion* there are similarities between the strings at one or more positions. To consider the number and form of the possible relevant similarities we consider a schema [2] or similarity template; a similarity template is simply a string over our original alphabet {1,0} with the addition of a wild card or don't care character *. For example, with string length $\ell = 7$ the schema 1*0**** represents all strings with a 1 in the first position and a 0 in the third position. A simple counting argument shows that while there are only 2^ℓ strings, there are 3^ℓ well-defined schemata or possible templates of similarity. Furthermore, it is easy to show that a particular string is itself a representative of 2^ℓ different schemata. Why is this interesting? The interesting part comes from considering the effect of reproduction and crossover on the multitude of schemata contained in a population of n strings (at most $n \cdot 2^\ell$ schemata). Reproduction on average gives exponentially more samples to the observed best similarity patterns (a near-optimal sampling strategy if we consider a k-armed bandit problem). Second, crossover, combines schemata from different strings so that only very long defining length schemata (relative to the string length) are interrupted. Thus, short defining length schemata are propagated generation to generation by giving exponentially increasing samples to the observed best, and all this goes on in parallel with little explicit bookkeeping or special memory other than the population of n strings. How many of the $n \cdot 2^\ell$ schemata are usefully processed per generation? Using a conservative estimate, Holland has shown that $0(n^3)$ schemata are usefully sampled per generation. This compares favorably with the number of function evaluations (n), and because this processing leverage is so important (and apparently unique to genetic algorithms) Holland gives it a special name, implicit parallelism.

The leverage of implicit parallelism permits genetic algorithms to be applied in a variety of optimization problem domains. A recent conference [3] saw application of genetic

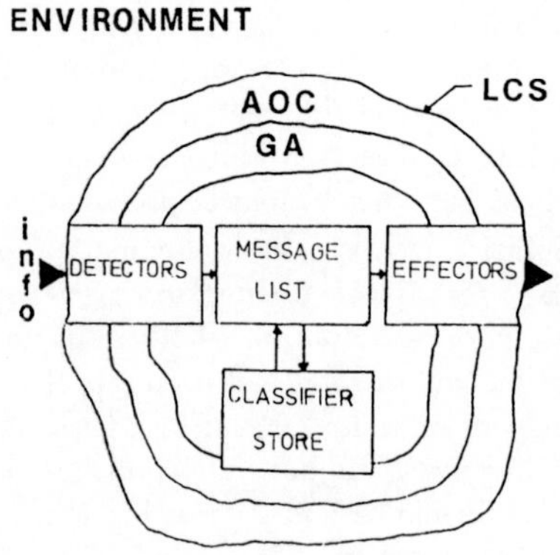

Figure 1: Schematic - Learning Classifier System .

algorithms in parameter optimization problems, combinatorial optimization problems and multi-objective optimization problems. Furthermore, because GA's are designed specifically to exploit similarities in arbitrary codings, they are one of the few known search methods capable of searching through a search space of computer programs. In the next section, we briefly outline one system, a learning classifier system (LCS), which uses a genetic algorithm as its underlying rule discovery heuristic.

3. A Learning Classifier System

A learning classifier system (LCS) is an artificial system that learns rules, called classifiers, to guide its interaction in an arbitrary environment. It consists of three main elements:

1. Rule and Message System
2. Apportionment of Credit System
3. Genetic Algorithm

A schematic of an LCS is shown in Figure 1. In this schematic, we see that the rule and message system receives environmental information through its sensors, called *detectors*, which decode to some standard *message* format. This environmental message is placed on a *message list* along with a finite number of other internal messages generated from the previous cycle. Messages on the message list may activate *classifiers*, rules in the *classifier store*. If activated, a classifier may then be chosen to send a message to the message list for the next cycle. Additionally, certain messages may call for external action through a number of action triggers called *effectors*. In this way, the rule and message system combines both external and internal data to guide behavior and the state of mind in the next state cycle.

In an LCS, it is important to maintain simple syntax in the primary units of information, messages and classifiers. In the current study messages are ℓ-bit (binary) strings and classifiers are 3ℓ-position strings over the alphabet $\{0, 1, \#\}$. In this alphabet, # is a wild card, matching a 0 or a 1 in a given message. Thus, we maintain powerful pattern recognition capability with simple structures.

In traditional rule-based expert systems, the value or rating of a rule relative to other rules is fixed by the programmer in conjunction with the expert or group of experts being emulated. In a rule learning system, we don't have this luxury. The relative value of different rules is one of the key pieces of information which must be learned. To facilitate this type of learning, Holland has suggested that rules coexist in a competitive service economy. A competition is held among classifiers where the right to answer relevant messages goes to the highest bidders with this payment serving as a source of income to previously successful message senders. In this way, a chain of middlemen is formed from manufacturer (source message) to message consumer (environmental action and payoff). The competitive nature of the economy insures that the good rules survive and that bad rules die off.

In addition to rating existing rules, we must also have a way of discovering new, possibly better, rules. This, of course, is the appropriate role for our genetic algorithm. In the learning classifier system application, we must be less cavalier about replacing entire string populations each generation, and we should pay more attention to the replacement of low performers by new strings; however, the genetic algorithm adopted in the LCS is very similar to the simple tripartite algorithm described earlier.

Taken together, the learning classifier system with a computationally complete and convenient rule and message system, an apportionment of credit system modeled after a competitive service economy, and the innovative search of a genetic algorithm, provides a unified framework for learning. Classifier systems and classifier-like systems have been applied to maze running tasks [4,5], pipeline control [6], search for food in 2-D domains [7,8], and poker playing [5]. Classifier systems are also under consideration for modeling adaptation in the immune system [9] and for modeling sub-committee and coalition formation in legislative bodies [10]. Despite the growing number of applications, there is a need for basic empirical and theoretical investigation of genetic algorithms and the learning systems which use them to make them more useful in ever more complicated applications. In the next section, we outline some of the salient research needs.

4. Research Needs in Genetic Algorithms and Classifier Systems

Although genetic algorithms have proven useful in a growing number of optimization and machine learning applications, theoretical and empirical studies are necessary to broaden their impact in increasingly nonlinear and nonstationary environments. In this section, we discuss some of the fundamental research needs in genetic algorithm and classifier system work.

Further work is required to study the importance of dominance, ordering, and crowding operators in genetic algorithms. In most genetic algorithm studies, the artificial chromosome has been represented as a single (haploid) string, whereas many complex organisms carry a pair of strings (diploid) where certain alleles dominate over others. Holland [2] has suggested that this redundancy is useful because it holds currently low-ranking schemata in abeyance, thus maintaining the possibility of resorting to recessive gene solutions when environmental shifts give the recessive sufficient selective pressure. This hypothesis should be checked in a number of nonstationary environments using a simple model of dominance and dominance shift. Ordering operators are especially important to the ultimate success of genetic algorithm methods. The primary hypothesis of a simple genetic algorithm (reproduction-crossover-mutation) is that tightly linked, highly fit building blocks lead to optimal or near-optimal solutions. While Bethke's Walsh function work [11] has recently added rigor to the study of GA-hard functions, the ultimate answer to poor building blocks lies in the ability to look for better string orderings. Studies are needed which investigate the use of ordering operators like inversion [2] and PMX (partially matched crossover) [12]. Further work is also needed in the area of crowding operators. In resource-limited environments, it is necessary to allocate resources among competing hypotheses according to their merit. Without this, even strings with small advantages are magnified and soon dominate all available positions. More work is needed along the lines suggested by De Jong [13] and Booker [7] to help promote a more ecological distribution of population slots, especially in multimodal function optimization and machine learning applications.

Work in classifier systems should proceed along several lines. The syntax of early classifier systems has been severely restricted to obtain explicit pattern matching capability with minimal alphabets. These restrictions should be carefully relaxed to allow more sophisticated string manipulations and thus permit more interesting rules and data manipulations. For example, the addition of a single character shift would permit the formation of messages which were equivalent to time histories, thus permitting the recognition of temporal patterns with one or two rules. More detailed simulations of simple bucket brigade performance are needed, especially in cases with intermittent or performance-driven reward. Additionally, more careful connections should be drawn between the bucket brigade algorithm and theoretical work in learning automata. Finally, extensions to classifier syntax should be considered which permit the genetic algorithm and apportionment of credit mechanism to be coded themselves in classifier system form. Holland originally suggested such a system with his Broadcast language proposal [2]; however, what is needed now is a minimal extension of classifier system syntax to accomplish LCS self-description [14].

5. Conclusion

Genetic algorithms have received growing attention and application in both optimization and machine learning problems. In this paper, a simple genetic algorithm has been outlined, and its application as the fundamental rule discovery heuristic in a larger rule learning system has been examined.

Despite early analytical and empirical successes, more work is needed to demonstrate genetic algorithm robustness and practicality. Specifically, studies are needed to investigate the roles of dominance, ordering, and crowding operators. Classifier system syntax and bucket brigade structure also require careful examination. Efforts in these areas will help extend the usefulness of genetic algorithms in ever more complex, nonlinear and nonstationary environments.

Acknowledgments

This material is based upon work supported by the National Science Foundation under Grant ENG-8451610.

References

[1] J. Hadamard, The Psychology of Invention in the Mathemnatical Field, Princeton University Press, Princeton, 1945.

[2] J.H. Holland, Adaptation in Natural and Artificial Systems, University of Michigan Press, Ann Arbor, 1975.

[3] J.J. Grefenstette (ed.), "Proceedings of an International Conference on Genetic Algorithms and Their Applications," held at Carnegie- Mellon University, Pittsburgh, Pa., July 24-26, 1985.

[4] J.H. Holland and J.S. Reitman, "Cognitive Systems Based on Adaptive Algorithms," in Pattern-Directed Inference Systems, D.A. Waterman and F. Hayes-Roth (eds.), pp. 313-329, Academic Press, New York, 1978.

[5] S.F. Smith, "A Learning System Based on Genetic Adaptive Algorithms," Ph.D. Dissertation, University of Pittsburgh, Pittsburgh, 1980.

[6] D.E. Goldberg, "Computer-Aided Pipeline Operation using Genetic Algorithms and Rule Learning," Ph.D. Dissertation, University of Michigan, Ann Arbor, 1983.

[7] L.B. Booker, "Intelligent Behavior as an Adaptation to the Task Environment," Ph.D. Dissertation, University of Michigan, Ann Arbor, 1982.

[8] S.W. Wilson, "Knowledge Growth in an Artificial Animal," in Proceedings of an International Conference on Genetic Algorithms and Their Applications," J.J. Grefenstette (ed.), Carnegie-Mellon University, Pittsburgh, Pa., July 24-26, 1985, pp. 16-23.

[9] J.D. Farmer, N.H. Packard and A.S. Perelson, "The Immune System and Artificial Intelligence," paper presented at an International Conference on Genetic Algorithms and Their Applications, Carnegie-Mellon University, Pittsburgh, Pa., July 24-26, 1985.

[10] S. Taylor, private communication, 1985.

[11] A.D. Bethke, "Genetic Algorithms as Function Optimizers," Ph.D. Dissertation, University of Michigan, Ann Arbor, 1981.

[12] D.E. Goldberg and R. Lingle, "Alleles, Loci, and the Traveling Salesman Problem," in Proceedings of an International Conference on Genetic Algorithms and Their Applications, J.J. Grefenstette (ed.), Carnegie-Mellon University, Pittsburgh, Pa., July 24-26, 1985, pp. 154-159.

[13] K.A. De Jong, "Analysis of the Behavior of a Class of Genetic Adaptive Systems," Ph.D. Dissertation, University of Michigan, 1975.

[14] D.E. Goldberg, "Control of Dynamic Systems Using Rule Learning and Genetic Algorithms," research proposal submitted to NSF, November 1984.

Knowledge Growth in an Artificial Animal

Stewart W. Wilson
Rowland Institute for Science
Cambridge, Ma. 02142

This paper describes work using an artificial, behaving, animal model (termed an "animat") to study intelligence at a primitive level. The motivation for our somewhat unusual approach is the view that the essence of intelligence is exhibited by animals surviving in real environments. Therefore, insight into intelligence should be obtainable from simulated animals and environments, even simple ones, provided the simulations suitably reflect the animal's survival problems. The starting point for the research is an explicit definition of intelligence which guides model construction. In experiments, a particular animat is placed in an environment and evaluated as to its rates of improvement in performance and perceptual generalization. Learning is central, because we wish to provide the animat with adaptive mechanisms which yield rapid and solid improvement but themselves contain minimal *a priori* information.

Animat research is one possible bridge between the fields of artificial intelligence and adaptive control. Because the environment may be complicated, the animat must employ a rudimentary perception to classify and simplify raw sense data. The resulting symbols and symbol processing are the stuff of AI and many techniques are potentially relevant. At the same time, the animat must learn to choose appropriate actions depending upon its objectives and the environmental state: in this the work is related to adaptive control.

1. A Definition of Intelligence

A good definition should be relatively simple and yet cover most of the things we regard as belonging to the concept and few we regard as not belonging. The psychological literature offers a number of useful similar efforts but the best definition of intelligence we have found is the following, from the physicist van Heerden:

> Intelligent behavior is to be repeatedly successful in satisfying one's psychological needs in diverse, observably different, situations on the basis of past experience [1].

This definition (vH) is suitable for the computer study of intelligence because it is comprehensive and its terms are not difficult to define computationally for experimental purposes. A high rate of receipt of certain reward quantities can correspond to "repeatedly successful in satisfying one's psychological needs" (on the simplest level, somatic needs). To "diverse, observably different, situations" can correspond sets of distinct sensory input "vectors" with each set having a particular implication for optimal action. To "past experience" can correspond a suitable internal record of earlier interactions with the environment, and their results.

2. The Animat Model

To define our model, we abstract four basic characteristics of simple animals:

1) The animal exists in a sea of sensory signals. At any moment only some signals are significant; the rest are irrelevant.

2) The animal is capable of actions (e.g., movement) which tend to change these signals.
3) Certain signals (e.g., those attendant on consumption of food), or certain signals' absence (e.g., absence of pain) have a special status for him.
4) He acts, both externally and through internal operations, so as approximately to optimize the rate of occurrence of the special signals.

An animal's sensory-motor situation is described in very general terms by (1) and (2). Characteristics (3) and (4) are assumptions which provide a way of making definite the notion of "needs" and their satisfaction. Together, the four characteristics define the animat model.

We take as the animat's overall problem the generation of rules which associate sensory signals with appropriate actions so as to achieve the optimization of (4), above. For this, the major questions are adaptive, namely:

1) How to discover and emphasize rules that work,
2) Get rid of those that don't (since memory space is limited and noise is undesirable),
3) Optimally generalize the rules that are kept (since space is limited).

There is some previous work along these lines. Notable were Grey Walter's: *machina speculatrix*, which was a sort of sub-animat which chose actions based on needs and the sensory situation, but did not adapt its rules; and *m. docilis*, which could be taught a conditioned response [2]. More recently, Holland and Reitman [3] exhibited successful performance by a rule-adaptive animat-like system which optimized its rate of satisfaction of two distinct needs. Booker [4] experimented with an animat-like "hypothetical organism" which adapted its rules in a simple environment that contained both attractive and aversive stimuli; he also provides a review of earlier systems. The present investigation is indebted to the last two works.

3. Implementation

Within the above framework we make the model definite by defining the animat's environment, sensory channels, repertoire of actions, its association rules, and then its performance and adaptive algorithms.

Environment

A rectangle on the computer terminal screen, 18 rows by 58 columns and continued toroidally at its edges, defines the environmental space. Alphanumeric characters at various positions represent objects; the animat itself is denoted by *. Some, possibly many, positions are just blank.

Sensory Channels

In studies so far, * has been given the ability to pick up sensory signals from objects which happen to be one step (row and/or column) away, in any of the eight (including diagonal) directions; nothing is detected from more distant objects. Thus the "sense vector" has eight positions. With * located, for example, as shown below left, the sense vector would be as shown at the right:

$$\begin{matrix} TT \\ {*F} \end{matrix} \qquad\qquad T\,T\,F\,b\,b\,b\,b\,b\,,$$

where b stands for blank. To form the sense vector, the circle of positions surrounding * is mapped, clockwise starting at 12 o'clock, into a left-to-right string.

But this vector is not the final sensory input. We imagine that an object is ultimately sensed as the outcome of measurements upon it by one or more feature or attribute detectors.

Without loss of generality we assume each detector produces either a 0 or 1 output. If there are d detector types, an object translates into a binary string d bits in length. The sense vector as a whole thus translates into a "detector vector" of $8d$ bits.

Detector translations or encodings of objects are fixed in *'s "low-level" sensory hardware. They are assigned at the beginning of an experiment. For example, in experiments discussed here, "*F*" (food) is encoded as "11"; "*T*" (tree or obstacle) as "01"; and "*b*" (open space) as "00". [The first bit might be thought of as the output of a "food smell" detector; the second, of an "opacity" detector.] Thus the above sense vector translates into the detector vector:

01 01 11 00 00 00 00 00 .

The associative apparatus takes the detector vector as input.

Repertoire of Actions

*'s actions are restricted to single step moves in each of the eight directions. The directions are numbered 0-7 starting at 12 o'clock and proceeding clockwise; for example, a move in direction 3 would be south-easterly.

The animat may move, or attempt to move, to a position occupied by an object. The environment's response for each kind of object is pre-defined. In present experiments, if the move is into a position whose encoding is 00 (the blank object), there is no response (though the new sense vector will in general be different). If * steps into a space occupied by an object whose encoding has the first bit equal to 1, * is regarded as having eaten the object and receives a reward signal. If * tries to step toward an adjacent object whose encoding is 01, the step is not permitted to occur (a collision-like banging may be displayed).

The foregoing establish a semi-realistic situation in which sensory signals carry partial, but uncertain, information about the location of food, and available actions permit exploration and approach. Environmental predictability can be varied through the choice and arrangement of the objects. The number of object types which may be experimented with is limited only by the number of bits in the detector encoding scheme.

Association Rules

For its association rules, the animat uses a rudimentary form of Holland's [5] "classifier" rule. The animat's rules each consist of a "taxon" and an "action". The taxon is a sort of template capable of matching a certain set of detector vectors. The action is some one of the available actions. The animat's classifier says, in effect, "if my taxon matches the current detector vector, then consider taking this action." It is a kind of hypothesis about what to do given a certain sensory situation (class of detector vectors). An example of a classifier would be:

0# 01 1# 0# 00 00 0# 0# / 2

The matching rule requires that for any taxon position having a 0 or 1, the same value must occur in the detector vector; taxon positions with # ("don't care") match unconditionally. Because of the #'s, which confer a kind of generality on the classifier, the above taxon, for example, will match 32 possible detector vectors, including the one discussed earlier.

It is worth making a few further observations about this classifier. First, it is a pretty good one because if food is present in direction 2 and the classifier matches the detector vector, the action recommended is to move in direction 2 and not some other direction! Second, in directions 0, 3, 6, and 7, the taxon only requires that the object be, in effect, non-food, it being irrelevant whether these directions have obstacles or are blank. Directions 1, 4, and 5 have not been so generalized. Broadly speaking, a classifier is more useful to the animat to the extent it is general (matches many detector vectors) without being so general that it makes too many errors (i.e., that in certain matching situations its recommended action is inappropriate).

Besides taxon and action, each classifier possesses a "strength", a quantity serving as the principal measure of a classifier's value to the animat.

The animat keeps a classifier population [P] of fixed size. Usually, [P] is initialized by filling all the taxa with 1, 0, and # according to some random rule; actions are similarly filled in. As the animat's CRT "life" evolves, the classifier population changes, as will be described.

4. Performance Algorithm

*'s basic cycle is one "step", within which events having purely to do with immediate behavior are very simple. First, the current detector vector is calculated. Second, [P] is searched for classifiers which match it; these classifiers form the "match set" [M]. Third, a classifier is selected from [M] using a probability distribution over the strengths of [M]'s classifiers; that is, the probability of selection of a particular classifier is equal to its strength divided by the sum of strengths of classifiers in [M]. Fourth, * moves according to the action of the selected classifier, or tries to. The environment's response to the move will be as described earlier.

It can be seen that *'s move choice tends to be the one having the greatest total strength among the classifiers of [M] advocating it. Thus, overall, * first asks which classifiers of [P] "recognize" the current sensory situation, then from these tends to pick the move with the greatest associated strength. The subset of [M] consisting of classifiers whose action is the same as the chosen action is called the "action set" [A].

5. Adaptation Algorithm

The adaptation algorithm has three distinct aspects: 1) reinforcement of classifier strengths; 2) "genetic" operations on classifiers yielding new classifiers; and 3) direct creation of classifiers.

Reinforcement

As discussed in the last section, a classifier's strength is a major determinant of its ability to influence *'s action and therefore performance. We therefore want strength to reflect the performance which tends to result when this classifier is in [A]. That would be straightforward if every step were rewarded: we could, for example, adjust the classifier's strength by an amount proportional to the reward. Classifiers which got bigger rewards would be stronger, thus more likely to be in [A], etc.

Realistically, however, it is usually the case that only some of an organism's actions receive a definite reward from the environment. Actions leading up to, or setting the stage for, a rewarded action are themselves not directly rewarded. But they must somehow be encouraged or the final payoff will not occur. Holland [6] addressed this problem in proposing a "bucket-brigade" algorithm in which, very briefly, 1) classifiers make payments out of their strengths to classifiers which were active on the preceding cycle, and 2) the same classifiers later correspondingly receive payments from the strengths of the next set of active classifiers. External reward goes to the final active set in the chain. In effect, a given amount of external reward will eventually flow all the way back through a reliable chain, reinforcing every precurser classifier.

Our basic implementation of this idea is as follows: On each step:

1) all classifiers in [A] have a fraction e of their strengths removed;
2) the total strength thus removed from [A] is distributed to the strengths of any classifiers in $[A-1]$, defined as the action set in the previous step;
3) * then moves and if external reward is received it is distributed to the strengths of [A]; if external reward is not received, the classifiers of [A] replace those of $[A-1]$.

Thus every [A] participates in general in two transactions, one paying out, the other receiving. We can write

$$S'_A = S_A - eS_A + p,$$

where S_A is $[A]'s$ total strength on one step, S'_A its total on the next, and p is the total payoff received (either external reward or from the next [A]). If p is the same over time, S_A approaches a constant value given by p/e, so that under reasonably steady payoff conditions, S_A is an estimator of typical payoff. Similarly, the strength of any individual classifier is an estimator of its typical payoff.

The total payoffs to [A] or [A − 1] are in the simplest case shared equally by the recipient classifiers. This has the consequence that the more classifiers are in, say, [A], the less payoff each gets.

Genetic Operations

Consider two classifiers which match similar situations:

0# 01 1# 0# 00 00 0# 0# / 2

and

0# 0# 11 01 00 0# 0# 0# / 2 .

Each is good, but each still lacks something in generality since, for example, the matching requirements for 01 in bits 2-3 and 6-7, respectively, of each are perhaps unnecessarily restrictive. Suppose we make a new classifier by combining bits 5-9 of the first with bits 0-4 and 10-15 of the second. The result would be the slightly more general classifier:

0# 0# 1# 0# 00 0# 0# 0# / 2 .

The above operation on two classifiers resembles a kind of crossing-over or recombination of chromosome parts in genetics. It is an operation in which two "parent" classifiers produce an offspring that is possibly an improvement over both of them. Another "genetic" operation, this time using just one parent, would first clone the parent, then mutate one or more of the clone's taxon positions. Other types of operations on classifier structure can be imagined (one will be discussed later). In each case the attempt is to use existing classifiers as the starting points for improved classifiers.

But the crossover points above were chosen quite carefully; otherwise the offspring might have been no improvement, or even a retrogression (to a classifier more specific than either parent). We do not expect the animat to know where best to cut or mutate. How can we expect genetic operations to be of any use?

Holland [7] presents a mathematical theory showing that a population of individual symbol strings, in which each string can be assigned a numerical worth, will progressively increase in average worth as its members undergo reproduction, genetic operations on or among the offspring, and deletion of individuals to maintain constant population size. The key requirement is that an individual's probability of reproduction be proportional to its worth. Holland extended the theory to include classifier systems. In employing genetic operations, our animat constitutes an exploration and test of the theory.

The specific algorithm employed is as follows:

1) A first classifier $c1$ of [P] is selected with probability proportional to its strength;
2) If $c1$ is merely to be reproduced, a copy of it is made and added to [P]. To make room, some classifier is deleted;
3) If $c1$ is to be crossed with another classifier, a second, $c2$, is selected, also with probability proportional to its strength, but from the subset of [P] of classifiers having the same action as $c1$. Two cut points are chosen as above, but at random, and an offspring $c3$ constructed out of the parts. $C3$ is added to [P] and some classifier is deleted.

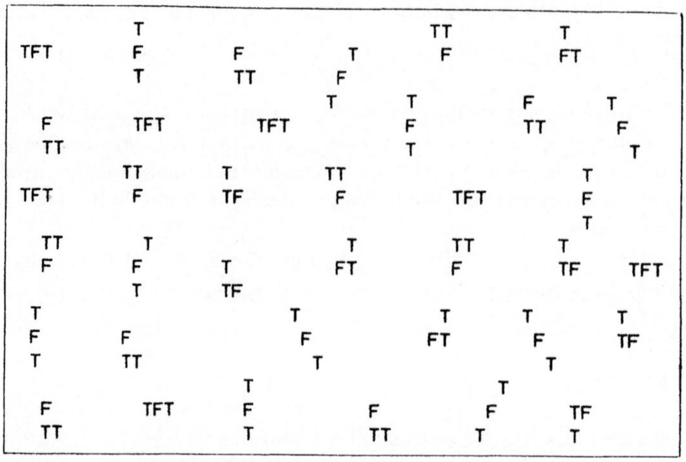

Figure 1: The Environment "WOODS7".

"Create" Operations

Occasionally, as * executes the performance algorithm, a detector vector may occur that no classifier of [P] matches, i.e., the situation is unrecognized. The animat's response is to create a new, matching classifier. A taxon is made by adding some *#'s* at random to the detector vector; an action is chosen randomly. The created classifier is added to [P] and one is deleted. The new classifier immediately matches the previously unrecognized situation and action occurs by the normal mechanism.

6. Experimental Procedure

The animat model was designed with the vH-intelligence definition as a guide. In experiments with the model we are interested in finding procedures and parameter values that seem to give * greater rather than less vH-intelligence. For this two measures have been adopted. One is a performance measure: given an environment, how many steps does * take, on average, to find food objects? The other is a generality measure: does * evolve classifiers each tending to be useful in a number of distinct situations? Generality is important because it suggests that a high level of performance developed in one environment will carry over to a somewhat different environment.

The experimental procedure is to fix *'s methods and parameters, then have him do a large number of "problems" in a particular environment E. The measures of performance and generality are tracked. A "problem" always consists of starting * at a randomly selected blank position in E; then * moves until he eats some food, at which point the problem ends. The number of steps between start and food is recorded; a moving average of this quantity over the previous 50 problems is the performance measure, STPSAV.

To track generality, we calculate a histogram over the "periods" of all classifiers in [P]. The period of a classifier is a moving average of the number of steps by * between occurrences in [A] of this classifier. Thus a frequently used classifier will have a low period. [P] will then be general to the extent the histogram of periods is largest at low period. As [P] evolves we expect the histogram peak to move toward lower period, if *[P]'s* generality is increasing.

An environment used for many of the experiments is "WOODS7", shown in Fig. 1. Although WOODS7 may look easy, it actually contains a total of 92 distinct sense vectors, so *'s need to discover and generalize is substantial. To obtain performance baselines, we can start * randomly, then have him move completely randomly until food (F) is bumped into. For WOODS7, this takes about 41 steps on average, giving us a measure of worst case

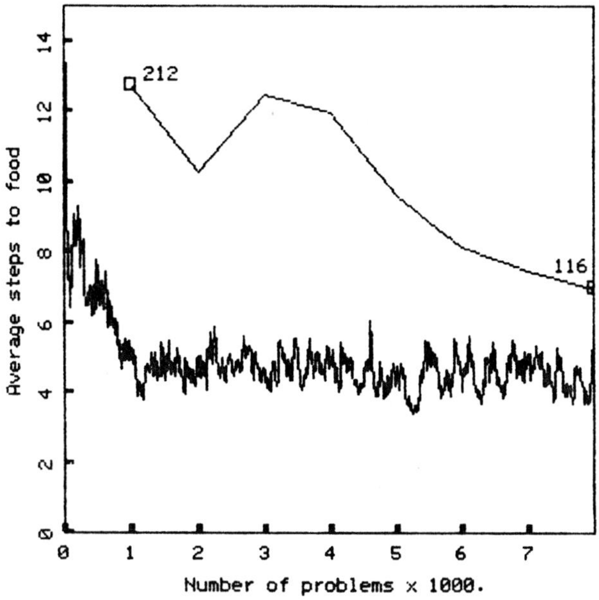

Figure 2: STPSAV (ragged line) and Period Average (broken line) for * to 8000 problems. Period values as marked.

performance. We may also ask [8]: What is the best conceivable performance (for, say, an animat with human capabilities)? For every starting position, the number of steps to the nearest F can be found and averaged over all starting positions. The result for WOODS7 is 2.2 steps.

7. Results and Discussion

Fig. 2 shows a performance curve for a combination of procedures and parameter settings that is among the best so far found. There is an initial rapid improvement within the first 1000 problems (untypically good during the first 100 problems, where STPSAV usually stays above 15), followed by very gradual improvement thereafter. The performance at 8000 problems, between 4 and 5 steps, is quite respectable compared with "perfect" (2.2 steps), especially since * has no information whatsoever until he is next to a non-blank object.

For the same animat, Fig. 3 shows the histogram of periods of $[P]$ at 8000 problems. There is a definite bulge for low periods; the average period is 116. For comparison, the broken line in Fig. 2 shows the trend of the period averages at earlier epochs, indicating gradual generalization in the sense we have defined.

Qualitatively, a * such as this one gives the impression of "knowing" the Woods quite well. When next to F, * nearly always takes it directly; occasionally he will move one step sideways and take it from that direction. When next to one or more $T's$, but with no F immediately in sight, * quite reliably steps around the obstacle(s) and finds the F. When * is "out in the open", i.e., the sense vector consists of blanks, he has no information about the best way to go, as in a thick fog. One might expect *'s behavior to resemble a random walk but this is not the case. Instead, the movements look more like a general "drift" in some direction, with some superimposed randomness. After several problems the drift may shift to another direction.

Distance Estimation

Performance in earlier animat experiments was below the level of Fig. 2. One defect

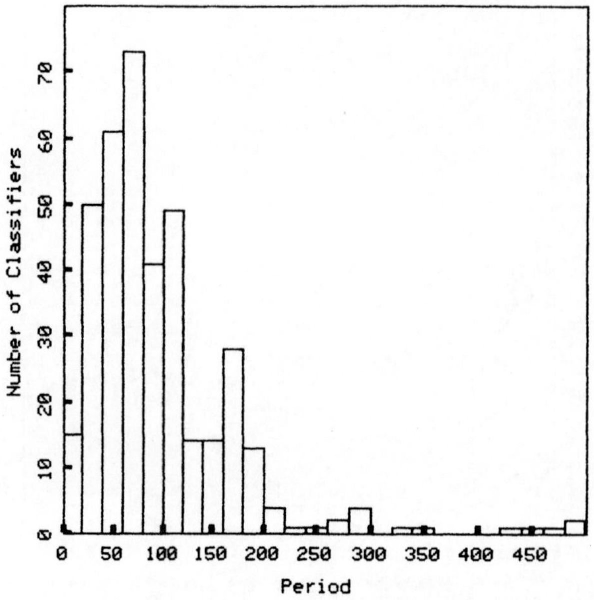

Figure 3: Histogram of classifier periods for the * of Fig. 2 at 8000 problems.

was a kind of "dithering" in which while * would tend toward $F's$, the path would have unnecessary sidesteps and wanderings. It was soon realized that the basic reinforcement algorithm does not care whether a path from point A to food is long or short; there is nothing which preferentially reinforces the most expeditious classifiers. Any path, even a looping one, will come to equilibrium at a high strength level in its constituent classifiers.

The solution had to be more subtle than simply penalizing long paths. What is required is a technique that, at every position, tends to prefer the most direct of several possible moves, but does not prevent the setting up of a long path if that is actually the shortest path available. Our solution was twofold. First, each classifier was made to keep an estimate of its distance (in steps) to food. This did not require elaborate look-ahead. Instead each classifier in $[A-1]$ adjusted its distance estimate according to an average of the distance estimates of $[A]$; when reward was received, the members of $[A]$ were similarly adjusted, using the quantity 1. This technique, with each estimate an average over the last few updates, is quite satisfactory.

The distances are employed as follows: In the performance cycle, selection from $[M]$ is based on probability proportional to strength/distance instead of just strength. Consequently, a move tends to be selected that is not only strong, but also "short". Now comes the second part of the solution. At the same time as $[A]$ is formed, the set $NOT[A]$ of the remaining classifiers in $[M]$ is taxed by a small amount: the "longer" classifiers thus tend to incur a loss by not being selected. This "lateral inhibition" induces a sort of catastrophe in which the shorter classifiers become even more likely to be picked and the longer become ever weaker, and can disappear entirely. Note that the competition is purely local and does not work against the setting up of minimal long paths.

This technique is very effective against "dithering"; the progressive takeover of a match set by a discovered shorter move has been repeatedly observed. Our solution is not perfect, however, because to suppress the special case of occasional looping situations we had to impose a small tax on $[A]$. Since $[A]$ is the set which receives payoff, the tax has little effect except if a loop is taking place, and then the tax is soon very effective. Still, in principal even a small tax on $[A]$ reduces the strength flow in very long chains, putting them at a reproductive disadvantage. This residual problem may be an indication that as paths grow, they should be "condensed" into units of behavior longer than one step.

Extensions to "Create"

A second area of changes which improved performance had to do with the "Create" operations. As discussed, Create at first only occurred when $[M]$ was empty. It was found that * sometimes also got stuck looping among situations with non-empty $[M]'s$. The tax on $[A]$ enabled recognition of these loops because the total strengths in each $[A]$ would tend to zero. We put in a threshold that triggered Create if the strength of any $[M]$ got too low. This suppressed looping dramatically and improved performance.

It was also found important to trigger Create randomly, at a very low rate. * is engaged in path construction, using the best available current evidence. This can lead to good but nevertheless suboptimal paths which might be improved if * would only try something different. Random Creates are one way to introduce a new move direction. Usually the new classifier is no improvement. But when it is, and it gets tried (gets in $[A]$), it will be (often heavily) reinforced and therefore given a good chance at eventual reproductive success.

Effect of Genetic Operations

Finally, we shall discuss what the experiments suggest about the role of the genetic operations. To begin, it is helpful to define a "concept" as a set of classifiers from $[P]$ having exactly the same taxon and action, and for which there is no other classifier in $[P]$ with that taxon and action. The basic effect of *'s genetic operations then appears to be to exert a pressure tending to increase the generality of $[P]'s$ concepts. That is, with time, the periods of the concepts in $[P]$ tend to decrease. The pressure is restrained by the requirement that the concepts be more or less correct (* must get the food expeditiously). The precise point of balance appears to depend on the parameter regime.

An important experiment is to turn off the genetic operations and evolve an animat with reinforcement and Create going as usual. The result is a performance almost as good as Fig. 2. But significant generalization does not occur; $[P]$ remains a population of specialists. There thus appears to be a division of effort: Create introduces the new material, the specific examples to be evaluated; and the genetic operations produce more general concepts from the examples.

While general classifiers appear to have a selective advantage, this is of no use unless such classifiers can be formed and introduced in the first place. Crossover is adequate for some types of generalization. But a natural operation for the purpose is obviously intersection. We have implemented this operation as follows. Two parents are chosen and a new taxon is formed by intersecting copies of the parents' taxa over a randomly selected interval. In that interval, if the parents differ at a position, the new taxon gets a #; if not, the new taxon gets the common value. Outside the interval, the new taxon is filled in from parent 1.

Intersection is a "hot" operation which should be used cautiously because it can introduce $\#'s$ at a high rate. Nevertheless, our results show increased generalization with little performance loss when crossover and intersection are both available to *.

8. Conclusion

In its simple way, * meets the definition of intelligence stated at the beginning. * becomes good at satisfying its need for food in a Woods of diverse object configurations on the basis of experience. Though not yet tested, *'s rule generalization over time suggests that performance would be maintained in a somewhat different Woods, or if the Woods slowly changed.

The present animat consists essentially of a population of mutable, competing, and also cooperating hypotheses. As such it represents just one species within the much broader genus outlined at the beginning. Yet the notion of a population of hypotheses seems intuitive and basic. Using it, our experiments have suggested how the organism can 1) efficiently allocate memory resources, 2) establish and improve performance, and 3) form appropriate generalizations. Future work will use more complicated environments to find the limits of this structure. But we shall also be investigating 1) interaction among separate animats, 2) adaptive sensory communication between a human being and an animat, and 3) physical animat realizations. Since a substantial fraction of an animat's processing is parallelizable, the area may be an interesting one for the application of highly parallel computers.

Acknowledgment

The author wishes to acknowledge valuable conversations with C. G. Shaefer of the Rowland Institute, the support of the Institute, and stimulating interactions over several years with John Holland.

References

[1] P.J. van Heerden, The Foundation of Empirical Knowledge, Wistik, Wassenaar, The Netherlands, 1968.

[2] W.G. Walter, The Living Brain, Norton, New York, 1953.

[3] J.H. Holland and J.S. Reitman, "Cognitive Systems Based on Adaptive Algorithms," in Pattern-Directed Inference Systems, D.A. Waterman and F. Hayes-Roth (eds.), Academic Press, New York, 1978.

[4] L. Booker, "Intelligent Behavior as an Adaptation to the Task Environment," Ph.D. Dissertation (Computer and Communication Sciences), The University of Michigan, 1982.

[5] J.H. Holland, "Adaptation," in Progress in Theoretical Biology, 4, R. Rosen and F.M. Snell (eds.), Plenum, New York, 1976.

[6] J.H. Holland, "Genetic Algorithms and Adaptation," in Adaptive Control of Ill-Defined Systems, O.G. Selfridge, E.L. Rissland and M.A. Arbib (eds.), Plenum, New York, 1984.

[7] J.H. Holland, Adaptation in Natural and Artificial Systems, University of Michigan Press, Ann Arbor, 1975.

[8] Martha Gordon, personal communication.

CONTROL OF FLEXIBLE SPACE STRUCTURES

Simplify, simplify.

Henry David Thoreau

The analysis and control of large space structures are invariably based on the disciplines of materials technology, structural dynamics, and control theory, since strength, flexibility and performance are inseparable in such problems. The complex distributed models which describe these structures, together with stringent performance criteria regarding their pointing, shaping, and vibration suppression characteristics, make their study truly challenging. Since any feedback controller which is to be implemented on-line using a digital computer with a finite number of actuators and sensors must necessarily be finite dimensional and operate in discrete time, one of the fundamental problems that confronts the designer is the stability of the hybrid closed loop system. The question of most relevance to us, in the context of this book, is the role that adaptation and learning can be expected to play. The diverse viewpoints that exist in the control community are reflected in the four papers included in this section.

The first paper by Kosut and Lyons provides a comprehensive review of theoretical and practical issues involved in developing a design methodology. While both adaptive and non-adaptive methods are discussed, the authors claim that the former may be indispensable in situations where uncertainties in disturbance spectra and plant characteristics limit the performance obtainable with fixed gain, fixed order control. In such cases the disturbance and plant models have to be identified accurately on-line to make adaptive control effective. A similar position is taken in the paper by Sundararajan and Montgomery. Their methodology consists of obtaining a model on-line, validating the model, designing a controller based on the model, and finally engaging the control system. Since large flexible structures are inherently stable though underdamped, and colocated feedback can be used to improve their stability, the authors feel that this approach is ideal.

In contrast to the two foregoing papers, Longman and Lindberg argue that the large space structure control problem may not warrant the use of adaptive control. Playing devil's advocate, Longman proposes that disturbances such as gravity gradient, thermal distortion, solar radiation and internal disturbances due to moving parts, are reasonably predictable so that they can be compensated for by feedforward control or gain scheduling. Disturbances such as crew motion, he feels, are not sufficiently predictable in the short term to justify the use of adaptive control. He concedes, however, that since the dynamics of the space craft may not be known well enough before launch, self-tuning may be resorted to in orbit, but adds that once the parameters are tuned there may be no further need for adaptation.

The editor's position regarding the matter is closer to that taken in the first two papers. As performance specifications become more severe, the models of the disturbances discussed by Longman may not be sufficiently accurate for gain scheduling. This, together with variations in dynamic characteristics of the plant due to structural modifications and component failures, in our opinion, will call for adaptive methodologies in which the control system evolves with the structure.

The controversy regarding the need for approximation, and precisely at what stage such an approximation should be used in the control of large space structures, is a tradition of the field. Many control engineers feel that some form of approximation of the overall system is unavoidable if practical controllers are to result. In the last paper, Wen and Balas take a different stand on this issue and suggest that a model reference adaptive control method which has been suggested for finite dimensional plants can be rigorously extended to infinite dimensional plants as well.

Issues in Control Design for Large Space Structures

Robert L. Kosut and Michael G. Lyons
Integrated Systems Inc
101, University Avenue
Palo Alto, Ca. 94301

Abstract

The development of a design methodology for the control of Large Space Structures (LSS) involves many different issues. In this paper we present a selective discussion of the theoretical and practical issues that seem most relevant. The discussions cover various types of control design procedures, including both robust (non-adaptive) as well as adaptive, with an emphasis on their practical use.

1. LSS Control Problem Setting

1.1 Control Design Objectives

Problems associated with vibration control and accurate pointing of LSS systems typically involve a combination of the following control-performance objectives.

1. modal damping augmentation to enhance transient settling or improve quasi-static vibration propagation behavior,
2. stabilization of the attitude control system,
3. eigenvector modification to reject narrow band steady-state disturbances, and
4. maneuver load management to minimize structural loads or modal excitation (transient or steady-state).

1.2 Modeling

The basis for selecting a control strategy must include an adequate description of the relevant structural dynamics together with a description of how system performance is to be measured. Initially, continuum models were suggested as the basis for proper system design since discretization of the model could be postponed or eliminated. Unfortunately, practical spacecraft configurations do not present simple boundary conditions or simple shapes, hence partial differential equation (p.d.e.) representations are nearly impossible to write. However such continuum models have provided useful insight into appropriate discrete representations. Finite element models can provide adequate fidelity, at least over the frequency range needed for the control design model, and are supported with sophisticated software tools easily adapted to the needs of control design [1].

1.3 Two-Level Control Architecture

The natural structural properties of LSS systems compel the use of a two-level control system architecture as shown in Figure 1. The two levels are a colocated rate-damping control system and a noncolated high performance control system. The colocated system consists typically of rate damping devices, either active or passive, and requires a coarse knowledge of system dynamics. These are inherently robust but yield low performance. They essentially provide a wide-band, Low-Authority Control (LAC) and are often referred to as the LAC-system.

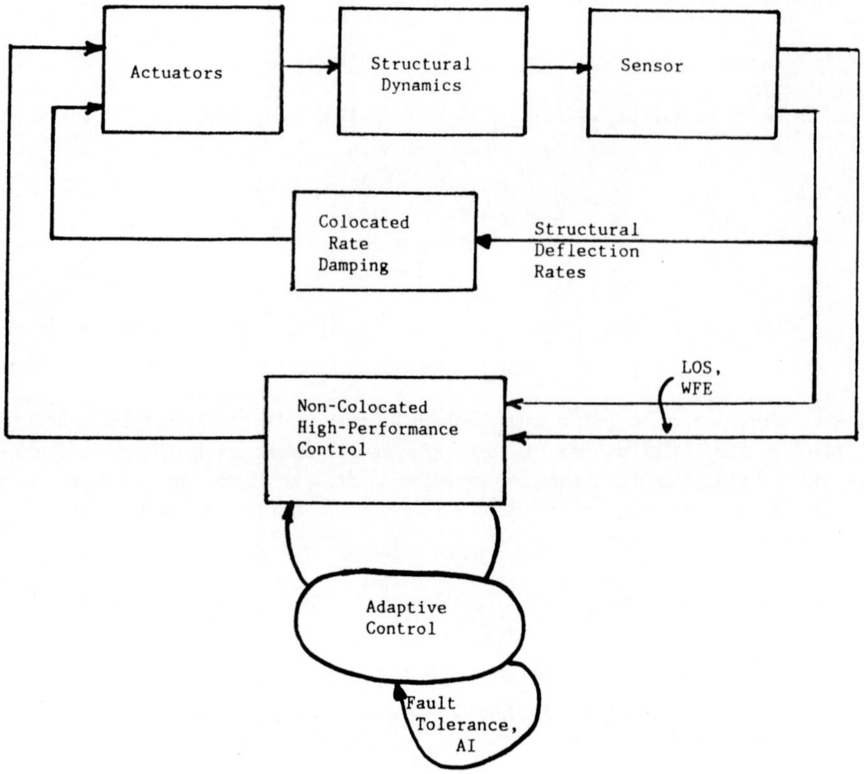

Figure 1: Two Level Control Architecture for LSS Systems.

The high performance control is non-colocated and requires accurate knowledge of critical modes, and hence, is very sensitive to disturbance and structural parameter variations. This controller system is essentially a narrow-band, High-Authority Control (HAC), and is referred to as the HAC-system. Typically, HAC provides high damping and mode shape adjustment in selected modes to meet performance requirements.

LAC synthesis principally involves passivity methods and rate feedback mechanizations, usually with colocated actuators and sensors [2].

HAC synthesis, in addressing performance goals associated with dynamic wavefront and line-of-sight error suppression, requires high modal damping and mode shape changes. Hence, HAC is dependent on accurate narrow-band models. For such requirement, it is essential that control design techniques manage both dependence on model fidelity and system gain in regions where model fidelity is poor. This has generally been accomplished using fixed-gain robust control theory, [4]. With this architecture it is likely that only the HAC would be tuned by an adaptive system since the LAC is inherently robust.

1.4 Adaptive Techniques

In general, uncertaintites in both disturbance spectra and system dynamical characteristics limit the performance obtainable with fixed gain, fixed order control, e.g. HAC system. The use of an adaptive control mechanization where disturbance and/or plant dynamics are identified prior to or during control, gives system designers more options for minimizing the risk in achieving performance benchmarks.

In the case of LSS systems, the performance levels are extremely high. Hence it is necessary that disturbance and plant models are accurately known. Since model data obtained from ground testing is unlikely to sufficiently match the actual on-orbit system, it follows

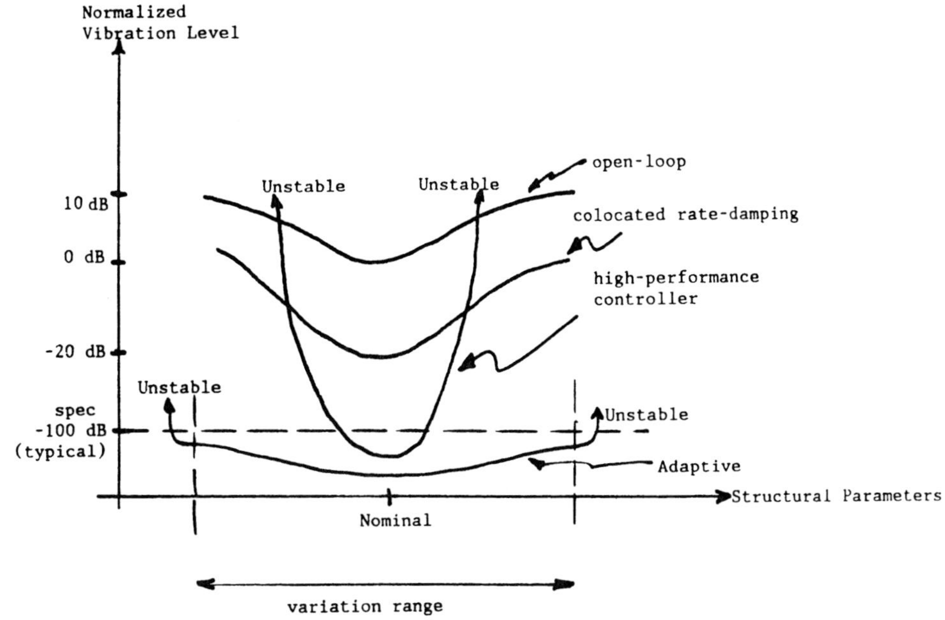

Figure 2: Closed-loop System Performance vs. Structural Parameter Variations.

that on-line procedures are needed for identification and control.

The generic properties of closed-loop system performance vs. structural parameter variations are depicted in Figure 2.

2. Control Design for HAC/LAC Architecture

In this section we wil discuss the steps involved in control design for the HAC/LAC architecture. Although the architecture is specialized, the control design methodology is not and can be quite general. We will discuss three methodologies for design: (1) an LQG based methodology whose genesis is the ACOSS/VCOSS programs, and (2) a more recent approach involving what is known as "Q-parametrization" and H_∞-optimization". These latter methods are frequency domain oriented rather than state-space oriented like the LQG approach. (3) We will also discuss an adaptive control strategy which can be utilized for online self-tuning. We refer to this approach as "adaptive calibration".

2.1 Limitations of Design

Independent of the design method, the defining characteristic of the vibration control problem is that there are an infinite number (theoretically) of elastic modes, with low natural damping, and the controller bandwidth extends over a significant number of these modes (Figure 3). The low frequency modes interact not only with the attitude controller but contribute directly to the deformation geometry of the structure which itself may require accurate control. Proper control synthesis requires that performance criteria be precisely formulated or the control problem is ill-posed.

The control design approach must properly handle the poorly known higher frequency modes by not destabilizing them while controlling the low frequency modes. Indeed, no matter where the controller roll-off frequency is situated, the infinite nature of the modal

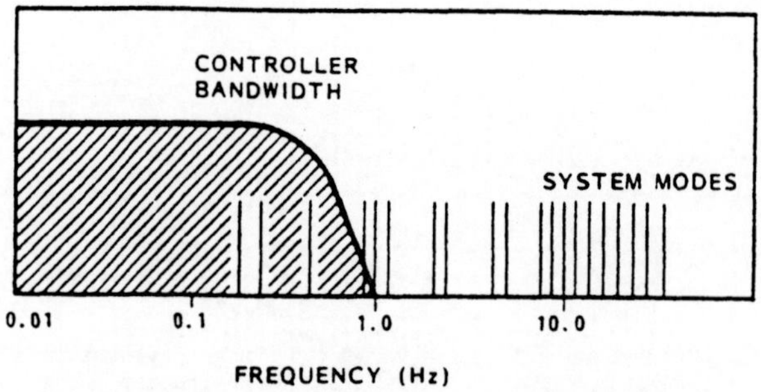

Figure 3: Flexible Structure Mode Location and Controller Bandwidth.

spectrum implies that there will be modes within and beyond the roll-off region. Furthermore, destabilization is likely and almost certain to occur in the roll-off region, a situation which can only worsen for closely packed modes and low natural damping. This phenomenon sometimes referred to as "spillover" is one of the most crucial problems faced by the control designer. In more general terms, spillover can be viewed as an aspect of the problem of robust control design; this will be discussed more in a later section.

2.2 Modeling of Flexible Spacecraft

A central issue in the active control of space structures is the development of "correct" mathematical models for the open and closed loop dynamical plants. Programs such as NASTRAN and SPAR are the primary current tools for generating dynamical models of conceptual spacecraft whose structure cannot be idealized by simple models of beams, plates, and beams with lumped masses.

Finite element structural programs generally provide the control designers with a set of modal frequencies and a set of mode shapes (eigenvectors) corresponding to appropriate boundary values (e.g. free-free modes). These eigenvectors are given in discretized form, i.e. a set of modal displacements in the x, y and z directions at each nodal station. In some cases, modal rotations are also required. In addition, coordinates and a "map" of the structure's nodes must be provided to allow the reconstruction of physical displacements in terms of their modal expansions.

The important point here is that, for any nontrivial flexible satellite configuration, the volume of information is so large that the data handling must remain entirely within the computer and its mass-storage facilities. Development of this database, in a form usable by control synthesis software, is a fundamental necessity for the synthesis and evaluation of complex control which require modal truncation, actuator/sensor location and type changes, and evaluation of system performance for parameter and system order changes. Preparation of a structure for controls is a major part of the overall effort required to develop structural control systems.

2.3 Nonlinear Models

For single-body monolithic structures, the fine-pointing attitude dynamics are subsumed in the rotational rigid body modes included in the modal matrix. When only "small" motions of a space structure are being considered, the conventional linear structural dynamics analyses (NASTRAN and SPAR) are adequate, and the rigid-body modes are formally handled together with the elastic modes, even though the actuators necessary to control them will be different, in general, from those used to control elastic vibrations. When larger attitude angles need to be considered, if the angular rates remain small, the linear equations are still applicable provided that the rigid-body modes are now given in terms of three attitude angles which then constitute the first three modal coordinates. The displacements are then

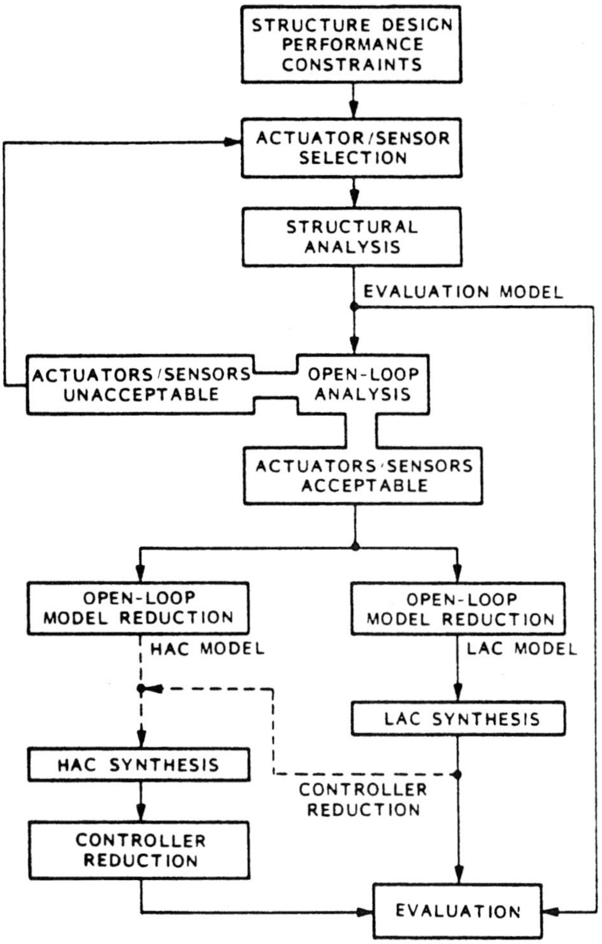

Figure 4: Analytical Control-Design Procedure.

interpreted as the linear deformations of the structure with respect to the rotated frame. This procedure removes the kinematic nonlinearities resulting from the linear stretching of the structure under the classical rigid-body modes. However, for large angular rates, non-linear dynamic effects have to be modeled, even though structural deformations can still be represented by linear equations.

2.4 Two-Level Control Design: The HAC/LAC Methodology

The two-level approach consists of a wide-band, low-authority control (LAC) and a narrow-band, high-authority control (HAC). HAC provides high damping or mode-shape adjustment in a selected number of modes to meet performance requirements. LAC, on the other hand, introduces low damping in a wide range of modes for maximum robustness. Figure 4 shows the control design procedure with integrated LAC and HAC designs.

LAC is usually implemented with colocated sensors and actuators. However, the theory, based on the work of Aubrun, is applicable to multiple actuators/sensors with cross-feedback and possible filters [2].

HAC uses a collection of sensors and actuators not necessarily colocated. Selecting the increase in damping ratio is realized by any number of methods including LQG with frequency shaping, Q-parametrization, or H_∞-optimization. These methods provide roll-off over desired frequency regions. HAC may destabilize modes not used in the design. LAC is,

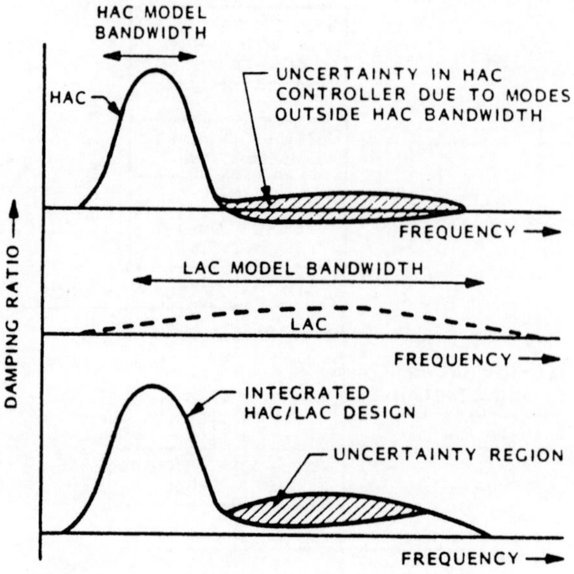

Figure 5: Need to Integrate HAC and LAC.

therefore, necessary to "clean up" problems created by HAC.

The need to integrate HAC with LAC is shown in Figure 5. HAC is based on models valid over a limited frequency region. It produces large increases in damping ratio and disturbance rejection in the frequency range of interest. The effect of the HAC controller on modes not used in the control design and outside the controller bandwidth may be stabilizing or destabilizing. LAC is designed to provide protection such that adequate damping is provided in the mode most adversely perturbed by HAC. With reference to Figure 5, the LAC moves the entire uncertainty region above the zero level damping ratio.

In the next few sections, a more in-depth discussion of the blocks in Figure 4 will be presented, in particular, actuator/sensor location, model and controller reduction methods, and HAC/LAC synthesis. These methodologies rely on certain properties of feedback control: this raises the issue of robust control design which is fundamental to the whole design philosophy of feedback, especially for LSS, and this will be discussed first.

2.5 Robust Control Design

This section will describe how to evaluate the robustness of a control design. The evaluation is independent of the methodology used to achieve a particular design. To illustrate the technique we will consider the robust control problem of vibration suppression with unmodeled high frequency dynamics. Figure 6 shows the control system where $P(s)$ is the plant transfer function matrix from actuator inputs to LOS sensor measurements, and where $C(s)$ is the controller transfer function matrix. Neglecting the rigid body modes in $P(s)$ and assuming infinite bandwidth sensors and actuators,

$$P(s) = \sum_{k=1}^{\infty} G_k(s)$$

where

$$G_k(s) = \frac{1}{s^2 + 2\xi_k \omega_k s + \omega_k^2} M_k.$$

Suppose that n of the modes are known. Let $P_n(s)$ denote the known part of $P(s)$.

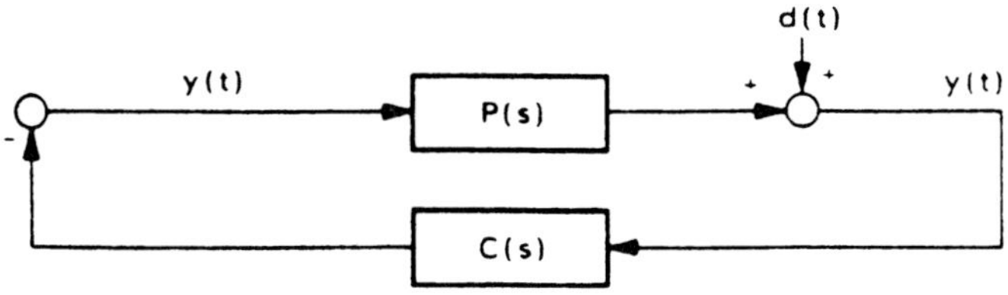

Figure 6: Vibration Suppression Control System.

For example, $P_n(s)$ can be obtained from $P(s)$ by modal truncation, i.e., the first n-modes of $P(s)$ are retained. One can ask the question: is this the best choice for a given model order n? In general, it depends on what is meant by "best". For closed loop control, it is usually better to retain those n-modes which most affect the closed-loop performance. How to select these modes will be discussed in the section on model reduction.

Assuming the modes have been selected, define model error as

$$\delta(s) = P(s) - P_n(s) = \sum_{k \in \Omega_n} G_k(s).$$

Observe that $\delta(s)$ is stable because both $P(s)$ and $P_n(s)$ are stable. Hence, it can be shown that the closed loop system is stable if

$$\overline{\sigma}\left[\delta(j\omega)\right] < \delta_{sm}(j\omega) = 1/\overline{\sigma}\left[Q_n(j\omega)\right]$$

where $Q_n(s)$ is given by $Q_n(s) = C(s)[I + P_n(s)C(s)]^{-1}$ and $\overline{\sigma}(\cdot)$ denotes the maximum singular value of the matrix argument. The quantity $\delta_{sm}(\omega)$ is referred to as the "stability margin", hence, the subscripts "sm". (See [3,4].)

The stability robustness test depends on the location of uncertainty. Additive perturbations such as those just discussed result in the test as shown. The table in Figure 7 shows a variety of stability margins corresponding to generic forms of model error. In Figure 7, P = plant, C = control, M nominal model, and δ = model error. The stability margin is expressed as a function of C and M which are known quantities. Examples of some model error testts are shown in Figure 8 for the CSDL #2 VCOSS model.

2.6 Performance Robustness

The stability robustness tests can be extended to evaluate performance robustness to model error. The evaluation is determined by how performance is measured. Consider the closed loop system

$$y(t) = H(s)d(t)$$

where $H(s)$ is the closed loop transfer function. Although $d(t)$ is not precisely known, it can be considered as the output of a weighting filter $W(s)$ driven by "noise" $w(t)$ so that $d(t) = W(s)w(t)$.

Typical performance bounds depend directly on the frequency dependent quantity $[H(j\omega)W(j\omega)]$. A natural frequency domain performance criterion is then

$$\overline{\sigma}[H(j\omega)W(j\omega)] \leq \rho(\omega)$$

where $\rho(\omega)$ is selected on the basis of power, energy, and magnitude specifications on the output signals. In terms of model error, performance specification is satisfied if

$$\overline{\sigma}[\delta(j\omega)] \leq \delta_{pm}(\omega)$$

where $\delta_{pm}(\omega)$ is the performance margin given by

$$\delta_{pm}(\omega) = [1 - \rho_n(\omega)/\rho(\omega)]\delta_{sm}(\omega)$$

GENERIC FORM OF MODEL ERROR	SOURCE OF MODEL ERROR IN SPACECRAFT SYSTEM		STABILITY MARGIN	
P: = ACTUAL PLANT M: = MODEL Δ: = MODEL ERROR			GUARANTEED STABILITY IF $\bar{\sigma}(\Delta) < \delta_{SM}$	CONTROLLED SPACE 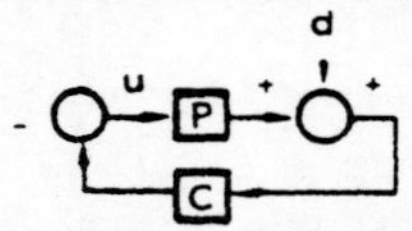
ADDITIVE $P = M + \Delta$	• NEGLECTED RESIDUAL MODES, e.g., – SPILLOVER – REDUCED ORDER MODELING – UNCERTAIN INTERACTING STRUCTURAL MODES		$\delta_{SM} = 1/\bar{\sigma}\,[C(I + MC)^{-1}]$	
OUTPUT MULTIPLICATIVE $P = (I + \Delta)M$	• SENSOR ERRORS – MISALIGNMENTS – BANDWIDTH – SCALE FACTORS	• NEGLECTED HIGH FREQUENCY PHENOMENA, e.g., – MODEL APPROX-IMATIONS – FRICTION – STICTION	$\delta_{SM} = 1/\bar{\sigma}\,[(I + MC)^{-1}MC]$	
INPUT MULTIPLICATIVE $P = M(1 + \Delta)$	• ACTUATOR ERRORS – BANDWIDTH – NONLINEARITIES – PARASITICS – QUANTIZATION		$\delta_{SM} = 1/\bar{\sigma}\,[(I + CM)^{-1}CM]$	

Figure 7: Source of Model Error in Spacecraft System.

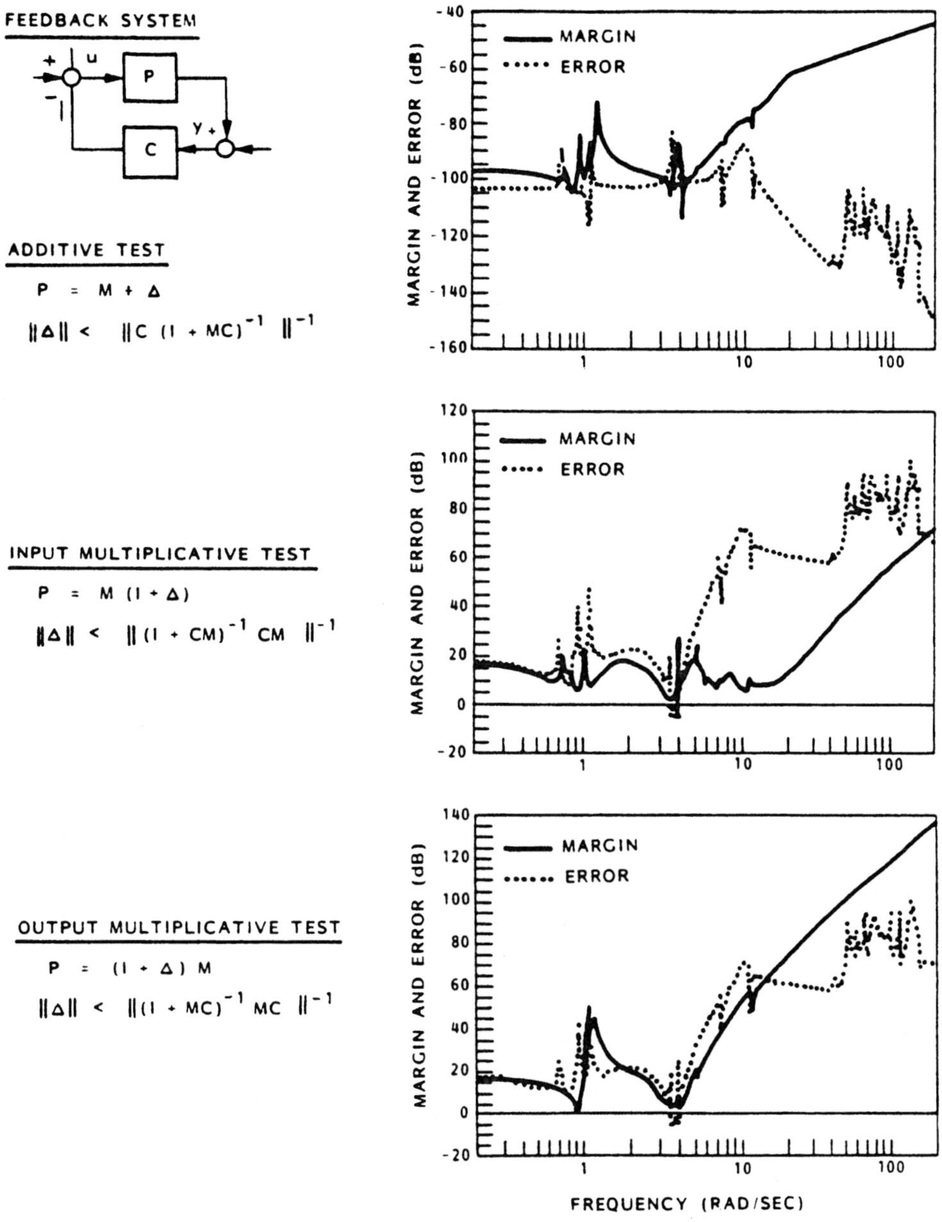

Figure 8: Results of Model Error Tests.

and $\rho_n(\omega)$ is the performance of the nominal closed loop system $H_n(s)$ with no model error. Then,

$$\rho_n(\omega) = \overline{\sigma}[H_n(j\omega)W(j\omega)]$$

which must always be smaller than $\rho(\omega)$ in order for $\delta_{pm}(\omega)$ to be meaningful. Note that $\delta_{pm}(\omega) > \delta_{sm}(\omega)$ as would be expected since performance includes stability. As before, the location of uncertainty modifies the calculation of $\delta_{pm}(\omega)$.

2.7 Usefulness of Stability/Performance Robustness Tests

The stability/performance robustness tests are indispensible in obtaining a realistic preliminary design. They are used in a number of places in the design cycle to establish the HAC/LAC gains, effect of actuator/sensor dynamics, and the criteria for model and controller reduction, which will be discussed in the next section. The tests are also invaluable in establishing criteria for online system identification and control, which will be discussed later on in this section.

2.8 Model Reduction

In general, the requirements for model reduction for active control of large space structures must include the following:

1. The reduced model should be suitable for control design and synthesis. It should incorporate all features critical for the selection of a feedback structure and control gains.
2. The reduced model should accurately incorporate actuator effectiveness, sensor measurements and disturbance distribution [1].
3. The dynamical characteristics of interest in the structure should be represented in the reduced model.

A basic methodology for model reduction which has been used successfully in ACOSS/VCOSS and a number of other programs such as internal balancing, is now described. Other approaches also exist which will be discussed in the sequel.

2.9 Internal Balancing

To determine the most important modes for control design, many criteria must be considered including controllability, disturbability, observability in performance, and observability in the measurements. Any mode which is highly controllable, observable, and disturbable must clearly be included in the design model: however highly controllable-but-unobservable modes, for example, are difficult to judge. Moore [5] has developed an "internal balancing" approach whereby asymptotically stable linear models are transformed to an essentially unique coordinate representation for which controllability and observability rankings are identical. The definition of internally balanced coordinates follows:

Definition: An asymptotically stable model

$$\begin{aligned} \dot{x} &= Ax + Bu \\ y &= Cx \end{aligned}$$

is *internally balanced* over $[0, \infty]$ iff

$$\int_0^\infty e^{At}BB^Te^{A^Tt}dt = \int_0^\infty e^{A^Tt}C^TCe^{At}dt = \sum{}^2$$

where

$$\sum{}^2 = diag[\sigma_1^2\ \sigma_2^2\ \ldots \sigma_n^2] \quad i \geq j,\ \sigma_i^2 \geq \sigma_j^2.$$

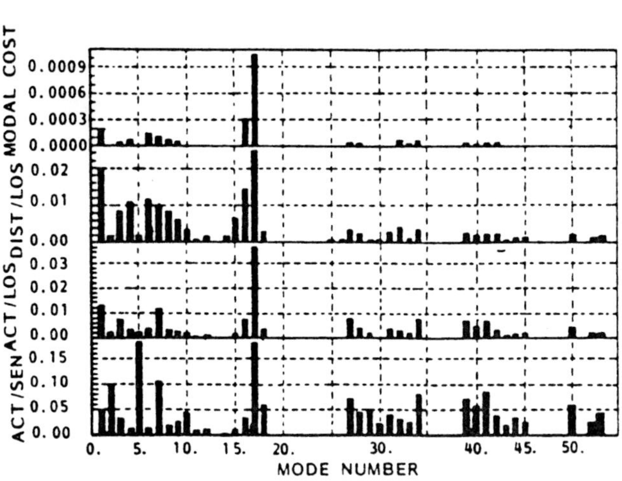

Figure 9: Open Loop Modal Analysis.

Notice that the balanced representation is such that the controllability Gramian and observability Gramian are equal and diagonal. The σ_i's are termed "second-order modes." In general, the required transformation "scrambles" the original coordinate system such that the physical meaning of the states is lost.

However, for lightly damped structural models with decoupled dynamics, the internally balanced coordinate representation is approximately equal to a scaled representation of the model states. Thus it is possible to write approximate formulae for the states in terms of the original model. Three modal rankings are considered:

- disturbance inputs to LOS
- actuator inputs to LOS
- actuator inputs to sensor outputs

These "second-order modes" rankings give important evaluations about which modes to retain and validity of a actuator/sensor placement. These rankings are shown in Figure 9 along with LOS modal cost [6] computed using the colored noise disturbance.

Here the absolute values of the modal costs (for the VCOSS 1 model) are used. The RMS second-order modes and modal costs are plotted versus mode number in Figure 9. Immediately evident is the clustering of these modal phenomena. The disturbance effect as seen through the line-of-sight is constrained to clusters of modes as is the ability to measure and control the model. The coincidence of the controllable clusters and disturbable clusters indicates a favorable actuator/sensor configuration for the problem.

2.10 Frequency Weighted Balanced Realizations

Balanced realization model reduction can be extended to finding a reduced model $P_n(s)$ of a high order model $P(s)$ such that

$$\sup_{\omega} \overline{\sigma}\left\{W_o(j\omega)[P(j\omega) - P_n(j\omega)]W_i(j\omega)\right\} < 1$$

where $W_o(s)$ and $W_i(s)$ are output and input frequency dependent weighting matrices. These can be chosen to reflect closed-loop requirements on model error, vis a vis, frequency domain stability and performance margins. For example, stability of the closed loop system with $C(s)$ designed from $P_n(s)$ is guaranteed if

$$\begin{aligned} W_o(s) &= I \\ W_i(s) &= C_n(s)[I + P_n(s)C_n(s)]^{-1}. \end{aligned}$$

The problem is that $W_i(s)$ is dependent on $P_n(s)$ which is unknown. The let out is that its shape is partially determined by the performance specifications, thus, we can make an initial guess. This technique is referred to as "advanced loop shaping." This involves an iterative problem which is solvable via successive approximation.

2.11 Compensator Order-Reduction

An alternative to plant order reduction is to design a high order compensator and then reduce the compensator order. Let $C(s)$ denote a high order compensator of order N designed to control $P(s)$ of order N or larger. Let $C_n(s)$ denote a reduced version of $C(s)$ of order $n < N$. Motivated by the stability robustness theory, view $C(s) - C_n(s)$ as a perturbation. Hence, the closed loop system with $P(s)$ and $C_n(s)$ is stable if

$$\sup_{\omega} \overline{\sigma} \{W(j\omega)[C(j\omega) - C_n(j\omega)]\} < 1$$

where $W(s) = (I + P(s)C(s))^{-1}P(s)$.

The weight $W(s)$ is stable because the high order control $C(s)$ stabilizes the closed loop system. In this case $W(s)$ is known and we can apply internal balancing to find $C_n(s)$. The disadvantage to this method is that it is necessary to find a high-order compensator. The advantage is that once it is found, internal balancing applies immediately since the weights are known. On the other hand, direct plant order reduction does not involve control design for the high order plant, but does involve an iterative process since the weights are functions of the (unknown) reduced model.

2.12 Low-Authority Control Design

LAC systems, when applied to structures, are vibration control systems consisting of distributed sensors and actuators with limited damping authority. The control system is allowed to modify only moderately the natural modes and frequencies of the structure. This basic assumption, combined with Jacobi's root perturbation formula, leads to a fundamental LAC formula for predicting algebraically the root shifts produced by introducing a LAC structural control system. Specifically, for an undamped, open-loop structure, the predicted root shift $(d\lambda_n)_p$ is given by

$$(d\lambda_n)_p \approx \frac{1}{2} \sum_{a,r} C_{ar} \phi_{an} \phi_{rn} \tag{1}$$

where the coefficient matrix C_{ar} is a matrix of (damping) gains, and ϕ_{an}, ϕ_{rn} denote respectively the values of the nth mode shape at actuator station a and sensor station r.

Equation (1) may also be used to compute the unknown gains C_{ar} if the $d\lambda_n$ are considered to be desired root shifts or, equivalently, desired modal dampings. While an exact "inversion" of equation (1) does not generally exist, weighted least-squares type solutions can be devised to determine the actuator control gains C_{ar} necessary to produce the required modal damping ratios. This determination of the gains is the synthesis of LAC systems.

For structures which already have some damping or control systems in which sensor, actuator, or filter dynamics can either be ignored or are already embedded in the plant dynamics, the root perturbation techniques and cost function minimization methods above can similarly be used to synthesize low-authority controls.

2.13 Robustness of LAC Systems

When sensors and actuators are colocated (i.e. $a = r$), are complementary, and only rate feedback is used, formula (1) reduces to

$$d\lambda_n = -\xi_n \omega_n \approx \frac{1}{2} \sum_{a} C_a \phi_{an}^2$$

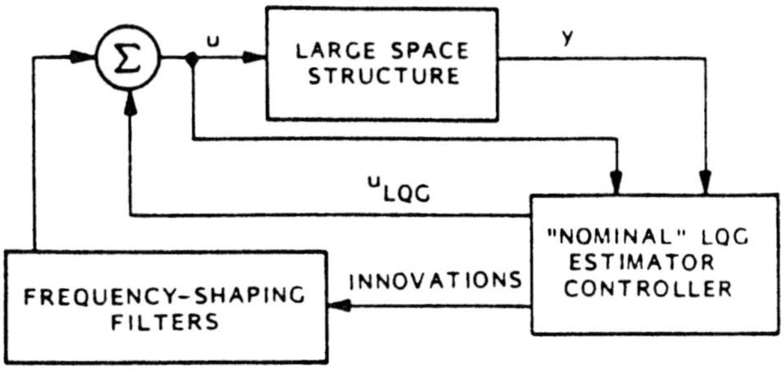

Figure 10: LQG Control With Frequency-Shaping Filters.

which shows that the root shifts are always towards the left of the j-axis if all the gains are negative. This robustness result is obviously based on the assumption that both sensors and actuators have infinite bandwidth, and also that the structure was initially undamped. Several departures from this idealization occur in the actual practical implementation of the LAC systems. The most severe of these results from the finiteness of the actuators' bandwidths. More precisely, the second-order roll-off introduced by the actuator dynamics will always destabilize an undamped structure. However, when some natural damping is present in the structure, or when a passive damper is mounted in parallel with the actuator, additional active damping can be obtained without destabilizing the structure.

2.14 High-Authority Control Design

The HAC control design procedure can be based on any number of multivariable design methods, e.g. LQG, Q-parametrization, H_∞-optimization, etc. Increased penalties in the LQG cost functional are placed at those frequencies where less response is desired. The concept of frequency-shaped cost functionals was introduced prior to ACOSS [7].

The frequency shaping methods are useful in several areas of large space structures control. Three principal applications are important: (1) robustness (spillover avoidance), (2) disturbance rejection, and (3) state estimation.

2.15 Management of Spillover

Spillover in closed loop control of space structures is managed by injecting minimum control power at the natural frequencies of the unmodeled modes. Procedures for controlling spillover at high frequencies are usually discussed, although similar techniques are applicable for other regimes.

The high frequency spillover may be controlled by modifying the state or the control weighting. Conversion to the frequency domain gives the following performance index:

$$R(j\omega) = \left(\frac{(\omega^2 + \omega_0^2)}{\omega_0^2} \right) R$$

The problem of robustness (spillover management) is solved by making Q and R functions of frequency. Figure 10 depicts the modification to the nominal LQG controller. Observe that frequency shaping adds filters whose inputs are the innovation outputs of the state-estimator in the LQG controller.

2.16 Summary

The application of frequency-shaping methods to large space structures leads to a linear controller with memory. However, additional states are needed to represent frequency-dependent weights, hence, there is an increase in the controller order. The software needed for these controller designs is similar to that for standard LQG problems.

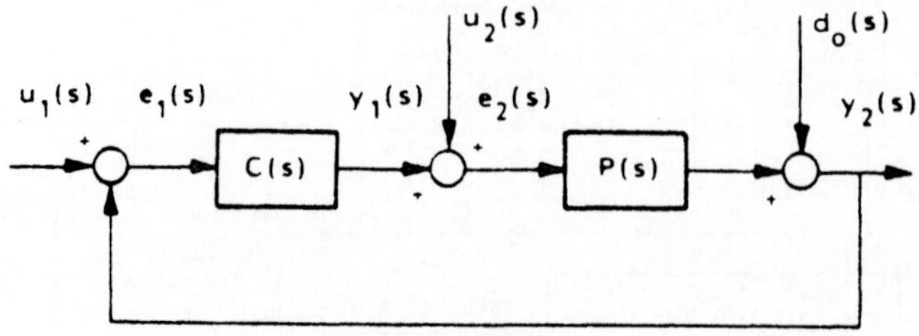

Figure 11: The Unity-Feedback Structure.

3. Controller Design Using Q-Parametrization and H_∞ Optimization

During the last decade, mathematical theories of servo design have been based mainly on quadratic minimization of the Wiener-Hopf-Kalman type, usually applied to state-space models, e.g. LQG controls. However, despite the academic success of these methods, classical frequency response techniques relying on "lead-lag compensators" to reduce sensitivity have continued to dominate industrial servo design. One reason is that quadratic design tends to have poor sensitivity. On the other hand, the frequency domain description has proven to be more suitable to characterize uncertainties which arise in the plant approximation/identification, and frequency domain technique usually results in more robust design, e.g. frequency-shaped LQG can be viewed as an indirect frequency-domain design approach.

Two direct multivariable frequency domain design techniques have become popular in recent years: the Q-parametrization technique and the H_∞-optimal sensitivity.

3.1 Q-Parametrization Design

Consider the linear unity-feedback systems shown in Figure 11 where $P(s)$ is the given linear time-invariant plant. $C(s)$ is the linear compensator, u_1 is the reference input, u_2, and d_0 are respectively the plant-input disturbance and plant-output disturbance, and y_2 is the plant output.

The closed loop system input-output transfer function is given by

$$H_{yu} = \begin{bmatrix} C(I+PC)-1 & -C(I+PC)^{-1}P & -C(I+PC)^{-1} \\ PC(I+PC)-1 & P(I+CP)^{-1} & (I+PC)^{-1} \end{bmatrix}$$

(For simplicity, we drop the argument s in $P(s), C(s)$ etc. in this section.)

By introducing the parameter (transfer function)

$$H_{yu} = \begin{bmatrix} u_1 \\ u_2 \\ d_o \end{bmatrix} \rightarrow \begin{bmatrix} y_1 \\ y_2 \end{bmatrix}$$

$$Q = C(I+PC)^{-1},$$

H_{yu} can be rewritten as

$$H_{yu} = \begin{bmatrix} Q & -QP & -Q \\ PQ & (I-PQ)P & I-PQ \end{bmatrix}$$

Note that the closed loop input-output transfer function, for the given plant P, is completely specified by the parameter Q in a very simple manner: it involves only sums and products of P and Q.

In a typical control system design problem, the two most important closed loop transfer functions are $H_{y_2u_1}$ and $H_{y_2d_o}$: $H_{y_2u_1}$ is the transfer function from reference input u_1 to output y_2 and $H_{y_2 d_o}$ is the transfer function from plant-output disturbance d_o to output y_2. They specify respectively the servo-performance and regulator performance of the feedback system S. The two transfer functions are given by

$$H_{y_2 u_1} = PQ$$

and

$$H_{y_2 d_0} = I - PQ$$

Therefore the control design problem reduces to choosing the parameter Q so that the closed loop system S is stable and that $H_{y_2 u_1}$ and $H_{y_2 d_o}$ are "satisfactory". After the parameter Q is chosen, the corresponding compensator C can be obtained by the formula

$$C = Q(I - PQ)^{-1}$$

Hence, there is a one-to-one correspondence between C and Q. Consequently, for each parameter Q chosen, there is a unique compensator C which achieves the specified Q.

The selection of the parameter Q in the design process raises several questions: What are the conditions on Q so that the resulting compensator C is realizable (e.g. proper)? What is the class of all Q's which result in a stable feedback system? How is an "optimal" Q chosen?

Realizability: If the plant P is realizable, then the compensator C is realizable if and only if the parameter Q is realizable. Note that a physical plant is always realizable.

Global Parametrization: If the open loop plant P is stable, then the closed loop system S is stable if and only if Q is stable, since sums and products of stable transfer function matrices are stable. Consequently, the class of all stabilizing compensators is given by

$$\{Q(I - PQ)^{-1} \mid Q \text{ is stable}\}$$

and the class of all achievable stable input-output transfer matrix $H_{y_2 u_1}$ and the class of all achievable stable disturbance-to-output transfer matrix $H_{y_2 d_o}$ are given respectively by

$$\{PQ \mid Q \text{ is stable}\}$$

and

$$\{I - PQ \mid Q \text{ is stable.}\}$$

These sets give global parametrization of all stabilizing compensators, and all achievable I/O characteristics in terms of a stable proper transfer matrix Q. In other words, the class of all "feasible" designs are parametrized by Q.

If the open loop plant P is not stable, additional constraints have to be added to the choice of Q, in addition to stability and realizability of Q. For example, Q must contain right half plane zeros to cancel the unstable poles of P. Currently, there are three approaches to obtain global parametrization of a given unstable plant: (i) Factorization representation theory [8]; (ii) Direct approach [9]; (iii) Two-step compensation [9].

Optimality: The Q-parametrization alone does not quantatively address the issue of optimal design. The designer selects Q from the class of "feasible" designs, on the basis of the desired input-output response, a priori knowledge of external disturbances, bandwidth, dynamic range and uncertainty of the plant, etc.

Optimal design based on the Q-parametrization and fractional representation framework has become very popular in the research community. The H_∞-optimal sensitivity design is among the results available.

3.2 H_∞-Optimal Sensitivity Design

The H_∞-optimal sensitivity design is an extension of the Q-parametrization technique to include a quantitative performance measure of the closed loop system and achievable optimality based on the performance measure. Roughly speaking, the H_∞ design problem is

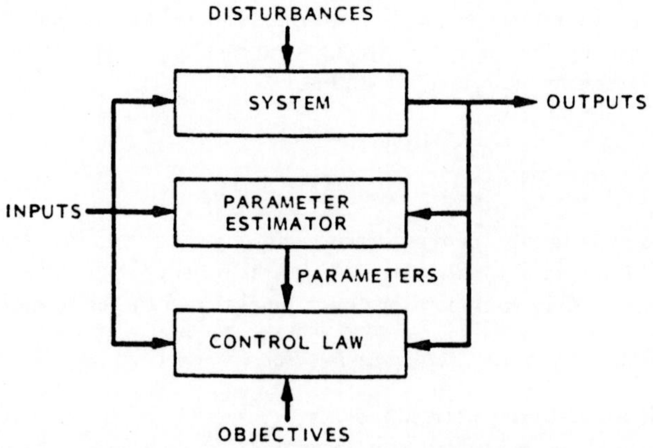

Figure 12: Adaptive Control System.

the following: Given an open loop plant $P(s)$ and a low pass weighting function $W(s)$, find the compensator $C(s)$ so that the H_∞-norm of the weighted sensitivity $(I + PC)^{-1}W$ is minimized subject to the stability of the closed loop system.

Using the Q-parametrization formulation, the problem is equivalent to the following: Find a Q in H_∞ such that the closed loop system is stable and that $(I - PQ)W$ is minimized. Since the weighted sensitivity function is affine in Q, the equivalent problem is easier to solve than the original problem.

Solution to the H_∞-Optimal Sensitivity Problem: Based on the fractional representation (coprime factorization) formulation, several solutions have been proposed and algorithms given. However, all the proposed algorithms are conceptual in nature, suitable only for simple text book example. More effort is needed towards a numerically robust synthesis procedure.

4. Adaptive Control Techniques

Uncertainties in both disturbance spectra and system dynamical characteristics will limit the performance obtainable with fixed gain, fixed order controls. The use of adaptive type control, where disturbance and/or plant dynamics are identified prior to or during control, gives system designers more options for minimizing the risk in achieving performance benchmarks. For the case of LSS systems where performance levels are extremely high, it is absolutely necessary that disturbance and plant models be equally accurate. Since data from ground tests do not usually represent the flight condition accurately, it follows that an on-line procedure for identification and control is necessary.

The need to identify modal frequencies, for example, in high performance disturbance rejection systems has been shown in [1]. The deployment of high performance optical or RF systems may require on-line identification of critical modal parameters before full control authority can be exercised. Parameter sensitivity, manifested by performance degradation or loss of stability (poor robustness) may be effectively reduced by adaptive feedback mechanizations.

Most adaptive control algorithms can be described in the form shown in Figure 12. For example, one could select from the folowing catalogs of major areas:

Model	Control Design	Adaptation
ARMAX	Model Reference	Gradient
State Space	Self-Tuning	Recursive Least Squares
	Pole-Placement	Recursive Max Likelihood
		Extended Kalman Filter

The schemes also differ in terms of update rates. Typically the outer control loop is at a fast rate, whereas the parameters from identification are updated more slowly. Adaptive schemes are referred to as ***recursive*** if the identification rate is a fixed multiple of the controller rate. If identification is used when necessary for calibration the scheme is referred to as ***adaptive calibration***.

Although a great deal of research results are available about adaptive control and identification, unmodeled dynamics and broadband disturbances will significantly upset most algorithms.

4.1 Adaptive Calibration:

The use of a "slow" adaptive control, which is more practical than recursive adaptive control in most space applications is described in this section. It is referred to as a method of adaptive calibration. The term "slow" means that there is sufficient time to run batch identification before the control system is modified. The methodology provides a guaranteed level of performance given an "identified" model of the system together with the model error between the system and the identified model. In fact, the methodology generates performance versus model error tables (to be stored in the computer) from which the control design is immediately obtained. Moreover, the order of the control design is determined strictly on the basis of model error and performance demand.

4.2 Application of Adaptive Calibration:

The basic problem with control based on identified models is that without a measure of model error it is very easy to destabilize the system - particularly when the goal is high performance - as in LSS systems. Adaptive calibration is an approach which incorporates a measure of model error with robust control design in an iterative way so that identification is performed only where it is needed. A proposed adaptive calibration system is shown in Figure 13 with test results, using the CSDL #2 model, shown in Figure 14. The adaptive calibration procedure involves the following steps:

1. The model $M(s)$ is a 10-mode model which has been obtained from I/O data.
2. Estimate $\delta(\omega)$ = model error versus frequency using FFT. This is the dashed curve in Figure 15.
3. Using the identified model $M(s)$ and the model error $\delta(\omega)$, synthesize a robust control (section 2).
4. Calculate δ_{sm} - stability margin. This is the dashed curve in Figure 15. Compare to model error δ both plotted in Figure 16. If acceptable go to Step 6 and implement controller. Otherwise go to Step 5.
5. Modify filter windows, number of parameters (e.g. number of modes), or input spectrum and then repeat Step 1 to obtain new ID model. Figure 16 shows result of identification after one mode is added in the frequency domain region where the test fails.
6. Implement controller.

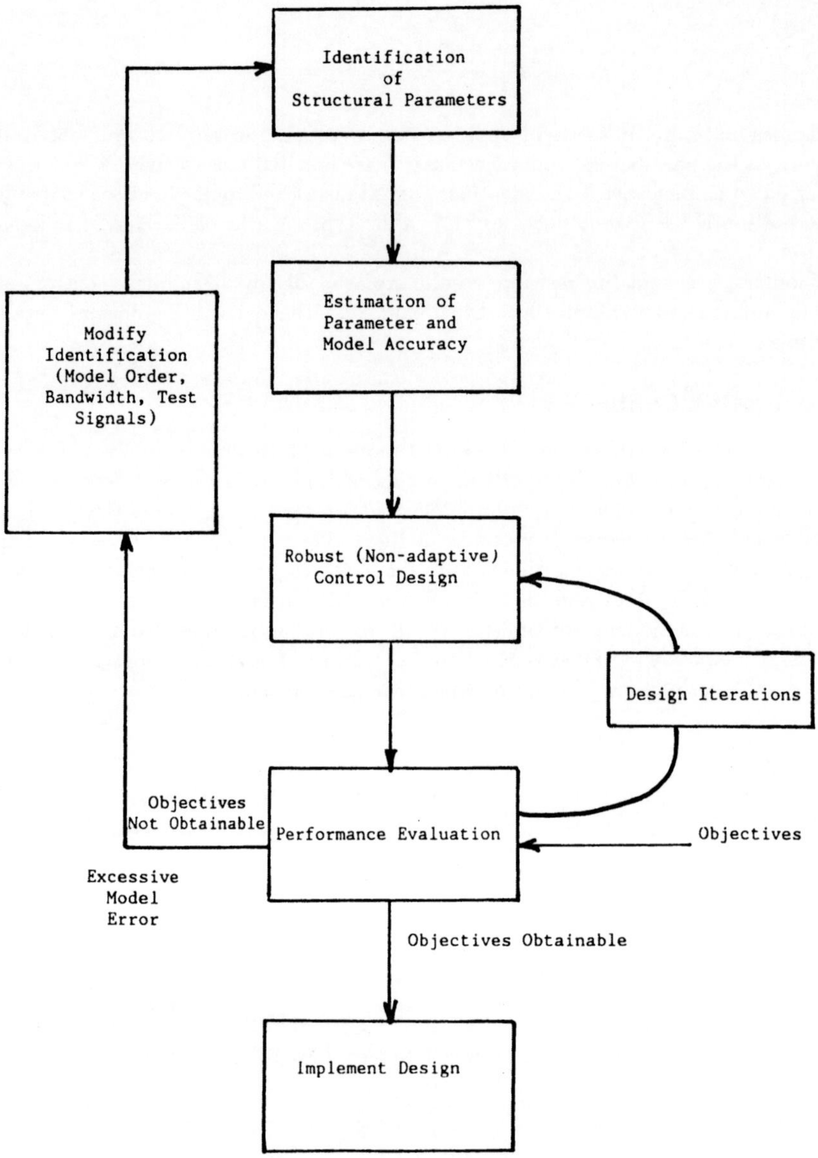

Figure 13: An Adaptive Calibration System.

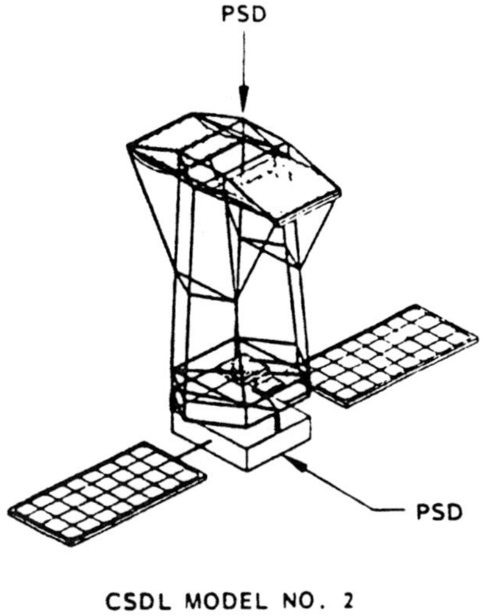

Figure 14: Draper Simulation System.

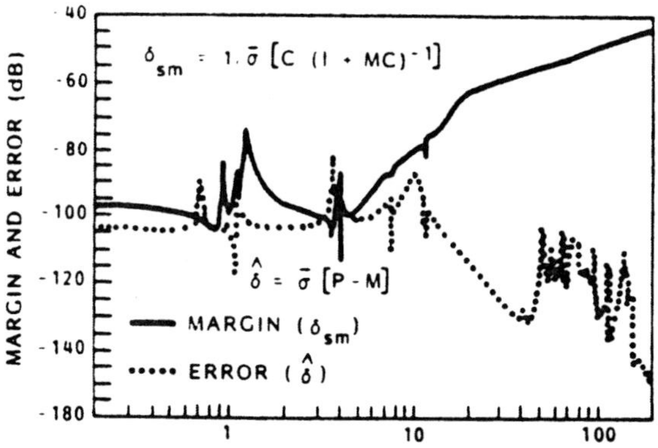

Figure 15: Comparison of Margin and Model Error.

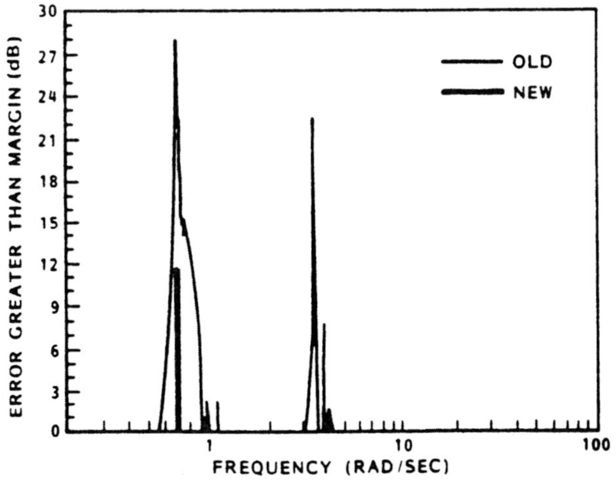

Figure 16: Data Before and After ID Cycle Plots Show Where Error is Margin.

Acknowledgments

The work reported here was sponsored by the Air Force Office of Scientific Research (AFSC) under contracts F49620-84-C-0054 and F49260-85-C-0094.

References

[1] "Active Control of Space Structures (ACOSS)," Phase 1A final report, Prepared by LMSC for Darpa under contract F30602-80-C-0087, Aug. 1981.

[2] J.N. Aubrun, et al., "Stability Augmentation for Flexible Space Structures," *Proceedings of the IEEE CDC*, Hollywood, Fl., Dec. 1979.

[3] J.C. Doyle and G. Stein, "Multivariable Feedback Design: Concepts for a Modern/Classical Synthesis," *IEEE Trans. on Autom. Control*, vol. 26, No. 1, pp. 4-17, Feb. 1981.

[4] R.L. Kosut, H. Salzwedal, and A. Emami, "Robust Control of Flexible Spacecraft," *AIAA J. of Guidance and Control*, vol. 6, No. 2, Mar-April 1983.

[5] B.C. Moore, "Principal component Analysis in Linear Systems: Controllability, Observability, and Model Reduction," *IEEE Trans. on Autom. Control*, AC-26, Feb. 1981.

[6] C.Z. Gregory, Jr., "Reduction of Large Flexible Spacecraft Models Using Internal Balancing Theory," *AIAA J. of Guidance and Control*, vol. 7, No. 6, Nov-Dec. 1984.

[7] N.K. Gupta, "Frequency Shaped Cost Functionals: Extensions of LQG Methods," *AIAA J. of Guidance and Control*, pp. 529-535, Nov-Dec. 1980.

[8] C.A. Desoer, R.W. Liu, J. Murray and R. Saeks, "Feedback System Design: The Fractional Representation Approach to Analysis and Synthesis," *IEEE Trans. on Autom. Control*, vol. 25, No. 3, pp. 399-412, June 1980.

[9] G. Zames, "Feedback and Optimal Sensitivity: Model Reference Transformation, Multiplicative Seminorms, and Approximate Inverses," *IEEE Trans. on Autom. Control*, vol. 26, No. 2, pp. 301-310, April 1981.

[10] J.C. Doyle, "Structured Uncertainty in Control Design," *IFAC Workshop on Model Error Compensation*, Boston, Ma., June 1985.

[11] R.L. Kosut and M.G. Lyons, "Adaptive Techniques for Control of Large Space Structures," ISI Report 50 for AFOSR, under contracts F49620-83-C-0107, F49620-84-C-0054.

[12] R.L. Kosut and C.R. Johnson, Jr., "An Input-Output View of Robustness in Adaptive Control," *Automatica*, vol. 20, No. 5, Oct. 1984.

Progress in Adaptive Control of Flexible Spacecraft Using Lattice Filters

N. Sundararajan
Old Dominion University Research Foundation
Hampton, Va.
and
Raymond C. Montgomery
NASA Langley Research Center
Hampton, Va.

Abstract

This paper reviews the use of the least square lattice filter in adaptive control systems. Lattice filters have been used primarily in speech and signal processing, but they have utility in adaptive control because of their order-recursive nature. They are especially useful in dealing with structural dynamics systems wherein the order of a controller required to damp a vibration is variable depending on the number of modes significantly excited. Applications are presented for adaptive control of a flexible beam. Also, difficulties in the practical implementation of the lattice filter in adaptive control are discussed.

1. Introduction

For large flexible spacecraft, design models will probably not be adequate. Hence, an adaptive control system is highly desirable. Early research into adaptive vibration control of large flexible structures is reported in [1]. Therein, adaptive control of a spinning angular momentum control device (AMCD) was studied. That scheme consisted of simultaneous identification and control with the objective of regulating the out-of-plane deflections of the spinning AMCD. Some of the disadvantages of the method were the requirement of selecting the number of modes to be used for controller design, the use of analytically predicted mode shapes, and the coupling between modes due to inhomogenities in the system. Lattice filter adaptive control is a new method which attempts to overcome these problems. It is, hence, well suited for the adaptive control of flexible spacecraft.

The least square lattice filter has been used extensively in the field of speech and signal processing [2]. In these applications the filter is designed based on a predetermined estimate of system order. Reference [3] is a comprehensive tutorial on this subject. Concerning adaptive control, [4] proposes a self-tuning controller configuration using lattice filters. This scheme requires computing the polynomial coefficients for the plant and controller at each iteration and enforcing a known feedback structure for the controller. Reference [5] proposes inverting the transfer function of the plant for general adaptive control. This idea, with the least mean squares (LMS) algorithm, was utilized in [6] to obtain adaptive control. Reference [7] proposes a similar approach using lattice forms instead of the LMS algorithm. Reference [8] takes this approach but uses a lattice model instead of an autoregressive, moving average with exogenous variables (ARMAX) model where familiar controller techniques could be used. All of these schemes attempt simultaneous identification and control or direct adaptive control. For each case stability questions are not resolved analytically; neither are simulation results available in the open literature.

As opposed to simultaneous identification and control, the scheme discussed herein consists of conducting tests to obtain a design model, validating the model, designing a controller

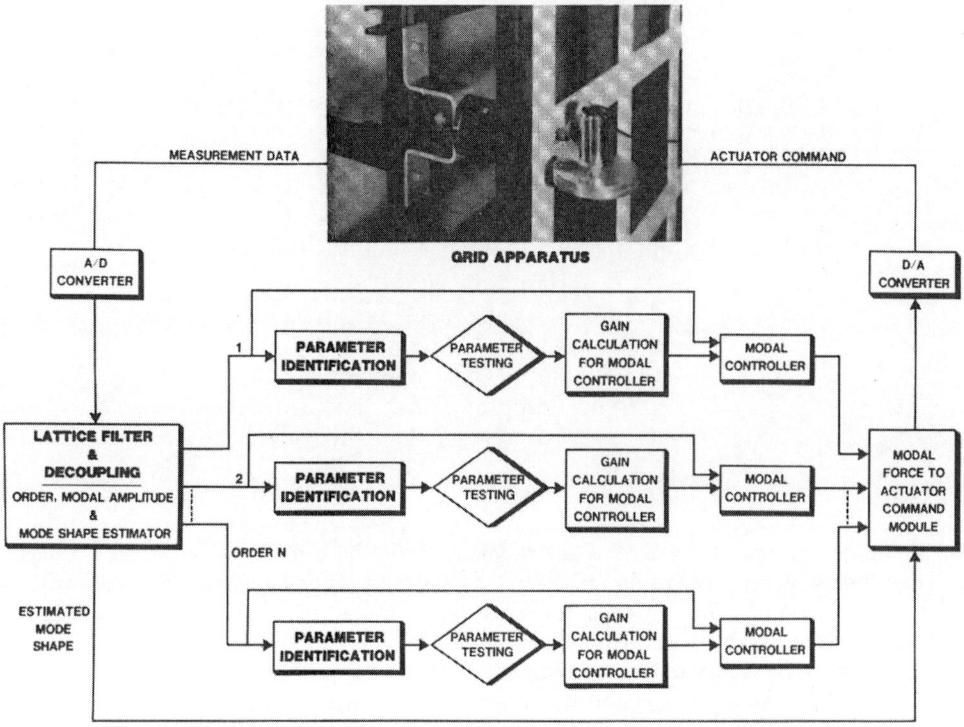

Figure 1: Adaptive Control with Lattice Filters .

based on the validated model, and finally, engaging the control system. This approach is ideally suited to the control of large flexible spacecraft because of the passive environment of outer space and the potential of relaxation to a controller that is known to be stable - that of collocated rate feedback. It was originally presented in [9] and represented the first use of a recursive variable order structure for adaptive control. Therein, the lattice filter was used to provide an on-line estimate of the system order, mode shapes, and modal amplitudes to provide a validated modal control design model. After the identified model parameters are validated through a series of test procedures, they are used in a modal pole-placement control law design. Figure 1 shows the adaptive control scheme using lattice filters.

The purpose of this paper is to assess progress in using lattice filters in adaptive control of flexible spacecraft and to highlight problem areas for further research. First, lattice filter theory and order determination is summarized following the original development of [10]. Then, their use in adaptive control is discussed along with applications to the vibration control of a beam. Finally, difficulties arising in the practical implementation are discussed.

2. Summary of Lattice Filter Theory and Order Determination

For application considered herein, we assume that the ith measurement sample is of the form

$$y_i^T = [y_1(i), y_2(i), \ldots, y_{NS}(i)]$$

where NS represents the number of sensors. It is assumed that y is generated from a model system wherein

$$y_i = \Phi \cdot \Psi_i + v_i \tag{1}$$

Here, Φ is an $NS \times NM$ mode shape matrix, Ψ_i is the $NM \times 1$ modal amplitude vector, v_i is a Gaussian-random variable with zero mean and covariance matrix R. NM represents the number of modes in the system or order of the system.

Reference [10] presents a derivation of the equations that relate any order, n, and time, i, recursions for the normalized forward and backward residuals as well as the least square estimate of the measurement vector. These equations are listed below:

$$\begin{aligned} \underline{e}_{i,n+1} &= (1 - k_{i,n+1}^2)^{-1/2}(\underline{e}_{i,n} - k_{i,n+1}\underline{r}_{i-1,n}) \\ \underline{r}_{i,n+1} &= (1 - k_{i,n+1}^2)^{-1/2}(\underline{r}_{i-1,n} - k_{i,n+1}\underline{e}_{i,n}) \end{aligned}$$

wherein

$$k_{i,n+1} = < \underline{e}_{i,n}, \underline{r}_{i-1,n} >$$

and <> represents an inner product. The symmetry of the recursion formulae is apparent. The equations are coupled by the term $k_{i,n+1}$ which is customarily called the "reflection coefficient." The estimate of the measurement [10] at sample i for a model of order n is

$$\hat{y}_{i,n} = \sum_{j=0}^{n-1} \mathcal{E}\,(\underline{e}_{i,j} | \underline{r}_{i-1,j})$$

where $\mathcal{E}$ represents an orthogonal projection operator. Hence,

$$\hat{y}_{i,n} = [r_{i-1,0}, r_{i-1,1}, \ldots, r_{i-1,n-1}] \begin{bmatrix} k_{i,1} \\ k_{i,2} \\ \cdot \\ \cdot \\ \cdot \\ k_{i,n} \end{bmatrix}$$

so that

$$y_i = \Phi_L \cdot \Psi_i + \underline{e}_{i,n} \tag{2}$$

where Φ_L is an orthonormal $NS \times n$ matrix $[\underline{r}_{n-1,0}, \ldots, \underline{r}_{n-1,n-1}]$ generated from the lattice filter, and Ψ_i is the n dimensional vector of reflection coefficients and $\underline{e}_{i,n}$ is the NS dimensional estimation error vector.

Clearly, in this approach one may "fit the noise" by continually increasing the order of the system; however, once the order of the estimator has increased beyond the correct order, then the residual errors should lie within a noise band which can be predicted *a priori* based on assumed noise characteristics. A threshold value can be selected based on this predicted noise band and order determined by a test of whether or not the residuals have been reduced to lie within the noise band. The residuals will generally consist of signal and noise parts - the signal part being reduced as the correct order is reached until the residuals essentially consist only of noise. This test is carried out based on a data window of NW samples. Thus, assuming that the data can fit a linear model and that the noise process is Gaussian, for i large enough,

$$E(\sum_{i=1}^{NW} \underline{e}_{i,n}^T \underline{e}_{i,n}) = NW\ tr\ E(v_i v_i^T) = NW \sum_{j=1}^{NS} \sigma_j^2 \tag{3}$$

where E is the expectation operator. This can be used as the one sigma threshold for the order determination test. In the last equation σ_j is the standard deviation of the noise process for the jth sensor. Reference [10] documents experience in choosing the data window size NW and the threshold level based on simulations.

3. Adaptive Control Using Lattice Filters

Independent Modal Space Control (IMSC) [11] is a control scheme specifically designed to deal with flexible spacecraft in a modal form amenable to control law design. Unfortunately, it requires natural modes and not the orthonormal basis provided by the lattice filter. Consequently, in order to interface the lattice filter outputs with the target adaptive control scheme (Figure 1) and to make comparisons with finite element analysis predictions of natural modes, a method is needed to obtain natural mode shape estimates from the lattice filter basis. The filter updates the NM basis vectors at every sample instant. While the order estimate NM remains constant, the updated basis vectors are related by a mere rotational transformation. The assumption of the target adaptive control scheme is that the system motions can be modelled by a constant and finite set of natural modes and their associated modal amplitudes over a reasonably long time interval. Hence, when the estimated system order is constant, the basis elements used to derive the modal amplitude time series required by the target adaptive control scheme are not changed.

The transformation from the lattice filter to a natural mode basis should satisfy

$$\hat{y} = \Phi_L \Psi_L = \Phi_N \Psi_N$$

wherein the subscript L refers to the lattice filter and N refers to the natural modes. A nonsingular matrix T, will satisfy this condition provided

$$\Phi_L T = \Phi_N; \qquad \Psi_N = T\Psi_L \tag{4}$$

Since the order estimate is assumed constant, this matrix can be approximately determined on-line using the digital Fourier transform (DFT). Herein, this is accomplished as follows. Since the lattice filter uses the current measurement sample as its first basis element, the corresponding modal amplitude time series contains NM frequencies. Hence, the DFT spectrum of this series will contain NM peaks corresponding to these frequencies. The frequencies $(\omega_1, \omega_2, \ldots, \omega_{NM})$ can thus be identified by searching this spectrum for these peaks. Assuming that the motion is comprised of undamped structural vibrations, the matrix T, which produces the desired transformation can be calculated as

$$T = \begin{bmatrix} Re[\Psi_L^1(\omega_1)] & \cdots\cdots & Re[\Psi_L^1(\omega_{NM})] \\ \cdot & & \cdot \\ \cdot & & \cdot \\ \cdot & & \cdot \\ Re[\Psi_L^{NM}(\omega_1)] & \cdots\cdots & Re[\Psi_L^{NM}(\omega_{NM})] \end{bmatrix}$$

wherein, $[\Psi_L^1(\omega), \ldots, \Psi_L^{NM}(\omega)]$ is an NM dimensional vector of the modal amplitude transform. Using this matrix, the digital Fourier transform of each component of Ψ_N will be zero at the discrete frequencies, $\omega_j, j \neq i$. One item which degrades this approximation is the error in using DFT instead of the true Fourier transform. Still another is the assumption that the motion is made up of undamped structural oscillations. In spite of these items, [12] shows that this approach produces good estimates of the natural modes for the beam used herein.

The decoupled modal amplitude time series, $\Psi_N(i)$, as obtained above in equation (4), is then analyzed, for each mode, to identify the parameters of its autoregressive, moving average (ARMA) model. The inputs to each ARMA modal model are the generalized forces and hence, each model takes on the form:

$$\Psi_N(i) = A_1\Psi_N(i-1) + A_2\Psi_N(i-2) + B_1 f(i-1) + B_2 f(i-2) \tag{5}$$

where the f represents the generalized forces. Given the Ψ_N and $f's$, the parameters A and B above are identified and used in the control law design process. Thus, the ARMA model output error is

$$e(i-1) = \Psi_N(i-1) \quad -[\hat{A}_1(i-1)\Psi_N(i-2) + \hat{A}_2(i-1)\Psi_N(i-3) + \hat{B}_1(i-1)f(i-2) + \hat{B}_2(i-1)f(i-3)] \tag{6}$$

The gradient technique of [1] is used to identify the parameters $p = (A_i, B_i)$ using the iteration sequence

$$p(i) = p(i-1) + e(i-1)[W_1\Psi_N(i-2), W_2\Psi_N(i-3), W_3f(i-2), W_4f(i-3)] \tag{7}$$

As indicated in [1], the weights W must be selected consistent with the relation

$$W_1\Psi_N^2(i-2) + W_2\Psi_N^2(i-3) + W_3f^2(i-2) + W_4f^2(i-3) < 2$$

and the inputs to the algorithm, Ψ_N and f, must be sufficiently varying and large if the parameters are to converge to their correct value.

For the identification and control scheme explained above to work satisfactorily in a closed loop environment, it is necessary to validate the design model. Three tests are suggested herein which check the following: 1) model fit error; 2) parameter convergence; and, 3) signal information. These tests have been used successfully in simulation and experimental work. The fit error test uses a fixed parameter set to calculate an estimated modal displacement for the past NT samples.

$$\sigma_{fit} \quad > \sum_{n=0}^{NT} \hat{\Psi}_N(i-n) - \{A_1\hat{\Psi}_N(i-n-1) + A_2\Psi_N(i-n-2) + \hat{B}_1f_N(i-n-1) + \hat{B}_2f(i-n-2)\},\ k\ >\ NT$$

If the absolute sum of the error between the modeled displacement and the displacement calculated by the lattice filter exceeds a given threshold, the fixed parameter set is updated with the present identified parameter set. This process is repeated until the parameter set fits the data. The convergence test runs concurrently with the fit test. It simply checks the magnitude of the changes in successfive estimated parameters.

$$\sigma_{conv} > \sum_{n=0}^{NT} |p_n - p_{n-1}| \qquad for \qquad p^T = [\hat{A}_1,\ \hat{A}_2,\ \hat{B}_1,\ hatB_2]$$

If the absolute sum of ten successive parameter estimates changes is above a specified level, a logical switch is set to indicate failure. The third and final test is on information content of the estimated modal amplitude signals from the lattice filters. The purpose of this test is to check whether enough information is present in the signal for proper identification of the parameters. If this test fails, the controller gains are not updated based on the identified parameters, but are frozen at the last values before the test failed. Here, the estimated modal amplitudes and velocities from the lattice filter are checked for sufficient excitation by summing over ten samples.

$$\sigma_{inf} < \sum_{n=0}^{NT} |\Psi_N| + |\Psi_N - \Psi_{N-1}|$$

The second term in the above equation represents a measure of velocity estimates. If the sum is below a threshold, σ_{inf}, the updating of the control gains based on the identified parameters is stopped. The information and fit error tests constitute one test for each mode and the convergence and reasonability tests constitute four tests for each mode. Thus, six tests must be passed before control is applied to a given mode. The actual stability and performance of the controller is directly affected by the criteria chosen for passing a test. If the test criteria are too stringent, system noise and nonlinearities may preclude initiation of control. However, if the tests are not adequate, it is possible that an error in the estimated parameters could result in gain calculations which produce an unstable system.

Now, consider the philosophy to be used when the tests described above pass or fail. When all the tests for parameters of a given mode have passed, control gains are calculated according to a previously developed pole placement scheme [1]. The control force commands are then calculated using these control gains. Considering the philosophy used when the tests fail, two cases were studied. In the first case, when the tests failed, control was turned off and the control forces were made zero. In the second case, when the tests failed, updating of the control gains was stopped and they were frozen at their values prior to the test failure. In this case, the control forces were not made zero and were computed using the frozen control gains. From a detailed study of both cases, it was found that the performance of the adaptive control system in the first case was superior to that of the second case.

4. Application to a Flexible Beam

The closed-loop adaptive control scheme of Figure 1 has been tested in the digital simulation for the 12-foot, flexible free-free beam located at NASA Langley Research Center. The simulation contains the mathematical model of the beam apparatus in modal form. For this study, the simulation contains one rigid-body mode, the first three flexible modes, nine deflection sensors, and four actuators for control purposes. The initial conditions on the modal displacements were set to .05 and the modal velocities were set to zero. The modal damping was also set to zero. A digital sampling rate of 32 Hz was selected for the simulation, and the standard deviation for all measurement noise was assumed to be .005 based on observed noise in the available hardware. The lattice filter estimates were based on a data window size of 4 [10]. The testing procedures were all carried out based on data window (NT) of ten samples. Initial parameters estimates were offset from the mathematically correct values to test and verify the rapid convergence of the identification algorithm. An arbitrary delay of 2 seconds was added between the time identification starts and when the control would be applied to show the behavior of the identification scheme.

At the start of the simulation, the lattice filter determines the number of modes in the simulation along with the mode shapes. Modal amplitude time histories are then generated. From the lattice filter mode shapes and modal amplitudes, natural modes and modal amplitudes are obtained through a linear transformation explained in the earlier section. The application of the transformation is delayed for 2 seconds because the online transformation technique of reference 12 requires 2 seconds of data for a digital Fourier transform data base to obtain the required transformation. The natural modal amplitudes are input to the equation-error parameter identifier which identifies the ARMA parameters. The identification results are then tested using the test procedures described above. When the tests are passed, the control is turned on. Results of the simulations are presented in Figures 2-4.

Figure 2 shows the estimated modal displacement for the first lattice filter mode. The order estimate plot shows that the correct order of 4 is obtained in .3 seconds. After the parameter identification, when all the tests are passed, the control is turned on at 5.5 seconds and the modes are damped. The result of the adaptive control on the natural modes is shown in Figure 3. It is evident that when the identification is validated by passing the tests and control turned on, vibration suppression is achieved. When the modes are damped out the lattice filter order estimate drops from 4 to 1 indicating the flexible modes are damped out. Although the lattice filter order decreased, the control design order was maintained at 4 throughout the time interval when control was on. Allowing the order to vary in real time and updating the control order is a topic for further study.

The main results of the identification and the test procedures are summarized in Figure 4. For the first flexible mode, the figure shows the time histories of identified frequency parameter A_1 the fit error, a parameter that indicates algorithm convergence, and a parameter that indicates information content of the measurements. When all the tests are passed, the corresponding pass parameter (plotted as a binary logical variable) is set to one. The various thresholds for the tests are also marked to indicate when the tests pass. These thresholds were determined based on detailed sensitivity studies of the modal control scheme

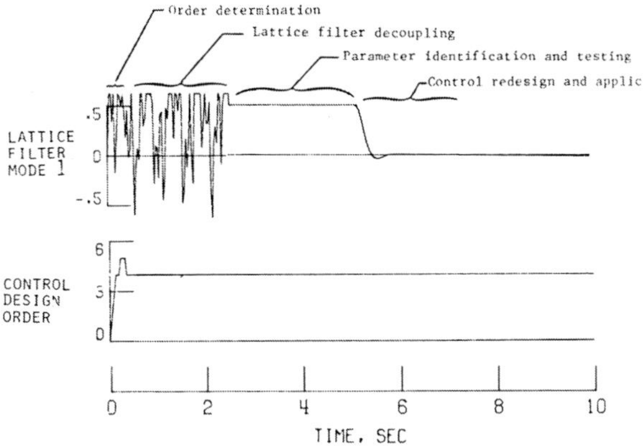

Figure 2: Typical time history of an adaptive control run using identification, testing, and control design.

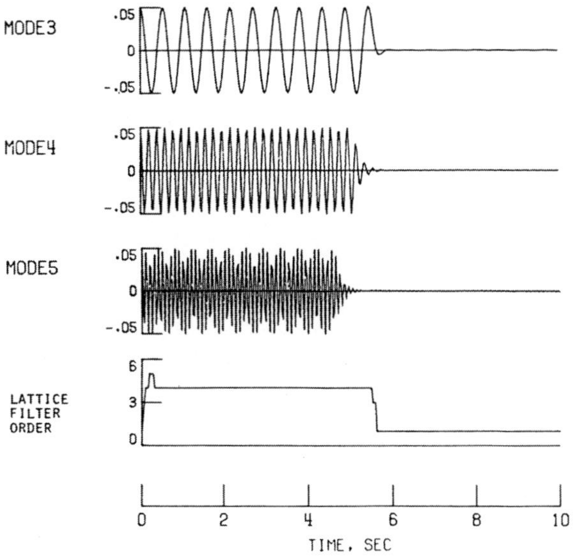

Figure 3: Time histories of three natural modes with the lattice filter order indicated .

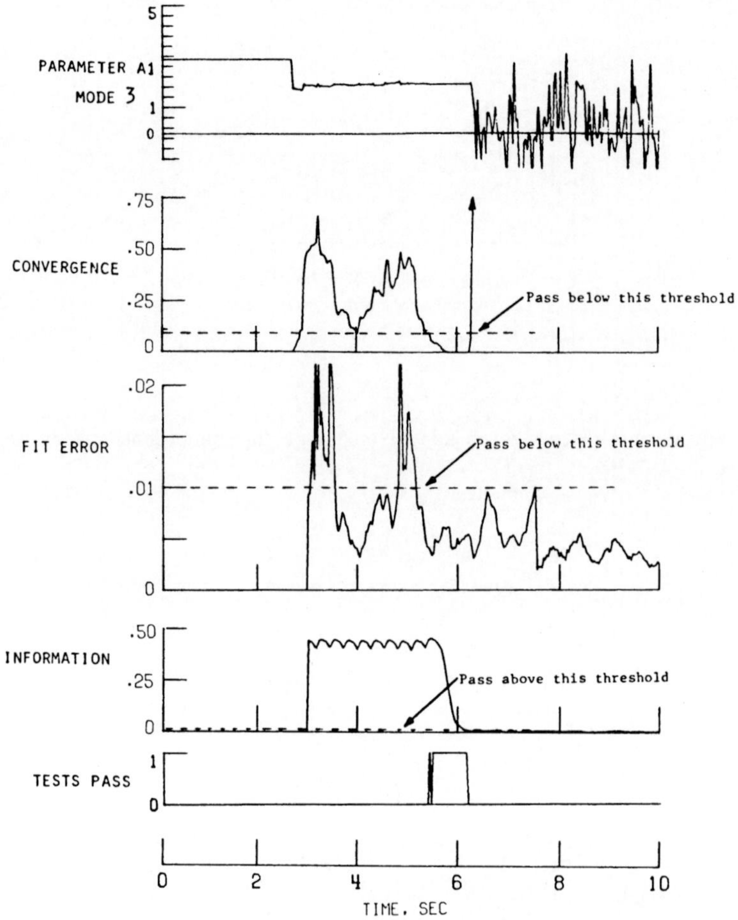

Figure 4: Time histories of the test variables for one mode with the test thresholds and logic sum of the tests indicated.

for the beam [13]. An error was intentionally put on the initial estimate of A_1 so that the convergence of the estimates to the correct value could be observed. When the identifier is turned on, the estimate converges to the true value of 1.8 from 3. The thresholds indicate that the fit error test is passed first and then the convergence test. With enough signal in the measurements the information test is always passed. When all the tests are passed at 5.5 sec, the control is turned on. When control is fully effective, that is when the modes are damped out, the measurement data will contain only the noise and the information test will fail. This is immediately seen from the history of A_1 as it starts oscillating with large amplitude indicating that the modal amplitude signal contains mainly noise. Also, if the parameter excursions are large, the convergence tests will also fail indicating a failure for the binary variable pass. Once this happens, the control gain updating is stopped, and control forces were made zero.

5. Problems in Practical Implementation

The adaptive control scheme of Figure 1 is good from the engineering point of view since only validated models are used for control system design. A natural question arises as to the course of action when validation tests fail. The operating environment for large flexible spacecraft is, fortunately, benign and a system designed to suppress vibrations can be shut down at the expense of having to conduct relatively long term maneuvers. Another saving feature of large flexible spacecraft is that collocated rate feedback is stable and relaxation of the system to this mode of operation is also possible, again, with corresponding degradation in performance. Therefore, two options that can be evoked are; one, to shut down the control system and the other, to revert to a robust control system design which insures stability.

At first glance one may wish to use the ARMA model generated by the lattice filter directly in the design process rather than using IMSC with its requirement of generating natural modes. Unfortunately, the current online design capability for controllers of vector ARMA processes is not adequate. Having selected IMSC, one must obtain natural modes from the vector ARMA model or from the measurement time series. Here the same problem arises, that is, the current capability of eigenvalue/vector analysis for vector ARMA processes is inadequate for online implementation. Hence, a time series analysis using a DFT has been selected. The accuracy of the process of extracting natural modes is directly affected by the number of data points processed. Hence, there is a tradeoff to be made between the higher complexity in control computations versus the error in the natural modes using the DFT approach. Also, significant computational saving results if the approximation of zero damping can be evoked. If this approximation cannot be made, then one must work with complex modes.

Since several approximations are required by the system, a method of validating the models used in the online controller design is essential. Analytic methods of validating models based on statistical error analysis (e.g. Cramer-Rao bounds) are not adequate. Currently, tests on fit error, algorithm convergence, information content of the measurements, and reasonability have been used. The thresholds and design constants for these tests can be determined only by exhaustive simulation and/or hardware tests and is not an online procedure.

6. Conclusion

This paper reviews the use of the least square lattice filter in adaptive control systems. Emphasis is placed on the integration of the lattice filter into a practical parameter adaptive control system. One novel feature of the recommended system is the inclusion of a design model validation scheme based on model fit error, algorithm convergence, and signal information content. An application is presented for adaptive control of a flexible beam. These results indicate that the lattice filter adaptive scheme is practical for vibration control of large flexible spacecraft. Difficulties in the practical implementation of the lattice filter in adaptive control are also discussed. These are centered around the computational burden of transforming lattice filter modes into natural modes and the selection of the thresholds for online validation tests.

References

[1] C.R. Johnson and R.C. Montgomery, "A Distributed System Adaptive Control Strategy," *IEEE Trans. on Aerospace and Electronic Systems*, vol. AES-15, No. 5, pp. 601-612, 1979.

[2] *IEEE Trans. on Acoustics, Speech and Signal Processing, Special Issue on Adaptive Signal Processing*, vol. ASSP-29, No. 3, June 1981.

[3] B. Friedlander, "Lattice Filters for Adaptive Processing," Proc. of IEEE, vol. 70, No. 8, pp. 829-867, August 1982.

[4] B. Friedlander, "Recursive Lattice Forms for Adaptive Control," Proc. JACC, San Francisco, California, 1980.

[5] J. Martin-Sanchez, "A New Solution to Adaptive Control," Proc. IEEE, vol. 64, No. 8, pp. 1209-1218, August 1976.

[6] B. Widrow, J.M. McCool and B.P. Medoff, "Adaptive Control by Inverse Modelling," Proc. of 12 ASILOMAR Conference on Circuits, Systems and Signal Processing, 1978, pp. 90-94.

[7] S. Olcer and M. Morf, "Adaptive Control by Ladder Forms," Proc. of American Control Conference, San Diego, June 1984, pp. 1265-1270.

[8] S.C. Shah, R.A. Walker and H.A. Saberi, "Multivariable Adaptive Control Algorithms and Their Mechanizations for Aerospace Applications," Proc. of IEEE Conference on Decision and Control, Las Vegas, Nevada, December 1984, pp. 381-386.

[9] N. Sundararajan, J.P. Williams and R.C. Montgomery, "Adaptive Modal Control of Structural Dynamics Systems Using Recursive Lattice Filters," *J. of Guidance, Control and Dynamics*, vol. 8, No. 2, pp. 223-229, 1985.

[10] N. Sundararajan and R.C. Montgomery, "Identification of Structural Dynamics System Using Least Square Lattice Filters," *J. of Guidance, Control and Dynamics*, vol. 6, No. 5, pp. 374-381, September-October 1983.

[11] L. Meirovitch and H. Oz, "Modal Space Control of Distributed Gyroscopic Systems," *J. of Guidance and Control*, vol. 3, No. 2, pp. 140-150, January-February 1980.

[12] N. Sundararajan and R.C. Montgomery, "Decoupling and Structural Mode Estimated Using Recursive Lattice Filters," Proceedings of the 21st IEEE Conference on Decision and Control, December 1982, pp. 998-999.

[13] J.P. Williams and R.C. Montgomery, "Simulation and Testing of Digital Control on a Flexible Beam," Proceedings of the AIAA Guidance and Control Conference, August 1982, pp. 403-409.

The Search for Appropriate Actuator Distribution Criteria in Large Space Structures Control

Richard W. Longman*and Robert E. Lindberg
Naval Research Laboratory
Washington, D. C. 20375

Abstract

In a series of previous publications the authors have defined various concepts of the degree of controllability, together with a unifying framework for such definitions. These concepts can be used as control system design tools, with applications that include actuator placement, model reduction, and the quantification of the degree of coupling between subsystems. These works were motivated by the actuator placement question in the shape control of future large flexible spacecraft. Previous use of these concepts have all pointed toward placement of the actuators at the corners of the structure where the modes have the largest deflections and slopes. Various physical situations suggest that this is not always the most desirable location, and this paper demonstrates that the degree of controllability concepts can give optimal locations distributed throughout the spacecraft when the time given to accomplish the control is relatively short.

1. Comments on Adaptive Control of Large Space Structures

The technical contribution of this paper is in the resolution of certain issues associated with placement of actuators in the control of large space structures. However, since this paper appears in a volume emphasizing adaptive control, it is perhaps appropriate to start with a section which comments on the future use (or lack of use) of adaptive control theory to accomplish attitude and shape control of future large flexible spacecraft. In these comments I don't wish to discourage people from investigating this interesting application–the comments might best be viewed as playing devil's advocate.

First, consider the fact that large flexible spacecraft are governed by partial differential equations and hence are infinite dimensional systems, whereas adaptive control theorems seem to always need an upper bound on the dimension of the system in order to be able to prove convergence. A natural question is then: Will the government ever decide to use adaptive control if we cannot prove that it will converge? I can imagine two answers to this, and leave it to the reader to decide which is right. One answer is: No! We must play it safe. These spacecraft cost millions of taxpayers' dollars, and we cannot afford the chance of a failure. The other answer is: Yes! Who cares about theorems? Practicing engineers don't understand them. If you simulate the dynamic behavior of the adaptive control system for a hundred cases representing all the reasonable situations, and it works every time, then use it!

Perhaps we should take a step back and ask whether there is a need for adaptive control in the large space structures problem. Adaptive control is useful in two types of situations. First is the case of disturbance dynamics. If the disturbances to a system change too quickly to be predicted long term, but do not change so quickly that short term identification and

*Also, Professor, Columbia University, N. Y., N. Y., 10027

prediction are impossible, then adaptive control can be very useful in improving the performance of a control system. An excellent example of this is the use of adaptive control in position control of undersea vehicles subjected to disturbance currents. The second situation is one in which the system parameters are slowly changing, for example in aircraft control systems where the aircraft have very different dynamic behavior at subsonic, transonic, and supersonic speeds.

Do either of these situations apply to the large space structures control problem? There is no reason to expect important changes in the parameters. There can be changes in parameters due, for example, to outgassing of water vapor from graphite epoxy parts, but these changes are very long term and therefore not appropriate for real time adaptation. Turning our attention to disturbance dynamics, the major disturbances can be listed as: 1) gravity gradient torque, 2) thermal distortion, 3) solar radiation and residual atmospheric pressures, and 4) internal disturbances. Disturbances 1, 2, and to a lesser extent, 3, are all reasonably predictable and long term. Also, the usual actuators envisioned for spacecraft shape control (such as proof mass actuators, control moment gyros, reaction wheels, etc.) may not be appropriate for these long term effects, suggesting that they might better be handled by a separate controller using, for example, tension control in structural members, perhaps employing feedforward control or gain scheduling. This leaves item 4. If the internal disturbances are in fact crew motion, then the disturbances are not sufficiently predictable even in the short term to justify use of adaptive control. If the internal disturbances are from moving parts, then perhaps again they become so predictable that adaptive control is not appropriate. So it appears that the large space structures control problem does not have the properties that make adaptive control appropriate. However, one can imagine that in the space station design one would want the individual control systems for pointing of each experiment to be as autonomous as possible, and this might suggest adaptive control rather than feedforward control to limit the disturbances on one experiment due to motion of neighboring experiment packages.

The most fundamental difficulty in attitude and shape control of large flexible spacecraft is the fact that one may not know the spacecraft dynamics well enough before launch to design a good control system. Many of these structures cannot support themselves under the load of gravity, and therefore cannot be experimentally tested on the ground. The system parameters are constant, but we don't know what they are. Therefore, the logical thing to do is to perform tests in orbit for purposes of system identification, and then tune the controller once the results are obtained. The tuned controller would be used from then on, without any need for adaptive control. (Reference [1] gives an analysis to determine how many sensors and actuators would be needed to be certain that one could identify the system mass and stiffness matrices after launch, and this number turns out to be large. On-line operation of an adaptive controller may have similar requirements.)

One other phase of large space structures operations, is the construction phase in orbit that applies to all structures that are too large to fit on a single shuttle flight. During construction the structural dynamics equations are constantly changing. But the pointing requirements during this phase are very lax, which suggests the use of a simple low authority controller such as a positive real controller that stabilizes all modes without requiring detailed knowledge of the modal dynamics. Here again, there seems to be no strong incentive to try to apply adaptive control.

From the above arguments, much as I would like to say otherwise, I conclude that adaptive control will not be used in any general sense in the large space structures attitude and shape control problem, but there may be certain specialized situations in which adaptive control is the logical approach. Now let us turn to the main subject of this paper, the search for actuator distribution criteria in large space structures (LSS).

2. Introduction

Initial Considerations

Various concepts of the degree of controllability have been generated by the authors in previous publications. The original motivation was to generate methods of guiding the control system designer in choosing locations for actuators in the LSS control problem, but they also apply to actuator placement in chemical process control, for example. Additional uses for the concepts include model reduction, and for understanding the degree of coupling between parts of systems for use in hierarchical or decentralized control design.

The seminal paper which spawned the series of publications is a two part work appearing in [2] and [3] (and in their conference proceedings versions in [4] and [5]). The introduction to that work follows some of the original train-of-thought that led to the ultimate formulation, and a few words on this subject might be appropriate here. The most commonly stated test for controllability is rank of the controllability matrix $Q = [B \; AB \cdots A^{n-1}B]$. A simplistic attempt at defining a degree of controllability is to determine how near the matrix Q is to being rank deficient by looking at the minimum eigenvalue of QQ^T. The manner in which such a simplistic approach fails, sheds light on the characteristics that a workable definition must have. In particular we note the following:

1) A workable definition must have the property that the degree of controllability is zero if and only if the system is uncontrollable.
2) The control effort must be standardized or restricted in some way, since the degree of controllability must in some manner indicate how much control authority is obtained for a given amount of control effort.
3) The control objective, such as in the regulator problem or the tracking problem, should influence the definition.
4) The definition should depend on how much time T is allowed to accomplish the control objective.

The original definition in [2] and [4] addresses these issues by considering saturation constraints on the controllers ($|u_i| \leq 1$ after suitable normalization) and considering the control objective to be to return the state to the origin after a disturbance, i.e. the regulator problem which is appropriate for shape control of an antenna spacecraft. Controllability requires the existence of a control function which can transfer any initial state to any final state in finite time. With the regulator control objective, the degree of controllability should be related to the size or shape of the volume of initial conditions (or states resulting from disturbances) which can be returned to the origin in a prescribed amount of time T using the standardized control effort. References [2] and [4] took some comfort in the fact that for pure modal systems without damping the parameter T reduces to a scale factor in the resulting degree of controllability, provided T is long relative to the period of the slowest mode of structural vibration, and hence did not appear to play a fundamental role.

In an uncontrollable system the volume of the initial states described above will degenerate in the sense that there will be at least one direction in the state space for which initial conditions cannot be returned to the origin. For a controllable system which is nearly uncontrollable, only initial conditions which are very near the origin in the above mentioned direction can be returned to the origin in time T using the bounded controls. Hence, the degree of controllability was based on the minimum distance from the origin to the boundary of this initial condition region. The coordinates of a state space will usually have different physical dimensions, which suggests the need for a normalization to make a unitless set of coordinates. But the need for a normalization of the space is more fundamental, since a ranking of the degree of controllability of two systems will depend on comparison of distances in different directions, and hence the normalization must be such that the need for recovery from an initial condition of unit displacement from the origin be considered the same for all directions from the origin.

The degree of controllability definition in [2,4] can then be summarized as follows. Let the normalized state equations be

$$\dot{x}(t) = Ax(t) + Bu(t) \qquad |u_i| \leq 1 \tag{1}$$

Definition 1: The recovery region for time T for normalized system (1) is the set

$$\mathcal{R} = \{x(0) |\, \exists u(t),\; t \in [0,T],\; |u_i(t)| \leq 1 \forall i,\; s.t. x(T) = 0\}$$

Definition 2: The degree of controllability in time T of the $x = 0$ solution of normalized system (1) is defined as

$$\rho = \inf_{x(0) \notin \mathcal{R}} \|x(0)\| \tag{2}$$

where $\|\cdot\|$ represents the Euclidean norm.

The Basic Structure of a Regulator D.O.C. Definition

By examining the above definition and the thinking that led to it, one can extract a basic structure that can be used to generate other possible concepts of a degree of controllability (D.O.C.). This has been done in detail in [6]. The corresponding dual concepts of the degree of observability are also considered, together with the combination of the associated degree of controllability and degree of observability in balanced forms.

For the purposes of this paper we only need to consider the following four steps in making a degree of controllability definition:

(i) Normalize the state space and the control vector.

(ii) Specify a performance criterion.

(iii) Define a recovery region.

(iv) Specify a scalar measure of the recovery region (one which reduces to zero when the system is uncontrollable) and this is the degree of controllability.

Above, the recovery region was the set of states that could be returned to the origin in time T with the bounded controls. Its boundary is then the set of points whose time optimal control to the origin uses time T–hence, buried in the development of this degree of controllability is a time optimal performance criterion. The value of this criterion, i.e. time T, determined the definition of the recovery region. The scalar measure of the recovery region used was the minimum distance from the origin to the boundary, but other possible scalar measures include the volume of the region or a combination of these two.

The Range of Possible D.O.C. Definitions

Having recognized the time optimality implicit in the above definition, it is clear that there are many possible definitions associated with different performance criteria. These have been investigated in the references. Besides [6] which gives an overall understanding of the possibilities, [7] develops the energy optimal degree of controllability, while [8] and [9] consider a fuel optimal performance criterion. Reference [10] develops still other concepts in which certain possibilities are unified within a framework based on defining gain measures. The use of the usual quadratic cost criterion is considered in [6], but it is shown to fail to satisfy the requirement of a D.O.C. definition that it give zero whenever the system is uncontrollable, and is therefore discarded except when modified to include a fixed endpoint and a finite time interval.

Reference [11] moves out of the framework developed here, that puts emphasis on response to initial conditions, and considers concepts of the degree of disturbance rejection for classes of disturbances. Also, when tracking or maneuver problems are considered instead of regulator problems, the definitions must change, and this is discussed in [12] and [7].

3. The Problem Addressed

In order to compute the time optimal degree of controllability of Definition 2 one needs a method of computing the minimum distance to the boundary of the recovery region of Definition 1. References [2] and [3] develop an approximate method which is simple to use, especially for modal systems ([13] develops a more complicated, but more precise computational procedure, and an improved version is to appear). The approach finds the distance to the boundary of the recovery region in n carefully chosen directions which are guaranteed to show lack of controllability when it exists. These directions correspond to the real eigenvectors of the n dimensional system matrix together with the real parts and the imaginary parts of complex eigenvectors. (Things are more complicated for repeated eigenvalues, see [3] and [5].)

For large values of T the D.O.C. takes on a particularly simple form, not only for the time optimal case but also for fuel and energy as well (see Eq. 5 and Ref. [7]), when applied to modal systems. These approximations may be very useful for placement of actuators in such applications as chemical process control, or shape control of ground based structures that are, for example, simply supported. But when used on the attitude and shape control of flexible spacecraft which have free boundary conditions, the time optimal degree of controllability with its bounded actuators exhibits the following behavior:

(i) The optimal actuator locations are always at the ends or corners of the structure where the modal displacements and modal slopes are the largest.

(ii) When only two actuators are used to control a free-free beam, there is no difference between placing one actuator at each end, and placing both actuators at the same end. This property generalizes to more actuators, and to more complicated structures.

The same results are observed for energy optimal degree of controllability (which will be demonstrated in the next section), and for the approximate fuel optimal degree of controllability. However, there is a possible difference in the fuel optimal case in that it can occur that once a sufficient number of actuators are placed optimally by the fuel optimal D.O.C., any additional actuators may be placed anywhere on the structure (including the ends or corners) without changing the degree of controllability.

Several situations can be hypothesized for which the properties (i) and (ii) above seem to be intuitively unacceptable:

1) If the time allowed to achieve the control is short, an actuator on the perimeter may not be very effective in controlling a disturbance at the center, considering the wave propagation time in the structure.

2) If there is considerable damping, the corrective signal may be very attenuated by the time it travels from the perimeter of the structure to a disturbance at the center, suggesting a more even distribution of actuators.

A previous paper [14] looked for possible parameters which could be adjusted in order to have the degree-of-controllability concepts produce results consistent with the above two points. The candidates considered as possible ways to produce optimal actuator positions that are somewhat distributed throughout the structure were:

1) The time for control T, since short times would emphasize the delay from wave propagation.

2) The amount of damping considered in the structure, which could be either modal damping, or nonmodal damping which further dissipates the energy in any given mode.

3) The choice of the coordinates of the state space used for the normalization. If the state space contains a physical coordinate representing displacement at one point in the structure, and the normalization is chosen to make that coordinate all-important, then one would expect the best actuator location to be that point in the structure.

Reference [14] used the approximate degree of controllability computation method in [2-5], and investigated the above points. In order to eliminate time T as a factor, analytically evaluated integrals were used to find the approximate time optimal D.O.C. for any T. Arbitrary normalization matrices were carried throughout the development, with the normalization being made in any chosen coordinate system before transformation to modal coordinates. The following result was obtained:

Theorem: Given system (1) which is expressible in second order modal coordinates with all distinct non-zero eigenvalues for the system matrix A. If there exists a choice of actuator locations which simultaneously maximizes the magnitude of all elements of the B matrix, then that choice will maximize the approximate time optimal degree of controllability [2], irrespective of the time T allowed, the choice of coordinates, the normalization, or the modal damping of the system.

Large flexible spacecraft with free boundary conditions, for example, rods, beams, flat plates, and truss structures, exhibit the property that the maximum deflection and maximum slope of every mode is at the extremity of the structure, and therefore satisfies the assumptions of the theorem for placement of force or torque actuators. The theorem can be extended to include rigid body modes as well.

Hence, none of the suspected ways of making the approximate time optimal degree of controllability give optimal actuator locations distributed throughout the structure, did in fact produce the desired result. This leaves one main suspect, the fact that the computation method is still approximate in the sense that it finds the minimum distance to the boundary of the recovery region among n carefully chosen directions, but this does not necessarily produce the actual minimum distance. This paper addresses this last question, by studying the energy optimal degree of controllability as a function of T, since this definition of the degree of controllability is the only one which can be calculated exactly without any approximation.

One might note that the failing of the approximate time optimal degree of controllability computation for this class of problems is somewhat of a singular situation, since in most problems one would not be able to find an actuator location that simultaneously maximizes all the elements of the B matrix as required by the above theorem.

4. Energy Optimal D.O.C. Definition and Computation for Modal Systems

Using the basic structure of D.O.C. definitions given above, the energy optimal D.O.C. can be written as:

Definition 3: The recovery region for time T and energy E_c for normalized system (1) (with control saturation limits removed) is the set

$$\mathcal{R}_e = \{x(0) | \, \exists u(t), \; t \in [0,T] \text{ s.t. } J(u(\cdot),T,0) \leq E_c, \; x(T) = 0\}$$
$$J(u(\cdot),T,0) = 1/2 \int_0^T u^T R u dt, \quad R > 0$$

Definition 4: The energy optimal degree of controllability for time T and energy E_c for normalized system (1) is defined as

$$\rho_e = \inf_{x(0) \notin \mathcal{R}_e} \|x(0)\| \tag{3}$$

where $\|\cdot\|$ represents the Euclidean norm.

The reader may note that this special case of a degree of controllability definition is one that has appeared previously, in slightly altered form, without the benefit of the systematic framework discussed here. References [15] and [16] both use this general concept, although a reachable region is used in place of the recovery region used here. The recovery region is required from physical considerations in our actuator placement problem. Equation (5) below, and the discussion which follows it, give results in which the overall framework of degree of controllability definitions presented here is seen to put the above definition, together with its predecessors in [15] and [16], into a new perspective.

The energy optimal control to the origin [7] and the associated recovery region are expressed in terms of a controllability gramian

$$u(t) = -R^{-1}B^T e^{-A^T t} W^{-1}(T,0)x_o$$
$$\mathcal{R}_e = \{x_o | \tfrac{1}{2} x_o^T W_c^{-1}(T,0) x_o \leq E_c\}$$
$$W_c(T,0) = \int_0^T e^{-At} B R^{-1} B^T e^{-A^T t} dt$$

The recovery region is an ellipsoid whose shortest axis length is the degree of controllability which is given in terms of the eigenvalues of W_c as

$$\rho_e = \min_{i=1,2,\cdots,n} [2E_c \lambda_i (W_c(T,0))]^{1/2} \tag{4}$$

One way to compute this is to use a different controllability gramian $W(T,0)$ whose integrand contains $T - t$ in place of the $-t$ in the exponentials. This gramian can be computed from a Liapunov differential equation

$$\dot{W}(t,0) = AW(t,0) + W(t,0)A^T + BR^{-1}B^T$$

and is related to $W_c(t,0)$ by $W_c(t,0) = exp[-At]W(t,0)exp[-A^T t]$. This approach is convenient in that it involves use of standard computer programs, but in the case of modal systems we will generate a different method which does not require solution of a differential equation, but rather uses analytical evaluations of certain integrals, and this alternative method requires much less computation than the above when the number of modes is large.

Consider an undamped system of modal equations

$$\ddot{\eta}_i + \omega_i^2 \eta_i = \Gamma_i^T u \qquad i = 1, 2, \cdots, (n/2)$$

(although with a small increase in complexity we could extend the approach to include modal damping). The normalized state variables are chosen as $x_{2i-1} = \eta_i / N_i$, and $x_{2i} = \dot{\eta}_i / \bar{N}_i$ for rigid body modes, and $x_{2i} = \dot{x}_{2i-1} / \omega_i$ for vibration modes. Letting $R = diag[r_1, r_2, \cdots, r_m]$, we can write the gramian $W_c(T,0)$ in terms of two-by-two partitions of the form

$$W_{ij} = \frac{(\Gamma_i^T \Gamma_j)_r}{N_i N_j \omega_i \omega_j} \int_0^T \begin{bmatrix} sin\ \omega_i t\ sin\ \omega_j t & sin\ \omega_i t\ cos\ \omega_j t \\ cos\ \omega_i t\ sin\ \omega_j t & cos\ \omega_i t\ cos\ \omega_j t \end{bmatrix} dt$$

when both i and j are vibration modes. The factor $(\Gamma_i^T \Gamma_j)_r$ is defined as $\Sigma_{k=1}^m \Gamma_{ik} \Gamma_{jk} / r_k$ with Γ_{ik} being the kth element of Γ_i. When i and j are both rigid body modes

$$W_{ij} = (\Gamma_i^T \Gamma_j)_r \int_0^T \begin{bmatrix} \frac{1}{N_i N_j} t^2 & \frac{-1}{N_i \bar{N}_j} t \\ \frac{-1}{N_j \bar{N}_i} t & \frac{1}{\bar{N}_i \bar{N}_j} \end{bmatrix} dt$$

and when i is a vibration mode and j is a rigid body mode,

$$W_{ij} = (\Gamma_i^T \Gamma_j)_r \int_0^T \begin{bmatrix} \frac{1}{N_j N_i \omega_i} t\ sin\ \omega_i t & \frac{-1}{\bar{N}_j N_i \omega_i} sin\ \omega_i t \\ \frac{-1}{N_j N_i \omega_i} t\ cos\ \omega_i t & \frac{1}{\bar{N}_j N_i \omega_i} cos\ \omega_i t \end{bmatrix} dt$$

and the integrand is transposed when i is the rigid body mode. The above partitions of W_c contain 13 types of integrals, each of which can be evaluated analytically allowing one to obtain $W_c(T,0)$ analytically instead of by numerical integration of the Liapunov equation, together with the state transition matrix evaluation and multiplication, of the more general method.

The above analytical expressions giving $W_c(T,0)$ were used in a computer program to evaluate the optimal actuator locations as a function of T. The program evaluated the 13 types of integrals associated with all types of mode products in the gramian, then evaluated $(\Gamma_i^T\Gamma_j)_r$ for the prescribed control weighting R and any chosen actuator locations. This is assembled to form $W_c(T,0)$ from which ρ_e is determined by an eigenvalue routine. This can be used to evaluate the effect of changing actuator locations, changing the time for control T, and changing the normalizations N_i and R as desired.

Note that the result quoted in the previous section, i.e. that the optimal actuator locations are at the corners of a flexible spacecraft when T is large and the time optimal D.O.C. actuator placement criterion is used, can be seen to hold for the energy optimal criterion as well. Considering shape control only, so that only the first type of W_{ij} above is needed, and assuming that ω_i and ω_j are not comensurate for any $i \neq j$, then the associated products of the trigonometric functions will average to zero. However, the diagonal terms in $W_c(T,0)$ involve the square of the sine or cosine, and for large T will approach the value $\frac{1}{2}T$ times the coefficient in front of the integral. Therefore

$$\rho_e \cong \sqrt{E_c T}\,\min_i\,[||\Gamma_i||_{r2}/(N_i\omega_i)] \tag{5}$$

where $||\Gamma_i||_{r2}$ is the weighted L_2 norm $(\Sigma_{k=1}^m \Gamma_{ik}^2/r_k)^{1/2}$. Since all elements of Γ_i are maximized for every mode when the actuator is placed at the end, this position is found to be optimal. As a point of interest, the time and fuel optimal degrees of controllability are found to yield a similar form for large T with the coefficient in front of the minimization changed, and the norm changed to a weighted L_1 norm and a weighted L_∞ norm, respectively, in place of the weighted L_2 norm.

5. Optimal Actuator Locations as a Function of T for Energy Optimal D.O.C.

The optimal placement for two torque actuators on a free-free beam was computed for various values of time-for-control T, using the energy optimal D.O.C. criterion. Planar attitude and shape control of the beam was treated, so that the first mode is rigid body rotation, and there are four flexible body modes considered. The frequencies of the beam were normalized so that the modulus of elasticity E, the moment of inertia I, and the mass of the beam M, are all unity as is the length of the beam L. Then the angular frequencies of the four bending modes are π^2 times the square of 1.506, 2.500, 3.500, and 4.500 [17]. Starting with the rigid body rotation which has one node, each mode has the samenumber of nodes as the mode number. To avoid an uncontrollable system at least one of the actuators for each mode must be located at a point where the mode shape derivative is nonzero.

The actuator weighting matrix R in the energy optimality criterion was taken as the identity matrix. The position and velocity normalizations for the rigid body mode was set to unity, while the flexible body mode normalizations were scaled approximately as $1/\omega^2$ and were 0.2, 0.025, 0.007, and 0.002 respectively. This reflects the decreased amplitude of higher frequency modes for the same energy per mode, and also reflects a choice of relative importance of flexible vs. rigid body mode control (how this choice is best made is still a question for research).

The mode shapes of the beam come from solving the beam equation

$$d^4Y(x)/dx^4 - \beta^4 Y(x) = 0\ ;\ \beta^4 = \omega^2 m/EI$$
$$d^2Y/dx^2(0) = d^2Y/dx^2(L) = 0$$
$$d^3Y/dx^3(0) = d^3Y/dx^3(L) = 0$$

which dictates that β satisfy $cos\beta L\ cosh\beta L = 1$ to obtain nontrivial solutions, and this produces the frequencies cited above. The mode shapes, excluding the transverse displacement mode which is not part of the attitude and shape control problem, are

$$\begin{aligned} Y_1(x) &= A_1(x - L/2) \\ Y_r(x) &= A_r[(cos\beta_r L - cosh\beta_r L)(sin\beta_r x + sinh\beta_r x) \\ &\quad -(sin\beta_r L - sinh\beta_r L)(cos\beta_r x + cosh\beta_r x)] \end{aligned}$$

The derivatives of these shape functions determine the input influence coefficients Γ_i when evaluated at each torque actuator location. The values of A_1 and A_r have been arbitrarily set to the value that makes the maximum deflection of the mode unity.

For simplicity, as well as for determination of global optimality, no search routine was used to locate optimal (or local optimal) actuator locations, but rather the set of possible actuator locations was restricted to 21 evenly spaced positions on the beam, producing 441 possible combinations of two actuator positions (allowing both actuators to be at the same position).

When the time for control T was set to values $T >> 1$ (if the units are chosen so that angular frequencies ω_i are in rad/sec, then the time units are seconds), we obtained the same results as the approximate time optimal degree of controllability, i.e. the optimal actuator locations are at the ends of the beam, and there is no indication of a difference in control authority when both actuators are placed at the same end and when one actuator is placed at each end.

When T was set to 1 unit, the optimal actuator locations remain at the ends of the beam, but now there is a $1\frac{1}{2}\%$ improvement in the energy optimal degree of controllability when one actuator is at each end when compared to having both actuators at the same end. Some symmetry was lost, but without any reason to think otherwise, one might attribute the difference to numerical computation effects.

The value of T was next set to 0.5. Note that the period of the slowest mode is 0.28, so that this T is still several periods of oscillation, while the other modes have periods of .102, .052, and .031. The best actuator positions are still at the ends of the beam, but this time putting one actuator at each end is 37% better than putting both at the same end. When T reaches .25 the difference has increased to a factor of 5 to 1.

When $T = .1$ the best location of the actuators is no longer the ends of the beam, but at 0.15 and 0.80 instead. Actually there are four equivalent solutions, one at .2 and .85 and then the same two solutions with the identical actuators interchange. Three percent of the solutions in the 441 element array are better than having one actuator at each end. At $T = 0.05$ the best location has moved to 0.15 and 0.55 although it is only somewhat better than the 0.15 and 0.80 locations. The locations in between are significantly worse. One should note that the propagation speed of disturbances at the fundamental frequency is 22.43 beam lengths per second (twice this rate applies to the group velocity) so that propagation of a corrective signal from the actuators to all points of the beam is still possible with these actuator locations.

When $T = .01$ the eigenvalue solver exhibits ill-conditioned behavior which should be related to the fact that a harmonic disturbance at the fundamental frequency will travel only $.22L$ in .01 sec. indicating that there is no pair of actuator locations such that corrective action from an actuator can propagate to all parts of the beam in $T = .01$ sec. The wave propagation information says that the full distributed parameter system is not controllable to the origin in $T = .01$, but all finite dimensional modal representations are controllable in any period of time, no matter how small, and hence an exact computation of the D.O.C. should give some nonzero, but very small, answer for this T.

6. Conclusions

Considerable effort was expended in [2-5] to develop a very simple way to compute the time optimal degree of controllability, and this was extended to fuel optimal D.O.C. in [8,9]. This method may be sufficiently precise for many classes of problems, but it is seen not to be good enough for the problem of shape control of flexible spacecraft when one wants to accomplish control within a length of time equal to a few periods or less of the lowest frequency vibration mode. (We assume that the true D.O.C. for time or fuel would show the same expected characteristics which were demonstrated here in the case of the energy optimal D.O.C.)

This indicates that better methods of computation are needed for time and fuel optimal D.O.C., and work is in progress on this problem pursuing the general approach in [13]. Until these results are available, we conclude that the only computationally practical approach to actuator placement in the large flexible spacecraft shape control problem using D.O.C. concepts, is the energy optimal D.O.C. This paper developed an efficient method to compute this using analytical evaluations of integrals in order to eliminate the numerical integration of the Liapunov equation, and the state transition matrix, and in order to alleviate the dimension problem in many mode systems.

References

[1] S.W. Sirlin, R.W. Longman and J.N. Juang, "Identifiability of Conservative Linear Mechanical Systems," *The Journal of the Astronautical Sciences*, Special Issue on Structural Modeling and System Identification of Flexible Space Structures, pp. 95-118, January-March 1985.

[2] C.N. Viswanathan, R.W. Longman and P.W. Likins, "A Degree of Controllability Definition: Fundamental Concepts and Application to Modal Systems," *Journal of Guidance, Control and Dynamics*, vol. 7, pp. 222-230, March-April 1984.

[3] C.N. Viswanathan, R.W. Longman and P.W. Likins, "A Degree of Controllability Definition–Repeated Eigenvalue Systems and Examples of Actuator Placement," *The Journal of the Astronautical Sciences*, to appear.

[4] C.N. Viswanathan, R.W. Longman and P.W. Likins, "A Definition of the Degree of Controllability–A Criterion for Actuator Placement," Proceedings of the Second VPI&SU/AIAA Symposium on Dynamics and Control of Large Flexible Spacecraft, Blacksburg, Va., June 1979.

[5] C.N. Viswanathan and R.W. Longman, "The Determination of the Degree of Controllability for Dynamic Systems with Repeated Eigenvalues," *Advances in the Astronautical Sciences*, vol. 50, Part II, pp. 1091-1111, 1983.

[6] R.W. Longman, S.W. Sirlin, T. Li and G. Sevaston, "The Fundamental Structure of Degree of Controllability and Degree of Observability," AIAA/AAS Astrodynamics Specialist Conference, Paper 82-1434, San Diego, Calif., Aug. 1982.

[7] R.W. Longman and K.T. Alfriend, "Energy Optimal Degree of Controllability and Observability for Regulator and Maneuver Problems," Proceedings of the Sixteenth Annual Conference on Information Sciences and Systems, Princeton, N.J., March 1982.

[8] R.A. Laskin, R.W. Longman and P.W. Likins, "A Definition of the Degree of Controllability for Fuel-Optimal Systems," Proceedings of the Third VPI&SU/AIAA Symposium on Dynamics and Control of Large Flexible Spacecraft, Blacksburg, Va., June 1981.

[9] R.A. Laskin, R.W. Longman and P.W. Likins, "Actuator Placement in Modal Systems Using Fuel Optimal Degree of Controllability Concepts," Proceedings of the 20th

Annual Allerton Conference on Communication, Control and Computing, Monticello, Ill., Oct. 1982.

[10] G. Sevaston and R.W. Longman, "Gain Measures of Controllability and Observability," *International Journal of Control*, vol. 41, No. 4, pp. 865-893, 1985.

[11] R.A. Laskin, R.W. Longman and P.W. Likins, "A Definition of the Degree of Disturbance Rejection," AIAA/AAS Astrodynamics Specialist Conference, AIAA/AAS Paper No. 82-1466, San Diego, Calif., Aug. 1982.

[12] R.W. Longman and K.T. Alfriend, "Actuator Placement from Degree of Controllability Criteria for Regular Slewing of Flexible Spacecraft," *Acta Astronautica*, vol. 8, pp. 703-718, July 1981.

[13] G. Klein, R. Lindberg and R.W. Longman, "Computation of a Degree of Controllability via System Discretization," *Journal of Guidance, Control and Dynamics*, vol. 5, pp. 583-588, Nov.-Dec. 1982.

[14] R.E. Lindberg and R.W. Longman, "The Influence of Time and Normalization on Actuator Placement by Degree of Controllability," *Advances in the Astronautical Sciences*, Proceedings of the AAS/AIAA Astrodynamics Specialist Conference, Lake Placid, N.Y., August 1983.

[15] R.E. Kalman, Y.C. Ho and K.S. Narendra, "Controllability of Linear Dynamical Systems," *Contributions to Differential Equations*, vol. 1, No. 2, pp. 189-213.

[16] B.C. Moore, "Principal Component Analysis in Linear Systems: Controllability, Observability and Model Reduction," *IEEE Transactions on Automatic Control*, vol. AC-26, No. 1, pp. 17-32, Feb. 1981.

[17] L. Meirovitch, Analytical Methods in Vibrations, MacMillan, pp. 161-166, 1967.

Direct Model Reference Adaptive Control in Infinite-Dimensional Hilbert Space

John Ting-Yung Wen
Jet Propulsion Laboratory
California Institute of Technology
Pasadena, CA 91109
and
Mark J. Balas
Department of ECSE
Rensselaer Polytechnic Institute
Troy, NY 12181

Abstract

Though great advances have been reported in adaptive control of single-input/single-output (SISO) systems and some multi-input/multi-output (MIMO) systems, some precise a priori structural information of the plant (at least the order) is needed for most of the methods proposed. This is unsatisfactory in some applications because of unmodeled dynamics and structure and noisy operating environment. In fact, in many high performance control system designs, as for example the control of large space structures, the distributed nature of the plant must be taken into account. These distributed parameter systems are frequently modeled by partial differential equations. Therefore, they must be analyzed in the appropriate infinite-dimensional state space.

A particular approach based on model reference adaptive control (MRAC) with command generator tracker (CGT) concepts, adopts a set of assumptions that are not dependent on the system dimension. The method has been applied successfully to some finite-dimensional systems and shows promise for the infinite-dimensional state space generalizations as well. In this paper, the scheme is modified in order to make the transition of this theory from finite dimensions to the infinite-dimensional Hilbert Space, mathematically rigorous. Four main technical difficulties for such a transition are discussed: coercivity of the solution in the Lyapunov equation, application of the Invariance Principle in infinite dimensions, the strict positive realness condition, and the existence and the uniqueness of solutions. We investigate some of the ramifications and the remedies of these issues. Robustness with respect to state and output perturbations and parameter variations is also discussed to demonstrate the practicality of this controller operating under realistic conditions.

1. Introduction

As stated in [1], the control of distributed parameter systems (DPS) pose a real challenge to control engineers. From the theory side, functional analytic tools have to be employed to tackle mathematical issues normally taken for granted in lumped parameter systems (LPS). From the practical viewpoint, the control of an infinite-dimensional system represents a tremendous load on computation and instrumentation. In many applications, the DPS setting arises naturally, such as in chemical processes, aerospace systems, and magneto-hydrodynamic systems. These days, with the advance of space transportation systems, control of large space structures (LSS) has attracted increasing attention. LSS involve a high level of mechanical flexibility, and therefore, should be considered as DPS [2]. In many cases, due to the lack of sufficient knowledge of the system, for example, insufficient modal data in LSS, an effective non-adaptive controller is difficult to find. Adaptive control of DPS therefore, is an area that merits investigation.

Adaptive control can be classified into two types: indirect and direct adaptive control. The indirect scheme identifies the plant with an adaptive identifier and then adjusts the controller parameters based on the identified model. The direct scheme adjusts the controller gain directly. The former approach was adopted in [3,4], and the latter in [5,6]. Results with explicit consideration of the infinite-dimensional state space have been scarce. In [7], sufficient conditions have been given to ensure the stability of an adaptive identifier using the reduced order model (ROM) obtained via projection method. Utility of this result in indirect adaptive control is limited by the multiequilibria issue and the input richness condition. Scheme also exist [8] for identifying DPS with a finite unknown parameter set. Several algorithms to perform self-tuning regulation for DPS with finite unknown parameter set have been proposed [9,10]. Only simulation results but not stability proofs have been given. In [11,13], some of the possible directions of investigation and the main areas of difficulty were surveyed.

The CGT theory was first proposed in [14] for the model following problem with known parameters. An adaptive control algorithm was subsequently developed in [15,16] using the CGT theory under a plant/model structural matching condition. Strict positive realness of the closed-loop plant is shown to imply asymptotic tracking. This condition is relaxed to positive realness in [17]. The potential generalization of this method to infinite dimensions has been discussed in [11,12]. The appeal of this method is that it does not require the reference model to be of the same order as the plant and the knowledge of the plant order is not needed; in fact, the plant order can even be infinite. In other words, the plant dynamics can be governed by a partial differential equation or delay differential equation. The reason for considering an algorithm which is not critically dependent on the assumption of plant order is two-fold. First, all systems are by nature distributed, and hence infinite-dimensional. Secondly, it has been shown in [18] and [19] that unmodeled dynamics can cause instability even for simple systems.

This method is naturally suitable for application to high-order systems since its main structure is the adjustment of a low-order feedback gain matrix. The drawback is that instead of the dimensionality requirement, there are other conditions to be satisfied. While the generalizations of the CGT technique to the infinite-dimensional systems is not straight forward, it is very promising. There are four difficulties that are encountered: 1. The solution of the Lyapunov equation may not be coercive. 2. The Invariance Principle cannot be immediately applied. 3. Strict positive realness is never attainable without positive feedthrough. 4. Existence and uniqueness of a continuous solution in closed loop operation must be shown. The objective of this paper is to delineate the difficulties and their partial solutions, in applying this method to infinite-dimensional systems in Hilbert Space.

The paper is organized as follows. Section 2 states the problem formulation and the mathematical preliminaries. Section 3 reviews the adaptive control with CGT concept in finite dimensions and points out the problem areas in the generalization to the infinite dimensions. Sections 4 to 7 discuss each of these problem areas in detail and state the proposed solution. Section 8 investigates the robustness of the scheme with respect to state and output perturbations and parameter variations. Finally, the concluding remarks are given in section 9. This paper is based on [21] where further details and mathematical proofs can be found.

2. Problem Formulation and Mathematical Preliminaries

The plant under consideration is modeled in the state space form

$$\begin{aligned} \dot{x}(t) &= Ax(t) + Bu(t); \quad x(0) = x_0 \in D(A), \quad t \geq 0 \\ y(t) &= Cx(t) \end{aligned} \tag{1}$$

where $x(t) \in X$, is the state vector, X is a Hilbert Space with inner product $< \cdot, \cdot >$ and corresponding norm $\|.\|$, $D(A)$ is the domain of A, dense in X, $A : D(A) \rightarrow X$ may be an unbounded operator and A generates a C_o-semigroup of bounded operators $U(t)$ on X, $B : \mathcal{R}^m \rightarrow X$ is the bounded linear input operator with rank m, $C : X \rightarrow \mathcal{R}^m$ is the bounded linear output operator with rank m, $u(t) \in \mathcal{R}^m$ is the input vector, and $y(t) \in \mathcal{R}^m$ is the output vector.

We consider the solution of (1) in the mild sense. We use the mild formulation because dynamic equations arising from physical situations are usually in integral form and the condition for existence and uniqueness of solutions are weaker.

$$\begin{aligned} x(t) &= U(t)x_0 + \textstyle\int_0^t U(t-\tau)Bu(\tau)d\tau; \quad x(0) = x_0 \in X, \quad t \geq 0 \\ y(t) &= Cx(t) \end{aligned} \tag{2}$$

where $U(t)$ is the C_o-semigroup of bounded operator generated by A on X. The assumption of A generating a C_o-semigroup is a natural one, if we are considering a physically meaningful dynamic system.

If the control objective is regulation, our goal is to drive $y(t)$ to zero with a bounded control signal. If the control objective is model following, the goal is to drive $y(t) - y_m(t)$ to zero with a bounded control where $y_m(t)$ is the output of a reference model given by

$$\begin{aligned} \dot{x}_m(t) &= A_m x_m(t) + B_m u_m(t); \quad x_m(0) = x_{m0}, \quad t \geq 0 \\ y_m(t) &= C_m x_m(t) \\ x_m(t) \in & \ \mathcal{R}^{n_m}, u_m(t) \in \mathcal{R}^{p_m}, y_m(t) \in \mathcal{R}^m \end{aligned} \tag{3}$$

It should be noted that the only requirement on the reference model is that the model output has the same dimension as the plant output. The order of the model may be much smaller than the order of the plant. This feature is the greatest appeal of this method, since no large-dimensional or infinite-dimensional model needs to be constructed. Since the dimensions of X and $\mathcal{R}^{n_m}$ are not equal in general, we cannot create the error signal by directly subtracting x_m from x. We follow the CGT approach in [17], and assume there exists an intermediate system that is "ideal" in some sense:

$$\begin{aligned} \dot{\overline{x}}(t) &= A\overline{x}(t) + B\overline{u}(t); \quad \overline{x}(0) = \overline{x}_0, \quad t \geq 0 \\ \overline{y}(t) &= C\overline{x}(t) \end{aligned} \tag{4}$$

where $\overline{u}(t)$ is some input signal such that (4) has a unique mild solution $\overline{x}(t)$. In [17], (4) is required to satisfy the following exact model matching condition:

Assumption 1
There exist bounded, linear operators $S_{11}, S_{12}, S_{21}, S_{22}$ such that

$$\begin{bmatrix} \overline{x}(t) \\ \overline{u}(t) \end{bmatrix} = \begin{bmatrix} S_{11} & S_{12} \\ S_{21} & S_{22} \end{bmatrix} \begin{bmatrix} x_m(t) \\ u_m(t) \end{bmatrix} \tag{5}$$

and

$$\overline{y}(t) = y_m(t) \qquad \forall t \geq 0. \tag{6}$$

We have thus gotten around the problem of dimensionality mismatch between the plant and the reference model by assuming the existence of an ideal system (4), the state space of which is X but looks the same as the reference model from the output. The idea is similar to MRAC with a reference model of the same order as the plant in which such an ideal system is shown to exist that not only appears the same as the reference model from the output but in fact has the same transfer function [20]. Clearly, we do not have the freedom to make such a strong assumption when $X \neq \mathcal{R}^{n_m}$. The following error system can be created:

$$\begin{aligned} e(t) &= x(t) - \overline{x}(t) \\ \dot{e}(t) &= Ae(t) + Bu(t) - B\overline{u}(t); \quad e(0) = x_0 - \overline{x}_0, \quad t \geq 0 \\ e_y(t) &= y(t) - \overline{y}(t) = Ce(t) \end{aligned} \tag{7}$$

The control objective for MRAC is to find a bounded control signal that drives $e_y(t)$ to zero asymptotically and keeps $e(t)$ uniformly bounded. This is equivalent to our previously stated objective since $e_y(t) \to 0$ implies $y(t) \to y_m(t)$.

The CGT condition is difficult to verify since the ideal trajectory and control, $\overline{x}$ and $\overline{u}$ respectively, are not actually available. A sufficient condition is given in the following proposition:

Proposition 1 [7]
Assume there exists bounded linear operators $S_{11}, S_{12}, S_{21}, S_{22}$, that satisfy

$$\begin{bmatrix} A & B \\ C & 0 \end{bmatrix} \begin{bmatrix} S_{11} & S_{12} \\ S_{21} & S_{22} \end{bmatrix} = \begin{bmatrix} S_{11} & 0 \\ 0 & 1 \end{bmatrix} \begin{bmatrix} A_m & B_m \\ C_m & 0 \end{bmatrix} \tag{8}$$

Suppose $u_m(t)$ is constant for $t \geq 0$. Then Assumption 1 is satisfied.

The assumption that u_m is constant can be immediately relaxed to piecewise constant, since the derivative of piecewise constant u_m is zero almost everywhere with respect to the Lebesgue measure. In [22], the condition on $u_m(t)$ is further relaxed to the output of a linear filter with piecewise constant input. (8) may be an easier condition to check than Assumption 1, but when the knowledge of A, B, C is absent or fuzzy, it is still difficult to verify. Further manipulation of (8) gives the condition

$$||S_{11}|| ||A_m|| < 1 \tag{9}$$

Since $||S_{11}||$ is fixed for a given plant, (9) means that if $||A_m||$ is sufficiently small, Assumption 1 is satisfied. However, this may contradict the practical design requirement. The discussion so far may give the impression that the CGT condition is very stringent. However (8) is by no means necessary, as demonstrated by the following example. Let

$$A = \begin{bmatrix} -1 & 0 \\ 3 & 1 \end{bmatrix} \quad B = \begin{bmatrix} 1 \\ 0 \end{bmatrix} \quad C = [1 \; 1] \tag{10a}$$

$$A_m = -2 \quad B_m = 1 \quad C_m = 1. \tag{10b}$$

(8) cannot be satisfied since we simultaneously require $S_{11}^{(1)} + S_{11}^{(2)} = 0$ and $S_{11}^{(1)} + S_{11}^{(2)} = 1$. However, it can be shown that Assumption 1 is satisfied [21]. A closer look at the CGT condition reveals [21] that it is not difficult to satisfy provided the state space dimension of the reference model is not bigger than the dimension of a completely controllable and completely observable subsystem of the plant under piecewise constant u_m. In [23], it has been shown that a stabilizable, detectable (possibly infinite-dimensional) system can be decomposed into

$$\begin{aligned} \dot{\xi} &= \begin{bmatrix} A_1 & 0 \\ 0 & A_2 \end{bmatrix} \xi + \begin{bmatrix} B_1 \\ B_2 \end{bmatrix} u \\ y &= [C_1 \; C_2] \xi \end{aligned} \tag{11}$$

where $\xi = P^{-1}x$ is a linear coordinate transformation, (A_1, B_1, C_1) is a finite-dimensional minimal system and A_2 generates an exponentially stable C_o-semigroup. The CGT condition can be justified as follows. Suppose the dimension of A_1 is the same as that of A_m, and A_m has distinct eigenvalues. Assume u_m is a constant input. Choose $u = G\xi_1 + v$, where G is a constant gain matrix, such that $\sigma(A_1 + B_1 G) = \sigma(A_m)$($\sigma(F)$ means the spectrum of an operator F). A further coordinate change gives

$$\begin{aligned} \dot{\eta} &= \begin{bmatrix} \Lambda_1 & 0 \\ 0 & \Lambda_2 \end{bmatrix} \eta + \begin{bmatrix} \beta_1 \\ \beta_2 \end{bmatrix} v \\ y &= [\gamma_1 \; \gamma_2] \eta \end{aligned} \tag{12}$$

where $\sigma(\Lambda_1) = \sigma(A_1 + B_1 G)$ and $\sigma(\Lambda_2) = \sigma(A_2)$. Choose $\eta_2(0) = -\Lambda_2^{-1}\beta_2 v$. Then $\eta_2 = -\Lambda_2^{-1}\beta_2 v$ for all $t \geq 0$. Therefore, y is a linear combination of $e^{-\lambda_i t}$, $\lambda_i \in \sigma(A_m)$, and a constant. The CGT condition can be satisfied with proper choice of v and $\eta_1(0)$ provided that y_m does not contain any modes that do not appear in y (i.e. those modes annihilated in the null space of γ_1). The condition on y_m may sometimes be removed by introducing an observer on the subsystem (A_1, B_1, C_1). This is due to the additional signals appearing in ξ_1, the dynamics of which is determined by $\sigma(A_1 + KC_1)$. We can select $\sigma(A_1 + KC_1) = \sigma(A_m)$ such that this extra signal may account for some of the missing modes. The piecewise constant condition on u_m can be replaced by u_m being generated by a linear filter with piecewise constant input. The dimensionality condition for CGT must then be relaxed to the dimension of the largest minimal subsystem of the plant exceeding the sum of the orders of the model and the filter.

The concept of positive realness plays an important role in the finite dimension theory. We define the positive realness in infinite dimensions via a frequency condition similar to that in the finite dimensions case.

Definition 1
Given a system

$$\begin{aligned} \dot{x} &= Ax + Bu \\ y &= Cx + Du \end{aligned} \tag{13}$$

where A, B, C are as defined in (1), D is a $\mathcal{R}^{m \times m}$ feedforward matrix. A constant $k \in \mathcal{R}$ is called the positive realness index $(PRI)_\Sigma$ for the quadruplet $\Sigma = (A, B, C, D)$ if

$$k = \sup\{d \in \mathcal{R} : Re < (D + C(j\omega I - A)^{-1}B)w, w > \ \geq d\|w\|^2, \ \ \forall w \in \mathcal{C}^m, \text{and all } \omega \in \mathcal{R}\} \tag{14}$$

$\Sigma = (A, B, C, D)$ is called strictly positive real if there exists $\delta > 0$ such that $(PRI)_\Sigma \leq \delta$ and A generates an exponentially stable C_o-semigroup.

The definition for the positive real system differs slightly from the finite-dimensional definition where A is only required to be stable. The definition of strictly positive real systems is stronger than the finite dimension counterpart. Here, a "uniform" lower bound above zero is needed. As will be seen in Section 6, this has serious implications. We also need the following definition on the output feedback matrix G.

Definition 2
A matrix $G \in \mathcal{R}^{m \times m}$ is called k-stabilizing for (A, B, C, D) if $(A + BGC)$ is exponentially stable and $(PRI)_\Sigma = k$, $\Sigma = (A + BGC, B, C, D)$. Clearly, if G is k–stabilizing for (A, B, C, D) with $k > 0, (A + BGC, B, C, D)$ is strictly positive real and G 0-stabilizing implies $(A + BGC, B, C, D)$ is positive real. If $dim\ X < \infty$, we call G 0^+-stabilizing if $(A + BGC, B, C, D)$ is strictly positive real (in the sense of definition 1).

We shall progressively weaken the assumption of the existence of 0^+-stabilizing G to 0-stabilizing G, and to $-d$-stabilizing G where D is a small positive constant.

For finite-dimensional systems, positive realness is closely related to the Lur'e equations via the Kalman-Yakubovich Lemma. We need a similar result in the infinite dimensional case also.

Lemma 1[24,25]
Given a real Hilbert Space X, consider the following dynamic system

$$\begin{aligned} \dot{x} &= Ax + Bu; \ \ x(0) \in X, u(\cdot) \in \mathcal{L}_2(\mathcal{R}^+; \mathcal{R}^m) \\ y &= Cx + Du \end{aligned} \tag{15}$$

where A, B, C are as defined in (1), $D \in \mathcal{R}^{m \times m}$. If (A, B, C, D) is strictly positive real, then there exists $P \in \mathcal{L}(X), P > 0$, $Q \in \mathcal{L}(X, \mathcal{R}^m)$, $W \in \mathcal{R}^{m \times m}$ and $\epsilon > 0$ sufficiently small, such that

$$(A^*P + PA)x = -(\epsilon I + Q^*Q)x, \quad \forall x \in D(A) \tag{16a}$$

$$B^*P = C - W^*Q \tag{16b}$$

$$W^*W = (D + D^*)/2 \tag{16c}$$

(A^* means the adjoint operator of $A, Q \in \mathcal{L}(X, \mathcal{R}^m)$ means $Q : X \to \mathcal{R}^m$ is a bounded linear operator, $\mathcal{L}(X) \equiv \mathcal{L}(X, X)$.)

In the proof of Lemma 1, the strict positive realness assumption is needed to assert the invertibility of an operator. This then leads to the existence of a unique minimizer of a linear-quadratic minimization problem, the solution of which is directly related to the Lur'e equations (16a)-(16c). Using this proof technique, we do not have the result for positive real systems as in the finite dimensional case. This then necessitates the modification of the adaptive controller as will be seen in section 5.

3. Adaptive Control In Finite-Dimensional Space

Let us consider the case $X = \mathcal{R}^n$. Before we attempt to investigate the adaptive control problem, we need to see how the problem can be solved when the parameters are known.

Suppose there exists a static output feedback gain $G \in \mathcal{R}^{m \times m}$ such that the eigenvalues of the closed looop system matrix $(A + BGC)$ all have negative real part, i.e., $(A + BGC)$ is strictly stable. Assume that when (A, B, C) in (1) are known, we can find G and S_{21}, S_{22} in (5). Then the following simple control law can be constructed as a linear combination of the available measurements:

$$u(t) = G(y(t) - y_m(t)) + S_{21}x_m(t) + S_{22}u_m(t) \tag{17}$$

Substituting (17) into (7) yields

$$\dot{e}(t) = Ae(t) + BG(y(t) - y_m(t)) + B(S_{21}x_m(t) + S_{22}u_m(t) - \overline{u}(t)). \tag{18}$$

The CGT conditions (5) and (6) can be applied to (18) and we obtain

$$\dot{e}(t) = (A + BGC)e(t).$$

By assumption, $(A + BGC)$ is strictly stable. Hence, in the case of regulation, $x(t) \to 0$, and $e(t) \to 0$ as $t \to \infty$ for model following. Therefore, our control objective is met. However, computation of G, S_{21}, S_{22} requires explicit knowledge of the plant and may often be very difficult. Therefore, we use the following adaptive version of (17):

$$u(t) = G(t)(y(t) - y_m(t)) + S_{21}(t)x_m(t) + S_{22}(t)u_m(t) \tag{19}$$

where $G(t), S_{21}(t), S_{22}(t)$ are the adaptive estimates of G, S_{21}, S_{22} respectively. Define

$$\begin{aligned} \Delta G(t) &= G(t) - G & (20a)\\ L(t) &= S(t) - S & (20b)\\ w(t) &= [x_m(t)^T \;\; u_m(t)^T]^T & (20c)\\ S(t) &= [S_{21}(t) \;\; S_{22}(t)] & (20d)\\ S &= [S_{21} \;\; S_{22}] & (20e) \end{aligned}$$

Substituting (19) into (1) yields the closed-loop dynamic equations:

$$\begin{aligned} \dot{e}(t) &= (A + BGC)e(t) + B\Delta G(t)e_y(t) + BL(t)w(t) \\ e_y(t) &= Ce(t) \end{aligned} \tag{21}$$

A stronger requirement is now placed on the output feedback gain G that $((A+BGC), B, C, 0)$ is strictly positive real. It is a stronger condition on G since the Lyapunov equation (16a) implies $(A + BGC)$ is strictly stable. To select an adaptive strategy that will stabilize (21) we use Lyapunov's Direct Method. Choose the quadratic Lyapunov function candidate.

$$V(e, \Delta G, L) = e^T Pe + tr[\Delta G \Gamma_1^{-1} \Delta G^T] + tr[L \Gamma_2^{-1} L^T] \tag{22}$$

where $P > 0$ follows from (16a) with A replaced by $A + BGC$. Let $A_c \triangleq A + BGC$. Suppose the following adaptive control strategy is used:

$$\dot{G}(t) = \Delta \dot{G}(t) = -\Gamma_1 e_y(t) e_y^T(t) \tag{23a}$$

$$\dot{S}(t) = \dot{L}(t) = -\Gamma_2 e_y(t) w^T(t) \tag{23b}$$

Then

$$\dot{V} = -||Qe||^2 - \epsilon||e||^2 \leq 0. \tag{24}$$

Hence, V is bounded. By using the extension of LaSalle's theorem [26] to time varying systems [27], $e \to 0$ and therefore, $e_y \to 0$ as $t \to \infty$. Furthermore, $G(t), S_{21}(t), S_{22}(t)$ are uniformly bounded. Therefore, the control objective is met. For regulation, only (23a) is needed.

A slight modification is made in [22] to weaken the requirement on G to $(A_c, B, C, 0)$ being positive real. The control signal (19) becomes

$$u(t) = G(t)(y(t) - y_m(t)) + S_{21}(t)x_m(t) + S_{22}(t)u_m(t) + K(y(t) - y_m(t)) \tag{25}$$

where K is a positive definite constant matrix. Using the same quadratic Lyapunov function candidate as in (22), $\dot{V}$ becomes ($\epsilon = 0$ in this case)

$$\dot{V} = -||Qe||^2 - e_y^T K e_y. \tag{26}$$

Since $K > 0$, the generalization of LaSalle's theorem implies $e_y(t) \to 0$ as $t \to \infty$.

In the generalization of this method to the infinite-dimenisonal Hilbert Space, we run into some difficulties. In the stability proof, a term like $< e, Pe >$ is included in the Lyapunov function. However, under the condition that Q is bounded, P in (16a) is coercive if and only if A generates a C_o-group [28]. For example, for hyperbolic PDE, the C_o-group assumption is satisfied, but for parabolic PDE, the C_o-group assumption does not hold. The coercivity of P is important since the norm equivalence between $||e||_1 \equiv (< e, Pe >)^{\frac{1}{2}}$ and $||e||$ is needed in the stability proof. In section 4, we will show that without the C_o-group condition, asymptotic stability can still be achieved sometimes. Without the assumption, only stability can be concluded in an enlarged space X_1.

In the stability analysis in finite dimensions, generalization of the LaSalle's Invariance Principle is employed to show the convergence of the state error e or the output error e_y. This is from the fact that $\dot{V}$ is only negative semidefinite and not negative definite. Bounded sets are precompact in $\mathcal{R}^n$. Therefore, bounded positive orbits, as concluded from $\dot{V}$ being negative semidefinite, are precompact. This implies that the positive limit set $\Omega(x_0)$ is nonempty. Since $\Omega(x_0)$ belongs to $\{x \in X : \dot{V}(x) = 0\}$, the desired convergence property follows. When $dim\ X = \infty, X$ is no longer locally compact. Hence the bounded positive orbits may not be precompact and the positive limit set may be empty. The convergence property can, therefore, no longer be concluded. In Section 5, we will see that a potential remedy directly motivates our modified control algorithm, under which, interestingly, the Invariance Principle is no longer needed.

In the linear portion of the error equation (21), the feedforward matrix $D = 0$. The strict positive realness assumption now reads

$$Re < C(j\omega I - A_c)^{-1}B)w, w > \geq \delta||w||^2 \quad \forall w \in \mathcal{C}^m, \ \forall \omega \in \mathcal{R} \tag{27}$$

for some $\delta > 0$. From [29],

$$(j\omega I - A_c)^{-1}x = \int_0^\infty e^{-j\omega t} U_c(t)x\, dt, \quad \forall x \in X \tag{28}$$

where $U_c(t)$ is the exponentially stable C_o-semigroup generated by A_c. From the fact $U_c(t)x \in \mathcal{L}^1(\mathcal{R}^+; X)$ and the Riemann-Lebesgue Lemma [30]

$$(j\omega I - A_c)^{-1}x \to 0 \text{ as } \omega \to \infty \tag{29}$$

Thus (29) means that the strict positive realness assumption can never be satisfied. Furthermore, Lemma 1 does not apply to the positive real system for $\epsilon = 0$ as in the $\mathcal{R}^n$ case and hence, the control law (25) is of no value. This motivates a further modification of the control algorithm in Section 6, where the strict positive realness requirement is weakened to almost positive realness.

All the proofs of stability would be futile if a unique continuous solution does not exist in the closed loop. In finite dimensions, a well known condition exists (see for example, [31]). Section 7 assures that in the infinite dimensions we do have the existence and uniqueness of closed loop solution.

4. Coercivity of the Solution of the Lyapunov equation

When A in the Lyapunov equation (16a) does not generate a C_o-semigroup and Q is a bounded linear operator, it has been shown in [28] that P is *not* coercive. In other words, P does not have a bounded inverse [32]. This can cause difficulty, especially in light of the modified controller given in section 2.3, since bounded $< e, Pe >$ does not imply bounded $\|e\|$. There are two possible ways around this: a) Let Q in (16a) be an unbounded operator and investigate conditions under which P is coercive when A does not generate a C_o-group or, b) work in the larger space $(X_1, \|.\|_1)$, the completion of $(X, \|.\|_1)$, where $\|x\|_1 \equiv < x, Px >$ and try to deduce a weaker type of stability. It can be easily shown [33] that $\|.\|_1$ is a norm and X_1 is a Hilbert Space with the inner product $< x, y >_1 \equiv < x, Py >$.

No general result is known for the first case. However, there is a special case where the idea can be readily applied. Consider the following system with colocated sensors and actuators:

$$\begin{aligned} \dot{x}(t) &= Ax(t) + Bu(t); \quad x(0) = x_0 \in X, \quad t \geq 0 \\ y(t) &= B^* x(t). \end{aligned} \tag{30}$$

Assume that there exists $G : \mathcal{R}^m \to \mathcal{R}^m$ such that $(A + BGB^*)$ is strictly dissipative, i.e. there exists $\delta > 0$, called the dissipation constant, such that

$$< x, (A + BGB^*)x > \ \leq -\delta \|x\|^2, \quad \forall x \in D(A) \tag{31}$$

Then a solution of the Lur'e equations (16a-c), with $D = W = 0$, is given by

$$P = 1 \tag{32a}$$
$$\epsilon = \delta \tag{32b}$$
$$Q^*Q = -[(A + A^*)/2 + BGB^* - \delta I] \tag{32c}$$
$$W = 0 \tag{32d}$$

P and $(\epsilon I + Q^*Q)$ are both coercive. P is bounded while $(\epsilon I + Q^*Q)$ is not, in general. The method from Section 3 can then be applied. Note that B and B^* need no longer be bounded operators since Lemma 1 is not used.

There are many more systems that are feedback dissipative, rather than strictly dissipative, i.e., there exists $G \in \mathcal{R}^{m \times m}$ such that

$$< x, (A + BGB^*)x > \leq 0. \tag{33}$$

Then a solution of the Lur'e equations is given by

$$P = 1 \tag{34a}$$
$$\epsilon = 0 \tag{34b}$$
$$Q^*Q = -[(A + A^*)/2 + BGB^*] \tag{34c}$$
$$W = 0 \tag{34d}$$

The control law (25) can now be used to establish asymptotic stability of $e_y(t)$ and boundedness of $e(t)$.

For the second case, construct a new Hilbert Space as follows:

Definition 3
Let P be the solution of the Lyapunov equation (16a). For $x, z \in X$, denote the inner product

$$< x, z >_1 = < P^{\frac{1}{2}}x, P^{\frac{1}{2}}z > \tag{35}$$

and let X_1 be the Hilbert Space given by the completion of X in the X_1 inner product. Denote the norm generated by $< \cdot, \cdot >_1$ by $||.||_1$.

The square root $P^{\frac{1}{2}}$ exists and is positive definite and self-adjoint since P is positive definite and self-adjoint. (The positive definiteness and self-adjointness of $\epsilon I + Q^*Q$ imply that the same properties hold for P [24, Lemma 1]. P is therefore m-accretive and a unique positive definite, self-adjoint $P^{\frac{1}{2}}$ exists [34].) Clearly, $||.||$ and $||.||_1$ are equivalent norms if and only if P is coercive (which is a necessary and sufficient condition for invertibility). If $\epsilon I + Q^*Q$ is bounded in (16a), this is equivalent to A generating a C_o-group. Without the C_o-group assumption, we only have a one-sided inequality $||x||_1 \leq k||x||$. An example has been constructed in [21] in which asymptotic stability in X_1 does not even lead to the boundedness in X. One idea is, as in [33], when A does not generate a C_o-group, we shall consider stability in X_1 only. This approach has the obvious drawback of not being able to state the result in the original space. However, if the controller (25), (23) is used, convergence of e_y to zero can be concluded by using the infinite-dimensional generalization of the Invariance Principle (see Section 5), but the state is bounded in X_1 rather than in X.

5. Application of the Invariance Principle

In section 3 we have seen the application of the generalization of LaSalle's Invariance Principle in proving the convergence of $e(t)$ or $e_y(t)$. The application hinges on the fact that bounded positive orbits are precompact in the locally compact space, $\mathcal{R}^n$, and therefore, the positive limit set is nonempty. In infinite-dimensional spaces, bounded positive orbits may not be precompact. Hence the Invariance Principle is not directly applicable.

A closer examination of the proof of the Invariance Principle in [Theorem IV.4.2, 35] reveals that the potential trouble is that the positive limit set may be empty if the positive orbit is not precompact. From Theorem III.10.6 of [36], we know that a weakly convergent subsequence exists for each bounded positive orbit. Hence, under the weak topology, the positive limit set is nonempty. Under certain conditions [37], the Invariance Principle can be used to conclude the weak convergence of $e(t)$ to zero using (24) or the weak convergence of $e_y(t)$ to zero using (26). Weak convergence of $e(t)$ implies uniform boundedness in the strong topology using the Uniform Boundedness Principle [Theorem III.9.4 of 36]. Hence with the controller given by (19),(23), $e(t) \to 0$ weakly as $t \to \infty$ and $||e(t)||$ is bounded in X_1 for all $t \geq 0$. Weak convergence in finite dimensions corresponds to norm convergence. Therefore, with the controller (25), $e_y(t) \to 0$ in $\mathcal{R}^n$ as $t \to \infty$.

In [11], a different approach has been taken. We shall see that even though only Lagrange stability is obtained, the approach yields a modified controller that does not utilize the Invariance Principle in the stability analysis. The following theorem from [35] is needed:

Theorem 1 [35]
Let A_+ generate the linear C_o-semigroup $T_+(t)$ on X_+. Let $F: D(f) \subset X_+ \to X_+$, $D(F) \supset D(A_+ + F)$ be a bounded, continuous operator such that $(A_+ + F)$ generates a nonlinear semigroup on $D(A_+ + F)$. If either

(a) $T_+(t)$ is compact for every $t > 0$ or

(b) $T_+(t)$ is exponentially stable and F is compact,

then all bounded positive orbits of $\dot{x} = (A_+ + F)x$, $x(0) \in D(A_+ + F)$, are precompact.

In the proof of the above theorem, F is allowed to be time-varying. From Section 3,

$$X_+ = X \times \mathcal{R}^{m\times m} \times \mathcal{R}^{m\times(n_m+p_m)} \tag{36a}$$

$$A_+ = \begin{bmatrix} A_c & 0 & 0 \\ 0 & 0 & 0 \\ 0 & 0 & 0 \end{bmatrix} \tag{36b}$$

$$F(e, \Delta G, L) = \begin{bmatrix} B\Delta GCe + BLw \\ -\Gamma_2 Cew^T \\ -\Gamma_1(Ce)(Ce)^T \end{bmatrix} \tag{36c}$$

$$T_+(t) = \begin{bmatrix} U_c(t) & 0 & 0 \\ 0 & 0 & 0 \\ 0 & 0 & 0 \end{bmatrix} \tag{36d}$$

Clearly, if the reference model is stable, $D(F) = X_+$, $D(A + F) = D(A_+)$ and F is bounded and continuous, since $\Delta G, e$ are bounded from $\dot{V} \leq 0$ and w is bounded from the stable model assumption. In section 7, we shall see that $(A_+ + F)$ generates a nonlinear semigroup; hence, the assumptions of the theorem are satisfied. Condition (a) is satisfied by many practical systems, e.g., beam equation and heat equation. However, it also rules out many systems of interest, for example, convection over an infinite slab. Condition (b) is more general and appealing since F is compact from the fact that B is a bounded finite-rank operator, and hence, compact. However, $T_+(t)$ is obviously not exponentially stable. The natural modification of (23) to overcome this problem leads to

$$\dot{G}(t) = \dot{\Delta G}(t) = -\gamma_1 G(t) - \Gamma_1 e_y(t) e_y^T(t) \tag{37a}$$

$$\dot{S}(t) = \dot{L}(t) = -\gamma_2 S(t) - \Gamma_2 e_y(t) w^T(t) \tag{37b}$$

Suppose we use the same Lyapunov function candidate as in (22), $\dot{V}$ becomes

$$\dot{V} = -\|Qe\|^2 - \epsilon\|e\|^2 - 2\gamma_1 tr[\Delta G\Gamma_1^{-1}\Delta G^T] - 2\gamma_1 tr[G\Gamma_1^{-1}\Delta G^T] - 2\gamma_2 tr[L\Gamma_2^{-1}L^T] - 2\gamma_2 tr[S\Gamma_2^{-1}L^T] \tag{38}$$

Completing squares, we obtain

$$\begin{aligned} \dot{V} &= -\|Qe\|^2 - \epsilon\|e\|^2 - \gamma_1 tr[\Delta G\Gamma_1^{-1}\Delta G^T] - \gamma_2 tr[L\Gamma_2^{-1}L^T] \\ &\quad -\gamma_1 tr[(\Delta G + G)\Gamma_1^{-1}(\Delta G + G)^T] - \gamma_2 tr[(L+S)\Gamma_2^{-1}(L+S)^T] \\ &\quad +\gamma_1 tr[G\Gamma_1^{-1}G^T] + \gamma_2 tr[S\Gamma_2^{-1}S^T] \\ &\leq -\lambda V + \rho \end{aligned} \tag{39}$$

with

$$\lambda = \min(\epsilon\|P\|^{-1}, \gamma_1, \gamma_2) \tag{40a}$$

$$\rho = \gamma_1 tr[G\Gamma_1^{-1}G^T] + \gamma_2 tr[S\Gamma_2^{-1}S^T]. \tag{40b}$$

By using the Comparison Principle [38],

$$V \leq e^{-\lambda t}(V(0) - \frac{\rho}{\lambda}) + \frac{\rho}{\lambda}. \tag{41}$$

The stability has been weakened to Lagrange type and the ultimate bound on V equal to $\rho\lambda^{-1}$. We have the following situation:

$$< e(t), Pe(t) > +tr[G(t)\Gamma_1^{-1}G(t)^T]+tr[S(t)\Gamma_2^{-1}S(t)^T] \to \lambda^{-1}(\gamma_1 tr[G\Gamma_1^{-1}G^T]+\gamma_2 tr[S\Gamma_2^{-1}S^T]) \quad (42)$$

exponentially with rate λ. By choosing Γ_1, Γ_2 large, we can effectively control the size of the ultimate bound on $||e(t)||$. $G(t), S(t)$ are clearly uniformly bounded. The convergence rate is globally exponential with rate λ. The Invariance Principle and Theorem 1 can also be used to conclude Lagrange stability of e in X_1 by defining a new Lyapunov function. If $V(0) \geq \rho/\lambda$,

$$V_1 = V - \frac{\rho}{\lambda} \quad (43)$$

and, if $V(0) < \rho/\lambda$,

$$V_1 = e^{\lambda t}(V - \frac{\rho}{\lambda}). \quad (44)$$

Note that V_1 is continuous and bounded below and $\dot{V}_1 \leq 0$. Therefore, the Invariance Principle can be applied. With controller as in (25), (23), we can conclude that $||e_y|| \to 0$.

6. Strict Positive Realness Condition

As mentioned in Section 3, the strict positive realness condition as required in the infinite dimensions can never be satisfied for a strictly proper system. This causes no problem for finite-dimensional systems since, (i) the frequency condition for strict positive realness does not require a *uniform* bound above zero, and, (ii) the Kalman-Yakubovich Lemma applies to positive real systems also. Neither is true in the infinite dimensional case. Clearly, the scheme in Section 3 must be modified. A natural question to ask is: Suppose the plant is almost feedback positive real in the sense that there exists an output feedback G that is $-d$-stabilizing with $d \geq 0$, will the scheme in Section 3 work? And if it does not, where does it fail and how can one remedy the situation?

For the first question, we assume there exists G that is $-d$-stabilizing for $(A, B, C, 0)$. Then apply the controller (19), (23a,b) to the error equation (21) using the same quadratic Lyapunov function candidate (22) as before. The full Lur'e equations must be used:

$$(A_c^*P + PA_c)x = -(\epsilon I + Q^*Q)x, \quad \forall x \in D(A_c) = D(A) \quad (45a)$$

$$B^*P = C - (2d)^{\frac{1}{2}}Q \quad (45b)$$

Take the derivative of V along the solution and use (45a,b),

$$\dot{V} = -\epsilon||e||^2 - ||Qe||^2 + 2e_y^T\Delta Ge_y - 2(2d)^{\frac{1}{2}}(Qe)^T(|\delta Ge_y) + 2e_y^T Lw - 2(2d)^{\frac{1}{2}}(Qe)^T(Lw) + 2tr[\dot{\Delta G}\Gamma_1^{-1}\Delta G^T] + 2tr[\dot{L}\Gamma_2^{-1}L^T].$$

After applying the control law and completing squares,

$$\dot{V} = -\epsilon||e||^2 - ||2^{-\frac{1}{2}}Qe + 2d^{\frac{1}{2}}(\Delta Ge_y)||^2 - ||2^{-\frac{1}{2}}Qe + 2d^{\frac{1}{2}}(Lw)||^2 + 4d||\Delta Ge_y||^2 + 4d||Lw||^2. \quad (46)$$

The last two terms are positive semidefinite. Therefore, for any $d > 0$, they can possibly lead to instability. Suppose the same modification is made as in Section 5, i.e., (37) is used instead of (23), then (46) becomes

$$\dot{V} \leq -\lambda V + 4d||\Delta Ge_y||^2 + 4d||Lw||^2 + \rho \quad (47)$$

Assuming that a stable model and P^{-1} exists, this can be further simplified as

$$\dot{V} \leq -\lambda V + \eta V^2 + \rho. \quad (48)$$

For

$$4\eta_\rho < \lambda^2 \tag{49}$$

we can obtain Lyapunov stability from (48) [21]. Since λ depends on ϵ which in turn depends on d, the inequality (49) may not be satisfied even locally. Also, local stability is much less desirable than global stability. A natural modification is to add terms like $G(t)e_y(t)e_y^T(t)$ to the adaptive gain update equation, to introduce a negative semidefinite term $-\|\Delta G(t)e_y(t)\|^2$. Also, motivated by (25), the control signal is modified to

$$u(t) = G(t)e_y(t) + S(t)w(t) - he_y(t). \tag{50}$$

The following proposition then shows that if d is small enough, and the controller parameters are properly chosen, global Lagrange stability of e in X_1 and asymptotic stability of e_y result.

Proposition 2 [21]
Assume the following statements are true.

1. There exists $G \in \mathcal{R}^{m\times m}$ that is $-d$-stabilizing, $d > 0$.
2. Assumption 1 holds.
3. The reference model is stable.
4. The adaptive control law (50) and

$$\dot{G}(t) = -\gamma_1 G(t) - \Gamma_1 e_y(t)e_y^T(t) - \xi\Gamma_1 G(t)e_y(t)e_y^T(t) \tag{51a}$$

$$\dot{S}(t) = -\gamma_2 S(t) - \Gamma_2 e_y(t)w^T(t) \tag{51b}$$

   is applied to the error system (21). Let $\|G\| \leq g$ where g is known and $h = g, \xi = \frac{1}{g}, \gamma_2$ is chosen sufficiently large.

If d is sufficientlly small, then the overall closed loop system is Lagrange stable in the enlarged state space X_1. Furthermore, the output error is asymptotically stable.

The main merit of this controller is that it applies to almost positive real systems in infinite dimensions. The closed loop strict positive realness assumption can now be relaxed to closed loop almost positive realness (PRI of the closed loop system is small). In particular, a system is allowed to be closed loop positive real, since in that case, d is arbitrarily small. The precise meaning of d being sufficiently small is

$$dg < \frac{1}{4}\frac{c-1}{c} \tag{52}$$

where $c \in (1, \infty)$ is a constant. Large c implies a correspondingly large γ_2 must be chosen. In adaptive regulation, c can be chosen as ∞. There are two situations under which this inequality can be satisfied. The first case is when the plant is almost stable, hence, (52) may be satisfied by choosing g small and d appropriately large, for example,

$$W_p(s) = (s-1)(s^2+1)^{-1}. \tag{53}$$

The performance of the closed-loop system may be sluggish, however, due to the small damping term. The second case is when the plant is almost positive real under feedback. e.g., consider

$$W_p(s) = (s-0.1)(s^2+1)^{-1} \tag{54}$$

This result also has application to frequency perturbed systems of the form

$$W_p(s) = W_1(s) + \delta W_2(s) \tag{55}$$

where δ is the perturbation parameter, and $W_1(s)$ is closed loop positive real.

For δ sufficiently small, the controller (50),(51) that Lagrange-stabilizes the unperturbed plant will Lagrange-stabilize the perturbed plant as well. The above discussion is useful in both the context of regular perturbation and singular perturbation. Suppose the generator A is perturbed by a linear operator δA_1. Then the open loop transfer function is

$$W_p(s) = C(sI - A - \delta A_1)^{-1}B = C(sI - A)^{-1}B + \delta W_2(s) \tag{56}$$

which is in the same form as (55). If the plant is singularly perturbed, i.e.,

$$\begin{aligned} \dot{x} &= A_{11}x + A_{12}z + B_1 u \\ \delta\dot{z} &= A_{21}x + A_{22}z + B_2 u \\ y &= C_1 x + C_2 z \end{aligned} \tag{57}$$

then the transfer function can also be written in the form of (55) where $W_1(s)$ is the transfer function when $\delta = 0$.

7. Existence and Uniqueness of Closed Loop Solutions

None of the stability analysis is meaningful if a unique continuous solution does not exist under the closed loop operation. In this section, we show that such solution does in fact exist by using the infinite-dimensional version of a familiar finite-dimensional existence and uniqueness result. In the subsequent analysis, controller (19), (37) is used. We first state the definition of a solution.

Definition 4
Given

$$\begin{aligned} \dot{v}(t) &= A_+ v(t) + F(t, v(t)) \\ v(0) &= v_0 \in X_+ \end{aligned} \tag{58}$$

where A_+ is the infinitesimal generator of the C_o-semigroup $T_+(t)$ on a Hilbert Space X_+ and $F : X_+ \rightarrow X_+$. Then a continuous $v(t)$ solving

$$v(t) = T_+(t - t_0)v_0 + \int_{t_0}^{t} T_+(t - \tau)F(\tau, v(\tau))d\tau \tag{59}$$

is called a mild solution of (58). For our adaptive controller operating in closed loop,

$$\begin{aligned} A_+ &= diag\{A_c,\ -\gamma_1 I,\ -\gamma_2 I\} && (60a) \\ T_+ &= diag\{U_c(t),\ e^{-\gamma_1 t},\ e^{-\gamma_2 t}\} && (60b) \\ F &= [\{B(G(t) - G)Ce + B(S(t) - S)w\}^* \ - (\Gamma_1(Ce)(Ce)^T)^* \ - (\Gamma_2(Ce)w)^T]^* && (60c) \\ v &= [e^*(t)\ \ G^T(t)\ \ S^T(t)]^* && (60d) \\ X_+ &= X_1 \times \mathcal{R}^{m\times m} \times \mathcal{R}^{m\times(n_m+p_m)} && (60e) \end{aligned}$$

X_1 is considered in (60) since only stability in X_1 can be obtained in general (see Section 4). Note that we have already assumed the existence and uniqueness of $\overline{x}$ in X (therefore in X_1 also). We first note that F is locally Lipschitz.

Lemma 2[21]
$F(t, v)$ is continuous in t for $t \geq 0$, locally Lipschitz continuous in v, uniform in t on bounded intervals.

We are now ready to apply Theorem 6.1.4 of [29]. The theorem is a generalization of the familiar continuation of solution result for ordinary differential equation to the infinite-dimensional state space. We state that result as follows:

Lemma 3
Let $F : \mathcal{R}^+ \times X_+ \rightarrow X_+$ be continuous in t, for all $t \geq 0$ and locally Lipschitz continuous in v,

uniform in t on bounded intervals. Then for all $v_0 \in X_+$, there exists $t_{max} \leq \infty$ such that (58) has a unique mild solution v on $[0, t_{max})$. Moreover, if $t_{max} < \infty$, then $\lim_{t \to t_{max}} \|v(t)\| = \infty$.

Lemma 2 together with Lemma 3 imply the local existence and uniqueness of closed loop solutions. But from Section 5, we have shown that $e(t), G(t)$ and $S(t)$ are uniformly bounded in X_1 in $[0, t_{max})$ for all $t_{max} \in \mathcal{R}^+$. Hence, $t_{max} = \infty$, and we actually have global existence and uniqueness of a mild solution.

8. Performance Under Perturbations

In real world situations, the model of the physical system (1) is subject to external disturbances and the modeling errors. The effect of these on our adaptive scheme is investigated in this section. The cases of linear, regular, and singular perturbations have been considered in Section 6 where Lagrange stability is shown to be preserved when the perturbation is small . Here, we will examine the effect of state and output disturbances. The disturbance is respectively considered as bounded, tending to zero with polynomial rate and tending to zero with exponential rate. The sensitivity issue is also considered. The unperturbed plant satisfies the feedback positive realness requirement while the perturbed plant may not. This situation will be shown to be closely related to the robustness question already addressed in Section 6.

Suppose the actual plant dynamics is given by

$$\begin{aligned} \dot{x}(t) &= Ax(t) + Bu(t) + v_1(t); \quad x(0) = x_0 \in X, \quad t \geq 0 \\ y(t) &= Cx(t) + v_2(t) \end{aligned} \tag{61}$$

where v_1 and v_2 are state noise and output noise respectively. The error dynamics is given by

$$\begin{aligned} \dot{e}(t) &= Ae(t) + Bu(t) - BSw(t) + v_1(t) \\ e_y(t) &= Ce(t) + v_2(t) \end{aligned} \tag{62}$$

There are three cases that will be considered:

A. There exist $M_1, M_2 < \infty$ such that

$$\|v_1(t)\| \leq M_1 \tag{63a}$$

$$\|v_2(t)\| \leq M_2, \quad \forall t \geq 0 \tag{63b}$$

This case corresponds to persistent, bounded state and output perturbations. M_1 and M_2 are usually small; it is frequently used to model the plant operating in a noisy environment.

B. There exist $M_1, M_2 < \infty, \ 0 < t_1, t_2, n_1, n_2 < \infty$ such that

$$\|v_1(t)\| \leq M_1(t + t_1)^{-n_1} \tag{64a}$$

$$\|v_2(t)\| \leq M_2(t + t_1)^{-n_2}. \tag{64b}$$

This case corresponds to slowly decaying state and output disturbances and is used to model the parameter drift phenomenon.

C. There exist $M_1, M_2 < \infty, \ 0 < \beta_1, \beta_2 < \infty$ such that

$$\|v_1(t)\| \leq M_1 e^{-\beta_1 t} \tag{65a}$$

$$\|v_2(t)\| \leq M_2 e^{-\beta_2 t}. \tag{65b}$$

This case corresponds to fast decaying state and output disturbances. When $M_1, M_2, \beta_1, \beta_2$ are all large, it is used to model impulsive disturbances.

We first show that under case A, with similar assumptions as in Proposition 2, the closed loop system remains Lagrange stable in X_1. The disturbances affect only the size of the ultimate bound.

Theorem 2[21]
Assume that the following statements are true:

1. There exists $G \in \mathcal{R}^{m\times m}$ that is $-d-$stabilizing, $d > 0$.
2. Assumption 1 holds.
3. The reference model is stable.
4. The adaptive control law (50), (51) is applied to the error system (62) under Case A. Let $||G|| \leq g$ where g is known and $h = g, \xi = kg^{-1}$, $k \in (0,1)$, γ_1 and γ_2 are chosen sufficiently large.

If d is sufficiently small, then the overall system is Lagrange stable in the enlarged state space X_1.

The precise meaning of d being sufficiently small is similar to (52):

$$dg < \frac{k}{4}\left(\frac{c-1}{c}\right) \tag{66}$$

where $k \in (0,1)$, $c > 1$. To make (66) as weak as possible, we can increase c and make k close to 1. The penalty that must be paid is that γ_2 must be chosen proportionally large for the former and the ultimate bound will be large for the latter. With the aid of Theorem 2, stability under cases B and C can be analyzed with ease.

Corollary 1
Assume that all conditions in Theorem 2 hold except that $v_1(t), v_2(t)$ satisfy case B. Then the overall closed-loop system is Lagrange stable in X_1 and the ultimate bound on the Lyapunov function V is proportional to ρ where ρ is defined in (40).

By choosing $||\Gamma_1||, ||\Gamma_2||$ large, the ultimate bound can be made arbitrarily small. However, this means γ_1 and γ_2 must also be selected small and the convergence to the ultimate bound may be very slow. Case C follows similarly:

Corollary 2
Assume all conditions in Theorem 2 hold except that $v_1(t), v_2(t)$ satisfy case C. Then the overall closed-loop system is Lagrange stable in X_1 and the ultimate bound on the Lyapunov function V is proportional to ρ. Furthermore, the convergence to the residue set is with exponential rate.

In Section 4, and also in some large space structure applications [39,22], it has been suggested to use sensor-actuator colocation to obtain the closed-loop positive realness property. However, exact colocation is clearly an unreasonable requirement. Therefore, there are two issues to resolve: a) Is stability preserved under small misalignment? b) How much misalignment can be tolerated? Both of these questions can be answered using Proposition 2. Without loss of generality, we shall consider the perturbation of the output operator C only. Let us investigate from the frequency domain point of view. Suppose the actual output operator C_1 is a slightly perturbed version of the assumed C:

$$C_1 = C + \delta\Delta C; \quad \delta \geq 0 \tag{67}$$

where $||\Delta C|| = 1$. Then the closed loop transfer function evaluated on the $j\omega$-axis is

$$C(j\omega I - A_c - BG\delta\Delta C)^{-1}B + \delta\Delta C(j\omega I - A_c - BG\delta\Delta C)^{-1}B = C(j\omega I - A_c)^{-1}B + \delta Y(j\omega) \tag{68}$$

where

$$Y(j\omega) = \delta C(j\omega I - A_c)^{-1} BG\Delta C(j\omega I - A_c - BG\delta\Delta C)^{-1} B + \delta\Delta C(j\omega I - A_c)^{-1} B.$$

This is in the same form as (55). Therefore, provided δ is sufficiently small, Lagrange stability of the perturbed plant is obtained by using controller (50),(51).

9. Summary

In this paper, we have discussed some of the problems (and solutions) of the generalization of a finite-dimensional adaptive control scheme to an infinite-dimensional Hilbert Space. This endeavor is important since all physical systems are in fact distributed. The problem is particularly relevant in the context of active control of large structures in space due to the mechanically flexible nature of the systems. Four main problems areas were addressed: the solution of Lyapunov equation may not be coercive, the strict positive realness frequency condition can never be satisfied, the Invariance Principle is not directly applicable in a non-locally-compact state space, and the existence and uniqueness of the closed-loop solution may not be assured. We summarize below the results obtained in each of these areas:

1. The coercivity of P, which determines whether the spaces X and X_1 are equivalent, is quite critical, as shown in section 4. When the plant is output feedback dissipative, the Positive Realness Lemma is not needed. The P that solves the Lyapunov equation is just the identity operator; hence, stability is in the original space X. In general, the stability results are confined to the enlarged space, X_1, which may not be equivalent to X. When the convergence is asymptotic in X_1 and the error state *converges* in X, then the convergence in X is also asymptotic.
2. In Section 5, it is pointed out that Invariance Principle cannot be used directly in infinite-dimensional spaces, since bounded orbits may not be precompact. There are two possible approaches to this issue. It is first shown in Section 5 that the Invariance Principle can sometimes be used in the weak topology to show weak convergence of the error state. Alternatively, Theorem 5.1 states two sufficient conditions under which bounded orbits are precompact. The first condition requires that the closed loop C_o-semigroup generated by the linear dynamics be compact. Though many DPS examples arising from mathematical physics do satisfy the condition (for example, the diffusion equation and the linearly damped beam equation), it also rules out other examples of interest. The second condition requires that the closed loop C_o-semigroup be exponentially stable, which is not satisfied by the unmodified control law. If the control law is modified such that the condition is satisfied, we find that the Invariance Principle is no longer needed. The drawback, as mentioned in Section 5, is that the error state no longer converges to the origin, but to a residue set indeed. The drawback, as mentioned in Section 5, is that the error state no longer converges to the origin, but to a residue set instead. The saving grace is that the size of this set can be made arbitrarily small and the asymptotic stability of the output error is maintained. A further advantage of the modified control law is that the convergence to the residue set is with exponential rate, while the Invariance Principle only provides strong stability.
3. In Section 6, it is shown that the closed loop strict positive realness assumption can never be satisfied. Intuitively, we expect a "good" controller to be robust with respect to structural properties, and in this case, the closed loop positive realness index (PRI). Therefore, the controller from Section 5 is further modified in Section 6. As a result, if the closed loop PRI is $-d, d > 0$, so long as dg remains small, g being an upperbound on $\|G\|$, the overall system is Lagrange stable and the output error is asymptotically stable.
4. In Section 7, a theorem similar to the continuation of solution results for ordinary differential equations is used to show the existence and the uniqueness of a mild solution under the adaptive schemes proposed in Sections 5 and 6.

5. In Section 8, it is shown that the adaptive controller in Section 6 robustly stabilizes the plant under various state and output disturbances and parameter variations.

This paper is a first attempt to extend an adaptive scheme to the infinite- dimensional state space in a rigorous fashion. Among the four technical difficulties encountered in making the transition, the strict positive realness requirement is the most impeding. Although a partial answer is provided here in Lagrange stable solution, much research is still required to gain further insight into this problem; nevertheness, a stable finite-dimensional adaptive control scheme has been demonstrated for a large class of linear infinite-dimensional systems.

References

[1] M. Athans, "Toward a Practical Theory for Distributed Parameter Systems," *IEEE Transactions on Automatic Control*, vol. 15, pp. 245-247, April 1970.

[2] M.J. Balas, "Trends in Large Space Structure Control Theory: Fondest Hopes, Wildest Dreams," *IEEE Transactions on Automatic Control*, vol. 27, No. 3, pp. 522-535, 1982.

[3] K.J. Åström , B. Wittenmark, "On Self-Tuning Regulators," *Automatica*, vol. 9, pp. 185-199, 1973.

[4] G. Kreisselmeier, "Adaptive Control via Adaptive Observation and Asymptotic Feedback Matrix Synthesis," *IEEE Transactions on Automatic Control*, vol. 25, No. 4, pp. 717-722, 1980.

[5] H.Elliott, W.A. Wolovich, "A Parameter Adaptive Control Structure for Linear Multivariable Systems," *IEEE Transactions on Automatic Control*, vol. 27, No. 2, pp. 340-352, 1982.

[6] R.V. Monopoli, "Model Reference Adaptive Control with an Augmented Error Signal," *IEEE Transactions on Automatic Control*, vol. 19, No. 5, pp. 474-484, 1974.

[7] M.J. Balas, J.H. Lilly, "Adaptive Parameter Estimation of Large Scale Systems by Reduced Order Modeling," *Proceedings of the 20th IEEE CDC*, San Diego, CA., pp. 233-239, 1981.

[8] H.T. Banks, J.M. Crowley, K.Kunisch, "Cubic Spline Approximation Techniques for Parameter Estimation in Distributed Systems," *IEEE Transactions on Automatic Control*, vol. 28, No. 7, pp. 773-786, 1983.

[9] M.H. Hamza and M.A. Sheirah, "A Self-Tuning Regulator for Distributed Parameter Systems," *Automatica*, vol. 14, pp. 453-463, 1978.

[10] M. Vajta, Jr., L. Keviczky, "Self-tuning Controller for Distributed Parameter Systems," *Proceedingts of the 20th IEEE CDC*, San Diego, pp. 38-41, 1981.

[11] M.J. Balas, H. Kaufman and J.Wen, "Stable Direct Adaptive Control of Linear Infinite Dimensional Systems Using a Command Generator Tracker Approach," Workshop on Identification and Control of Flexible Space Structures, San Diego, Ca., June 1984.

[12] J.Wen, M.J. Balas, "Direct MRAC in Hilbert Space," *5th Symposium on Dynamics and Control of Large Space Structures*, Blacksburg, Va., June 1985.

[13] M.J. Balas, "Some Critical issues in Stable Finite-Dimensional Adaptive Control of Linear Distributed Parameter Systems," *Proceedings of the Third Yale Workshop on Applications of Adaptive Systems Theory*, New Haven, Ct., 1983.

[14] J. Broussard, M. O'Brien, "Feedforward Control to Track the Output of a Forced Model, *Proceedings of the 17th IEEE CDC*, pp. 1144-1155, Jan. 1979.

[15] K. Sobel, H. Kaufman and L. Mabius, "Implicit Adaptive Control Systems for a Class of Multi-Input Multi-Output Systems," *IEEE Trans. on Aeros. and Electr. Syst.*, vol. 18, No. 5, pp. 576-590, 1982.

[16] K. Sobel, *Model Reference Adaptive Control for Multi-Input Multi-Output Systems*, Ph. D. Thesis, RPI, Troy, NY, June 1980.

[17] I. Bar-Kana, H. Kaufman and M.J. Bals, "Model Reference Adaptive Control of Large Structural Systems," *AIAA J. Guidance and Control*, vol. 6, No. 2, pp. 112-118, 1983.

[18] P.A. Ioannou and P.V. Kokotovic, "Instability Analysis and Improvement of Robustness of Adaptive Control," *Automatica*, vol. 20, No. 5, pp. 583-594, 1984.

[19] C.E. Rohrs, L. Valavani, M. Athans, G. Stein, "Analytic Verification of Undesirable Properties of Direct Model Reference Adaptive Control Algorithms," *Proceedings of the 20th IEEE CDC*, San Diego, Ca., pp. 1272-1284, 1981.

[20] K.S. Narendra and L.S. Valavani, "Stable Adaptive Controller Design - Direct Control," *IEEE Transactions on Automatic Control*, vol. 23, No. 4, pp. 570-583, 1978.

[21] J. Wen, *Direct Adaptive Control in Hilbert Space*, Ph.D. Thesis, Electrical, Computer and Systems Engineering Department, RPI, Troy, Ny., June 1985.

[22] I. Bar-Kana, *Direct Multivariable Model Reference Adaptive Control with Application to Large Structural Systems*, Ph.D. Thesis, RPI, Troy, Ny., 1983.

[23] C. Jacobson, "Some Aspects of the Structure and Stability of a Class of Linear Distributed Systems," Masters Thesis, RPI, Troy, Ny., 1984; available as Robotics & Automation Laboratory Report, RAL31.

[24] D. Wexler, "On Frequency Domain Stability for Evolution Equations in Hilbert Space via the Algebraic Riccati Equation," *SIAM J. Math. Anal.*, vol. 11, pp. 969-983, 1980.

[25] A. Likhtarnikov, V. Yakubovich, "The Frequency Theorem for Equations of Evolution Type," Sibirskii Mathematicheskii Zhurnal, vol. 17, pp. 1069-1085, 1976.

[26] J.P. LaSalle, "Some Extensions of Lyapunov's Second Method," *IRE Transactions on Circuit Theory*, pp. 520-527, Dec. 1960.

[27] J. Hale, Ordinary Differential Equations, Wiley Interscience, New York, 1969.

[28] A. Pazy, "On the Applicability of Lyapunov's Theorem in Hilbert Space," *SIAM J. Math. Anal.*, vol. 3, No. 2, pp. 291-294, 1972.

[29] A. Pazy, Semigroup of Linear Operators and Applications to Partial Differential Equations, Springer-Verlag, New York, 1983.

[30] S. Lang, Real Analysis, John Wiley and Sons, 1983.

[31] Hirsche, Smale, Differential Equations, Dynamics Systems and Linear Algebra, Academic Press, New York, 1974.

[32] Naylor, Sell, Linear Operator Theory in Engineering and Science , Springer-Verlag, New York, 1982.

[33] M. Slemrod, "An Application of Maximal Dissipative Sets in Control Theory," *J. Math. Analy. and Appl.*, vol. 46, pp. 369-387, 1974.

[34] T. Kato, Perturbation Theory for Linear Operators, Springer-Verlag, New York, 1966.

[35] J. Walker, Dynamic Systems and Evolution Equations: Theory and Applications, Plenum Press, New York, 1980.

[36] A.E. Taylor and D.C. Lay, Introduction to Functional Analysis, John Wiley & Sons, Second Edition, 1980.

[37] J.M. Ball and M. Slemrod, "Feedback Stabilization of Distributed Semilinear Control Systems," *Appl. Math. Optim.*, 5, pp. 169-179, 1979.

[38] D.D. Siljak, *Large Dynamical Systems*, North-Holland, Amsterdam, The Netherlands, 1978.

ROBOTICS

There were free men in Greece
because there were slaves there.

Albert Camus

The field of robotics is a diverse, rapidly growing, and quickly evolving discipline which resists succinct characterization, much less adequate representation by any small group of practitioners. This section contains a collection of six papers which address the general problem of robot task encoding: the modeling and analysis of desired tasks resulting in the production of a set of commands which cause the robot to accomplish them. It would be fair to say that all the papers fall within two extremes which might be loosely referred to as "high level" and "low level" techniques. From the point of view of the high level task planning community a solution is a formal algorithm or heuristic, whose output is a plan of action that the robot must subsequently perform, leaving aside the details of how sensors and actuators should be monitored and controlled. From the point of view of the low level control community, the plan of action is assumed to be known and a solution amounts to designing an appropriate feedback controller. This section presents a sampling of perspectives between the two extremes.

The paper by Reif, representative of the high level path planning community, considers the intrinsic complexity of motion planning in a task domain modeled by linked polyhedra. He surveys results which suggest that a large class of seemingly modest tasks gives rise to computationally intractable problems, and proposes a number of new problems. Lumelsky offers a new algorithm for solving motion planning problems involving two degrees of freedom with complete uncertainty regarding obstacles, avoiding the computational burden described above by using environmental feedback generated on-line.

Intermediate between the path planning and the control communities are the ideas of Hogan and Khatib. The fact that the dynamic coupling between the work-piece and the robot may lead to poor performance or even instability motivates Hogan to address problems in which robots perform mechanical work. Using what he terms impedance control, he discards the conventional objective of requiring system state variables to evolve to some desired value in favor of making the manipulator converge to some desired impedance function, relating input motions to output forces. His paper presents an application of this methodology to deburring problems. Khatib employs a potential function to model the presence of obstacles in the environment and directly generates a feedback law based on the gradient of the function. He uses a two-level control architecture, consisting of a low rate parameter evaluation level and a high rate servo control level, to implement the control law directly in the operational space in which the task is described.

The final two papers are closer to the traditional realm of control theory. Arimoto assumes that a reference trajectory is specified and applies a learning method to converge towards a time-varying control input, which causes the robot manipulator to track the trajectory. His technique converges globally when the underlying dynamics are linear and locally for the general class of nonlinear dynamics typical of revolute robot arms. In contrast to this control methodology, which utilizes a reference trajectory, the paper by Koditschek discusses feedback control strategies arising from task description schemes similar to those of Hogan and Khatib. His analysis, based on nonlinear control theory, examines the convergence properties of the resulting closed loop systems. Koditschek feels that simple error driven algorithms may be successful in accomplishing a wide variety of robot tasks and hence deserve a better hearing from the robotics community.

A Survey on Advances in the Theory of Computational Robotics

John H. Reif
Aiken Computation Laboratory
Division of Applied Sciences
Harvard University
Cambridge, Massachusetts 02138

Abstract

This paper describes work on the computational complexity of various movement planning problems relevant to robotics. This paper is intended only as a survey of previous and current work in this area. The generalized mover's problem is to plan a sequence of movements of linked polyhedra through 3-dimensional Euclidean space, avoiding contact with a fixed set of polyhedra obstacles. We discuss our and other researchers' work showing generalized mover's problems are polynomial space hard. These results provide strong evidence that robot movement planning is computationally intractable, i.e., any algorithm requires time growing exponentially with the number of degrees of freedom. We also briefly discuss the computational complexity of four other quite different types of movement problems: (1) movement planning in the presence of friction, (2) minimal movement planning, (3) dynamic movement planning with moving obstacles and (4) adaptive movement planning problems.

1. Introduction: The Mover's Problem

The classical *mover's problem* in d-dimensional Euclidean space is:

Input: (O, P, p_I, p_F) where O is a set of polyhedral *obstacles* fixed in Euclidean space and P is a rigid polyhedron with distinguished initial position p_I and final position p_F. The inputs are assumed to be specified by systems of rational linear inequalities.

Problem: Can P be moved by a sequence of translations and rotations for p_I to p_F without contacting any obstacle in O?

For example, P might be a sofa[1] which we wish to move through a room crowded with obstacles. Figure 1 gives a simple example of a two-dimensional mover's problem.

The mover's problem may be *generalized* to allow P (the object to be moved) to consist of multiple polyhedra freely linked together at various distinguished vertices. (A typical example is a robot arm with multiple joints.) Again, the input is specified by systems of rational linear inequalities.

The paper is organized as follows: Section 2 concerns lower bounds for generalized mover's problems in 2D and 3D. Section 3 concerns efficient solution of restricted mover's problems. Section 4 concludes the paper with discussion of further problems in computational robotics.

2. Lower Bounds for Generalized Mover's Problems

In [1] (also appearing in [2]) we proved that the generalized mover's problem in three dimensions is polynomial space hard. That is, we proved that the generalized mover's problem

[1]The author first realized the nontrivial mathematical nature of this problem when he had to plan the physical movement of an antique sofa from Rochester to Cambridge.

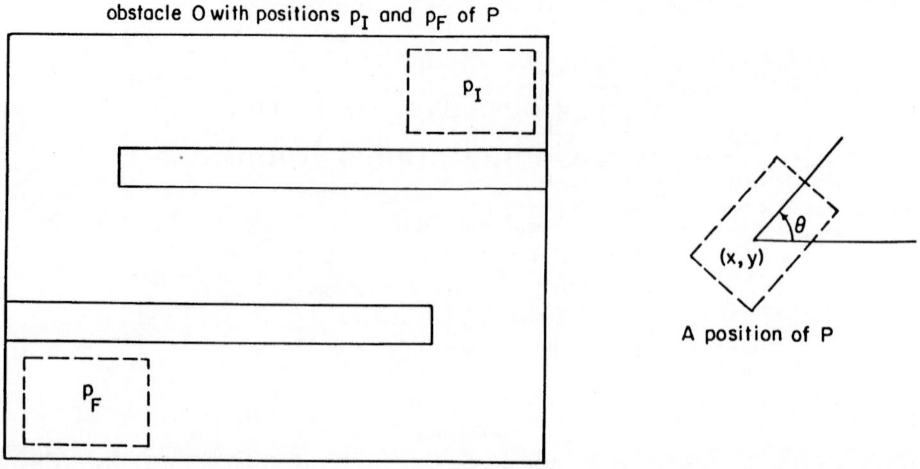

Figure 1: A 2-D Mover's Problem: Can rectangle P be moved from p_I to p_F without contacting an obstacle in O?

is at least as hard as any computational problem requiring polynomial space. (Polynomial space problems are at least as hard as the well known NP problems; see [3].)

This was the first paper investigating the inherent computational complexity of a robotics problem in Computational Geometry. Our proof technique is to use the degrees of freedom of P to encode the configuration of a polynomial space bounded Turing machine M, and to design obstacles which forced the movement of P to simulate the computation of M.

This work was originally motivated by application to robotics: the author felt it was important to examine *computational complexity issues in robots* given the recent development of mechanical devices autonomously controlled by micro and minicomputers, and the swiftly increasing computational power of these controllers. However, it took a number of years before computational complexity issues in robotics became of more general interest. Recently there have been a flurry of papers in the now emerging area which we might term *Computational Robotics.*

Recent investigations in lower bounds have provided some quite ingenious lower bound constructions for restricted cases of the generalized mover's problem. For example, [4] showed that the generalized mover's problem in three dimensions is also polynomial space hard, and [5] showed that the problem of moving a collection of disconnected polyhedra in a two-dimensional maze is polynomial space hard. The problem of moving a collection of disks in two dimensions is known to be NP-hard [6], but it remains open to show this problem polynomial space hard.

3. Upper Bounds for Mover's Problems

Our lower bounds for the generalized mover's problem provided evidence that time bounds for algorithms for movement planning must grow exponentially with the number of degrees of freedom. We next give a brief description of known algorithms for mover's problems. In our original paper [1] we also sketched a method for efficient solution of the classic mover's problem where P, the object to be moved, is rigid. In spite of considerable work on this problem by workers in the robotics fields and in artificial intelligence (for example [7-11]), no algorithm guaranteed to run in polynomial time had previously appeared. Our approach was to transform a classic mover's problem (O, P, p_I, p_F) of size n in d dimensions to an apparently simpler mover's problem (O', P', p'_I, p'_F) of dimensions d', where P' is a single part and d' is the number of degrees of freedom of movement in the original problem. The

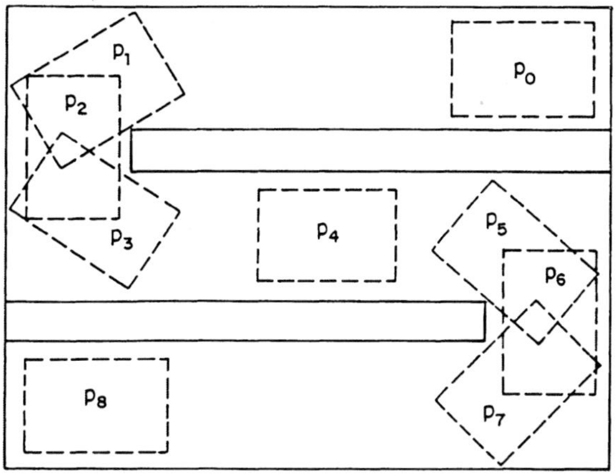

Figure 2: A solution to the 2-D Mover's Problem of Figure 1. P may be moved through positions $p_I = p_0, p_1, \ldots, p_8 = p_F$.

transformed problem is thus to find a path in d'-dimensional space avoiding the transformed obstacles O. The fundamental difficulty is that the induced obstacles may be nonlinear constraints. (In [11], Lozano-Pérez and Wesley did not construct O', but instead approximated the induced obstacles O' by linear constraints. Unfortunately, an exponential number of linear constraints were required to approximate even a quadratic constraint within accuracy 2^{-n}. Thus their method required exponential time (i.e., 2^{cn} time for some $c > 0$) even if the original mover's problem was two-dimensional.)

Example. Consider a classical mover's problem (O, P, p_I, p_F) restricted to *dimension* $d = 2$, with the obstacles O consisting of a set of line segments and P a single polygon. A *position* of P can be specified by a triple (x, y, θ) where (x, y) are the Cartesian coordinates of some fixed vertex of P and θ is the angle of rotation around this vertex. We define a mapping f from the position of P to 3-space. Let $f(x, y, \theta) = (x', y', z')$ where $y = z', tan(\theta) = x'/y'$, and $x = (x')^2 + (y')^2 - \alpha$, for some sufficiently large constant $\alpha \geq 0$. (α may be taken as the diameter of a circle enclosing P.) See Figure 3.

In this case, we define a *1-contact set* to be a maximal set of positions of P where a vertex of P contacts a line segment of O, or a vertex of O contacts a line segment of P. (See Figure 4.) The transformed obstacles O' are the union of these 1-contact sets. Thus each obstacle in O' is a quadratic surface patch which may be easily constructed from the input, there are at most $O(|O||P|)$ such obstacles and their $O(|O|^2|P|^2)$ intersections can easily be computed within accuracy $2^{n^{-c}}$ for any $c > 0$, by known polynomial time procedures [12] for intersection of quadratic surface patches. Hence in this simple example the connected regions bounded by O' can be explicitly constructed in polynomial time within accuracy 2^{-n^c} which is sufficient for solution of this mover's problem.

In the case of a classical mover's problem (O, P, p_I, p_F) of dimension $d = 3$, the transformed problem (O', P', p'_I, p'_F) has dimension $d' = 6$. In this case we define a 1-contact set to be a maximal set of positions of P where an edge of P contacts a face of O or an edge of O contacts a face of P. Again, the 1-contact sets are constant degree polynomial. The transformed obstacles O' are the union of the 1-contact sets. The connected regions defined by O' can again be explicitly constructed by intersecting these constraints. In [1], we briefly suggested a method for this construction, but the full credit should be given to [13] who later gave a complete detailed description of a method for explicit construction of such a transformed mover's problem in 3 dimensions in polynomial time. (In [14] O'Dunlaing et al. further improved this construction by observing that movement of P can be restricted to be

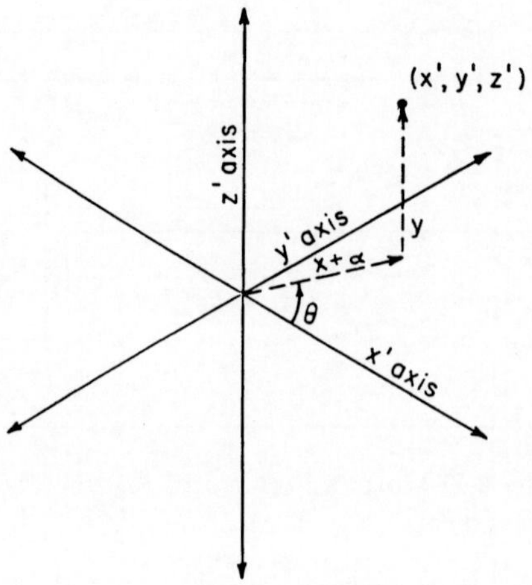

Figure 3: The mapping $f(x,y,\theta) = (x',y',z')$.

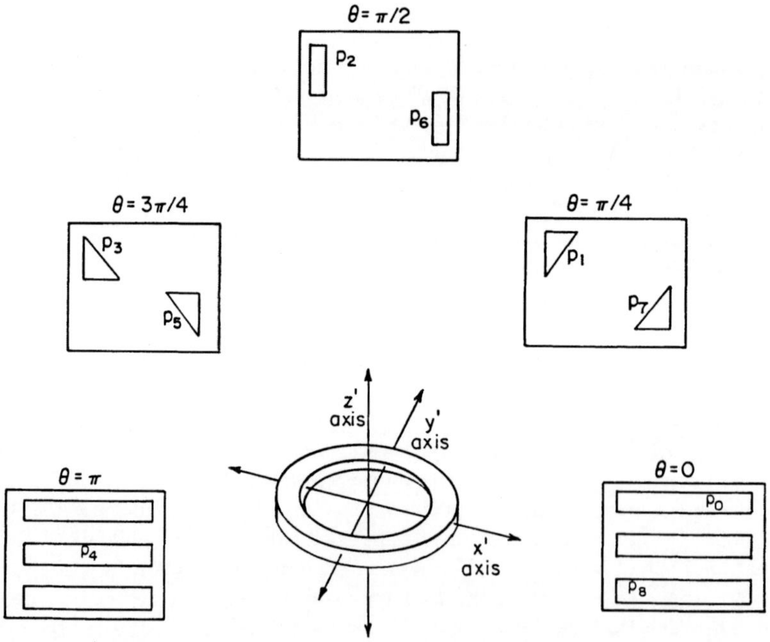

Figure 4: Transformed mover's problem from Figure 1. The obstacles of the transformed problem define a torus with cross-sections illustrated for $\theta = 0, \pi/4, \pi/2, 3\pi/4, \pi$. P may be moved through positions $p_I = p_0, p_1, \ldots, p_8 = p_F$. as in Figure 2.

equidistant from the obstacles.)

This approach was extended in [15] to solve any generalized mover's problem of input size n with d' degrees of freedom in time $n^{2^{O(d')}}$. They make use of the algebraic decomposition of [16] (previously used to decide formulas of the theory of real closed fields) to construct the connected regions bounded by O'. Note that their upper bounds grow doubly exponentially with d', whereas our polynomial space lower bounds suggest only single exponential time growth with d'. It remains a challenging problem to close the gap between those lower and upper bounds for generalized mover's problems. Further progress will likely depend on improvements to decision algorithms for the theory of real closed fields; recently Ben-Or et al. gave a single exponential space decision algorithm [17].

4. Further Problems in Computational Robotics

There are some very challenging problems remaining in the field of Computational Robotics beyond the complexity of the mover's problems and some recent progress.

(1)Frictional Movement. The problem here is to plan movement for (O, P, p_I, p_F) in the case contact is allowed in the presence of friction between surfaces. In [18], Rajan and Schwartz give the first known decision algorithm in the case that O is a cylindrical hole and P is a peg. In [19], Miller and Reif prove undecidability of planning frictional movement. What natural subclass of frictional movement problems is decidable?

(2)Minimal Movement. The problem is, given a set of k polygonal obstacles in d space defined by a total of n linear constraints, and points p_I, p_F find a minimal length path from p_I to p_F avoiding the obstacle O. [20] gives a $O(n \log n)$ algorithm in the case $d = 2$ and $k = 1$. [21] give a $2^{2^{O(n)}}$ algorithm for $d = 3$. Recently, Reif and Storer [22] gave a $O(nk \log n)$ algorithm for $d = 2$ and $n^{k^{O(1)}}$ time and $n^{O(\log k)}$ space algorithms for $d = 3$. Is there a $n^{O(1)}$ algorithm for $d = 3$?

(3)Dynamic Movement. The problem is to plan the movement of a polygon in d dimensions with bounded velocity modulus between points p_I and p_F, so as to avoid contact with a set O of k polygonal obstacles (defined by a total of n linear constraints) moving with fixed, known velocity. [22] give the first known investigation of the computational complexity of planning dynamic movement. They show that the problem of planning dynamic movement of a single ($k = 1$) disk P in $d = 3$ dimensions is polynomial space hard. (This result is somewhat surprising, since P in this case has only 3 degrees of freedom. Our key new idea is to use time to encode a configuration of a polynomial space bounded Turing machine.) Is this problem polynomial space hard for dimension $d = 2$?

Asteroid avoidance problems are a natural subclass of dynamic mover's problems where each obstacle is convex and does not rotate. In [22] Reif and Sharir give a polynomial time algorithm for dimension $d = 2$ with a bounded number $k = O(1)$ of obstacles and give $2^{n^{O(1)}}$ time and $n^{O(\log n)}$ space algorithms for dimension $d = 3$ with an unbounded number k of obstacles. Is the asteroid avoidance problem polynomial in the case $d = 3$?

(4)Adaptive Movement Planning. The problem is to do dynamic movement planning in the case where the obstacles make unpredicted movements in real time. This problem requires some sort of adaptive response to the changes of obstacles' trajectories, and appears considerably more difficult than the dynamic movement problem where the obstacles are assumed to make predictable movements. Although no previous work has been done in this area, it seems to be of central importance.

Acknowledgment

This work was supported by the Office of Naval Research Contract NOOO14-80-C-0647 and NSF Grant DCR-85-03251.

References

[1] J.H. Reif, "Complexity of the Mover's Problem," *Proc. 20th IEEE Symposium on Foundations of Computer Science*, San Juan, Puerto Rico, pp. 421-427, 1979, also appearing in Planning, Geometry and Complexity of Robot Planning, J. Schwartz, ed., Ablex Pub., Norwood, N.J. 1985.

[2] J.H. Reif and M. Sharir, "Motion Planning in the Presence of Moving Obstacles," *26th IEEE Symposium on Foundations of Computer Science*, Portland, Oregon, October 1985.

[3] M.R. Garey and D.S. Johnson, Computers and Intractability: A Guide to the Theory of NP-Completeness, Freedman and Co., San Francisco, 1979.

[4] J.E. Hopcroft, D.A. Joseph and S.H. Whitesides, "On the Movement of Robot Arms in 2-dimensional Bounded Regions," *Proc. 23rd IEEE Symposium on Foundations of Computer Science*, Chicago, Il., pp. 280-289, 1982.

[5] J.E. Hopcroft, J.T. Schwartz and M. Sharir, "On the Complexity of Motion Planning for Multiple Independent Objects: PSPACE Hardness of the Warehouseman's Problems," TR-103, Courant Institute of Mathematics, Feb. 1984.

[6] P. Spirakis and C. Yap, "Strong NP-Hardness of Moving Many Discs," to appear, *Information Processing Letters*, August 1985.

[7] N.J. Nilsson, "A Mobile Automation: An Application of Artificial Intelligence Techniques," Proceedings IJCAI-69, pp. 509-520, 1969.

[8] R. Paul, "Modelling Trajectory Calculation and Servicing of a Computer Controlled Arm," Ph.D. Thesis, Stanford University, Nov. 1972.

[9] S. Udupa, "Collision Detection and Avoidance in Computer Controlled Manipulators," Ph.D. Thesis, Cal. Inst. Tech., 1977.

[10] C. Widdoes, "A Heuristic Collision Avoider for the Stanford Robot Arm," Stanford CS Memo 227, June 1974.

[11] T. Lozano-Pérez and M. Wesley, "An Algorithm for Planning Collision-Free Paths Among Polyhedral Obstacles," *CACM*, vol. 22, pp. 560-570, 1979.

[12] P.G. Comba, "A Procedure for Detecting Intersections of Three-Dimensional Objects," *J. ACM*, vol. 15, No. 3, pp. 354-366, July 1968.

[13] J.T. Schwartz and M. Sharir, "On the Piano Mover's Problem: I. The Special Case of a Rigid Polygonal Body Moving Amidst Polygonal Barriers," *Comm. Pure Applied Mathematics*, vol. XXXVI, pp. 345-398, 1983.

[14] C. O'Dunlaing, M. Sharir and C.K. Yap, "Retraction: A New Approach to Motion Planning," Proc. 15th ACM Symposium on the Theory of Computing, Boston, Ma., pp. 207-220, 1983.

[15] J.T. Schwartz and M. Sharir, "On the Piano Mover's Problem: II. General Techniques for Computing Topological Properties of Real Algebraic Manifolds," *Adv. Applied Mathematics*, vol. 4, pp. 298,351, 1983.

[16] G.E. Collins, "Quantifier Elimination for Real Closed Fields by Cylindric Algebraic Decomposition," Proc. 2nd GI Conference on Automata Theory and Formal Languages, Springer-Verlag, LNCS 35, Berlin, pp. 134-183, 1975.

[17] M. Ben-Or, D. Kozen and J.H. Reif, "Complexity of Elementary Algebra and Geometry," 16th Symposium on Theory of Computing, 1984, also to appear in *J. Computer and System Sciences*, 1985.

[18] V.T. Rajan and J.T. Schwartz, work in progress, 1985.

[19] G. Miller and J.H Reif, "Robotic Movement Planning in the Presence of Friction is Undecidable," to appear, 1985.

[20] B. Chazelle, "A Theorem on Polygon Cutting with Applications," *Proc. 23rd IEEE Symposium on Foundations of Computer Science*, Chicago, Il., pp. 339-349, 1982.

[21] M. Sharir and A. Schorr, "On the Shortest Path in Polyhedral Spaces," *Proc. 16th ACM Symposium on the Theory of Computing*, Washington, D.C., pp. 144-153, 1984.

[22] J.H. Reif and J. Storer, "Shortest Paths in Euclidean Space with Polyhedral Obstacles," Center for Research in Computing Technology, Harvard University, TR-05-85, May 1985.

[23] M. Brady, J.M. Hollerbach, T.L. Johnson, T. Lozano-Pérez and M.T. Mason (eds.), Robot Motion: Planning and Control, M.I.T. Press, 1983.

[24] H.R. Lewis and C.H. Papadimitriou, "Symmetric Space Bounded Computation," *Theor. Comput. Sci.*, vol. 19, pp. 161-187, 1982.

[25] W.J. Savitch, "Relationships Between Nondeterministic and Deterministic Tape Complexities," *J. Computer Sci.*, vol. 4, pp. 177-192, 1970.

[26] J.T. Schwartz and M. Sharir, "On the Piano Mover's Problem: III. Coordinating the Motion of Several Independent Bodies: The Special Case of Circular Bodies Moving Amidst Polygonal Barriers," *The International Journal of Robotics Research*, vol. 2, No. 3, pp. 46-75, fall 1983.

Continuous Robot Motion Planning In Unknown Environment

Vladimir J. Lumelsky
Center for Systems Science
Department of Electrical Engineering
Yale University

Abstract

An overview is given of a class of non-heuristic algorithms for planning collision-free motion for robotic systems (such as an autonomous vehicle or a manipulator arm) operating in an environment with obstacles, when any point of the (multi-link) robot body can be subject to collision. In this approach (called Continuous Path Planning, CPP), obstacles are considered to be unknown, and the system has a feedback providing it with information about its immediate surroundings. Such local information is shown to be sufficient to guarantee reaching a global goal, while generating reasonable (if, in general, not optimal) paths. No constraints on the shape of the robot or of the obstacles are imposed. The general idea is to reduce motion planning to the analysis of simple closed curves on the surfaces of appropriate two-dimensional manifolds. The approach is readily compatible with real-time and sensory feedback applications, and thus presents an attractive alternative to the Piano Mover's approach [1,3] where full information about the environment is assumed. In this paper, versions of CPP algorithms are reviewed and examples are given for the cases of a point automaton and of planar and three-dimensional manipulator arms.

1. Introduction

The task of moving an automaton (a body) in an environment with obstacles includes designing such continuous trajectories for one or more of the automaton's points that the automaton could pass, collision-free, from the starting position to a predefined target position. In the sequel, a distinction will be made between those approaches where full information about the environment (the scene) is given (the Piano Mover's model, [1,3]) and approaches where an element of uncertainty is assumed. Also, our emphasis is on non-heuristic ("exact") algorithms - that is, on such path planning procedures which guarantee either reaching the target, or concluding that the target cannot be reached.

The complexity of the Piano Mover's problem for cases of rigid or hinged bodies has been extensively studied in recent years; a number of algorithms have been described. The approach was shown to be computationally prohibitive [3,13]. A two-dimensional (2D) case is studied in [1,10,12]; cases where the objects to be moved are polygons (polyhedra) or discs (spheres) moving amidst polygonal (polyhedral) obstacles are considered in [1,2,3,10,20,22]. The case of moving an object with a number of free-hinged links was studied in 1968 by Pieper [27] in the context of robot arm's control. Exact algorithms [3,12,13] as well as algorithms involving various heuristics [17,19] have been suggested for this problem.

The problem of path planning in an uncertain environment has been considered mainly in works related to 2D navigation of an autonomous vehicle; a number of path planning heuristics were introduced [14,15,16,21]. Typically, obstacles are approximated by polygons, and a piece of the path is formed from the edges of the connectivity graph resulting from studying the surrounding area for which information is available (for example, from a vision module). Within such limited areas, the problem is usually treated as one with full information.

Instead of assuming full information about the environment and computing the whole path at once (which is typical in the Piano Mover's approach), we attempt to use local algorithms in which the path is computed one step (point) at a time, based on the automaton's current position and on local information about its immediate surroundings coming, for example, from a sensory system.[1] (As an example, recall that normally one does not need the full layout of a building in order to go around it.) To emphasize the fact that local algorithms generate paths continuously, based on the incoming information, the approach is referred to as *Continuous Path Planning* (CPP).

Sensory feedback is a typical source of local information in robotic systems, and so local algorithms present an attractive alternative to the Piano Mover's approach, especially in real-time and sensor-based applications. Other advantages include independence of algorithms on the shape of obstacles and of the robot body (robot arm links). Obstacles may be convex or concave and may be formed by any linear or nonlinear (physically realizable) surfaces (lines). Computational characteristics of local algorithms are good since very little information has to be processed at each step. Also, because the processing at each path point depends only on its immediate surroundings and not on the rest of the scene, the solution requires time linear in the complexity of the scene.

The CPP approach exploits a known topological fact that following a simple closed curve ("obstacle") lying in an orientable 2-dimensional surface is a uniquely defined operation - it can be done only clockwise or counter-clockwise. A point automaton moving in the plane presents the simplest case. As for an arm manipulator with a fixed base, an attempt is made, using natural constraints imposed by the system kinematics, to present the system as a point moving along some 2D manifold, with the transformed obstacles forming combinations of simple closed curves. In this case, the arm feedback is assumed to be similar to that of a human arm: when the manipulator touches an obstacle, the point(s) of contact (in the arm reference system) is known.[2] Also,a notion of a desired trajectory is introduced (for example, a straight line trajectory for the arm endpoint) which helps in limiting the automaton's search.

We consider only those arm manipulators that obey the principle of *arm separability* and have the minimum number of degrees of freedom (dof) (in other words, are not redundant)[3] In a separable arm [6], all the links and joints can be divided into the *major linkage* responsible for positioning of the arm in the required area of the work space, and the *minor linkage* (wrist) responsible for the orientation of the end effector. As a rule, existing robot systems have separable linkages.

In a 2D arm with no redundancy, the major linkage consists of the first two links, with the third link forming the wrist; in a 3D arm, the major linkage consists of the first three links. We assume (as is commonly done in the work on path planning for arm manipulators [17,19,23,28]) that, from the standpoint of the gross motion, wrist dimensions are negligible, and the problem of path planning can be solved taking into account only the major linkage; then, in the vicinity of the target, both linkages can be used for the final orientation of the arm end effector.

Below, an overview of the CPP approach, along with examples of performance of a num-

[1] An interesting issue of the value of local means in reaching a global goal has been actively studied in the last twenty years (although, not in the geometric context) in the theory of automata and the theory of games (see, e.g., [29]).

[2] Although, for the sake of clarity, a model of "tactile" sensory feedback is being used here, the approach is applicable to a non-contact sensory feedback as well. Moreover, it can be shown that increasing the distance at which the automaton detects the obstacle, improves the path and, in the limit, leads to locally optimal paths.

[3] The minimum number of dof is what is required to guarantee that the arm end effector can reach an arbitrary point in the work space (taking into account both the position and the orientation). In a redundant arm, the number of dof exceeds this minimum. For a planar arm, the minimum is three dof, and for a 3D arm, it is six dof [4]. In general, in an arm with the minimum number of dof, a given path point corresponds to a unique set of values for the arm joints, whereas infinite number of such sets appear in a redundant arm.

ber of CPP algorithms, are given. For proofs of convergence and other details, references are provided. We start with a CPP algorithm for a point automaton traveling in the plane (Section 2). This algorithm forms a basis for a series of CPP procedures for various robot arms. Although the general idea of continuous path planning is the same for all cases, some important details of the procedures turn out to be dependent on the arms kinematics. CPP algorithms for two types of planar arms are considered in Section 3, and a case of a 3D arm is considered in Section 4.

2. Point Automaton in Plane

The theory for this case was developed in [5,7]. The goal of the automaton (for example, a mobile robot) is to reach a point T (Target), starting at a point S (Start), in an environment with unknown obstacles. Each obstacle is a simple closed curve homeomorphic to a circle. There is a finite number of obstacles in the scene; any "reasonable" trajectory or a disk drawn in the scene will have a finite number of intersections with the obstacles. The automaton knows its current coordinates as well as the coordinates of the point T and is capable of: (i) moving along a desired trajectory (called the *main line* or *M-line* - this can be, for example, a straight line), (ii) "feeling" a contact with the obstacle, and (iii) following the obstacle boundary. The strictly local information thus available to the automaton was shown in [5] to be sufficient to guarantee, using the suggested ***basic path planning procedure***, reaching the target in a finite time (or concluding that the target cannot be reached), while producing reasonable, if not optimal, paths. Also, an upper bound on the length of paths generated by the basic procedure had been produced, as well as a lower bound on the length of paths generated by any algorithm operating in unknown environment.[4]

In the basic procedure, when the automaton, while moving along M-line, meets an obstacle, it defines on it a ***hit point***, H_j; when, in order to start moving again along M-line, it leaves the obstacle, it defines on it a ***leave point***, L_j. Except Start and Target, hit and leave points are numbered in pairs, in order of their occurrence; L_o =Start. The distance is Euclidean distance in the plane. Since no apriori information is given for deciding on the direction for passing around an obstacle, both directions (when facing the obstacle, these are right and left) are equally acceptable. A *local direction* is defined as a once and for all decided upon direction for passing around the obstacle; it can be either right or left. The procedure does the following (refer to Figure 1).

1. Set $j = 1$. Go to Step 2.
2. From point L_{j-1}, the automaton moves along M-line until one of the following occurs:
   (a) Target is reached. The procedure stops.
   (b) An obstacle is encountered and a hit point, H_j, is defined. Go to Step 3.
3. Using the accepted local direction, the automaton follows the obstacle boundary until one of the following occurs:
   (a) Target is reached. The procedure stops.
   (b) M-line is met at a distance d from T such that $d < d(H_j, T)$; point L_j is defined. Increment j. Go to Step 2.
   (c) The arm returns to H_j (and thus completes a closed curve along the obstacle boundary) without ever meeting M-line. The target is trapped and cannot be reached. The procedure stops.

[4]We are aware of two distinct ideas on how an exact local algorithm for a point automaton can be designed. The basic procedure above is one of them; another scheme is described in [5,7]. Interestingly, not all seemingly similar strategies guarantee termination. For example, the Pledge algorithm [26] can help one to find a way out of a maze, but cannot help in reaching a specific point in the maze.

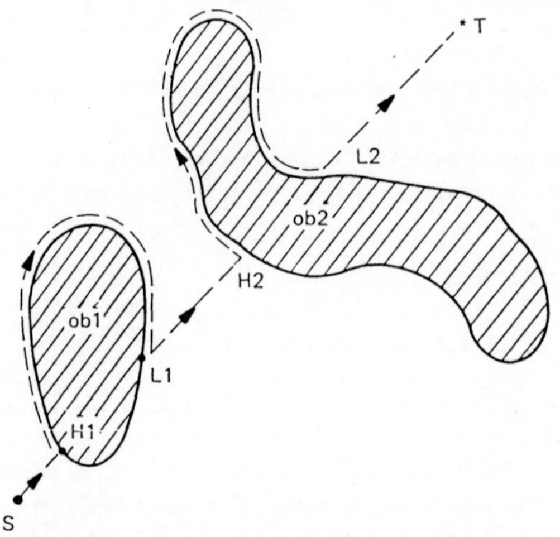

Figure 1: Basic path planning procedure .

It has been shown in [7] that in some special cases this procedure creates local cycles in which the automaton returns to some of the points that it has passed before. One characteristic of scenes where local cycles develop is that one or both points S and T lie inside a convex hull of one of the obstacles. The number of local cycles can be shown to be always finite, and so the convergence of the algorithm is not affected. An example of a scene which produces local cycles is shown in Figure 2. Although scenes which create local cycles in the considered case may seem to be rather unusual, studying local cycles turns out to be important. This is primarily because local cycles can appear in cases of arm manipulators even when they operate in simple scenes with convex obstacles (see the following sections).

3. Planar Arm Manipulators

We consider two planar arms whose kinematics are depicted in Figure 3; these and some other types are studied in detail in [11,18,25]. In the rest of the cases considered below the shape of the arm links and of the obstacles (e.g., their convexity or concavity) is of no importance to the CPP algorithms. Solely for better visualization, line segment links are used.

3.1 Arm of Figure 3a.

The arm consists of two links, l_1 and l_2, and of two joints, J_0 and J_1; joint J_0 is fixed. A link l_i (i=1,2) is a straight line segment of the length l_i which can rotate indefinitely about the corresponding joint producing an angle (a joint value) θ_i. If no obstacles are present, the arm endpoint can reach any point within the boundaries of its *work space (W-space)* formed by a circle with the radius $(l_1 + l_2)$ and by a circular "dead zone" whose radius is $|l_1 - l_2|$. A pair of values (θ_1^p, θ_2^p), or a set of coordinates of the link endpoints a_p and b_p, corresponding to a given position P of the arm endpoint in W-space, represent an arm solution (arm position) for P. In general, any position of the arm endpoint in W-space, except for points along the W-space boundaries, corresponds to two arm solutions.

An obstacle in W-space is a closed curve of finite length homeomorphic to a circle. There may be only a finite number of obstacles present; any "reasonable" curve or a disk passing through W-space intersects a finite set of obstacles. Being rigid bodies, obstacles cannot

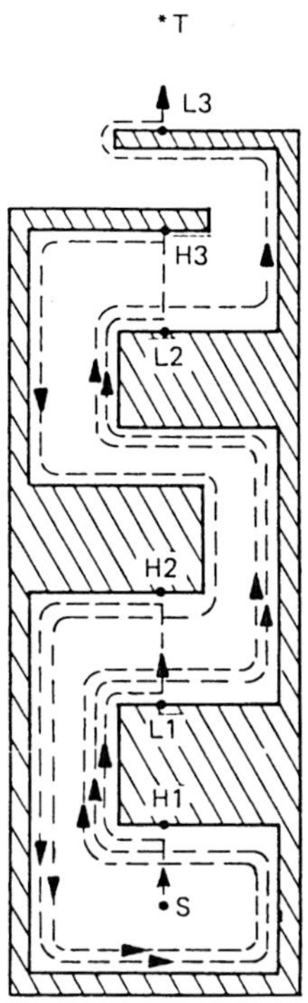

Figure 2: Local cycles created by the basic path planning procedure. The automaton passes points H_1 and L_1 three times, and points H_2 and L_2 twice.

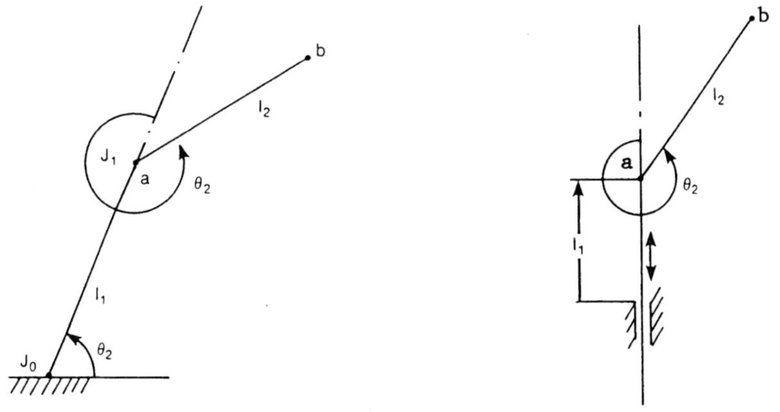

a) b)

Figure 3: Two planar arms: (a) with two revolute joints; (b) with one sliding and one revolute joint.

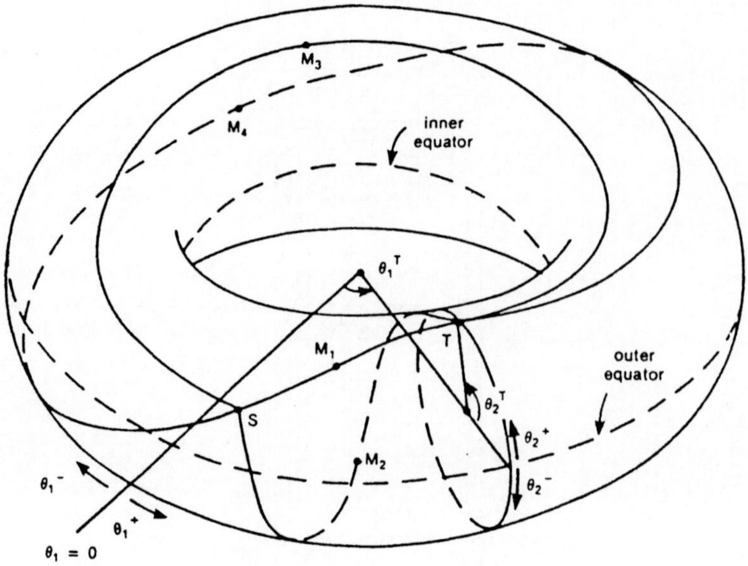

Figure 4: I-space torus.

Zeros and positive and negative directions for both angles, θ_1 and θ_2 are shown. For a given θ_1, point $\theta_2 = 0$ lies at a corresponding point of the outer equator; for example, coordinates of point T are (θ_1^T, θ_2^T). M_1, M_2, M_3, M_4 are middle points of four "shortest" routes between points S and T.

intersect. Two (or more) obstacles may touch each other, in which case the arm treats them as one obstacle. Only such configurations of sets of obstacles are considered for which, at any position of the arm, at least some arm motion is possible. Only continuous motion of robot links is allowed.

At any given moment, the arm knows its current coordinates θ_1 and θ_2, and the coordinates of the target position T. Position S is known to be ***reachable***; that is, when the arm is in the position S, no arm links intersect (interfere with) any obstacle. It is not known whether position T is reachable, and, if so, whether T can be reached from S. The arm is assumed to be capable of performing the following actions:

1. Moving the arm endpoint through a prescribed simple curve (M-line) connecting points S and T.
2. Identifying the point(s) of contact on the arm body when the arm hits an obstacle.
3. Following the obstacle boundary.

The arm is said to be moving in free space when it has no contacts with obstacles. When the arm touches an obstacle, it has only two directions for maneuvering around the obstacle while keeping in contact with it. *Passing around an obstacle* is such a continuous motion of the arm, at any moment of which the arm is in contact with some obstacle(s). When facing the obstacle's boundary, the point of contact can move along the boundary either right or left; this motion does not have to be continuous neither along the arm body nor along the boundary of the actual obstacle.

We introduce an *image space (I-space)* as the surface of a common two-dimensional torus defined by two independent angular variables, θ_1 and θ_2 [8]. An arm position P in W-space corresponds to a point P on the torus surface with coordinates θ_1^p and θ_2^p. A simple closed curve in I-space has its simple closed curve counterpart in W-space [9]. Given an M-line in W-space and the corresponding arm positions, there is an M-line image in I-space. A geodesic curve connecting points S and T on the torus surface corresponds to a straight line in the plane of variables (θ_1, θ_2) and thus can be used as the "shortest" M-line between positions S and T. Because of the torus topology, in general, four such "shortest" M-lines can appear. Two segments, M_k and M_m, $k, m = 1,2,3,4$, $k \neq m$, are said to be complementary over the angle θ_i, i=1,2, if the corresponding changes in θ_i from S to T for both segments add to 2π. Positive and negative directions and zero points for both variables of I-space are shown in Figure 4 along with four M_k-lines and their middle points.

In the example in Figure 5a, some intermediate positions of link l_2 during its passing around the obstacles A and B are shown. (Dashed lines represent the boundary of W-space and the locus of points of the endpoint of link l_1.) Note that there appear areas (shaded) which cannot be reached by the arm endpoint and which thus present to the arm obstacles as real as the actual obstacles. These areas (called obstacle *shadows*) form, together with the corresponding actual obstacles, *virtual obstacles*. A *virtual line* is a curve in W-space which the arm endpoint follows when the arm is passing around an obstacle. The virtual line forms the boundary of the virtual obstacle in W-space. An actual obstacle may form disconnected shadows; actual obstacles may interact in forming shadows - for example, when two arm points touch two or more actual obstacles simultaneously (Figure 5a).

A virtual obstacle in W-space has its counterpart in I-space; its boundary is called the *virtual boundary* and is defined by the virtual line with the corresponding link positions added. In Figure 5b, the virtual boundary of the obstacle (A+B) forms a single simple closed curve whereas the virtual boundary of the obstacle C forms two simple closed curves. It can be shown that, first, only simple closed curves may form virtual obstacles in I-space and, second, a virtual boundary may contain no more than two simple closed curves. The proof [25] uses Jordan Curve Theorem and the fact that the *first connectivity number* of the common torus (which defines the maximum number of closed cuts that can be made on a closed surface without dividing it into separate domains) is two [9]. Therefore, if, on its way around an obstacle, the arm endpoint completes a full circle, it does not necessarily mean that the whole virtual boundary has been traversed. There may be another, yet unobserved closed curve which limits the virtual obstacle "from the other side". To take this fact into account and to assure algorithm convergence for the arm of Figure 3a, some modifications of the CPP procedure of Section 2 are necessary; these are described below.

All possible ways to form virtual boundaries are shown schematically in Figure 6. An obstacle is of *Type I* if its virtual boundary is formed by a single closed curve; it is of *Type II* if its virtual boundary is formed by two closed curves. In the algorithm, testing for the obstacle type is done by introducing two counters, T_1 and T_2, corresponding to θ_1 and θ_2, respectively (for details and examples, see [25]). When the arm follows a closed curve of the virtual boundary, each counter integrates its angle, considering the sign. When the arm completes a closed curve, the result must be $n \cdot 2\pi$, $n = 0, 1, 2, \ldots$. For a given closed curve, the resulting values of the pair (T_1, T_2) define its *arm joints range* (or, simply, *range*). For a Type I obstacle, its range is (0,0). For a Type II obstacle, its range is $(n_1 \cdot 2\pi, n_2 \cdot 2\pi)$, with either $n_i = 0; n_{3-i} = 1$, or $n_i = 1; n_{3-i} = 1, 2, \ldots; i = 1, 2$.

If, after having completed a closed curve, the range is (0,0) (a Type I obstacle), this means that the whole virtual boundary of the obstacle in question has been observed, and, therefore, the target cannot be reached. If the range of the closed curve is different from (0,0) (a Type II obstacle), then there must be another closed curve somewhere corresponding to the same obstacle. In this case, a properly chosen complementary M_k-line is used in an attempt to reach the target "from the other side". Altogether, no more than two M_k-lines are needed to complete the task. Although any simple line can be used as an M-line, the algorithm becomes somewhat simpler if geodesics on the I-space torus are used (as shown in Figure 4).

Now, the whole path planning procedure for the planar arm of Figure 3a can be formulated. In the procedure, the hit points (positions), H_j, and the leave points, L_j, are numbered

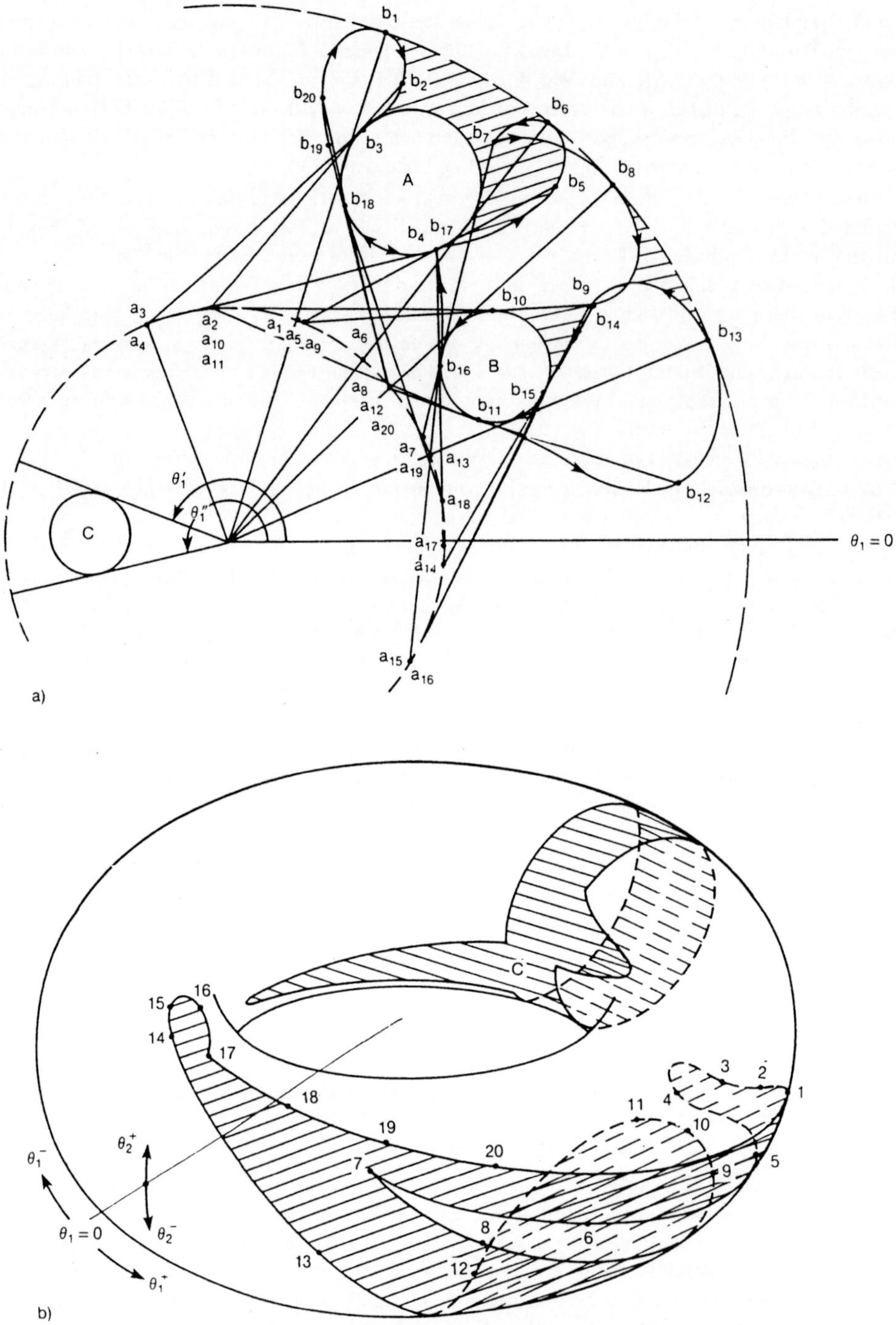

Figure 5: Example for the arm of Figure 3a.
(a) W-space; A,B,C-actual obstacles. (b) I-space image of the same obstacles.

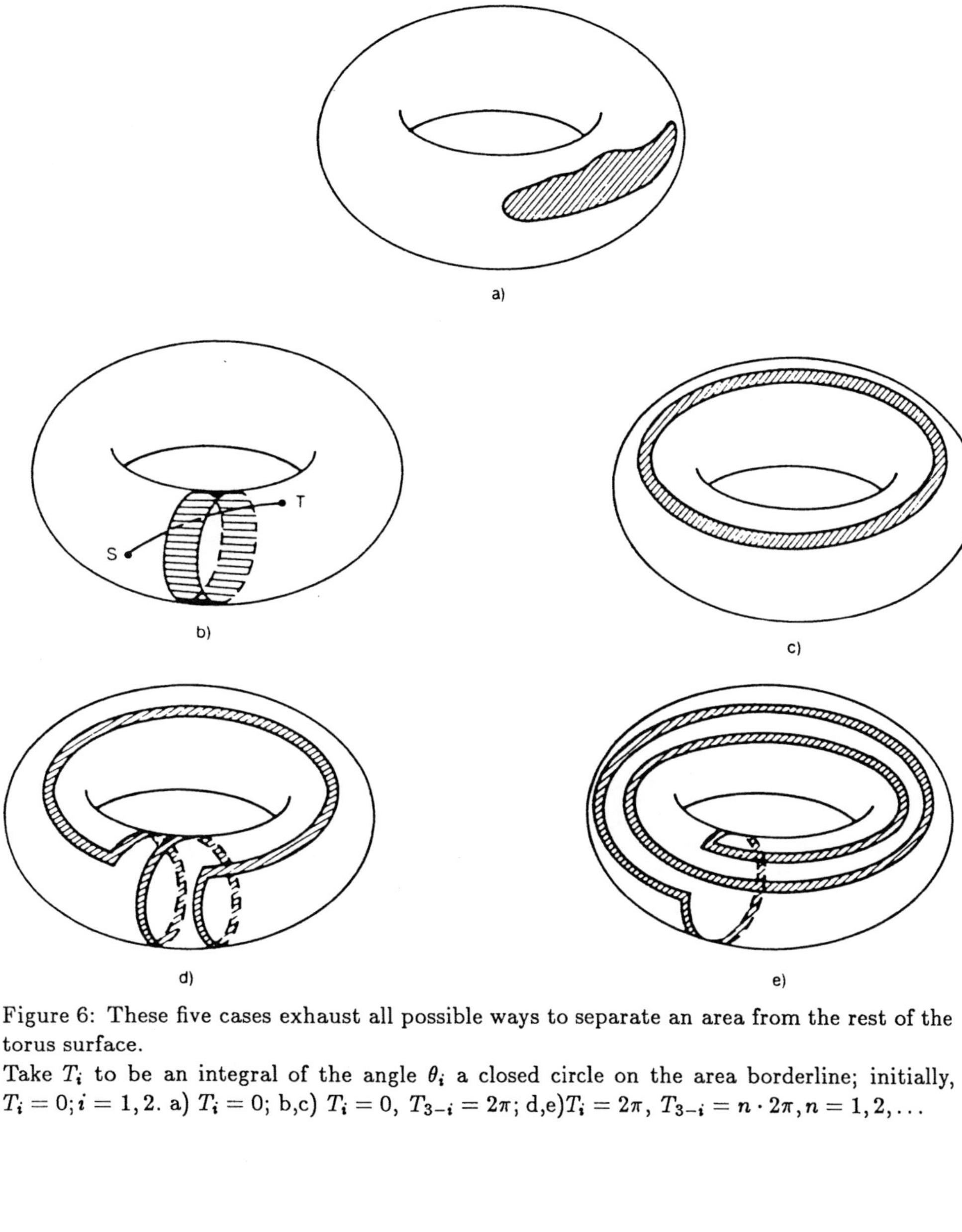

Figure 6: These five cases exhaust all possible ways to separate an area from the rest of the torus surface.

Take T_i to be an integral of the angle θ_i a closed circle on the area borderline; initially, $T_i = 0; i = 1, 2$. a) $T_i = 0$; b,c) $T_i = 0$, $T_{3-i} = 2\pi$; d,e)$T_i = 2\pi$, $T_{3-i} = n \cdot 2\pi, n = 1, 2, \ldots$

in the order of their occurrence, respectively; L_o=Start. If a new M-line is introduced, the arm starts again at point Start, and the numbering starts again. Distance d(P,Q) between arm positions P and Q in I-space is the Euclidean distance in the plane (θ_1, θ_2); the length of an M-line segment is defined similarly. A flag is used to indicate that (in case of a Type II obstacle) one of the two closed curves of the current virtual boundary has been processed. The algorithm includes the following steps.

1. All four complementary M_k-lines, $k = 1, 2, 3, 4$, are defined as geodesics on the I-space torus[5] (see Figure 4), and then ordered as follows: M_1 is the shortest of the segments, M_2 complements M_1 over the angle θ_2; M_3 complements M_1 over θ_1; M_4 complements M_1 over both θ_1 and θ_2. Go to Step 2.
2. M_1-line is designated as M-line. Set the flag down. Set j=1. Go to Step 3.
3. Counters T_1 and T_2 are set to zero. From point L_{j-1}, the arm moves along M-line until one of the following occurs:
   (a) Target is reached. The procedure stops.
   (b) An obstacle is encountered and a hit point, H_j, is defined. Go to Step 4.
4. Counters T_1 and T_2 are turned on. The arm follows the virtual boundary until one of the following occurs:
   (a) Target is reached. The procedure stops.
   (b) M-line is met at a distance d from T such that $d < d(H_j, T)$; point L_j is defined. Increment j. Go to Step 3.
   (c) The arm returns to H_j (which completes a closed curve along the virtual boundary) without ever meeting M-line. Go to Step 5.
5. Examine the obstacle range accumulated in the counters T_1 and T_2. One of the following takes place:
   (a) The range is (0,0) (i.e., this is a Type I obstacle). Target cannot be reached. The procedure stops.
   (b) The range is not (0,0) and the flag is up (i.e., this is the second closed curve of the virtual boundary of a Type II obstacle). Target cannot be reached. The procedure stops.

   Three cases below, (c,d,e), relate to the situation when the range is not (0,0) and the flag is down (i.e., the first closed curve of the virtual boundary of a Type II obstacle is being processed).

   (c) The range is $(0, n_2)$; $n_2 \geq 1$ - an integer; designate the shorter of M_3 and M_4 as M-line. Go to Step 6.
   (d) The range is $(n_1, 0)$; $n_1 \geq 1$; designate the shorter of M_2 and M_4 as M-line. Go to Step 6.
   (e) The range is (n_1, n_2); $n_1, n_2 \geq 1$; designate the shortest of M_2, M_3, and M_4 as M-line. Go to Step 6.
6. The arm moves back to Start. Set the flag up. Set j=1. Go to Step 3.

An "industrial-like" example shown in Figure 7 contains convex and concave obstacles A,B,C,D whose boundaries are formed by straight lines and curves. The desired M-line (dashed line) for the arm endpoint is the shortest path in the plane (θ_1, θ_2). Some intermediate positions of link l_2 along the path, (1,1), (2,2),..., (12,12), are shown. Note that although M-line plays a crucial role in the algorithm, the arm actually spends little time

[5] The corresponding formulas can be found in [11].

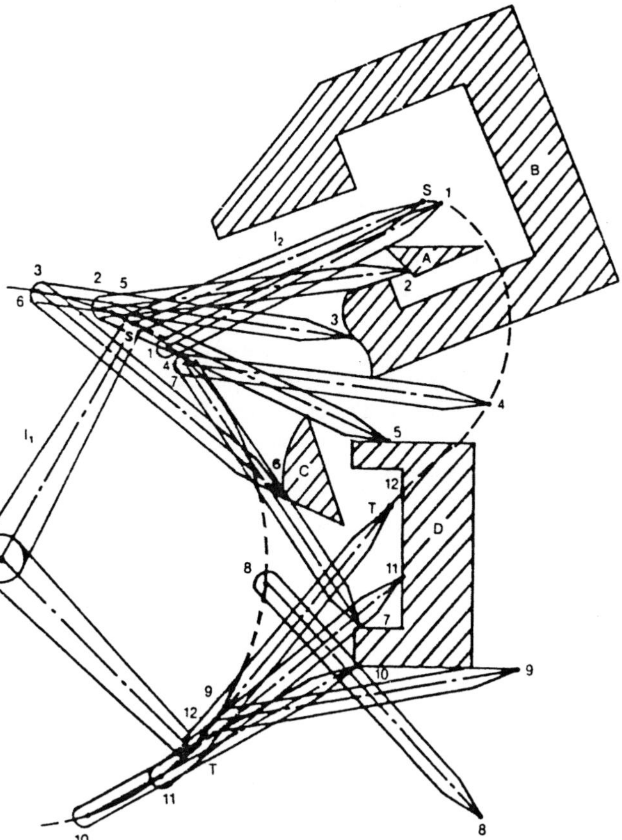

Figure 7: Example of path generation.

The arm moves from position (S,S)= Start to position (T,T)=Target. The work space is filled with obstacles A,B,C,D. M-line is the shortest path between S and T in the plane of variables θ_1 and θ_2. Some intermediate positions of the link l_2 are shown-(1,1),(2,2) ... (12,12). (The last of these is shown by its endpoints only).

moving along M-line but spends most of the time in passing around obstacles. Also, in spite of the fact that the path has been generated with no apriori knowledge about the obstacles, it is quite close to the optimum path (if such a path were somehow generated given the full information on the obstacles).

The described CPP procedure can create local cycles in scenes with some special configurations of obstacles and S and T points, even if obstacles are convex; one such example is described in [11].

3.2 Arm of Figure 3b

Two joint values of this arm are, respectively, the variable length of the first link, l_1, $0 \geq l_1 \geq l_{1max}$, and the angle of rotation, θ_2, of the second link. The length of link l_2 is constant. W-space of this arm (Figure 8) is thus limited by a combination of a rectangle whose sides are equal to $2 \cdot l_2$ and l_{1max}, and of two half-disks of the radius l_2 attached to the opposite sides of the rectangle.

We assume that no obstacles may interfere with the part of link l_1 expanding outside W-space. Circles on the top and on the bottom of W-space (Figure 8) form its *limit areas*. Any point outside the limit areas of W-space has two possible arm solutions; any point inside the limit areas has only one arm solution. For an actual obstacle, its virtual obstacle, virtual line, and virtual boundary are defined similarly to those of the arm of Figure 3a.

I-space for this arm is the surface of a cylinder whose height is l_{1max}, and whose base corresponds to the second joint value, θ_2. Virtual obstacles may be formed in I-space by single closed curves; these are Type I obstacles (Figure 9, obstacle A). If a part of a virtual line coincides with one of the limit values of the first joint value, l_1, then this part cannot be traced by the arm endpoint (Figure 9, obstacles B and C). If this occurs at both limit values of l_1, then the virtual obstacle forms a swath on the surface of the I-space cylinder, and its virtual boundary forms two open simple curves; this is a Type II obstacle. Tracing such a virtual boundary involves walking along one of the open simple curves to one of its dead ends, returning back, and walking in the other direction until reaching the other dead end of the open simple curve; then, the same may have to be repeated with the other open simple curve. Corresponding motion in W-space results in actual maneuvering of the arm around a Type II obstacle. A somewhat different Type II obstacle may form a band around the I-space cylinder; this happens when the obstacle interferes with link l_1 (for details, see [18]).

Although any desired M-line can be used, motion planning becomes somewhat simpler if M-line is defined as a straight line in the plane of variables l_1 and θ_2 - that is, as a function $\theta_2 = p \cdot l_1 + q$, with coefficients p and q determined from the known positions S and T. This line is designated as M_1-line. If, because of obstacles, T cannot be reached from S using M_1-line, a complementary M-line (designated as M_2-line) may be needed; it's coefficients p and q are obtained by substituting $(\theta_2^T - 2\pi)$ instead of θ_2^T in the equation for M-line.

Now, the CPP procedure for the arm of Figure 3a can be formulated. A parameter F is set to +1 when the arm, while following a virtual boundary, reaches the upper limit of the joint value l_1, to -1 when, while following a virtual boundary, it reaches the lower limit of the joint value l_1, and to 0 when the arm starts following an open curve of a Type II virtual boundary. A flag is used to distinguish between the first and the second open curve of the virtual boundary of a Type II obstacle. A counter T_2 is used to handle closed curves of virtual boundaries. Lines M_1 and M_2 are defined above. Hit points, H_j, and leave points, L_j, are defined as in Section 3.1; L_o=Start. Distances are Euclidean distances along M-line in the plane (l_1, θ_2). Every time a hit point is defined, the first local direction for passing around the obstacle is "left". The procedure consists of the following steps.

1. M_1-line is designated as M-line. Set the flag down. Set j=1. Go to Step 2.
2. Set counter T_2 to zero. Set F=0. From point L_{j-1}, the arm moves along M-line until one of the following occurs:

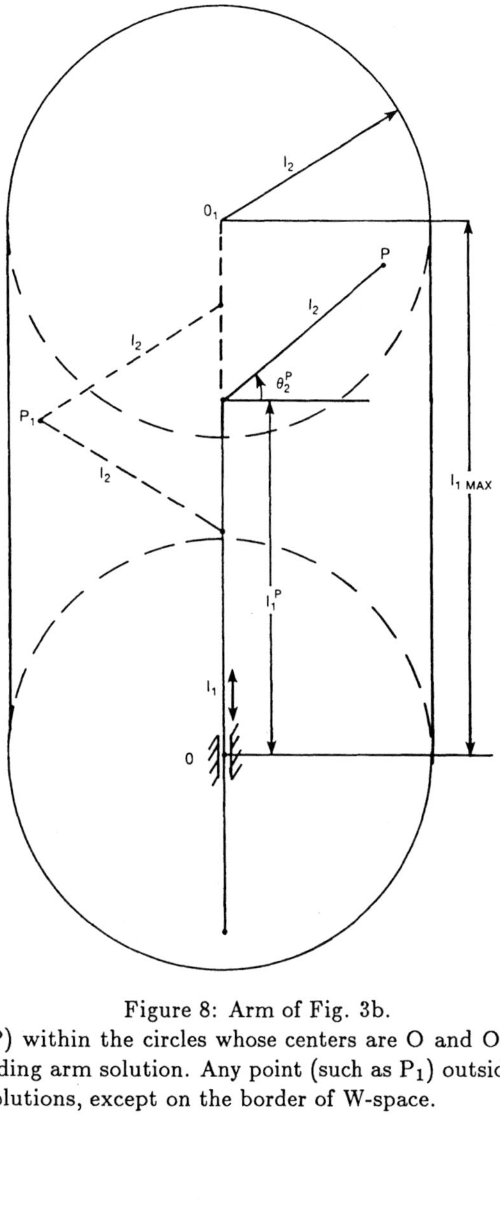

Figure 8: Arm of Fig. 3b.

Any point (such as P) within the circles whose centers are O and O_1, and whose radius is I_2, has one corresponding arm solution. Any point (such as P_1) outside these circles has two corresponding arm solutions, except on the border of W-space.

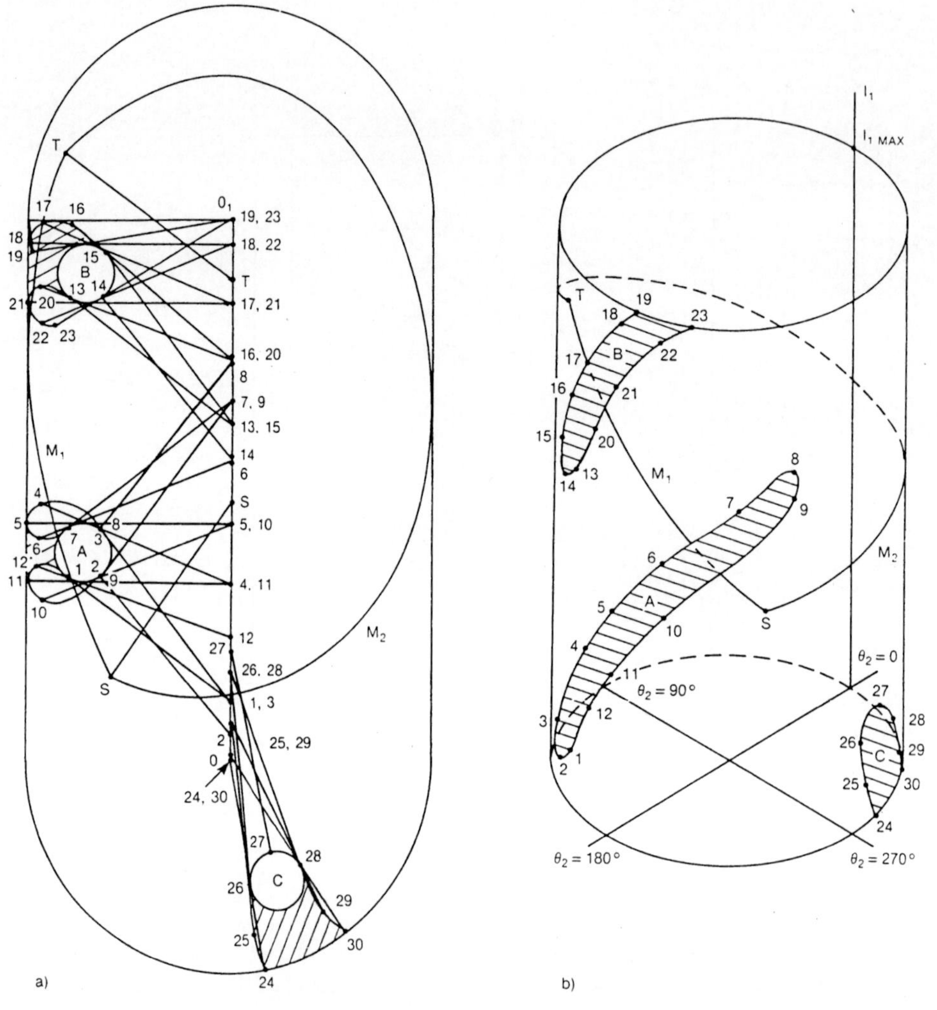

Figure 9: Example for the arm of Fig. 3b.
a) W-space; shown are M-lines M_1 and M_2, and link positions during passing around obstacles A,B, and C. b) I-space images of the same M-lines and of virtual obstacles.

(a) Target is reached. The procedure stops.

(b) An obstacle is encountered and a hit point, H_j, is defined. Choose the local direction "left". Turn on counter T_2. Go to Step 3.

3. The arm follows the virtual boundary until one of the following occurs:

(a) Target is reached. The procedure stops.

(b) M-line is met at a distance d from T such that $d < d(H_j, T)$. Define a leave point L_j. Set the flag down. Set j=j+1. Go to Step 2.

(c) The arm reaches one of the limits of l_1 (this corresponds to a dead end of an open curve of a virtual boundary) without ever meeting M-line. Go to Step 4.

(d) The arm returns to H_j (i.e., a closed curve along the virtual boundary has been completed) without ever meeting M-line. Target cannot be reached. The procedure stops.

4. Depending on the value F, on the flag condition, and on the current arm position, one of the following occurs:

(a) F=0. Set F to +1 or -1 according to the rule above. Return to the last hit point. Choose the local direction "right". Go to Step 3.

(b) The value of F corresponds to the current dead end (i.e., +1 for the upper limit and -1 for the lower limit of l_1); this means the whole obstacle has been investigated. Target cannot be reached. The procedure stops.

(c) The value of F does not correspond to the current dead end; the flag is down; this means the first open curve of a Type II obstacle has been investigated. Set j=1. Set the flag up. Designate M_2-line as M-line. Return to S. Go to Step 2.

(d) The value of F does not correspond to the current dead end; the flag is up; this means the second open curve of a Type II obstacle has been investigated. Target cannot be reached. The procedure stops.

4. Three-dimensional Cartesian Arm

The arm consists of three sliding mutually perpendicular links, l_1, l_2, l_3. In order not to clutter the picture, the arm is shown in Figure 10 in a position P outside the main "action area" of the scene. The arm endpoint can reach any point in the cubicle whose sides are l_{1max}, l_{2max}, and l_{3max}, respectively. The cubicle represents the arm's W-space. Assume that there are no obstacles outside W-space. An arm solution (arm position) corresponding to a given point P is defined by a set of three joint values (l_1^P, l_2^P, l_3^P). Any point in W-space has a unique arm solution. In Figure 10, equivalent coordinates of the link endpoints are indicated by letters a_i, b_i, c_i, i=1,2,3. Arm capabilities and available information are the same as those in the previous section.

Any obstacle in W-space is a volume homeomorphic to a 3D sphere. There may be a finite number of obstacles in the scene; a "reasonable" curve or a sphere drawn in W-space has a finite number of intersections with obstacles. No other constraints on the obstacle shapes are imposed. If two or more obstacles touch each other, they represent a single obstacle to the arm.

If a trivial case of possible interference of link l_1 with obstacles is ignored,[6] there are three important cases of arm-obstacle interaction during the arm motion: (1) only link l_2 has contacts with obstacles (this can occur only on the boundary of W-space); (2) only link l_3 has contacts with obstacles; (3) both links, l_2 and l_3, have contacts with obstacles. These cases are discussed in more detail below.

[6] If, while moving from S toward T, link l_1 meets an obstacle, then T is clearly out of reach. A link sliding along an obstacle toward its target point is not considered to be interfering with the obstacle.

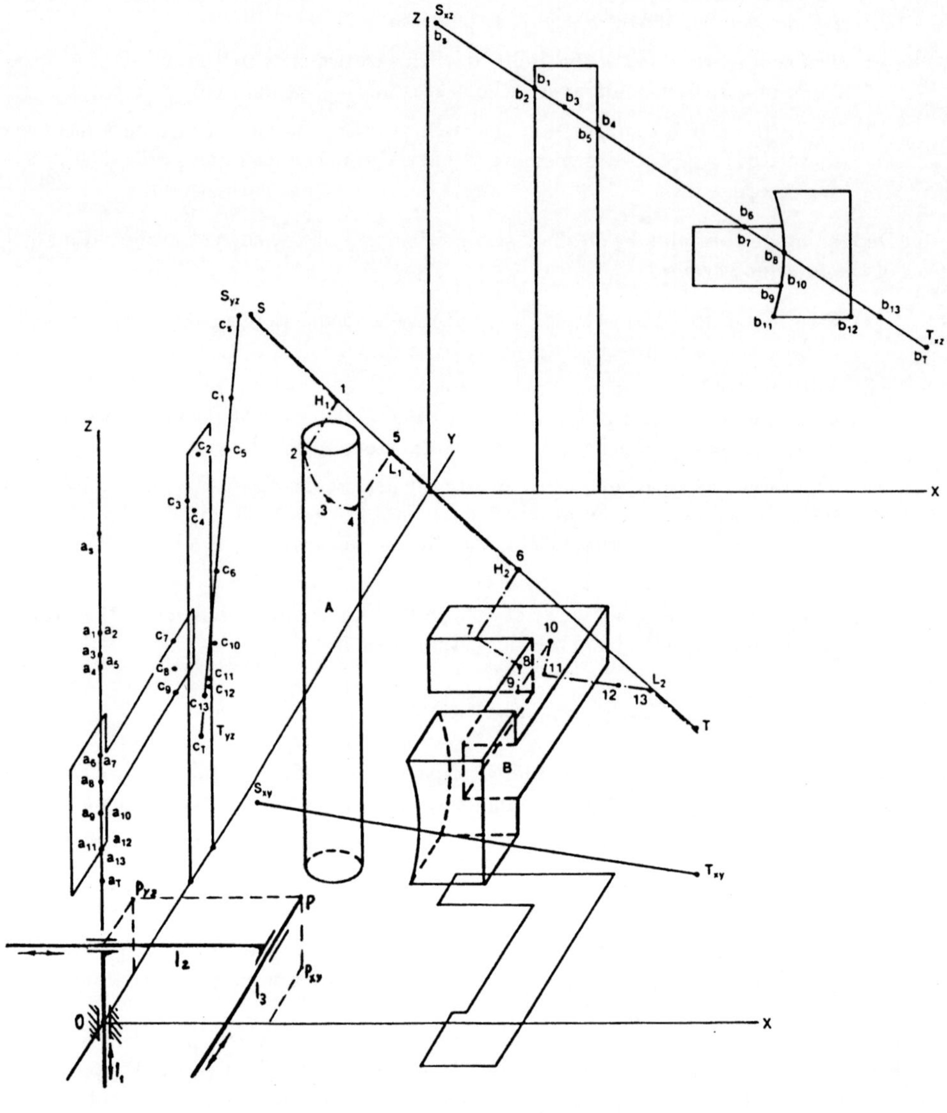

Figure 10: Example of path planning for a three-dimensional cartesian arm.

Case 1. If only link l_2 contacts an obstacle, then the arm motion can be planned effectively in the plane XOZ (Figure 10). This reduces the path planning problem to that of a planar cartesian arm; the corresponding procedure does converge [18]. It can be shown that if a path from S to T exists then there exists a path from S to T (points S and T excluded) such that, first, the arm endpoint moves in the plane XOZ and, second, the path corresponds to a monotonic change in the value l_1. Link l_3 is said to be "passive", in the sense that the value l_3 is not important and can be chosen arbitrarily. In the algorithm, l_3 is calculated so that it produces the shortest distance from the arm endpoint to M-line (in this position, the common perpendicular between link l_3 and M-line passes through the arm endpoint). This way, when the arm completes maneuvering around an obstacle according to the Case 1 strategy, the arm endpoint meets M-line.

Case 2. If the only contacts between the arm and the obstacles are with link l_3, then the motion of l_3 can be planned solely in the plane that contains link l_3 and points S and T (it is called below l_3ST-plane); this plane is perpendicular to the plane XOZ. This strategy converges because the following statement holds: if there is a path from S to T, then there is a path from S to T lying in l_3ST-plane and such that it corresponds to a monotonic change in the values l_1 and l_2. Although all three joint values are changing during the motion in this case, computationally path planning in l_3ST-plane is analogous to the case of a planar cartesian arm.

Case 3. In this combined case, both the links l_2 and l_3 come in contact with obstacles during the arm motion. Because of the "passive" behavior of link l_3 in Case 1, the combined strategy is designed such that contacts of link l_2 with obstacles always take precedence over contacts of link l_3. Whenever link l_3 contacts an obstacle, path planning is done according to the Case 2 strategy. If, in the middle of executing of a Case 2 path leg, link l_2 contacts an obstacle, path planning continues according to the Case 1 strategy. In this case, if free movement of link l_3 is obstructed by another obstacle, then links l_2 and l_3 are simultaneously following some obstacle surfaces, the computed values of l_3 are made as large as possible under the circumstances, and the distance between the arm endpoint and M-line does not necessarily correspond to the common perpendicular between link l_3 and M-line. It can be shown that the combination strategy presented by Case 3 cannot create cycles and terminates in a finite time. The generated paths are, in general, not planar.

In the resulting CPP procedure, two indicators, F_1 and F_2, are used to establish the *M-line directions of change* of joint values l_1 and l_2. The indicator F_i takes one of the values +1, 0, or -1, depending on whether the difference $(l_i^S - l_i^T)$ is positive, zero, or negative, respectively; i=1,2. A local direction for passing around an obstacle is chosen such that it produces the required M-line directions of change of one or both l_1 and l_2 (whatever is applicable in the case at hand). If M-line direction of change of a joint value cannot be determined because this value is constant in the vicinity of the path point currently under consideration, then such a local direction is chosen which corresponds to decreasing values of the joint value with the lower number (see [18] for details). In the procedure described below, H_j and L_j are hit and leave points, respectively, L_o=Start, and distance is Euclidean distance in 3D space.

1. Compute values F_1 and F_2. Compute coefficients of the l_3ST-plane. Set j=1. Go to Step 2.

2. From point L_{j-1}, the arm moves along M-line until one of the following occurs:

   (a) Target is reached. The procedure stops.

   (b) Link l_3 interferes with an obstacle. Define a hit point H_j. Choose a local direction in the l_3ST-plane such that it produces motion corresponding to the current value F_2. If locally the joint value l_2 is constant, choose a local direction that corresponds to decreasing values of l_3. Go to Step 3.

   (c) Link l_2 or both links, l_2 and l_3, interfere with obstacles. Define a hit point H_j. Choose a local direction in the plane XOZ such that it corresponds to the current value F_1. If locally the value l_1 is constant, choose a local direction that corresponds to decreasing values of l_2. Go to Step 4.

3. The arm follows the virtual obstacle, values l_1, l_2, and l_3 are calculated such that l_3 stays in the l_3ST-plane. (As a result, l_2 changes monotonically according to F_2 and l_1 changes monotonically according to F_1). This is done until one of the following occurs:
   (a) Target is reached. The procedure stops.
   (b) M-line is met at a distance d from T such that $d < d(H_j, T)$. Define a leave point L_j. Set j=j+1. Go to Step 2.
   (c) One or both of the current values l_1 or l_2 are outside the intervals (l_1^S, l_1^T) and (l_2^S, l_2^T), respectively. The target cannot be reached. The procedure stops.
   (d) Link l_2 interferes with an obstacle. Choose a local direction in the plane XOZ such that it produces motion corresponding to the current value F_1. If locally the joint value l_1 is constant, choose a local direction that corresponds to decreasing values of l_2. Go to Step 4.

4. Values l_1 and l_2 are calculated so as to allow passing around the obstacle in the plane XOZ. (As a result, values l_1 change monotonically according to F_1). Corresponding values of l_3 are chosen so as to produce the minimum distance from the arm endpoint to M-line. This is done until one of the following occurs:
   (a) Target is reached. The procedure stops.
   (b) M-line is met at a distance d from T such that $d < d(H_j, T)$. Define a leave point L_j. Set j=j+1. Go to Step 2.
   (c) Current value l_1 is outside the interval (l_1^S, l_1^T). Target cannot be reached. The procedure stops.

In the example shown in Figure 10, the task is to move the arm endpoint from point S to point T, preferably along the straight M-line (S,T). Two obstacles, A and B, formed by planar, curved, convex, and concave surfaces, prevent the arm from the straight line motion. Also shown in Figure 10 are projections of the obstacles and of M-line onto the planes XOY, YOZ, and XOZ; the point numbers indicate some of the intermediate positions of the arm endpoint. Under the CPP algorithm presented above, the arm, first, moves from S toward T along the M-line until, at point 1 (p. 1), link l_3 hits obstacle A. Here, link l_3 starts retracting, and the arm maneuvers around obstacle A while keeping in contact with it (points 2,3,4); during this motion, the arm endpoint stays in a plane that contains M-line and is perpendicular to the plane XOZ. At p. 5, the arm meets M-line and follows it in free space, until, at p. 6, link l_3 hits obstacle B. Similar maneuvering follows until, at p. 8, link l_2 touches the obstacle B. At this point, the role of link l_3 becomes passive; its relative positions are computed so as to produce, given l_1 and l_2 values, the minimum distance from the arm endpoint to M-line. As a result, the following arm positions (p. 9 to 13) are computed based on the interaction of l_2 with obstacle B. Finally, the arm meets M-line (p. 13), follows it in free space and eventually arrives at point T.

References

[1] J.T. Schwartz and M. Sharir, "On the 'Piano Movers' Problem. I. The Case of a Two-Dimensional Rigid Polygonal Body Moving Amidst Polygonal Barriers," *Comm. on Pure and Applied Mathematics*, vol. 34, 1983.

[2] C. O'Dunlaing, M. Sharir and C.K. Yap, "Retraction: a New Approach to Motion Planning," Proc. 15th Symposium on Theory of Computing, May 1984.

[3] J. Reif, "Complexity of the Mover's Problem and Generalizations," Proc. 20th Symposium of the Foundations of Computer Science, 1979.

[4] R. Paul, Robot Manipulators: Mathematics, Programming and Control, MIT Press, 1981.

[5] V.J. Lumelsky and A.A. Stepanov, "Navigation Strategies for an Autonomous Vehicle with Incomplete Information on the Environment," General Electric Company, Corporate Research Center, Tech. Report No. 84CRD070, April 1984.

[6] V. Milenkovic and B. Huang, "Kinematics of Major Robot Linkage," Proc. of 13th Intern. Symposium on Industrial Robots and Robots 7 Conference, Chicago, April 1983.

[7] V.J. Lumelsky and A.A. Stepanov, "Effect of Uncertainty on Continuous Path Planning for an Autonomous Vehicle," Proc. of the 23rd IEEE Conference on Decision and Control, Las Vegas, December 1984.

[8] W.S. Massey, Algebraic Topology, Harcourt, Brace & World, 1967.

[9] H. Behnke et al., (ed.) Fundamentals of Mathematics, Vol.II: "Geometry," Ch.16: "Topology," MIT Press, 1974.

[10] J.E. Hopcroft, J.T. Schwartz and M. Sharir, "On the Complexity of Motion Planning for Multiple Independent Objects; PSPACE Hardness of the 'Warehouseman's Problem'," *Int. Journal of Robotics Research*, vol. 3, No.4, 1984.

[11] V.J. Lumelsky, "On Non-Heuristic Motion Planning in Unknown Environment," Proc. of the IFAC Symposium on Robot Control (SYROCO'85), Barcelona, Spain, November 1985.

[12] J. Hopcroft, D. Joseph and S. Whitesides, "On the Movement of Robot Arms in 2-Dimensional Bounded Regions," Proc. of the IEEE Foundations of Computer Science Conference, Chicago, November 1982.

[13] J.T. Schwartz and M. Sharir, "On the 'Piano Movers' Problem. II. General Techniques for Computing Topological Properties of Real Algebraic Manifolds," New York University, Courant Institute, Tech. Report No.41, 1982.

[14] H. Moravec, "Obstacle Avoidance and Navigation in the Real World by a Seeing Robot Rover," Stanford AIM-340, Sept. 1980.

[15] B.Bullock, D. Keirsey, J. Mitchell, T. Nussmeier and D. Tseng, "Autonomous Vehicle Control: An Overview of the Hughes Project," Proceedings of IEEE Computer Society Conference 'Trends and Applications, 1983: Automating Intelligent Behavior', Gaithersburg, Maryland, May 1983.

[16] A.M. Thompson, "The Navigation System of the JPL Robot," Proceedings of 5th Joint International Conf. on Artificial Intelligence, Cambridge, Massachusetts, August 1977.

[17] S.M. Udupa, "Collision Detection and Avoidance in Computer Controlled Manipulators," Proceedings of IJCAI-5, Cambridge, Mass., August 1977.

[18] V.J. Lumelsky, "Continuous Path Planning for Various Planar Robot Arm Configurations," Yale University, Tech. Report No. 8506, April 1985.

[19] B. Faverjon, "Obstacle Avoidance Using an Octree in the Configuration Space of a Manipulator," IEEE Computer Society International Conference on Robotics, Atlanta, Georgia, March 1984.

[20] T. Lozano-Perez, "Automatic Planning of Manipulator Transfer Movements," *IEEE Transactions on Systems, Man and Cybernetics*, SMC-11, No. 10, October 1979.

[21] C.E. Thorpe, "Path Relaxation: Path Planning for a Mobile Robot," Carnegie-Mellon University, CMU-RI-TR-84-5, 1984.

[22] J. O'Rourke, "Convex Hulls, Voronoi Diagrams, and Terrain Navigation," Proc. Pecora IX Remote Sensing Symposium, Sioux Falls, South Dakota, 1984.

[23] R.A. Brooks, "Planning Collision-Free Motions for Pick-and-Place Operations," *Intern. Journal of Robotics Research*, vol.2, No.4, 1983.

[24] V.J. Lumelsky, "On Path Planning for a Planar Robot Arm with Uncertainty," *SIAM Conference on Geometric Modeling and Robotics*, Albany, July 1985.

[25] V.J. Lumelsky, "On Dynamic Path Planning for a Planar Robot Arm," Yale University, Tech. Report No. 8505, April 1985.

[26] H. Abelson and A. diSessa, Turtle Geometry, MIT Press, 1981.

[27] D.L. Pieper, "The Kinematics of Manipulators Under Computer Control," Ph.D. Thesis, Stanford University, 1972.

[28] L. Gouzenes, "Collision Avoidance for Robots in an Experimental Flexible Assembly Cell," IEEE Computer Society International Conference on Robotics, Atlanta, Georgia, March 1984.

[29] E.F. Moore, "The Firing Squad Synchronization Problem," in Sequential Machines: Selected Papers, (ed. E.F. Moore), Reading, Mass., 1964.

Impedance Control Applied to Automated Deburring

Neville Hogan
Department of Mechanical Engineering
Massachusetts Institute of Technology
77 Massachusetts Avenue, Room 3-449
Cambridge, Massachusetts 02139

Abstract

This paper considers some aspects of automatic deburring by robot. The strategy proposed is based on impedance control. Impedance control is a parameter-adaptive approach to manipulation aimed at modulating the effective dynamic behavior of a robot, rather than just its end-effector force or motion. To apply impedance control to robotic deburring, a simple model of the deburring process is developed. A target impedance for the robot is proposed which facilitates the identification of the parameters of the model of the deburring environment and leads to a straightforward parameter-adaptive control strategy for the robot. A method for selecting values for the parameters of an impedance appropriate for deburring is developed by describing the robot's task as an optimization problem. This method yields simple expressions relating the target impedance parameters to tolerances on the interface forces between the tool and the workpiece and tolerances on the desired motion of the tool.

1. Introduction

One of the most important applications of robot technology is the automation of manufacturing operations. To date, robots have been most successfully applied to manufacturing tasks such as spray painting, welding and materials handling (pick and place tasks). A common feature of these applications is that they may be performed by a robot which is essentially a programmable positioning system. The robot controller is designed to produce whatever actuator forces or torques are necessary to bring the robot through a desired trajectory or sequence of positions independent of disturbances or uncertainties.

However, many manufacturing operations require significant mechanical work to be performed by the robot; deburring, drilling, grinding and reaming are some examples. When a robot performs mechanical work on its environment, the robot and the workpiece are dynamically coupled. That dynamic coupling may have a profound effect on the performance of the robot, and may jeopardize its stability. A position controlled robot which can perform free motions without difficulty may develop serious instabilities when it contacts its environment or attempts to perform work on it. Attempts to implement force control in robots have been plagued by this problem [10].

Consequently, one of the foremost challenges in the development of robot technology is the automation of manufacturing operations which require the robot to perform mechanical work. Kramer et al. [9] have pointed out that of these, deburring is the most significant economically. Satisfactory robotic deburring will require a different approach than the usual positioning strategy used for robot control, and that is the topic of this paper.

2. Impedance Control

To deal with the problems of dynamic interaction between a robot and its workpiece a new approach to manipulation has been developed. The principal distinguishing feature of this approach, which has been termed impedance control [4-7] is in the objective of the controller. Whereas a conventional controller is structured to make some selected function of the system state variables (e.g. the position of the robot end-effector) converge to some desired value, an impedance controller attempts the more demanding task of making the entire dynamic behavior of the manipulator converge to some desired dynamic function relating input motions to output forces.

Several questions immediately arise: First of all, can this be done? The feasibility of imposing a desired impedance on a robot manipulator has been demonstrated and discussed in detail elsewhere [1,2,5,8]. It is a non-trivial problem: For example, if a linear model of the robot (valid for small deviations from an operating point) is considered, it can be shown [8] that achieving a desired impedance requires the control system to determine the closed loop eigenvectors as well as the eigenvalues. Even given full state feedback, arbitrary eigenvector assignment cannot be achieved. However, if a physically reasonable target impedance is chosen, Kazerooni [8] has shown that the required eigenstructure is, in fact, achievable. Furthermore, the feasibility of impedance control does not depend on a linear approximation. A nonlinear algorithm for imposing a desired impedance on a serial-link manipulator has been developed [2,5].

A second question: Is it worth doing? Yes; first of all, it has been shown [1,6] that impedance control provides a unified framework for coordinating free motions, obstacle avoidance, kinematically constrained motions and motions involving dynamic interaction. Secondly, a nonlinear impedance controller has been presented [2,5] which completely eliminates some of the more prominent computational problems associated with robot control; the inversion of the kinematic equations is unnecessary and (for a restricted class of target impedances) inversion of the Jacobian can be avoided. Thirdly, it has been shown [3,8] that the stability of an impedance controller is extremely robust. If a particular class of target impedances is achieved, then not only is the robot stable when decoupled from its environment, the stability of the robot is preserved when it is coupled to a very broad class of dynamic environments. In the nonlinear case, it has been shown [3] that this class includes any passive environment with an arbitrary number of degrees of freedom. Working with a linear model of the robot and the controller, Kazerooni [8] has shown that the environment may also include active energy sources.

A third question: How is an appropriate impedance to be chosen for a specific task? Reasoning about task-oriented motions is a formidable problem; reasoning about forces appears to be even more difficult and is a topic of intensive research. An impedance is fundamentally a more complex entity than a force or a motion, and therefore the problem of choosing a task-specific impedance may seem intractably difficult. However, one possible approach [6] is to use optimization theory to model the task. If the objective function describes a trade-off between interface forces and desired motions, optimization leads to a specification of an appropriate robot impedance. This paper will consider the application of that approach to automated deburring.

3. Modeling the Deburring Process

While one of the advantages of impedance control is that a detailed description of the environment is not required to perform some tasks, [1] nor to guarantee stability of the robot, [3,8] it should be clear that information about the environment can be used to improve performance. In fact, for some tasks the optimum impedance is expressed as a relation between the impedance of the robot and the admittance of the environment [6]. Accordingly, the first step in determining an appropriate impedance for deburring is to develop a reasonable description of the deburring environment.

The basic objective of deburring might be described as to generate a surface or an edge "hidden" beneath a burr. At first glance, this seems to be an ideal application for conventional position control which would be used to make the deburring tool describe the outlines of the desired "hidden" surface independent of the action of the "disturbing forces" required to remove the burr. The problem with such an approach is that to remove the burr mechanical work must be done on it by the robot, and failure to take the resulting dynamic interaction between robot and workpiece into account will result in unsatisfactory performance and possible instability. In fact, this application requires a more robust strategy such as impedance control.

4. Removing the Burr

Conceptually, it is useful to partition the deburring problem into two components, one dealing with motions along the normal into the desired "hidden" surface to be generated, the second dealing with motions along the tangent to that surface. Along the tangent to the desired surface or edge to be deburred, the principal concern is to remove the material of the burr. In this context the workpiece is basically an absorber or dissipator of mechanical work. Therefore, for a burr of given dimensions, the workpiece may be characterized by a static force-velocity relation, constrained so that power is absorbed and not generated.

$$F = D(V) \tag{1}$$

$$FV \geq 0 \tag{2}$$

where V is the velocity or feed rate of the deburring tool along the edge, F is the force required to move the tool along the edge and $D()$ is a static (algebraic) functional relation.

Although this force-velocity relation will be nonlinear in most applications, it is unlikely to be discontinuous (assuming the deburring tool is in contact with the workpiece). Consequently, in the neighborhood of any operating condition characterized by a given tool feed rate, V, the force-velocity relation for a burr of a given size may be linearized and the workpiece may be modeled as a linear viscous element.

$$F = B_e V \tag{3}$$

B_e is the effective viscous coefficient for the workpiece.

In practice, as the tool proceeds along an edge of the workpiece, the resistance offered to its motion changes [9] due to variations in the size of the burr, which may change the function $D()$ or the operating point. Strictly speaking, these variations should be represented as a function of distance travelled along the edge, but due to the irregular nature of the burr, the changes may be described in the model proposed above as an unpredictable variation of the viscous coefficient as a function of time.

$$F = B_e(t)V \tag{4}$$

t is time.

Given this description of the environment, what is an appropriate control strategy for the robot? One possibility is to give the manipulator the behavior of a mechanical Norton equivalent network [4] with a purely dissipative linear impedance.

$$F = B_m(V_o - V) \tag{5}$$

B_m is the target viscous coefficient for the robot and V_o is the commanded velocity.

One advantage of this behavior is that it is appropriate both when the tool is in contact with the workpiece and when it is not. When tool and workpiece are not in contact, the interface force is zero and the tool velocity, V, is the same as the commanded velocity, V_o. When tool and workpiece are in contact, increases in the resistance of the workpiece will act to reduce the rate at which the tool is fed, thereby reducing variations in the force exerted by the tool. This will reduce the probability of tool breakage and improve the finish of the deburred part [9].

5. Parameter Identification

A further advantage of this approach is that the simple behavior imposed on the manipulator facilitates identification of the parameters of the model of the environment when tool and workpiece are in contact. The dynamic coupling between the two generates no mechanical work, but merely transmits it from one system to the other without storage or loss. If the workpiece behaves as described by equation (3) and the robot behaves as described by equation (5), coupling the two systems means that the velocity, V, in these equations is common to both systems and the force, F, exerted by the manipulator is experienced by the environment. As a result the velocity, V, is related to the commanded velocity, V_o, as follows.

$$V = V_o B_m/(B_m + B_e) \tag{6}$$

Now assume that a sequence of measurements of the tool speed, V, are made at discrete time intervals. At the same points in time the commanded speed, V_o, and the target manipulator impedance, B_m, (which may vary with time) are assumed to be known. Denoting these sequences by the argument k, the effective viscosity of the workpiece may be estimated using the following formula derived from equation 6.

$$\hat{B}_e(k+1) = B_m(k)\left(\frac{V_o(k)}{V(k)} - 1\right) \tag{7}$$

where $\hat{B}_e(k+1)$ is the estimated effective workpiece viscosity at time interval $k+1$.

As an aside, this procedure could be used to provide real-time monitoring of the burr removal process, which might be valuable in its own right, e.g. for quality control purposes. For present purposes, it can be used as the basis of a parameter-adaptive control strategy which accommodates unpredictable changes in the resistance of the workpiece by periodically updating an estimate of the effective workpiece viscosity and then updating the target impedance of the manipulator to suit.

6. Choice of Impedance Parameters

Such a parameter adaptive strategy poses the question of what "rule" or "algorithm" should be used to select the target impedance for the manipulator. At this point a description of the objective of the task is required. Because the workpiece is absorbing power, one possible description of the task objective is to keep the actual mechanical work done on the workpiece close to some nominal value. In fact, if robot and workpiece behave as described by equations 3 and 5 and the effective viscous coefficient, B_e, of the environment is known, any desired rate of mechanical work can be enforced by appropriate choice of B_m. The mechanical power transmitted is:

$$P_d = FV = B_e B_m^2 V_o^2/(B_e + B_m)^2 \tag{8}$$

where P_d is the desired rate of mechanical power delivered to workpiece.

Solving for B_m,

$$B_m = B_e \frac{1 \pm \sqrt{B_e(V_o^2/P_d)}}{B_e(V_o^2/P_d) - 1} \tag{9}$$

However, this strategy has some unacceptable weaknesses. In principle, it would allow arbitrarily large forces or velocities, either of which would be undesirable. Although the principal task is to remove the material of the burr, the force exerted by the deburring tool must also be considered. Excessive variation in the interface force may result in an unacceptable finish on the deburred surface [9] and may increase the probability of tool breakage.

An alternative description of the task is as a tradeoff between maintaining the speed at which the tool traverses the edge and maintaining the force exerted by the tool at some nominal value. This task objective may be described by defining a quadratic penalty on deviations of force from a nominal value and velocity from a nominal value weighted by a force tolerance and a velocity tolerance.

$$Q = 1/2\left|\frac{F - F_o}{F_{tol}}\right|^2 + 1/2\left|\frac{V - V_o}{V_{tol}}\right|^2 \tag{10}$$

Q is the penalty function, F_o is the nominal interface force, F_{tol} is the force tolerance, V_o is the nominal (commanded) velocity and V_{tol} is the velocity tolerance.

The penalty function Q is to be minimized by choice of the target manipulator viscous coefficient B_m. The minimizing value is inversely proportional to the workpiece viscosity.

$$B_m = \left[\left(\frac{F_{tol}}{V_{tol}}\right)^2 + \frac{F_o}{V_o}\right] \cdot \frac{1}{B_e} \tag{11}$$

This strategy is physically more appealing – the less resistance offered by the workpiece, the more unyielding the manipulator may be. Note that when the tool is not in contact with the workpiece and the resistance to its motion is zero, this strategy specifies an infinite target impedance for the robot. In practice, the maximum achievable impedance is limited by the robot hardware.

7. Generating the Edge

Given this strategy for controlling motions along the edge to be deburred, what is an appropriate strategy for motions along the normal into the "hidden" surface or edge to be generated? As before, the first step is to model the robot's environment. In this direction the effect of irregular variations in the size of the burr is to produce an essentially unpredictable interface force between tool and workpiece which will cause the tool to deviate from its desired path. The interface force directed along the normal has a frequency content determined by the product of the speed with which the tool traverses the edge and the spatial frequency content of the edge [9]. There are two identifiable components, one due to the shape of the "hidden" surface to be generated, the other due to the irregular shape of the burr to be removed. In general, the frequency content of these two components differ substantially, the burr generating the higher-frequency forces [9].

Consequently, if the tool traverses the edge with a given velocity, there exists an identifiable separating frequency, ω_s, determined by the tool feed rate and the spatial distribution of the burr irregularities. Above that separating frequency, the interface forces are due to the burr. Below it, any forces due to the burr are indistinguishable from those needed to generate the edge. As the forces generated by the burr are essentially unpredictable, they may be characterized with reasonable fidelity by a purely stationary random process.

Again, an effective strategy is to control the impedance of the robot. However, the appropriate impedance for motion along the normal differs from that needed for motions along the edge. One component of the target impedance must permit the robot to specify the desired position of the tool along the normal. An appropriate impedance is:

$$F = K_n(X_o - X) - B_n U - M\dot{U} \tag{12}$$

K_n is the target robot stiffness, B_n is the target robot viscous coefficient, X_o is the commanded tool position, X is the actual position of the tool, U is the velocity of the tool and M is the effective robot end-effector inertia. All of these quantities refer to motion along the normal.

This target impedance may be thought of as specifying the "disturbance response" of the robot. Alternatively, it may be regarded as specifying the dynamics of the "suspension system" which supports the tool. Consequently, the problem of choosing parameter values for this impedance is exactly analogous to the problem of designing a vehicle suspension system. The choice of parameters should be such that the motion of the deburring tool follows the desired "hidden" surface or edge but does not follow the irregular shape of the burr.

8. Choice of Impedance Parameters

In some implementations, (e.g. Kramer et al., [9]) the effective robot end-effector inertia, M, will be a fixed property of the tooling. If it is subject to modulation, e.g. through the use of force feedback or by exploiting kinematic redundancies of the manipulator [5], it may be chosen to determine the bandwidth of the robot impedance (for a given stiffness and viscosity) such that below the separating frequency, the impedance is dominated by the stiffness and damping, while above it the inertia is the dominant term.

To determine appropriate values for the stiffness, K_n, and viscosity B_n, for a given effective robot end-effector inertia, M, optimization theory may again be used. The task objective may be described as a trade-off between maintaining the interface force close to a nominal value, F_o, (required by the deburring process) and simultaneously maintaining the position of the tool close to a nominal value, X_o, prescribed by the "hidden" surface to be generated. In any particular application the nominal force, F_o, is a constant. Compared to the irregular forces due to the burr, the nominal position, X_o, of the tool along the normal varies slowly, hence it may also be regarded as constant. In steady state, the two are related by

$$F_o = K_n(X_o - X). \tag{13}$$

This shows that either a one-sided tolerance should be specified for the edge to be deburred such that the nominal force is zero when the tool is at the nominal position, or, alternatively, an offset should be added to the nominal position equal to the nominal interface force divided by the stiffness.

In either case the equations of motion for deviations from the steady state become:

$$\begin{bmatrix} \dot{x} \\ \dot{u} \end{bmatrix} = \begin{bmatrix} 0 & 1 \\ -K_n/M & -B_n/M \end{bmatrix} \begin{bmatrix} x \\ u \end{bmatrix} + \begin{bmatrix} 0 \\ 1/M \end{bmatrix} F_b \tag{14}$$

x is the deviation from steady-state position, u is the deviation from steady-state velocity and F_b is the variation in interface force due to burr.

As the interface forces due to the burr are essentially unpredictable, F_b is described by a stationary, zero-mean, purely random process of strength S.

$$E[F_b(t)] = 0 \tag{15}$$

$$E[F_b(t)F_b(t+\tau)] = S\delta(\tau) \tag{16}$$

$E[]$ is the expectation operator and $\delta(\tau)$ is the dirac delta function.

The task objective may be quantified by defining an integral quadratic penalty function as follows:

$$Q = \lim_{T\to\infty} 1/T \int_0^T \left[\left(\frac{F}{F_{tol}}\right)^2 + \left(\frac{x}{X_{tol}}\right)^2 \right] dt \tag{17}$$

To obtain an impedance appropriate for following the desired surface, only the contributions of the robot's target stiffness and viscosity to the interface force need be considered. They are the dominant terms at frequencies below the separation frequency.

$$F = K_n x + B_n u \tag{18}$$

Because of the presence of the random process the penalty function to be minimized is the expectation of equation (17). The minimization is to be performed with respect to the impedance parameters K_n and B_n.

By reasoning as above, the problem of selecting K_n and B_n has been recast as a well-known stochastic optimization problem. The steady-state solution to this problem is [6]:

$$K_n = \frac{F_{tol}}{X_{tol}} \tag{19}$$

$$B_n = \sqrt{2K_nM} \tag{20}$$

Summarizing, the robot's target stiffness, K_n, is determined from task-specific tolerances on force and position. The target viscosity, B_n, is selected to yield a damping ratio of 0.707. If the effective robot inertia, M, may be modulated, it is chosen so that the bandwidth of the target impedance corresponds to the frequency above which disturbances due to the burr are to be rejected. If the inertia may not be modulated, the bandwidth has already been determined by the choice of K_n. In that case, the separating frequency may be changed by an appropriate choice of the rate at which the tool is fed along the edge, V_o, so that ω_s again corresponds to the bandwidth of the target impedance.

9. Conclusion

This paper has considered the problem of automated deburring as a particular application of impedance control strategies. A method for selecting the parameters of a target impedance appropriate for deburring was presented. It is based on a description of the robot's task as an optimization problem. A feature of this method is the ease with which the target impedance parameters can be related to simple physical parameters of the task to be performed, in this case the tolerances on the interface forces between the tool and the workpiece and the tolerances on the desired motion of the tool.

To select an appropriate target impedance, the deburring process was modeled as two orthogonal manipulation problems, one for motions along the edge to be deburred, one for motions along a normal into the edge. Simple models of the mechanics of the deburring process were developed for each of these directions. Using a simple impedance as the target behavior for the robot facilitates the identification of the parameters of the model of the environment and leads to a straightforward parameter adaptive control strategy for the robot.

It has previously been demonstrated that robot tasks requiring free motions, obstacle avoidance and kinematically constrained motions can be described successfully and effectively in terms of a target impedance for the robot [1,6]. Automated deburring provides another example of a robot task which may be characterized in terms of a target impedance. In effect, impedances provide a convenient set of "primitives" with which to describe a broad set of useful behaviors for a manipulator.

References

[1] J.R. Andrews, "Impedance Control as a Framework for Implementing Obstacle Avoidance in a Manipulator," S.M. Thesis, Department of Mechanical Engineering, M.I.T., 1983.

[2] S.L. Cotter, "Nonlinear Feedback Control of Manipulator End-Point Impedance," S.M. Thesis, Department of Mechanical Engineering, M.I.T., 1982.

[3] N. Hogan, "Control Strategies for Complex Movements Derived from Physical Systems Theory," paper presented at the International Symposium on Synergetics, Bavaria, May 1985.

[4] N. Hogan, "Impedance Control: An Approach to Manipulation: Part I - Theory," *ASME Journal of Dynamic Systems, Measurement and Control*, vol. 107, pp. 1-7, March 1985.

[5] N. Hogan, "Impedance Control: An Approach to Manipulation: Part II - Implementation," *ASME Journal of Dynamic Systems, Measurement and Control*, vol. 107, pp. 8-16, March 1985.

[6] N. Hogan, "Impedance Control: An Approach to Manipulation: Part III - Applications," *ASME Journal of Dynamic Systems, Measurement and Control*, vol. 107, pp. 17-24, March 1985.

[7] N. Hogan, "Impedance Control of Industrial Robots," *Robotics and Computer Integrated Manufacturing*, vol. 1, pp. 97-113, 1984.

[8] H. Kazerooni, "A Robust Design Method for Impedance Control of Constrained Dynamic Systems," Sc.D. Thesis, Department of Mechanical Engineering, M.I.T., 1985.

[9] B. Kramer, J.J. Bausch, R.L. Gott and D.M. Dombrowski, "Robotic Deburring," submitted to the *International Journal of Robotics and Computer Integrated Manufacturing*, 1985.

[10] D.E. Whitney, "Force Feedback Control of Manipulator Fine Motions," *ASME Journal of Dynamic Systems, Measurement and Control*, pp. 91-97, June 1971.

The Potential Field Approach And Operational Space Formulation In Robot Control

Oussama Khatib
Artificial Intelligence Laboratory
Stanford University

Abstract

The paper presents a radically new approach to real-time dynamic control and active force control of manipulators. In this approach the manipulator control problem is reformulated in terms of direct control of manipulator motion in operational space, the space in which the task is originally described, rather than controlling the task's corresponding joint space motion obtained after geometric and kinematic transformation. The control method is based on the construction of the manipulator end effector dynamic model in operational space. Also, the paper presents a unique real-time obstacle avoidance method for manipulators and mobile robots based on the "artificial potential field" concept. In this method, collision avoidance, traditionally considered a high level planning problem, can be effectively distributed between different levels of control, allowing real-time robot operations in a complex environment. Using a time-varying artificial potential field, this technique has been extended to moving obstacles. A two-level control architecture has been designed to increase the system real-time performance. These methods have been implemented in the COSMOS system for a PUMA 560 robot arm. We have demonstrated compliance, contact, sliding, and insertion operations using wrist and finger sensing, as well as real-time collision avoidance with moving obstacles using visual sensing.

1. Introduction

Conventional manipulator control, providing only linear feedback compensation to control joint positions independently, cannot meet the high accuracy and performance needed in precision manipulator tasks. Addressing this problem, much research has been directed at developing and modelling the dynamic equations of joint motion. Typical models relate joint variables to generalized torques and by necessity force the resulting control scheme to have two levels:

- Task description transformation level to express tasks in terms of joint coordinates;
- Joint space dynamic control level to calculate generalized force commands.

This first stage of transforming the task description is time consuming and prone to problems near kinematic singularities. Additionally, dealing with the dynamic compensation problem leads to high computational complexity in real-time control. Furthermore, the very approach of joint space control is ill-suited for active force control, an ability which is crucial in robot assembly tasks.

Robot collision avoidance, on the other hand, has typically been a component of higher levels of control in hierarchical robot control systems. It has been treated as a planning problem, and research in this area has focused on the development of collision-free path planning algorithms [1-4]. These algorithms aim at providing the low level control with a path that will enable the robot to accomplish its assigned task free from any risk of collision.

From this perspective, the function of low level control is limited to the execution of elementary operations for which the paths have been precisely specified. The robot's interaction with its environment is then paced by the time-cycle of high level control, which is generally several orders of magnitude slower than the response time of a typical robot. This places limits on the robot's real-time capabilities for precise, fast, and highly interactive operations in a cluttered and evolving environment. We will show, however, that it is possible to greatly extend the function of low level control and to carry out more complex operations by coupling environment-sensing feedback with the lowest level of control.

Increasing the capability of low level control has been the impetus for the work on real-time obstacle avoidance that we discuss here. Collision avoidance at the low level of control is not intended to replace high level functions or to solve planning problems. The purpose here is to make better use of low level control capabilities in performing real-time operations. At this low level of control, the degree or *level of competence* [5] will remain less than that of higher level control.

The *operational space formulation* is the basis for the application of the potential field approach to robot manipulators. This formulation has its roots in the work on end-effector motion control and obstacle avoidance [6] that has been implemented for an MA23 manipulator at the Laboratoire d'Automatique de Montpellier in 1978. The operational space approach has been formalized by constructing its basic tool, the equations of motion in the operational space of the manipulator end-effector. Details of this work have been published elsewhere elsewhere [7,8,9]. In this paper, we review the fundamentals of the operational space formulation and the artificial potential field concept.

2. Operational Space Formulation

An *operational coordinate system* is a set $\mathbf{x}$ of m_0 *independent* parameters describing the manipulator end-effector position and orientation in a frame of reference $\mathcal{R}_0$. For a non-redundant manipulator, these parameters form a set of configuration parameters in a domain of the operational space and constitute, therefore, a system of generalized coordinates. The kinetic energy of the holonomic articulated mechanism is a quadratic form of the generalized velocities

$$T(\mathbf{x},\dot{\mathbf{x}}) = \frac{1}{2}\dot{\mathbf{x}}^T \Lambda(\mathbf{x})\dot{\mathbf{x}}; \tag{1}$$

where $\Lambda(\mathbf{x})$ designates the symmetric matrix of the quadratic form, *i.e.* the kinetic energy matrix. Using the Lagrangian formalism, the end-effector equations of motion are given by

$$\frac{d}{dt}\left(\frac{\partial L}{\partial \dot{\mathbf{x}}}\right) - \frac{\partial L}{\partial \mathbf{x}} = \mathbf{F}; \tag{2}$$

where the Lagrangian $L(\mathbf{x},\dot{\mathbf{x}})$ is

$$L(\mathbf{x},\dot{\mathbf{x}}) = T(\mathbf{x},\dot{\mathbf{x}}) - U(\mathbf{x}); \tag{3}$$

and $U(\mathbf{x})$ represents the potential energy of the gravity. $\mathbf{F}$ is the operational force vector. These equations can be developed [7,8] and written in the form

$$\Lambda(\mathbf{x})\ddot{\mathbf{x}} + \mu(\mathbf{x},\dot{\mathbf{x}}) + \mathbf{p}(\mathbf{x}) = \mathbf{F}; \tag{4}$$

where $\mu(\mathbf{x},\dot{\mathbf{x}})$ represents the centrifugal and Coriolis forces, $\mathbf{p}(\mathbf{x})$ the gravity forces.

The control of manipulators in operational space is based on the selection of $\mathbf{F}$ as a command vector. In order to produce this command, specific forces $\mathbf{\Gamma}$ must be applied with joint-based actuators. The relationship between $\mathbf{F}$ and the joint forces $\mathbf{\Gamma}$ is given by

$$\mathbf{\Gamma} = J^T(\mathbf{q})\ \mathbf{F}; \tag{5}$$

where $\mathbf{q}$ is the vector of the n joint coordinates, and $J(\mathbf{q})$ the Jacobian matrix.

The extension of the operational space approach to redundant manipulators is presented in [7,8,9].

3. End-Effector Dynamic Decoupling

While in motion, a manipulator is subject to highly nonlinear inertial, centrifugal, Coriolis, and gravity forces. Dynamic decoupling, which is achieved by compensating for these forces, requires their evaluation using the manipulator dynamic model. Considering the complexity of these models, real-time dynamic control of manipulators has been viewed as a computationaly expensive approach. This problem is yet more acute in operational space, since the corresponding equations of motion are relatively more complex than those of joint motion.

Addressing this problem, we designed a two-level control system architecture based on isolating the configuration dependent coefficents in the dynamic model. The load of real-time computation of these coefficents can then be paced by the rate of configuration changes, which is much lower than that of the mechanism dynamics.

Using the dynamic model (4), the decoupling of the end-effector motion in operational space is achieved by

$$\mathbf{F} = \Lambda(\mathbf{x})\mathbf{F}^* + \mu(\mathbf{x}, \dot{\mathbf{x}}) + \mathbf{p}(\mathbf{x}); \tag{6}$$

where $\mathbf{F}^*$ represents the command vector of the decoupled end-effector, which becomes equivalent to a *single unit mass*.

Let $[\dot{\mathbf{q}}\dot{\mathbf{q}}]$ and $[\dot{\mathbf{q}}^2]$ be

$$\begin{aligned} [\dot{\mathbf{q}}\dot{\mathbf{q}}] &= [\dot{q}_1\dot{q}_2 \;\; \dot{q}_1\dot{q}_3 \cdots \dot{q}_{n-1} \;\; \dot{q}_n]^T \\ [\dot{\mathbf{q}}^2] &= [\dot{q}_1^2 \;\; \dot{q}_2^2 \cdots \dot{q}_n^2]^T. \end{aligned} \tag{7}$$

The joint torque vector corresponding to the operational space command vector (6) can be developed [7,8] following the structure

$$\mathbf{\Gamma} = J^T(\mathbf{q})\Lambda(\mathbf{q})\mathbf{F}^* + \tilde{B}(\mathbf{q})[\dot{\mathbf{q}}\dot{\mathbf{q}}] + \tilde{C}(\mathbf{q})[\dot{\mathbf{q}}^2] + \mathbf{g}(\mathbf{q}); \tag{8}$$

$\tilde{B}(\mathbf{q})$, $\tilde{C}(\mathbf{q})$, and $\mathbf{g}(\mathbf{q})$ are the $n \times n(n-1)/2$, $n \times n$, and $n \times 1$ matrices of the joint forces under the mapping into joint space of the end-effector Coriolis, centrifugal, and gravity forces, respectively.

The dynamic decoupling of the end-effector can thus be obtained using the configuration dependent dynamic coefficients $\Lambda(\mathbf{q})$, $\tilde{B}(\mathbf{q})$, $\tilde{C}(\mathbf{q})$ and $\mathbf{g}(\mathbf{q})$.

Furthermore, the rate of computation of the end-effector position, a costly step since it involves evaluations of the manipulator geometric model, can be reduced by integrating an operational position estimator into the control system. Finally, the control system (see Figure 1) has the following architecture:

- A low rate *parameter evaluation level*: updating the end-effector dynamic coefficents, the Jacobian matrix, and the geometric model.
- A high rate *servo control level*: computing the command vector using the estimator and the updated dynamic coefficents.

4. Active Force Control

Precise control of applied end-effector forces is crucial to accomplish advanced robot assembly tasks. An extensive research effort has been devoted to manipulator force control problems. Accommodation [10], joint compliance [11], active compliance [12], passive compliance, and hybrid position/force control [13] are among the various methods proposed. Active force control has been generally based on kinematic considerations, and has been treated within the framework of joint space control systems. However, task specification for motion and applied forces, wrist and finger force sensing feedback, and dynamics, are closely linked to the end-effector. Active force control can be naturally addressed in the framework of operational space control systems.

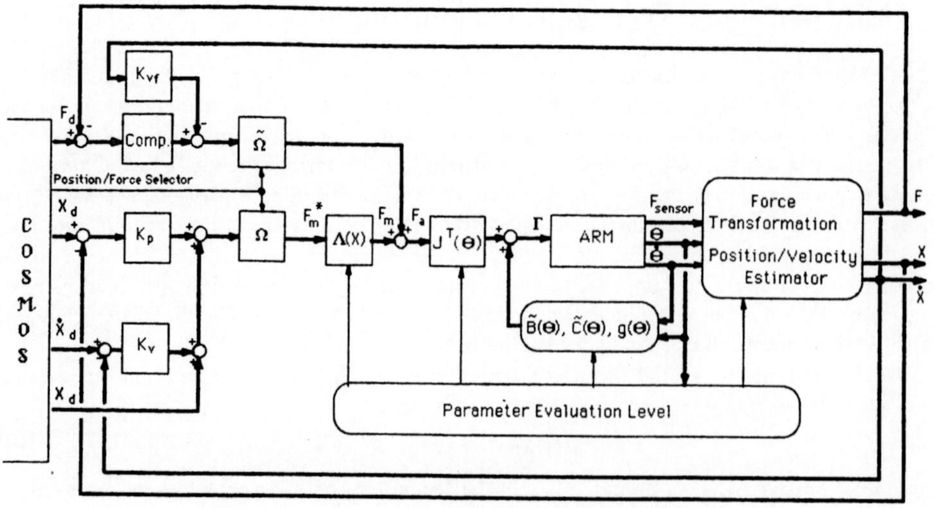

Figure 1: Operational Space Control System Architecture

A unified operational command vector can be used for end-effector dynamic decoupling, motion, and active force control [9]. The corresponding joint force vector is

$$\mathbf{\Gamma} = J^T(\mathbf{q})[\Lambda(\mathbf{q})(\Omega\mathbf{F}^*_m + \tilde{\Omega}\mathbf{F}^*_s) + \tilde{\Omega}\mathbf{F}^*_a] + \tilde{\mathbf{b}}(\mathbf{q}, \dot{\mathbf{q}}) + \mathbf{g}(\mathbf{q}). \tag{9}$$

where $\mathbf{F}_m$, $\mathbf{F}_a$ are the operational command vectors of motion and active force control. $\mathbf{F}^*_s$ represents the vector of end-effector velocity damping that acts in the direction of desired forces. Ω is *the generalized position and force specification matrix.* The control system architecture is shown in Figure 1, where k_{vf} is a velocity gain in $\mathbf{F}^*_s$.

5. The Artificial Potential Field Approach

We present this method in the context of manipulator collision avoidance. Its application to mobile robots is straightforward. The philosophy of the artificial potential field approach can be schematically described as follows:

The manipulator moves in a field of forces. The position to be reached is an attractive pole for the end-effector, and obstacles are repulsive surfaces for the manipulator parts.

Let us first consider the collision avoidance problem of a manipulator end-effector with a single obstacle $\mathcal{O}$. If $\mathbf{x}_d$ designates the goal position, the control of the manipulator end-effector with respect to the obstacle $\mathcal{O}$ can be achieved by subjecting it to the artificial potential field

$$U_{art}(\mathbf{x}) = U_{\mathbf{x}_d}(\mathbf{x}) + U_{\mathcal{O}}(\mathbf{x}). \tag{10}$$

This leads to the following expression of the potential energy in the Lagrangian (3)

$$U(\mathbf{x}) = U_{art}(\mathbf{x}) + U_g(\mathbf{x}); \tag{11}$$

where $U_g(\mathbf{x})$ represents the gravity potential energy. Using Lagrange's equations (2), and taking into account the end-effector dynamic decoupling (6), the command vector $\mathbf{F}^*_{art}$ of the decoupled end-effector that corresponds to applying the artificial potential field U_{art} (10) can be written as

$$\mathbf{F}^*_{art} = \mathbf{F}^*_{\mathbf{x}_d} + \mathbf{F}^*_{\mathcal{O}}; \tag{12}$$

with

$$\begin{aligned} \mathbf{F}^*_{\mathbf{x}_d} &= -\mathbf{grad}[U_{\mathbf{x}_d}(\mathbf{x})] \\ \mathbf{F}^*_{\mathcal{O}} &= -\mathbf{grad}[U_{\mathcal{O}}(\mathbf{x})] \end{aligned} \tag{13}$$

$\mathbf{F}^*_{\mathbf{x}_d}$ is an attractive force allowing the point $\mathbf{x}$ of the end-effector to reach the goal position $\mathbf{x}_d$, and $\mathbf{F}^*_O$ represents a *Force Inducing an Artificial Repulsion from the Surface* of the obstacle (FIRAS, from the French), created by the potential field $U_O(\mathbf{x})$. $\mathbf{F}^*_{\mathbf{x}_d}$ corresponds to the proportional term, *i.e.* $-k_p(\mathbf{x}-\mathbf{x}_d)$, in a conventional PD servo, where k is the position gain. The attractive potential field $U_{\mathbf{x}_d}(\mathbf{x})$ is simply

$$U_{\mathbf{x}_d}(\mathbf{x}) = \frac{1}{2}k_p(\mathbf{x} - \mathbf{x}_d)^2. \tag{14}$$

$U_O(\mathbf{x})$ is selected such that the artificial potential field $U_{art}(\mathbf{x})$ is a positive continuous and differentiable function which attains its zero minimum when $\mathbf{x} = \mathbf{x}_d$. The articulated mechanical system subjected to $U_{art}(\mathbf{x})$ is stable. Asymptotic stabilization of the system is achieved by adding dissipative forces proportional to $\dot{\mathbf{x}}$. Let k_v be the velocity gain; the forces contributing to the end-effector motion and stabilization are of the form

$$\mathbf{F}^*_m = -k_p(\mathbf{x} - \mathbf{x}_d) - k_v\dot{\mathbf{x}}. \tag{15}$$

This command vector is inadequate to control the manipulator for tasks that involve large end-effector motion toward a goal position without path specification. For such a task, it is better for the end-effector to move in a straight line, with an upper speed limit.

Rewriting equation (15) leads to the following expression, which can be interpreted as specifing a desired velocity vector in a pure velocity servo-control.

$$\dot{\mathbf{x}}_d = \frac{k_p}{k_v}(\mathbf{x}_d - \mathbf{x}). \tag{16}$$

Let V_{max} designate the assigned speed limit. The limitation of the end-effector velocity magnitude can then be obtained [14] by

$$\mathbf{F}^*_m = -k_v(\dot{\mathbf{x}} - \nu\dot{\mathbf{x}}_d); \tag{17}$$

where

$$\nu = min(1, \frac{V_{max}}{\sqrt{\dot{\mathbf{x}}_d^T\dot{\mathbf{x}}_d}}). \tag{18}$$

With this scheme the velocity vector $\dot{\mathbf{x}}$ is controlled to be pointed toward the goal position while its magnitude is limited to V_{max}. The end-effector will then travel at that speed, in a straight line, except during the acceleration and deceleration segments or when it is inside the repulsive potential field regions of influence.

6. FIRAS Function

The artificial potential field $U_O(\mathbf{x})$ should be designed to meet the manipulator stability condition and to create at each point on the obstacle's surface a potential barrier which becomes negligible beyond that surface. Specifically, $U_O(\mathbf{x})$ should be a non-negative continuous and differentiable function whose value tends to infinity as the end-effector approaches the obstacle's surface. In order to avoid undesirable perturbing forces beyond the obstacle's vicinity, the influence of this potential field must be limited to a given region surrounding the obstacle.

Using analytic equations $f(\mathbf{x}) = 0$ for obstacle description, the first artificial potential field function we used [6] was based on the values of the function $f(\mathbf{x})$

$$U_O(\mathbf{x}) = \begin{cases} \frac{1}{2}\eta(\frac{1}{f(\mathbf{x})} - \frac{1}{f(\mathbf{x}_0)})^2 & \text{if } f(\mathbf{x}) \leq f(\mathbf{x}_0); \\ 0 & \text{if } f(\mathbf{x}) > f(\mathbf{x}_0). \end{cases} \tag{19}$$

The region of influence of this potential field is bounded by the surfaces $f(\mathbf{x}) = 0$ and $f(\mathbf{x}) = f(\mathbf{x}_0)$, where $\mathbf{x}_0$ is a given point in the vicinity of the obstacle and η a constant gain. This potential function can be obtained very simply in real-time since it does not require any distance calculations. However, this potential is difficult to use for asymmetric obstacles, where the separation between an obstacle's surface and equipotential surfaces can vary widely.

Using the shortest distance to an obstacle $\mathcal{O}$, we have proposed [7] the following artificial potential field

$$U_{\mathcal{O}}(\mathbf{x}) = \begin{cases} \frac{1}{2}\eta(\frac{1}{\rho} - \frac{1}{\rho_0})^2 & \text{if } \rho \leq \rho_0 \text{ ;} \\ 0 & \text{if} \rho > \rho_0. \end{cases} \tag{20}$$

where ρ_0 represents the limit distance of the potential field influence and ρ, the shortest distance to the obstacle $\mathcal{O}$. The selection of the distance ρ_0 will depend on the end-effector operating speed V_{max} and on its deceleration ability. End-effector acceleration characteristics are discussed in [15].

Any point of the robot can be subjected to the artificial potential field. A *Point Subjected to the Potential* is called a PSP. The control of a PSP with respect to an obstacle $\mathcal{O}$ is achieved using the FIRAS function

$$\mathbf{F}^*_{(\mathcal{O},psp)} = \begin{cases} \eta(\frac{1}{\rho} - \frac{1}{\rho_0})\frac{1}{\rho^2}\frac{\partial \rho}{\partial \mathbf{x}} & \text{if } \rho \leq \rho_0 \text{ ;} \\ 0 & \text{if } \rho > \rho_0 \text{ ;} \end{cases} \tag{21}$$

where $\frac{\partial \rho}{\partial \mathbf{x}}$ denotes the partial derivative vector of the distance from the PSP to the obstacle

$$\frac{\partial \rho}{\partial \mathbf{x}} = [\frac{\partial \rho}{\partial x} \; \frac{\partial \rho}{\partial y} \; \frac{\partial \rho}{\partial z}]^T. \tag{22}$$

The joint forces corresponding to $\mathbf{F}^*_{(\mathcal{O},psp)}$ are obtained using the Jacobian matrix associated with this PSP. Observing (6) and (12), these forces are given by

$$\Gamma_{(\mathcal{O},psp)} = J^T_{psp}(\mathbf{q})\Lambda(\mathbf{x})\mathbf{F}^*_{(\mathcal{O},psp)}. \tag{23}$$

7. Obstacle Geometric Modeling

Obstacles are described by the composition of *primitives*. A typical geometric model base includes primitives such as a point, line, plane, ellipsoid, parallelepiped, cone, and cylinder. The first artificial potential field (19) requires analytic equations for the description of obstacles. For primitives such as a *parallelepiped, finite cylinder*, and *cone*, we have developed analytic equations representing envelopes which best approximate the primitives' shapes.

The surface, termed an *n-ellipsoid*, is represented by the equation

$$(\frac{x}{a})^{2n} + (\frac{y}{b})^{2n} + (\frac{z}{c})^{2n} = 1; \tag{24}$$

and tends to a parallelepiped of dimensions (2a,2b,2c) as n tends to infinity. A good approximation is obtained with $n = 4$, as shown in Figure 2.

A cylinder of elliptical cross section $(2a, 2b)$ and of length $2c$ can be approximated by the so-called *n-cylinder* equation

$$(\frac{x}{a})^2 + (\frac{y}{b})^2 + (\frac{z}{c})^{2n} = 1. \tag{25}$$

The analytic description of primitives is not necessary for the artificial potential field (20), since the continuity and differentiablity requirement is on the shortest distance to the obstacle. The primitives above, and more generally all convex primitives, comply with this requirement.

Determining the orthogonal distance to an n-ellipsoid or to an n-cylinder requires the solution of a complicated system of equations. To avoid this costly computation, a variational procedure for the distance evaluation has been developed. The distance expressions for other primitives are presented in [16].

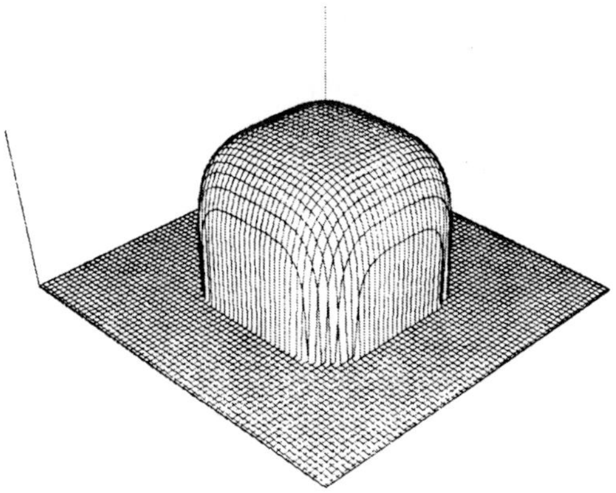

Figure 2: An n−ellipsoid with $n = 4$

8. Robot Obstacle Avoidance

An obstacle $\mathcal{O}_i$ is described by a set of primitives $\{\mathcal{P}_p\}$. The superposition property (additivity) of potential fields enables the control of a given point of the manipulator with respect to this obstacle by using the sum of the relevant gradients

$$\mathbf{F}^*_{\mathcal{O}_i,psp} = \sum_p \mathbf{F}^*_{(\mathcal{P}_p,psp)}. \tag{26}$$

Control of this point for several obstacles is obtained using

$$\mathbf{F}^*_{psp} = \sum_i \mathbf{F}^*_{(\mathcal{O}_i,psp)}. \tag{27}$$

It is also feasible to have different points on the manipulator controlled with respect to different obstacles. The resulting joint force vector is given by

$$\mathbf{\Gamma}_{obstacles} = \sum_j J^T_{psp_j}(\mathbf{q})\Lambda(\mathbf{x})\mathbf{F}^*_{psp_j}. \tag{28}$$

Specifying an adequate number of PSPs enables the protection of all of the manipulator's parts. An example of a dynamic simulation for a redundant 4 *dof* manipulator operating in the plane [Khatib and Le Maitre 1978] is shown in the display of Figure 3. The artificial potential field approach can be extended to *moving obstacles*, since stability of the mechanism persists with a continuously time-varying potential field.

The manipulator obstacle avoidance problem has been formulated in terms of *collision avoidance of links*, rather than points. Link collision avoidance is achieved by continuously controlling the link's closest point to the obstacle. At most, n PSPs then have to be considered. Additional links can be artificially intoduced or the length of the last link can be extended to account for the manipulator tool or load. In an articulated chain, a link can be represented as the line segment defined by the Cartesian positions of its two neighboring joints. In a frame of reference $\mathcal{R}$, a point $m(x, y, z)$ of the link bounded by $m_1(x_1, y_1, z_1)$ and $m_2(x_2, y_2, z_2)$ is described by the parametric equations

$$\begin{aligned} x &= x_1 + \lambda(x_2 - x_1) \\ y &= y_1 + \lambda(y_2 - y_1) \\ z &= z_1 + \lambda(z_2 - z_1) \end{aligned} \tag{29}$$

The problem of obtaining the link's shortest distance to a parallelepiped can be reduced to that of finding the link's closest point to a vertex, edge, or face. The analytic expressions of

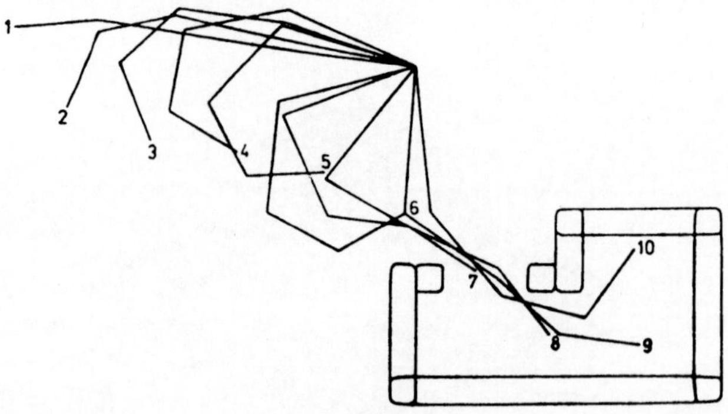

Figure 3: Displacement of a 4 dof manipulator inside an enclosure .

the link's closest point, the distance, and its partial derivatives for a parallelepiped, cylinder and cone are given in [16].

9. Joint Limit Avoidance

The potential field approach can be used to satisfy the manipulator internal joint constraints. Let $\underline{q}_i$ and $\overline{q}_i$ be respectively the minimal and maximal bounds of the ith joint coordinate q_i. q_i can be kept within these boundaries by creating barriers of potential at each of the hyperplanes $(q_i = \underline{q}_i)$ and $(q_i = \overline{q}_i)$. The corresponding joint forces are

$$\gamma_{\underline{q}_i} = \begin{cases} \eta(\frac{1}{\underline{\rho}_i} - \frac{1}{\underline{\rho}_{i(0)}})\frac{1}{\underline{\rho}_i^2} & \text{if } \underline{\rho}_i \le \underline{\rho}_{i(0)}; \\ 0 & \text{if } \underline{\rho}_i > \underline{\rho}_{i(0)}; \end{cases} \tag{30}$$

and

$$\gamma_{\overline{q}_i} = \begin{cases} -\eta(\frac{1}{\overline{\rho}_i} - \frac{1}{\overline{\rho}_{i(0)}})\frac{1}{\overline{\rho}_i^2} & \text{if } \overline{\rho}_i \le \overline{\rho}_{i(0)}; \\ 0 & \text{if } \overline{\rho}_i > \overline{\rho}_{i(0)}; \end{cases} \tag{31}$$

where $\underline{\rho}_{i(0)}$ and $\overline{\rho}_{i(0)}$ represent the distance limit of the potential field influence. The distances $\underline{\rho}_i$ and $\overline{\rho}_i$ are defined by

$$\begin{aligned} \underline{\rho}_i &= q_i - \underline{q}_i \\ \overline{\rho}_i &= \overline{q}_i - q_i \end{aligned} \tag{32}$$

10. Level of Competence

The potential field concept is indeed an attractive approach to the collision avoidance problem, and much research has recentely been focused on its applications to robot control [17-19]. However, the complexity of tasks that can be achieved with this approach is limited. In a cluttered environment, local minima can occur in the resultant potential field. This can lead to a stable positioning of the robot before reaching its goal. While local procedures can be designed to exit from such configurations, limitations for complex tasks will remain. This is because the approach has a *local* perspective of the robot environment.

Nevertheless, the resulting potential field does provide the global information necessary, and a collision-free path, if attainable, can be found by linking the absolute minima of the potential. Linking these minima requires, however, a computationally expensive exploration of the potential field. This goes beyond the real-time control we are concerned with here, but can be considered as an integrated part of higher level control. Work on high level collision-free path planning based on the potential field concept has been investigated in [20].

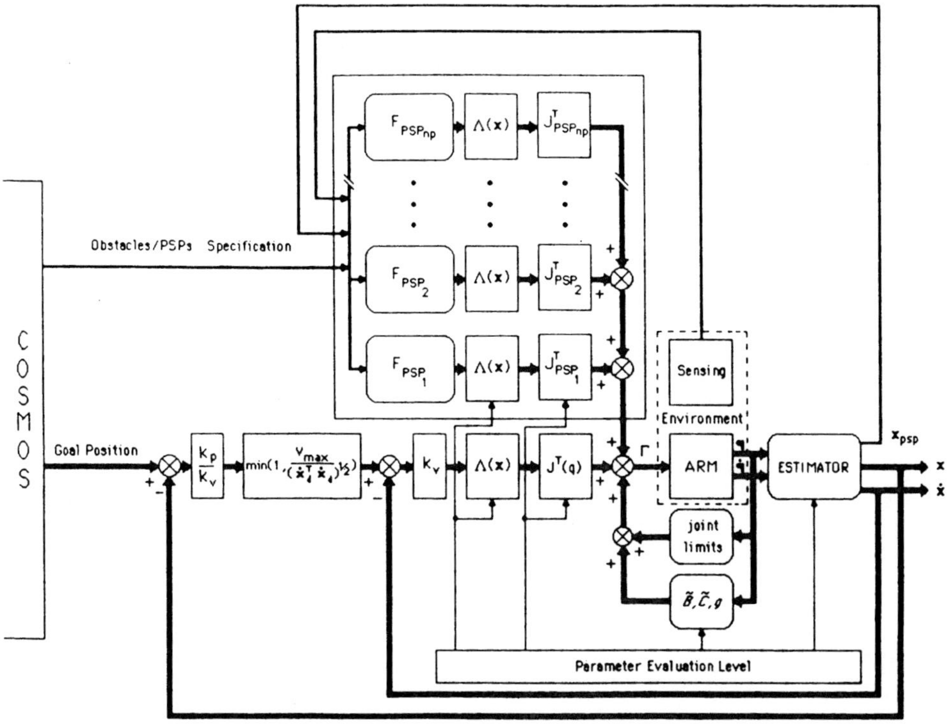

Figure 4: Control System Architecture.

11. Real-Time Implementation

Finally, the global control system integrating the potential field concept with the operational space approach has the following structure

$$\Gamma = \Gamma_{motion} + \Gamma_{obstacles} + \Gamma_{joint-limit}; \tag{33}$$

where

$$\Gamma_{motion} = J^T(q)\Lambda(q)F_m^* + \tilde{B}(q)[\dot{q}\dot{q}] + \tilde{C}(q)[\dot{q}^2] + g(q); \tag{34}$$

The control system architecture is shown in Figure 4 where np represents the number of PSPs. The Jacobian matrices $J_{psp_j}^T(q)$ have common factors with the end-effector Jacobian matrix $J^T(q)$. Thus, their evaluation does not require significant additional computation.

12. Applications

This approach has been implemented in an experimental manipulator programming system COSMOS (Control in Operational Space of a Manipulator-with-Obstacles System). Using a PUMA 560 and a Machine Intelligence Corporation vision module, demonstration of real-time collision avoidance with links and moving obstacles [21] have been performed.

We have also demonstrated real-time end-effector motion and active force control operations with the COSMOS system, using wrist and finger sensing. These include contact, slide, insertion, and compliance operations [22].

In the current multiprocessor implementation (PDP 11/45 and PDP 11/60), the rate of the servo control level is 225 Hz while the coefficient evaluation level runs at 100 Hz.

13. Summary and Discussion

We have presented our operational space formulation of manipulator control which provides the basis for this obstacle avoidance approach, and have described the two-level architecture designed to increase the real-time performance of the control system. The operational space formulation has been shown to be an effective means of achieving high dynamic performance in real-time motion control and active force control of robot manipulators for complex assembly tasks. In addition, the complex transformation of the task into joint coordinates, required in conventional joint space control approaches, is eliminated.

Further, we have described the formulation and the implementation of a real-time obstacle avoidance approach based on the artificial potential field concept. In this approach, collision avoidance, generally treated as high level planning, has been demonstrated to be an effective component of low level real-time control.

The integration of this low level control approach with a high level planning system seems to be one of the more promising solutions to the obstacle avoidance problem in robot control. With this approach, the problem may be treated in two stages:

- at high level control, generating a global strategy for the manipulator's path in terms of intermediate goals (rather than finding an accurate collision-free path);
- at the low level, producing the appropriate commands to attain each of these goals, taking into account the detailed geometry and motion of manipulator and obstacle, and making use of real-time obstacle sensing (low level vision and proximity sensors).

By extending low level control capabilities and reducing the high level path planning burden, the integration of this collision avoidance approach into a multi-level robot control structure will improve the real-time performance of the overall robot control system. Potential applications of this control approach include moving obstacle avoidance, collision avoidance in grasping operations, and obstacle avoidance problems involving multi-manipulators or multi-fingered hands.

References

[1] H.P. Moravec, "Obstacle Avoidance and Navigation in the Real World by a Seeing Robot Rover," Ph.D. Thesis, Stanford University, Artificial Intelligence Laboratory, 1980.

[2] T. Lozano-Perez, "Spatial Planning: A Configuration Space Approach," AI Memo 605, Cambridge, Mass., MIT Artificial Intelligence Laboratory, 1980.

[3] R. Chatila, Système de Navigation pour un Robot Mobile Autonome: Modélisation et Processus Décisionnels," Thése de Docteur-Ingénieur, Université Paul Sabatier, Toulouse, France, 1981.

[4] R. Brooks, "Solving the Find-Path Problem by Good Representation of Free Space," *IEEE Systems, Man and Cybernetics*, SMC-13, pp. 190-197, 1983.

[5] R. Brooks, "Aspects of Mobile Robot Visual Map Making," Second International Symposium of Robotics Research, Kyoto, Japan, Aug. 20-23, 1984.

[6] O. Khatib and J.F. Le Maitre, "Dynamic Control of Manipulators Operating in a Complex Environment," Proceedings Third CISM-IFToMM Symposium on Theory and Practice of Robots and Manipulators, Udine, Italy, 1978, pp. 267-282, (Elsevier 1979).

[7] O. Khatib, "Commande Dynamique dans l'Espace Opérationnel des Robots Manipulateurs en Présence d'Obstacles," Thése de Docteur-Ingénieur, École Nationale Supérieure de l'Aéronautique et de l'Espace (ENSAE), Toulouse, France, 1980.

[8] O. Khatib, "Dynamic Control of Manipulators in Operational Space," Sixth CISM-IFToMM Congress on Theory of Machines and Mechanisms, New Delhi, India, December 15-20, 1983, pp. 1128-1131.

[9] O. Khatib, "The Operational Space Formulation in Robot Manipulators Control," 15th International Symposium on Industrial Robots, Tokyo, Japan, Sept. 11-13, 1985.

[10] D.E. Whitney, "Force Feedback Control of Manipulator Fine Motions," *ASME, Journal of Dynamic Systems, Measurement, and Control*, pp. 91-97, June 1977.

[11] R.P. Paul and B. Shimano, "Compliance and Control," Proceedings of the Joint Automatic Control Conference, Purdue University, July 1976, pp. 694-699.

[12] J.K. Salisbury, "Active Stiffness Control of a Manipulator in Cartesian Coordinates," 19th IEEE Conference on Decision and Control, Albuquerque, New Mexico, December 1980.

[13] J.J. Craig and M. Raibert, "A Systematic Method for Hybrid Position/Force Control of a Manipulator," Proceedings 1979 IEEE Computer Software Applications Albuquerque Conference, Chicago, 1979.

[14] O. Khatib, M. Llibre and R. Mampey, "Fonction Decision-Commande d'un Robot Manipulateur," Rapport No. 2/7156, DERA/CERT, Toulouse, France, 1978.

[15] O. Khatib and J. Burdick, "Dynamic Optimization in Manipulator Design: The Operational Space Formulation," The 1985 ASME Winter Annual Meeting, Miami, November 1985.

[16] O. Khatib, "Real-Time Obstacle Avoidance for Manipulators and Mobile Robots," 1985 International Conference on Robotics and Automation, St. Louis, March 25-28, 1985.

[17] H.B. Kuntze and W. Schill, "Methods for Collision Avoidance in Computer Controlled Industrial Robots," 12th ISIR, Paris, June 9-11, 1982.

[18] N. Hogan, "Impedance Control: An Approach to Manipulation," 1984 American Control Conference, San Diego, June 6-8, 1984.

[19] B. Krogh, "A Generalized Potential Field Approach to Obstacle Avoidance Control. Robotics Research: The Next Five Years and Beyond," SME Conference Proceedings, Bethlehem, Pennsylvania, August 1984.

[20] C. Buckley, "The Application of Continuum Methods to Path Planning," Ph.D. Thesis (in progress), Stanford University, Department of Mechanical Engineering, 1985.

[21] O. Khatib, et al., "Robotics in Three Acts," (Film), Stanford University, Artificial Intelligence Laboratory, June 1984.

[22] O. Khatib, J. Burdick and B. Armstrong, "Robotics in Three Acts - Part II," (Film), Stanford University, Artificial Intelligence Laboratory, 1985.

Mathematical Theory of Learning with Applications to Robot Control

Suguru Arimoto
Faculty of Engineering Science
Osaka University
Osaka, Japan

Abstract

Fundamental forms of learning control law are proposed for linear and nonlinear dynamical systems which may be operated repeatedly at relatively low cost. Given a desired output $y_d(t)$ over a finite time duration $[0, T]$ and an appropriate input $u_0(t)$ for such a system, a general proposed law of learning control is described by a PID-type (Proportional, Integration, and Differentiation) iterative process: $u_{k+1}(t) = u_k(t) + \{\Phi + \Gamma d/dt + \Psi \int dt\}(y_d(t) - y_k(t))$, where u_k denotes the input at the kth trial, y_k the measured output when u_k excites the system, and Φ, Γ and Ψ are constant gain matrices. For a class of linear mechanical systems where x and $y(= dx/dt)$ stand for position and velocity vectors respectively, it is shown that a P-type or PI-type iterative learning control law with appropriate gain matrices Φ and Ψ is convergent in a sense that $y_k(t)$ approaches $y_d(t)$ pointwisely in $t \in [0, T]$ and $x_k(t)$ does $x_d(t)$ uniformly in $t \in [0, T]$ as $k \to \infty$. In case of using a D-type or DP-type iterative learning control law, an analogous conclusion is also proved for a class of nonlinear dynamical systems. Finally, proposed learning methods are applied to some problems of trajectory or path tracking control of robot manipulators.

1. Introduction

Human beings can learn much from experience. Great athletes exercise repeatedly to acquire a desired form of motion. Without the aid of human operators, are any mechanical systems like robots able to learn anything automatically from previous data and improve the performance at the next maneuvering? In response to this question the author and his colleagues have recently proposed learning control methods with simple iterative structures and applied them to the trajectory tracking control of mechanical manipulators ([1-4]).

In the present paper we recast proposed learning control laws in a form of PID-type iterative algorithm (see Fig. 1):

$$u_{k+1}(t) = u_k(t) + \{\Gamma \frac{d}{dt} + \Phi + \Psi \int dt\}(y_d(t) - y_k(t)). \tag{1}$$

This means that given a desired output $y_d(t)$ for a finite time duration $t \in [0, T]$, the next $(k+1)th$ input u_{k+1} is composed of the previous input u_k plus a PID modification of the present output error $e_k = y_d - y_k$, where y_k is the measured output when u_k excites the system at the kth operation. In the next section we first prove that for a class of linear time-invariant systems, a P-type learning law with an appropriate gain matrix Φ (where $\Gamma = \Psi = 0$) is convergent in the sense that $y_k(t)$ approaches the desired output $y_d(t)$ as $k \to \infty$ for any fixed $t \in [0, T]$. In Section 3 this result is extended to the case that the objective system is a linear time-varying mechanical system and the employed learning control law is of PI-type. In Section 4 it is shown that a D-type or DP-type learning control law is convergent for a class of nonlinear dynamical systems. In Section 5 these results are applied to a trajectory or path tracking control problem of robot manipulators. Finally it is concluded that even mechanical robots are able to improve autonomously their motion through self training.

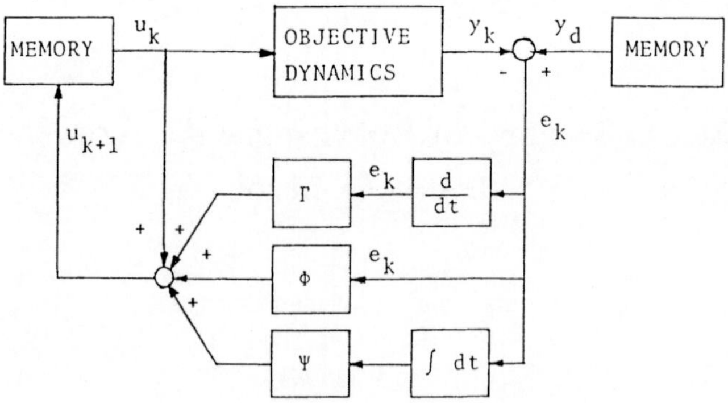

Figure 1: PID-type learning control law.

2. Learning Control for Linear Time-Invariant Mechanical Systems

Consider a linear time-invariant system that is subject to a set of linear differential equations

$$R\ddot{x} + Q\dot{x} + Px = u, \tag{2}$$

where

$$u,\ x \in R^n, \qquad R,\ Q,\ \text{and}\ P \in R^{n\times n}, \tag{3}$$

and $\dot{x} = dx/dt$ and $\ddot{x} = d\dot{x}/dt$. We assume that actual values of coefficient matrices R, Q, and P are not known but that all these are symmetric and positive definite. In such a case it is called a linear mechanical system, in which $x, \dot{x}$, and $\ddot{x}$ stand for position, velocity, and acceleration vectors respectively. Suppose that the velocity vector $\dot{x}$ can be measured and is set as the output, namely,

$$y(t) = \dot{x}(t). \tag{4}$$

First we deal with a P-type learning control law with an appropriate constant gain matrix Φ. Given a desired output $y_d(t)$ and an appropriate initial input $u_0(t)$ over a finite time duration $t \in [0,T]$, the learning process follows the recursive relation:

$$R\ddot{x}_k + Q\dot{x}_k + Px_k = u_k, \tag{5a}$$

$$y_k = \dot{x}_k, \tag{5b}$$

$$u_{k+1} = u_k + \Phi(y_d - y_k). \tag{5c}$$

It is also assumed that at every operation the initial position and velocity are set the same, namely

$$x_k(0) = x^o,\ \ \dot{x}_k(0) = \dot{x}^o = y_d(0). \tag{6}$$

Theorem 1. Assume that each component of $u_0(t)$ is continuous and each component of $y_d(t)$ is continuously differentiable, namely

$$u_0(t) \in C[0,T] \qquad \textit{and} \qquad y_d(t) \in C^1[0,T] \tag{7}$$

and gain matrix Φ is symmetric and positive definite and satisfies

$$2Q > \Phi > 0 \tag{8}$$

($2Q - \Phi$ is positive definite). Then the P-type learning control law defined by Eq.(5) is convergent in the sense that

$$e_k(t) = y_d(t) - y_k(t) \to 0 \qquad as \qquad k \to \infty \tag{9}$$

for any fixed $t \in [0,T]$.

Proof. It follows from Eq.(5) that

$$R(\ddot{x}_{k+1} - \ddot{x}_k) + Q(\dot{x}_{k+1} - \dot{x}_k) + P(x_{k+1} - x_k) = u_{k+1} - u_k = \Phi e_k. \tag{10}$$

Letting

$$d_k = x_{k+1} - x_k \tag{11}$$

and substituting this into Eq.(10), we obtain

$$R\ddot{d}_k + Q\dot{d}_k + Pd_k = \Phi e_k. \tag{12}$$

Note that by definition of e_k and d_k,

$$\dot{d}_k = e_k - e_{k+1} \tag{13}$$

from which it follows that

$$\begin{aligned} \textstyle\int_0^t e_{k+1}^T(\tau)\Phi e_{k+1}(\tau)d\tau &= \textstyle\int_0^t e_k^T(\tau)\Phi e_k(\tau)d\tau \\ &+ \textstyle\int_0^t \dot{d}_k^T(\tau)\Phi\dot{d}_k(\tau)d\tau - 2\int_0^t e_k^T(\tau)\Phi\ \dot{d}_k(\tau)d\tau. \end{aligned} \tag{14}$$

Substituting Eq.(12) into the last term of Eq.(14) and noting the initial condition of Eq.(6), we obtain

$$\begin{aligned} \textstyle\int_0^t e_{k+1}^T(\tau)\Phi e_{k+1}(\tau)d\tau &= \textstyle\int_0^t e_k^T(\tau)\Phi e_k(\tau)d\tau \\ &- \textstyle\int_0^t \dot{d}_k^T(\tau)(2Q - \Phi)\dot{d}_k(\tau)d\tau - \dot{d}_k^T(t)\ R\dot{d}_k(t) - d_k^T(t)Pd_k(t). \end{aligned} \tag{15}$$

Now it is important to note that conditions of Eq.(7) imply $x_0(t) \in C^2[0,T]$, $e_0(t) \in C^1[0,T]$, and hence $d_0(t) \in C^3[0,T]$, because the second derivative of $d_o(t)$ must be continuously differentiable according to Eq.(12). This means that $x_1(t) = (d_0 + x_0) \in C^2[0,T]$ and, thereby, $e_1(t) \in C^1[0,T]$. From mathematical induction, it thus follows that $d_k(t) \in C^3[0,T]$ and clearly $\ddot{d}_k(0) = 0$ for any k. Therefore it is possible to write the following equation:

$$R\dddot{d}_k + Q\ddot{d}_k + P\dot{d}_k = \Phi\dot{e}_k \tag{16}$$

with $\ddot{d}_k(0) = 0$. Then the same reasoning in the derivation of Eq.(15) gives rise to the following equation:

$$\begin{aligned} \textstyle\int_0^t \dot{e}_{k+1}^T(\tau)\Phi\dot{e}_{k+1}(\tau)d\tau &= \textstyle\int_0^t \dot{e}_k^T(\tau)\Phi\dot{e}_k(\tau) \\ &- \textstyle\int_0^t \ddot{d}^T(\tau)(2Q - \Phi)\ddot{d}_k(\tau)d\tau - \ddot{d}_k^T(t)\ R\ddot{d}_k(t) - \dot{d}_k^T(t)P\dot{d}_k(t). \end{aligned} \tag{17}$$

Now we define

$$\begin{aligned} a_k(t) &= \textstyle\int_0^t e_k^T(\tau)\Phi e_k(\tau)d\tau, \\ b_k(t) &= \textstyle\int_0^t \dot{e}_k^T(\tau)\Phi\dot{e}_k(\tau)d\tau. \end{aligned} \tag{18}$$

Since Φ, $2Q - \Phi$, R, and P are all positive definite, both $a_k(t)$ and $b_k(t)$ are monotonically non-increasing with increasing k and bounded from below for an arbitrarily fixed $t \in [0,T]$. This means

$$\begin{aligned} 0 \leqq \ddot{d}_k^T(t)R\ddot{d}_k(t) &\leqq b_k(t) - b_{k+1}(t), \\ 0 \leqq \dot{d}_k^T(t)R\dot{d}_k(t) &\leqq a_k(t) - a_{k+1}(t), \\ 0 \leqq d_k^T(t)Pd_k(t) &\leqq a_k(t) - a_{k+1}(t), \end{aligned} \tag{19}$$

which implies

$$\ddot{d}_k(t) \to 0,\ \ \dot{d}_k(t) \to 0,\ \text{ and } d_k(t) \to 0 \tag{20}$$

as $k \to \infty$ for any fixed $t \in [0,T]$. Substituting Eq.(20) into Eq.(12), we conclude that

$$e_k(t) \to 0 \text{ as } k \to \infty \tag{21}$$

for any fixed $t \in [0,T]$. This completes the proof.

This theorem assures only pointwise convergence of the velocity error $e_k(t) = y_d(t) - y_k(t) = \dot{x}_d(t) - \dot{x}_k(t)$. However, it is fortunate to see that the position vector $x_k(t)$ converges to the desired position trajectory $x_d(t)$ uniformly in $t \in [0,T]$ as $k \to \infty$. This is proved in the following theorem.

Theorem 2. Under the same conditions as in Theorem 1, the following uniform convergence holds:

$$x_k(t) \to x_d(t) \text{ uniformly in } t \in [0,T] \text{ as } k \to \infty \tag{22}$$

where

$$x_d(t) = x^o + \int_0^t y_d(\tau)d\tau = x^o + \int_0^t \dot{x}_d(\tau)d\tau. \tag{23}$$

Proof. As discussed in the derivation of Eq.(19), all terms of $\ddot{d}_k(t), \dot{d}_k(t)$, and $d_k(t)$ are uniformly bounded. In view of Eq.(12), this implies the uniform boundedness of $\{e_k(t)\}$. Let

$$E_k(t) = \int_0^t e_k(\tau)d\tau = x_d(t) - x_k(t). \tag{24}$$

Thus, the sequence $\{E_k(t)\}$ is uniformly bounded and equicontinuous in $t \in [0,T]$. In addition, it follows from Theorem 1 that $E_k(t) \to 0$ as $k \to \infty$ for any fixed $t \in [0,T]$. Now assume that the sequence $\{E_k(t)\}$ is not uniformly convergent. Then there exist $\epsilon_o > 0$, $k(m)$, and $t_m \epsilon [0,T]$ such that

$$|E_{k(m)}(t_m)| \geq \epsilon_o \tag{25}$$

where $|x| = max|x_i|$. Since the sequence $E_{k(m)}(t)$ itself is uniformly bounded, equicontinuous, and converges pointwise to zero, it follows from Ascoli-Arzela's theorem that the sequence $\{E_{k(m)}\}$ has a subsequence that converges to zero uniformly in $t \in [0,T]$. This contradicts Eq.(25), completing the proof.

3. Learning Control for Linear Time-Varying Mechanical Systems

Next we consider a class of linear time-varying mechanical systems described by

$$\begin{cases} R(t)\ddot{x}(t) & +Q(t)\dot{x}(t) + P(t)x(t) = u(t), \\ y(t) = & \dot{x}(t) = dx(t)/dt. \end{cases} \tag{26}$$

It is assumed that coefficient matrices $R(t)$, $Q(t)$, and $P(t)$ satisfy

$$\begin{cases} R(t) = & R^T(t) \geq R_o > 0, \; Q(t) + Q^T(t) \geq Q_o > 0, \\ P(t) + & P^T(t) \geq P_o > 0 \end{cases} \tag{27}$$

for all $t \in [0,T]$ and all their entries are continuously differentiable. Given a desired trajectory $y_d(t)$ that is continuously differentiable on $[0,T]$, and an initial input $u_0(t)$ that is continuous on $[0,T]$, we consider a PI-type learning control law with symmetric gain matrices Φ and Ψ, which is described by

$$\begin{cases} R\ddot{x}_k & +Q\dot{x}_k + Px_k = u_k, \\ y_k & = \dot{x}_k, \\ e_k & = y_d - y_k, \\ u_{k+1} & = u_k + (\Phi + \Psi \int dt)e_k. \end{cases} \tag{28}$$

Now we define the operator H as

$$(Hz)(t) = \Phi z(t) + \int_0^t \Psi z(\tau)d\tau \tag{29}$$

and the vector norm for the function $z(t)$ with property $z(0) = 0$ as

$$\begin{aligned}
|z(t)|_\rho = & \int_0^t e^{-\rho\tau}\{Hz(\tau)\}^T\Phi^{-1}\{Hz(\tau)\}d\tau \\
= & \int_0^t e^{-\rho\tau}\{z^T(\tau)\Phi z(\tau) + Z^T(\tau)\Psi\Phi^{-1}\Psi Z(\tau)\}d\tau \\
+ & \ 2\int_0^t e^{-\rho\tau}z^T(\tau)\Psi Z(\tau)d\tau \\
= & \ e^{-\rho t}Z^T(t)\Psi Z(t) \\
+ & \int_0^t e^{-\rho\tau}[z^T(\tau)\Phi z(\tau) + Z^T(\tau)\{\rho\Psi + \Psi\Phi^{-1}\Psi\}Z(\tau)]d\tau
\end{aligned} \tag{30}$$

where ρ is an arbitrarily chosen positive constant and

$$Z(t) = \int_0^t z(\tau)d\tau. \tag{31}$$

Since it follows from Eq.(28) that

$$\begin{aligned}
R(t)(\ddot{x}_{k+1} - \ddot{x}_k) & + Q(t)(\dot{x}_{k+1} - \dot{x}_k) + P(t)(x_{k+1} - x_k) \\
& = u_{k+1} - u_k = He_k,
\end{aligned} \tag{32}$$

we have

$$R(t)\ddot{d}_k + Q(t)\dot{d}_k + P(t)d_k = He_k \tag{33}$$

where we define

$$d_k = x_{k+1} - x_k, \quad \dot{d}_k = e_k - e_{k+1}. \tag{34}$$

Then, it is easy to see that

$$\begin{aligned}
|e_{k+1}(t)|_\rho & = |e_k(t) - \dot{d}_k(t)|_\rho \\
& = |e_k(t)|_\rho + |\dot{d}_k(t)|_\rho - 2\int_0^t e^{-\rho\tau}\{H\dot{d}_k(\tau)\}^T\Phi^{-1}\{He_k(\tau)\}d\tau \\
& = |e_k(t)|_\rho + |\dot{d}_k(t)|_\rho \\
& -2\int_0^t e^{-\rho\tau}\{H\dot{d}_k(\tau)\}^T\Phi^{-1}\{R(\tau)\ddot{d}_k(\tau) + Q(\tau)\dot{d}_k(\tau) + P(\tau)d_k(\tau)\}d\tau
\end{aligned} \tag{35}$$

which is reduced to

$$\begin{aligned}
|e_{k+1}(t)|_\rho & = |e_k(t)|_\rho - \int_0^t e^{-\rho\tau}[\dot{d}_k^T(\tau)S_1\dot{d}_k(\tau) + d_k^T(\tau)S_o d_k(\tau)]d\tau \\
& -e^{-\rho t}[\dot{d}_k^T(t)K_1\dot{d}_k(t) + d_k^T(t)K_o d_k(t)] \\
& + \int_0^t e^{-\rho\tau}\dot{d}_k^T(\tau)V_1 d_k(\tau)d\tau + e^{-\rho t}\dot{d}_k^T(t)L_1 d_k(t)
\end{aligned} \tag{36}$$

where

$$\left\{\begin{aligned}
S_1 = & \ \rho R(\tau) - \dot{R}(\tau) + Q^T(\tau) + Q(\tau) - \Phi - \Psi\Phi^{-1}R(\tau) + R(\tau)\Phi^{-1}\Psi, \\
S_o = & \ \rho\{P^T(\tau) + P(\tau) - \Psi\} - \Psi\Phi^{-1}\Psi - \dot{P}^T(\tau) - \dot{P}(\tau) + P^T(\tau)\Phi^{-1}\Psi + \Psi\Phi^{-1}P(\tau), \\
K_1 = & \ R(t), \\
K_o = & \ P^T(t) + P(t) - \Psi, \\
V_1 = & \ -2\rho R(\tau)\Phi^{-1}\Psi + 2\{P^T(\tau) + \dot{R}(\tau)\Phi^{-1}\Psi - Q^T(\tau)\Phi^{-1}\Psi\}, \\
L_1 = & \ -2R(t)\Phi^{-1}\Psi,
\end{aligned}\right. \tag{37}$$

Now we are in a position to state the central theorem of this paper:

Theorem 3. Assume that

$$u_o(t) \in C[0,T] \quad \text{and} \quad y_d(t) \in C^1[0,T] \tag{38}$$

and gain matrices Φ and Ψ are symmetric and satisfy

$$\Phi > 0, \ \|\Phi^{-1}\Psi\| = \lambda << 1, \ P_o - \Psi - 2\lambda^2 R(t) > 0 \tag{39}$$

where $\|A\|$ denotes the spectral radius of matrix A. Then the PI-type learning control law in Eq.(28) with the same initial condition as in Eq.(6), is convergent in the sense that $y_k(t) \to y_d(t)$ pointwisely in $t \in [0,T]$ and $x_k(t) \to x_d(t)$ uniformly in $t \in [0,T]$ as $k \to \infty$.

Proof. Applying Schwarz's inequality to the last two terms of Eq.(36), we see that

$$\begin{aligned} |e_{k+1}(t)|_\rho \ &\leq |e_k(t)|_\rho - \textstyle\int_0^t e^{-\rho\tau}\dot{d}_k^T\{\rho R - \gamma\rho R + 0(1)\}\dot{d}_k d\tau \\ &- \textstyle\int_0^t e^{-\rho\tau} d_k^T\{\rho(P^T + P - \Psi - \gamma^{-1}\Psi\Phi^{-1}R\Phi^{-1}\Psi) + 0(1)\}d_k d\tau \\ &-e^{-\rho t}\dot{d}_k^T(t)\{R(t) - \gamma R(t)\}\dot{d}_k \\ &-e^{-\rho t}d_k^T(t)\{P^T(t) + P(t) - \Psi - \gamma^{-1}\Psi\Phi^{-1}R(t)\Phi^{-1}\Psi\}d_k(t) \end{aligned} \tag{40}$$

where γ is an arbitrary constant satisfying $0 < \gamma < 1$ and $0(1)$ means a remaining term of matrices that are independent of ρ. If ρ is chosen sufficiently large and $\gamma = \frac{1}{2}$, then all brackets {} in Eq.(40) become positive definite due to Eq.(39). Hence,

$$|e_{k+1}(t)|_\rho \leq |e_k(t)|_\rho \tag{41}$$

which, together with Eq.(40), implies that

$$\dot{d}_k(t) \to 0, \ \ d_k(t) \to 0 \ \text{ as } \ k \to \infty$$

for any fixed $t \in [0,T]$. Next note that according to the condition of Eq.(38), $x_o(t) \in C^2[0,T]$, which implies $He_o(t) \in C^1[0,T]$ and furthermore $d_o(t) \in C^3[0,T]$, because the second derivative of $d_o(t)$ must be continuously differentiable due to Eq.(33). Therefore, with the aid of a similar argument to that given in the derivation of Eq.(16), we have

$$\frac{d}{dt}(R\ddot{d}_k + Q\dot{d}_k + Pd_l) = H\dot{e}_k. \tag{42}$$

Then we see that, similarly to Eqs.(35), (36), and (40),

$$\begin{aligned} |\dot{e}_{k+1}(t)|_\rho \ &= |\dot{e}_k(t)|_\rho + |\ddot{d}_k(t)|_\rho \\ &-2\textstyle\int_0^t e^{-\rho\tau}\{H\ddot{d}_k\}^T\Phi^{-1}\frac{d}{d\tau}(R\ddot{d}_k + Q\dot{d}_k + Pd_k)d\tau \\ &\leq |\dot{e}_k(t)|_\rho - \textstyle\int_0^t e^{-\rho\tau}\ddot{d}_k^T\{\rho R - \gamma\rho R + 0(1)\}\ddot{d}_k d\tau \\ &- \textstyle\int_0^t e^{-\rho\tau}\dot{d}_k^T\{\rho(P^T + P - \Psi - \gamma^{-1}\Psi\Phi^{-1}R\Phi^{-1}\Psi) + 0(1)\}\dot{d}_k d\tau \\ &- \textstyle\int_0^t e^{-\rho\tau}d_k^T\{0(1)\}d_k d\tau - e^{-\rho t}\ddot{d}_k^T(t)\{R(t) - \gamma R(t)\}\ddot{d}_k \\ &-e^{-\rho t}\dot{d}_k^T(t)\{P^T(t) + P(t) - \Psi - \gamma^{-1}\Psi\Phi^{-1}R(t)\Phi^{-1}\Psi\}\dot{d}_k(t) \end{aligned} \tag{43}$$

Addition of this to Eq.(40) yields

$$
\begin{aligned}
|e_{k+1}(t)|_\rho &+|\dot{e}_{k+1}(t)|_\rho \leqq |e_k(t)|_\rho + |\dot{e}_k(t)|_\rho \\
&- \int_0^t e^{-\rho\tau}\ddot{d}_k^T(\tau)\{\rho X(\tau) + 0(1)\}\ddot{d}_k(\tau)d\tau \\
&- \int_0^t e^{-\rho\tau}\dot{d}_k^T(\tau)\{\rho X(\tau) + \rho Y(\tau) + 0(1)\}\dot{d}_k(\tau)d\tau \\
&- \int_0^t e^{-\rho\tau}d_k^T(\tau)\{\rho Y(\tau) + 0(1)\}d_k(\tau)d\tau \\
&-e^{-\rho t}[\ddot{d}_k^T(t)\{X(t)\}\ddot{d}_k(t) + \dot{d}_k^T(t)\{X(t) + Y(t)\}\dot{d}_k(t) \\
&+d_k^T(t)\{Y(t)\}d_k(t)]
\end{aligned}
\tag{44}
$$

where

$$
\begin{cases} X(t) = & R(t) - \gamma R(t), \\ Y(t) = & P^T(\tau) + P(t) - \Psi - \gamma^{-1}\Psi\Phi^{-1}R(t)\Phi^{-1}\Psi. \end{cases}
\tag{45}
$$

Applying the same argument given in the derivation of Eq.(42) for Eq.(45), we conclude that

$$
\ddot{d}_k(t) \to 0, \quad \dot{d}_k(t) \to 0, \quad \text{and} \quad d_k(t) \to 0, \quad \text{as} \quad k \to \infty
\tag{46}
$$

for any fixed $t \in [0,T]$. Substituting this into Eq.(33), we have

$$
He_k \to 0 \text{ as } k \to \infty
\tag{47}
$$

for any fixed $t \in [0,T]$. Next we observe that the expression

$$
\Phi e_k(t) + \int_0^t \Psi e_k(\tau)d\tau = He_k(t)
\tag{48}
$$

leads to the integral inequality

$$
\|\Phi e_k(t)\| \leqq \|He_k(t)\| + \int_0^t \lambda\|\Phi e_k(\tau)\|d\tau,
\tag{49}
$$

where $\|\cdot\|$ denotes the euclidean norm. Applying Bellman-Gronwall's lemma to Eq.(49), and using Schwarz's inequality we obtain

$$
\begin{aligned}
\|\Phi e_k(t)\| &\leqq \|He_k(t)\| + \lambda\int_0^t \|He_k(\tau)\|e^{\lambda(t-\tau)}d\tau \\
&\leqq \|He_k(t)\| + \lambda[\int_0^t \|He_k(\tau)\|^2 d\tau]^{1/2}[\int_0^t e^{2\lambda(t-\tau)}d\tau]^{1/2}.
\end{aligned}
\tag{50}
$$

Then, for any fixed $t \in [0,T]$, Eq.(44) implies that

$$
\int_0^t \|He_k(\tau)\|^2 d\tau \to 0 \qquad as \qquad k \to \infty,
\tag{51}
$$

together with Eqs.(46) and (47), because $He_k(\tau)$ consists of a linear combination of $\ddot{d}_k(\tau)$, $\dot{d}_k(\tau)$, and $d_k(\tau)$ on $\tau \in [0,T]$ as seen in Eq.(33). Hence it follows from Eqs.(47),(50), and (51) that $\Phi e_k(t) \to 0$ pointwisely and thereby $e_k(t) \to 0$ pointwisely in $t \in [0,T]$ as $k \to \infty$, because of positive definiteness of Φ. Finally, note that $|\dot{e}_k(t)|_\rho$ is uniformly bounded according to Eq.(44). In view of Eq.(30) this shows the uniform boundedness of $\{e_k(t)\}$, which implies both uniform boundedness and equicontinuity of $\{E_k(t)\}$. Then the uniform convergence of $\{E_k\}$ to zero, namely $x_k(t) \to x_d(t)$, as $k \to \infty$ follows from the same argument as given in the proof of Theorem 2.

4. Learning Control for Robot Manipulators

It is well known (for example, see [5-7]) that the dynamics of serial-link manipulators with n degrees of freedom are described by

$$(J_o + H(q))\ddot{q} + f(q,\dot{q}) + g(q) = Kv, \tag{52}$$

where $q \in R^n$ is the vector of joint coordinates, H is a positive definite inertia matrix, J_o, is a nonnegative diagonal matrix that represents inertial terms of internal load distribution of actuators $f(q,\dot{q})$ is a vector-valued function of centrifugal, Coriolis, and viscous frictional forces, $g(q)$ is a vector due to gravity force, v is a vector of input voltages given to actuator servomotors, and K is a diagonal gain matrix. Given a desired output $q_d(t)$ over a finite time duration $t \in [0,T]$ which is of C^2-class, we consider a feedback control law

$$v = u + K^{-1}g(q) + \overline{A}(q_d - q) + \overline{B}(\dot{q}_d - \dot{q}), \tag{53}$$

where $\overline{A}$ and $\overline{B}$ are positive definite diagonal matrices. This type of control law for robot manipulators was first proposed by the author and his colleagues (see [8]), together with the proof of its stability and robustness in case of position control ($u \equiv 0$ and $q_d(t) \equiv$ const. in Eq.(53)). By substituting Eq.(53) into Eq.(52), the closed-loop system turns out to be subject to

$$(J_o + H(q))\ddot{q} + f(q,\dot{q}) + B(\dot{q} - \dot{q}_d) + A(q - q_d) = u, \tag{54}$$

where $A = K\overline{A}$ and $B = K\overline{B}$. At the first iteration we set $u(t) \equiv 0$. Then, by virtue of the stable and robust structure of the proposed feedback law (see [7]), it is reasonable to expect that the solution trajectory $\{q(t), \dot{q}(t)\}$ of Eq.(54) will remain in a neighborhood of the desired one $\{q_d(t), \dot{q}_d(t)\}$. Bearing this in mind, we define a deviation vector

$$x(t) = q(t) - q_d(t) \tag{55}$$

and rewrite Eq.(54) in terms of $x(t)$ in such a way as

$$(J_o + H(t))\ddot{x} + (B + G(t))\dot{x} + (A + F(t))x = h(t) + u, \tag{56}$$

where $H(t) = H(q_d(t)), G(t)$ and $F(t)$ are matrix-valued functions that are dependent on $q_d, \dot{q}_d$, and $\ddot{q}_d$, and $h(t)$ expresses remaining higher terms of x and $\dot{x}$. Note that the inertia matrix $J_o + H(t)$ is positive definite by definition and gain matrices A and B can be chosen so that both $A + F(t)$ and $B + G(t)$ are positive definite for any $t \in [0,T]$. Then, neglecting the higher term in Eq.(56), we finally obtain a linear time-varying system

$$R(t)\ddot{x} + Q(t)\dot{x} + P(t)x = u \tag{57}$$

where

$$R(t) = J_o + H(t),\ Q(t) = B + G(t),\ \text{and}\ P(t) = A + F(t). \tag{58}$$

Since all these coefficient matrices are positive definite for any $t \in [0,T]$, there are positive definite constant matrices R_o, Q_o, and P_o such that Eq.(27) is satisfied. Thus, it is possible to construct a PI-type learning control scheme as shown in Fig. 2.

Experimental results obtained by using an actual mechanical manipulator have shown the effectiveness of the proposed learning control method, improving remarkably the performance of trajectory tracking control (see [4]). It should be further remarked that this method can be extended to the case of force control or hybrid (force/position) control of robot manipulators (see [9]).

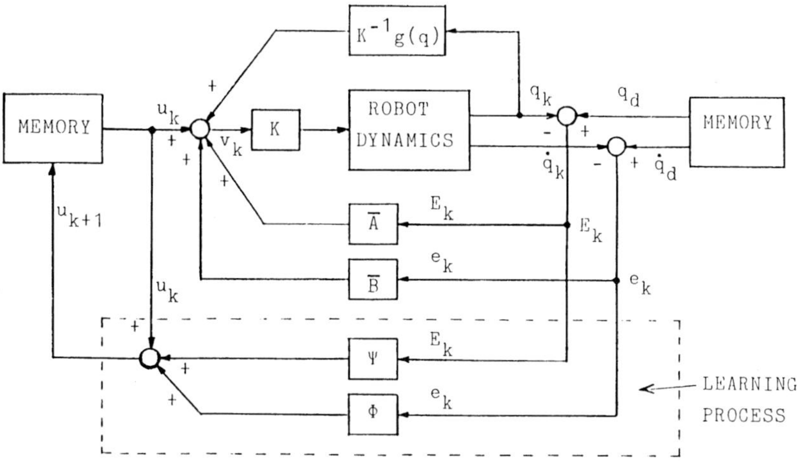

Figure 2: PI-type learning control law for robot manipulators.

5. Learning Control for Nonlinear Robot Dynamics

Previous discussions are made on the basis of the implicit assumption that the higher term $h(t)$ in Eq.(56) can be neglected. If this is not agreeable to the theoretical reasoning, it is necessary to apply an iterative learning scheme directly to the original nonlinear dynamics. Consider the dynamics with the feedback law of Eq.(53), which can be rewritten in the following state equation form:

$$\begin{cases} \dot{x} = & f(t,x) + D(x)u, \\ y = & [0, I]x = Cx, \end{cases} \tag{59}$$

where

$$x = \begin{bmatrix} x_1 \\ x_2 \end{bmatrix} = \begin{bmatrix} q \\ \dot{q} \end{bmatrix}, \quad D(x) = \begin{bmatrix} 0 \\ -(J_o + H(x_1))^{-1}K \end{bmatrix},$$

$$f(x) = \begin{bmatrix} x_2 \\ -(J_o + H(x_1))^{-1}\{f(x_1, x_2) + B(x_2 - x_{2d}) + A(x_1 - x_{1d})\} \end{bmatrix}. \tag{60}$$

It is assumed in this case that a desired velocity $y_d(t) = x_2(t) = \dot{q}_d(t)$ of C^1-class is given on $[0, T]$. Given an arbitrary constant $\lambda > 0$, we define the function norm

$$\|x\|_\lambda = \sup_{t \in [0,T]} e^{-\lambda t}|x(t)|. \tag{61}$$

Now, consider a PD-type learning control law first proposed by the author (see [1,2]), which is described by

$$\begin{cases} \dot{x}_k = & f(t, x_k) + D(x_k)u_k, \\ y_k = & Cx_k, \\ u_{k+1} = & u_k + (\Gamma\frac{d}{dt} + \Phi)(y_d - y_k), \end{cases} \tag{62}$$

in which at every k the initial state is set the same, namely

$$x_k(0) = x^o \text{ and } y_k(0) = y_d(0), \qquad k = 0, 1, 2, \dots . \tag{63}$$

Theorem 4. Assume that there are constants $\lambda_o > 0$ and $\alpha_o > 0$ such that the first trajectory $x_o(t)$ remains in a λ_o-neighborhood of trajectory $x_d(t)$ in $[0, T] \times R^{2n}$ as defined by

$$N(x_d; \lambda_o, \alpha_o) = \{(t, x) : \|x(t) - x_d(t)\|_{\lambda_o} < \alpha_o\}. \tag{64}$$

If the PD-type learning control law with $\lambda > 0$ and $\alpha > 0$ satisfies

$$\|I_n - CD(x)\Gamma\| = \|I_n - (J_o + H(q))^{-1}K\Gamma\| < 1 \qquad \text{for all } q \in R^n \tag{65}$$

then there exist constants $\lambda > 0$ and $\alpha > 0$ such that

$$x_k(t) \in N(x_d; \lambda, \alpha) \qquad \text{for } k = 1, 2, \dots . \tag{66}$$

Furthermore, the following uniform convergence holds:

$$y_k(t) \to y_d(t) \qquad \text{uniformly in } t \in [0, T] \text{ as } k \to \infty \tag{67}$$

The proof of this theorem is omitted. Instead we remark that a similar argument to that given in our previous paper [2] can be utilized in the proof.

6. Conclusion

A PID-type learning control law for linear and nonlinear dynamical systems is proposed for the purpose of improving the performance of tracking control. It has been shown that given a desired output $y_d(t)$ over a finite interval $t \in [0, T]$, the PI-type learning control law is uniformly convergent if the objective system is a linear time-invariant or time-varying system with positive definite coefficient matrices. It has also been pointed out that the PD-type learning control law becomes convergent for a class of nonlinear dynamical systems. The applicability of these methods to the control of robot manipulators is also discussed.

References

[1] S. Arimoto, S. Kawamura and F. Miyazaki, "Bettering Operation of Robots by Learning," *J. of Robotic Systems*, 1-2, pp. 123-140, 1984.

[2] Ibid., "Can Mechanical Robots Learn by Themselves?", in Robotics Research: The 2nd Inter. Symp., (H. Hanafusa and H. Inoue, ed.), pp. 127-134, MIT Press, Cambridge, Massachusetts, 1985.

[3] Ibid., "Bettering Operation of Dynamic Systems by Learning: A New Control Theory for Servomechanism or Mechatronics Systems," Proc. 23rd IEEE CDC, Las Vegas, Nevada, Dec. 1984.

[4] S. Kawamura, F. Miyazaki and S. Arimoto, "Iterative Learning Control for Robotic Systems," Proc. of IECON '84, Tokyo, Japan, Oct. 1984.

[5] R.P. Paul, Robot Manipulators, The MIT Press, Cambridge, 1981.

[6] P. Vukobratovic, Scientific Fundamentals of Robotics 1 and 2, Springer-Verlag, Berlin, 1982.

[7] S. Arimoto and F. Miyazaki, "Asymptotic Stability of Feedback Control Laws for Robot Manipulators," Proc. of IFAC Symp. on Robot Control '85, Barcelona, Spain, Nov. 1985.

[8] M. Takegaki and S. Arimoto, "A New Feedback Method for Dynamic Control of Manipulators," *Trans. ASME, J. Dynam. Syst. Measure. Control*, 103, pp. 113-125, 1981.

[9] S. Kawamura, F. Miyazaki and S. Arimoto, "Hybrid Position/ Force Control of Manipulators Based on Learning Method," Proc. of '85 Inter. Conf. on Advanced Robotics, Tokyo, Japan, Sept. 1985.

Automatic Planning and Control of Robot Natural Motion Via Feedback

Daniel E. Koditschek

Abstract

A feedback control strategy for the command of robot motion includes some limited automatic planning capabilities. These may be seen as sequential solution algorithms implemented by the robot arm interpreted as a mechanical analog computer. This perspective lends additional insight into the manner in which such control techniques may fail, and motivates a fresh look at requisite sensory capabilities.

1. Introduction

Much of classical control theory is concerned with the design of controllers possessing a "feedforward" structure which filters time-varying reference signals, forcing a closed loop plant compensated by the appropriate feedback structure. The task at hand might be said to be "encoded" by the reference signal (or class of signals) applied; the compensator is designed to force the plant to accomplish the task. While theoretically sound rules of thumb for the design of such controllers are available in the context of linear time invariant plants, there is, at present, no analytically tractable procedure for determining the response of the nonlinear dynamical systems defined by typical industrial robots to time varying inputs. In consequence, classical techniques cannot be applied to their control with confidence.

The last five years have seen a variety of nonlinear control methodologies proposed for the command of robot arm motion and force. It would be fair to say that in most cases, the conceptual basis for task encoding has remained fixed upon the paradigm of a reference trajectory, presumably the output of some arbitrary higher level planning algorithm, which the controller forces the plant to track. However, in the absence of theoretically informed choices for appropriate reference signals, even the best high level planning algorithm may produce poor command inputs: encoding a task in dynamical terms entirely unrelated to those of the plant may not be viable. Moreover, there is mounting evidence that the computational effort required to encode typical tasks (such as moving in a cluttered space) in terms of exact reference trajectories may be prohibitive [21], [24].

This paper will explore aspects of a rather different control methodology which replaces the role of such fedforward reference signals in favor of task characterization in terms of a feedback structure. "Natural Motion" refers to the resulting unforced response of the autonomous closed loop system. Specifically, we focus upon the mechanism by which the system generates a "plan of action" automatically, obviating the need for an independent reference signal. Earlier results of the author, [10],are re-interpreted to demonstrate their foundation upon two simple and very old ideas: that a set of algebraic equalities may be solved sequentially on an analog computer; and that a kinematic chain, in the absence of external torques, is a passive dynamical system. Their re-interpretation leads to a better understanding of how the algorithms may fail, and motivates a discussion concerning the trade-offs between sensing and computation. This understanding seems to be a first step along the way to a methodology of robot task encoding via feedback structures. In the long run, of course, robot controllers probably ought to be built with both feedforward precompensators and feedback. However, the capabilities of rather simple error-driven systems may afford the command of a surprisingly sophisticated range of robot tasks, hence, deserve a much better hearing in the robotics community than has yet been accorded.

The question of how particular robot tasks should be encoded, while not explicitly discussed in this paper, might be seen as a unifying theme for all the papers presented in this session of the Fourth Yale Workshop. The pure feedback based control structures discussed here represent a particularly simple implementation of tasks which have been encoded using potential functions - e.g. Hogan's impedance control methodology [6], or Khatib's obstacle avoidance methodology [9]. The extent to which other robot task paradigms can be implemented by pure feedback controllers is not at all clear, but seems a question well worth pursuing. Its pursuit is bound to shed more light on the general problem of how to translate abstract task goals into computationally efficient and dynamically sound controllers.

1.1 A Robot Model

The equations of motion of a kinematic chain have been extensively discussed in the robotics literature, and this paper will rely upon the standard rigid body model of an open chain with revolute joints. Thus, we consider a robot to be the dynamical system

$$M(q)\ddot{q} + B(q,\dot{q})\dot{q} + k(q) = \tau, \tag{1}$$

where B is linear in $\dot{q}$, and M, B, k all vary in q by polynomials of transcendental functions. This system is equivalent to

$$\begin{aligned} \dot{x_1} &= x_2 \\ \dot{x_2} &= M^{-1}(x_1)[B(x_1,x_2)x_2 + k(x_1) - \tau] \end{aligned} \tag{2}$$

where the generalized positions and velocities take values $x \triangleq \begin{bmatrix} x_1 \\ x_2 \end{bmatrix} \in \mathcal{P} \triangleq TJ$, *phase space*.

In this ideal world, the robot is a finite dimensional dynamical plant which accepts torque inputs,τ, and delivers joint positions and velocities, x_1, x_2. Missing from this model, evidently, is any account of actuator dynamics and nonlinearities (including absolute power constraints) as well as viscous and nonlinear friction at the joints. The omission of the former set of phenonmena from the model is problematic, but reflects the currently volatile state of experimentation in the robotics world with a variety of actuating technologies which promise to be "better" than the standard dc-servo with gear-train [2], [8]. The omission of the latter, seems to be more generally defensible on the grounds that, unlike the inverse-dynamics approach to arm control [5],[14], the reliance on the natural motion of an arm - a pure feedback stragtegy - requires no explicit knowledge of the underlying velocity coefficients. Although dramatic changes in damping over the course of motion are bound to degrade performance, it will be seen below that the feedback schemes being proposed are guaranteed to maintain *stability* at all times (assuming that any "negative" friction is entirely due to the coriolis term, B). As with any feedback scheme, however, the capabilities and deployment of sensors will be critical to the reliability of this analysis, and we will return to such questions in the last section.

The space of generalized positions, termed *joint-space,* (J in the sequel), is the cross product of a k dimensional torus with $\mathbf{R}^{n-k}$ where n is the total number of joints, and k is the number of revolute joints. There are two circumstances under which J may be accurately modelled as a simply connected subset of $\mathbf{R}^n$: if each revolute joint is mechanically constrained to prevent a full 360 degree revolution; or, if full revolution is mechanically possible, but joint sensors are available which transduce angular displacement with respect to some absolute position - e.g., from which a revolution and a half displacement is read as 540 degrees rather than 180 degrees. It will be seen in the sequel, e.g. the statement of Theorem 1, that such details may have significant consequences when pure feedback control is applied.

Of course, equation (1) is an incomplete model of a robot arm in the absence of an explicit output map from J to what we will term *work space*, or the set of *frames*. These, of course, represent the position and orientation of some orthonormal coordinate system fixed in the (rigid) end-effector with respect to an inertial system in the base of the arm. Appendix B introduces a notation for and develops some standard results from the algebra and calculus of frames which are modeled as points in $\mathcal{W} \triangleq \mathbf{R}^3 \times SO(3)$.

1.2 Inverse Dynamics

Before proceeding, it is worth pointing out that a conceptually very straight forward procedure exists which, if practicable, would bring robot control problems back into a domain entirely amenable to the well understood classical methodology. Robots are built with an actuator for each degree of freedom (assuming rigid links), hence, constitute a dynamical system which may be completely linearized and decoupled by state feedback. This possibility would suggest a return to the standard means of task encoding via reference trajectories.

Specifically, given a reference trajectory,q_r , and its derivatives, $\dot{q}_r, \ddot{q}_r$ the control strategy,

$$\begin{aligned} \tau &\triangleq [B - MK_2]\dot{q} + [k(q) - MK_1 q] + \tau_r \\ \tau_r &\triangleq M[\ddot{q}_r + K_2\dot{q}_r + K_1 q_r], \end{aligned} \tag{3}$$

$K_1, K_2 > 0$, may be shown *theoretically* to drive the errors, $e \triangleq \begin{bmatrix} q_r \\ \dot{q}_r \end{bmatrix} - x$ to zero, exponentially in time for arbitrary initial values. An analogous procedure, termed "resolved acceleration control" [14] can be used when the reference signals are expressed in work space coordinates avoiding explicit computation of the inverse kinematics.

This technique has been proposed by a variety of researchers in a diversity of guises [5], [14], [27] and is known most widely as the method of "computed torque" or "inverse dynamics". Since it "linearizes" the equations of motion by exact cancellation of thousands of nonlinear terms from the rigid body model of robot dynamics, the question arises as to whether the method may be applied in practice at all: can the computation be effected quickly enough; can it be effected accurately enough? A number of researchers have persuasively argued that the answer to the first question is (or may soon be) yes [7], [13], which conclusion we will accept with no further discussion here. In the absence of reported empirical experience or analytical study, however, the second question remains open. Caution seems justified in light of (i) the inaccuracies in the rigid body model upon which such computation would be based; (ii) uncertainty regarding the values of the link dynamical parameters, and (in general) total ignorance of the load dynamical parameters; and (iii) numerical inaccuracies in computation resulting from quantization effects.

2. Natural Motion of Robot Arms

Unlike other control methodologies that have emerged in the context of robotics, the feedback based algorithms described here admit a mathematical proof of global convergence with minimal a priori knowledge of the actual plant dynamics. The underlying mechanism of stability was understood as long as a century ago [25], but only introduced to the robotics community in 1981 by Arimoto [26] (and, independently, by this author in 1984 [10]). The idea of solving algebraic systems of equations by sequential techniques is much older. A direct consequence of the stability properties above is that a particular sequential technique may be modified for implementation on the second order nonlinear mechanical analog computer which we will interpret the robot arm to be. This interpretation of the author's earlier results [11] is partly suggested by a recent paper of Wolovich and Elliott [28].

2.1 Sequential Solution of a Set of Equations

Perhaps the most widely used general method for solving a set of nonlinear equations

$$0 = f(x)$$

is Newton-Raphson iteration

$$x_{n+1} = x_n - df_{x_n}^{-1} f(x_n) \tag{4}$$

which is known to converge to the *solution set*, $\mathcal{S} \triangleq f^{-1}[0]$, as long as x_0 is "close" to that set, and $|df|$ is bounded away from zero on the subsequent trajectory. Recent work of Smale [22] and Hirsch and Smale [23] has established variants of this technique which succeed for almost all initial points, x_0, arbitrarily far from $\mathcal{S}$, given mild assumptions on f. They point out that equation (4) is a discrete time version of the differential equation

$$\dot{x} = -df^{-1} f(x), \tag{5}$$

whose solutions are confined to curves in the inverse image $f^{-1}[\mathcal{L}]$ of the ray

$$\mathcal{L} \triangleq \{t f(x_0) \mid t > 0\}$$

connecting the origin of the target space to the image of the first guess, $f(x_0)$. They also mention a much older variant of this technique, the algorithm

$$\dot{x} = -df^{\mathrm{T}} f \tag{6}$$

which we will call the *gradient* method, since its trajectories follow the gradient of the scalar "error function",

$$\varepsilon \triangleq \|f\|,$$

in the target space. This system is also guaranteed to converge to $\mathcal{S}$ for all initial conditions which are sufficiently close to the solution set. Unfortunately, unlike algorithm (5) it cannot, in general, be modified to converge to $\mathcal{S}$ globally, even disregarding a set of measure zero, for reasons which will become apparent.

It should be mentioned that the gradient method has a long history in systems science as well as the general world of applied mathematics. For instance, the parameter adjustment algorithms of adaptive control and estimation schemes may be seen as a gradient solution to a set of linear algebraic equations [17]. In the field of robotics, Wolovich and Elliott [28] have recently proposed this algorithm as an off-line numerical solution technique to the same algebraic equations as we consider here - namely, the kinematic transformation of a robot arm. This paper, however, concerns an on-line implementation of the gradient algorithm implicit in the motion of the robot arm itself. In effect, it is proposed to perform the integration called for in equation (6) upon a mechanical analog computer formed by the robot arm itself, whose integrators obtain from its intrinsic dynamics. The arm solves its own inverse kinematics by moving to the desired target.

A number of limitations inhere from the beginning. In particular:

1. the technique is intrinsically time invariant - convergence is only guaranteed if $\frac{df}{dt} = 0$;
2. although the trajectories of system (5) can be shown to move toward their goal along straight lines in the target space, the gradient lines of the error magnitude, ε, depend entirely on properties of f itself, and, in general, will be curved in both the domain and target space;
3. while $\mathcal{S}$ is the positive limit set of almost every initial condition of system (5), (assuming a suitable adjustment in the sign of the vector field, as described in [23]) the same cannot be said of system (6);
4. while equation (6) calls for a set of first order integrators, robot arm analog computers, like every other mechanical system, obey Newton's laws, and come equipped only with a set of double integrators.

These limitations have some obvious and some less obvious implications for the performance of robot motion resulting from the feedback controls based upon the gradient algorithm.

The first caveat does not imply that a run-of-the mill tracking problem cannot be implemented in this fashion - simply that stability cannot be guaranteed. However, in the light of engineers' varied experiences with parametric excitation of forced second order systems caution is well advised. A determined application of this methodology will do better to rely as much as possible upon encoding robot tasks in terms of static geometric constructs rather than time varying reference signals. Recent work by this author [11], Hogan [6], and Khatib [9] begins to suggest that the set of tasks amenable to such construction is rich indeed.

The second caveat would be entirely irrelevant were we proposing simply another numerical procedure for the solution of inverse kinematics. In this context, solution trajectories describe the physical motion of a robot arm and we will be critically concerned with their position in the target space. The departure of the trajectories of system (6) from straight lines in the target space is determined locally by the eigenvectors and eigenvalues of $df^{\mathrm{T}}df$ near a limit point in $\mathcal{S}$, and, more generally, by the gradient lines of ε. These may be "shaped", then, by the construction of suitable error functions, ε, [11], [6].

The third caveat has more subtle origins and interpretation. System (6) has equilibrium states at the critical points of the error function, ε,

$$\mathcal{C} \stackrel{\triangle}{=} \{x \mid f(x) \in ker\ df^{\mathrm{T}}\}$$

and, in general, that set includes more than the desired $f^{-1}[0]$. Namely, there exist points $\mathcal{U} \stackrel{\triangle}{=} \mathcal{C} - \mathcal{S} \neq \emptyset$ if f has critical values other than 0. Since the hessian matrix of ε evaluated at a point in $\mathcal{C}$, (which will have a more complicated structure on $\mathcal{U}$ than $df^{\mathrm{T}}df$) specifies the linearized vector field of (6) around that equilibrium state, any non-zero valued local minima of $||f||$ define equilibrium states which are locally attractive and possess some open neighborhood within which initial conditions will converge to a useless value. This explains why the set of poor initial guesses does not, in general, have measure zero. If the error function, ε, has no local minima outside of $\mathcal{S}$ then $\mathcal{U}$ consists of unstable equilibrium states of system (6), and all initial conditions which do not lie on their stable manifold (necessarily a set of measure zero) converge to the solution - i.e. a point in $\mathcal{S}$.

Exploration of the final caveat is, of course, a main theme of this paper, and is now addressed explicitly.

2.2 Dissipative Mechanical Systems: Implementing a Sequential Algorithm with a Double Integrator

In light of the preceding discussion outlining the weaknesses of algorithm (6) relative to (5), it seems appropriate to inquire as to why we force the robot arm to integrate the former. For a broad range of mechanical systems, the Hamiltonian is an exact expression for total energy. In a conservative force field this scalar function is a constant (defines a first integral of the equations of motion) and, in the presence of the proper dissipative terms, it must decay [11]. If we now regard ε as a candidate potential function for the robot, and find a suitable dissipative function to match, then there is reason to hope for convergence results analogous to those of the previous section. Succintly, then, the vector field of system (6) always admits of a potential function, while that of system (5), in general, will not.

The dissipation of total energy, will not be of much practical use unless it has the properties of a Lyapunov Function. If we assume that kinetic energy is always quadratic in velocity and never zero in position, then any positive definite potential function will work. In the previous section, ε was a norm, and hence, certainly positive definite.

For example, suppose we are commanded to move the robot arm to a zero velocity point in the state space, $(q_d, 0)$. From the point of view of the previous section, we would like the robot to solve the trivial linear equation in $\mathcal{J}$,

$$f(q) \stackrel{\triangle}{=} q - q_d = 0.$$

To implement a version of algorithm (6), we first define a positive definite error function on the generalized coordinates,

$$\varepsilon \triangleq [q - q_d]^{\mathrm{T}} K_1 [q - q_d]$$

(K_1 is positive definite, and might specify the impedance characteristics of the manipulator [6]) and then form a scalar multiple of its gradient

$$d\varepsilon^{\mathrm{T}} = K_1 [q - q_d].$$

Defining a state feedback control law for (1) based upon removing the destabilizing gravitational field (which admits a "useless" indefinite potential function) and matching the desired gradient with a positive definite dissipative term,

$$\tau = k(q) - K_2 \dot{q} - df^{\mathrm{T}} K_1 f. \tag{7}$$

(K_2 is positive definite), yields the following result, stated here without proof.

Theorem 1 *(Takegaki and Arimoto [26], Koditschek [10]) Let J be a simply connected subset of $\mathbf{R}^n$. The closed loop system of equation (1), under the state feedback algorithm (7),*

$$\dot{x}_1 = x_2$$
$$\dot{x}_2 = -M^{-1}[(B + K_2)x_2 + K_1(x_1 - q_d)]$$

is globally asymptotically stable with respect to the state $(q_d, 0)$ for any positive definite symmetric matrices, K_1, K_2.

Since ε has a unique critical value, convergence to $(q_d, 0)$ is guaranteed from all initial conditions, including arbitary initial velocities. Note that the first condition of the hypothesis is required to avoid situations where topological properties of the manifold on which the dynamics is defined preclude the possibility of a smooth vector field with a unique equilibrium state.

There are two serious criticisms to be made of any robot controller based upon this result. First, the control law requires the exact cancellation of any gravitational disturbance. While $k(q)$ has a much simpler structure than the moment of inertia matrix, $M(q)$, or the coriolis matrix, $B(q, \dot{q})$, exact knowledge of the plant and load dynamical parameters would still be required, in general, to permit its computation. Since the dynamical parameters enter linearly in k, some progress has been made in the design of "adaptive gravity cancellation" algorithms [12] removing the need for any à priori information concerning the dynamical parameters. Second, this result affords very little understanding of the trajectory that the arm will take as it moves toward $(q_d, 0)$. While the singly integrated gradient system

$$\dot{q} = -K_1[q - q_d]$$

converges toward q_d along the dominant eigenvector of K_1, even a linear double integrator may drive the projection onto J of its state trajectory, (x_1, x_2) quite differently. Research addressing this question is also in progress.

As the title of this paper suggests, the real utility of this approach is not simply in commanding end-point control at the joint level. Any task encoded by means of a gradient in $\mathcal{W}$ may be commanded by simply composing its potential function with the kinematic map, and applying the same technique. The simplest example is provided by a point-to-point task. Suppose we are commanded to move the robot arm to the position $r_d \in \mathbf{R}^3$. The task may be encoded in terms of finding the solution to the set of nonlinear equations

$$f(q) \triangleq r_g(q) - r_d = 0,$$

where $r_g(q)$ is the position component of the kinematic transformation from J to $\mathcal{W}$ (see Appendix B). Again, we form a positive definite error function in the target space,

$$\varepsilon \triangleq f^{\mathrm{T}} K_1 f,$$

in order to use a scalar multiple of its gradient,

$$d\varepsilon^{\mathrm{T}} = dr_g^{\mathrm{T}} K_1 f,$$

as the input to the robot dynamics. Proceeding as before, we define a time invariant feedback control law

$$\tau \triangleq k(q) - K_2\dot{q} - dr_g^{\mathrm{T}} K_1(r_g(q) - r_d), \tag{8}$$

and state the following result, also without proof.

Theorem 2 *(Koditschek [11])*
The closed loop system of equation (1), under the state feedback algorithm (8),

$$\begin{aligned}\dot{x}_1 &= x_2\\ \dot{x}_2 &= -M^{-1}[(B + K_2)x_2 + [d_q r_g]^{\mathrm{T}} K_1(r_g - r_d)],\end{aligned}$$

has an attracting set contained in

$$\tilde{\mathcal{C}} = \{(q,0) \in \mathcal{P} : K_1(r_g - r_d) \in ker\ dr_g^{\mathrm{T}}\}.$$

for any positive definite symmetric matrices, K_1, K_2.

This result shows that the price paid for automatic inverse kinematics is the possible addition of undesirable equilibrium states, which inhabit the zero velocity hyperplane of $\mathcal{P}$ within the set of kinematic singularities,

$$\tilde{\mathcal{C}} \subset \{x \in \mathcal{P} : x_2 = 0,\ rank(dr_g) < n\}.$$

For all presently available commercial robots, kinematic singularities may be found in the inerior of the workspace, thus the problem is of practical concern. For a lucky choice of K_1 and r_d it might well turn out that $\tilde{\mathcal{C}}$ consists only of the points in the solution set,

$$\tilde{\mathcal{S}} \triangleq \{x \in \mathcal{P} : x_2 = 0,\ r_g(x_1) = r_d\},$$

but, in general, this cannot be expected. Given the existence of additional undersirable equilibrium states outside of the solution set,

$$\tilde{\mathcal{U}} \triangleq \tilde{\mathcal{C}} - \tilde{\mathcal{S}},$$

$\tilde{\mathcal{S}}$ is no longer a global attractor. However, there is some hope that the presence of *stall* points, attractive points of $\tilde{\mathcal{U}}$, can be ruled out. Research is now under way to determine a pragmatic approach to this problem: i.e., to determine when stall points may be ruled out, and what to do if they cannot. The failure to construct a globally asymptotically stable solution set, $\mathcal{S}$, for position control is seen to be a consequence of kinematic critical points in conjunction with the second order dynamics of the arm. When we pose the complete problem involving position and orientation simultaneously we may run into a still larger set of non-solution critical values which are a consequence of the intrinsic topology of $\mathcal{W}$. It is well known that more sophisticated tasks encoded as gradients in task space - e.g. the obstacle avoidance potentials of Khatib - almost inevitably give rise stall points in $\mathcal{W}$. This is probably an intrinisic limitation in the "intelligence" of feedback controllers, and must be taken into consideration at a higher level. A more complete characterization of "stall" will be required for this to be possible.

3. Feedback Measurements for Natural Control

It is hoped that the ideas presented here will afford the command of initially simple robot tasks - e.g. picking and placing objects of unknown mass, possibly with impedance control, possibly along a specified curve - at much higher speeds and less planning than currently possible. We are setting out in the Robotics Lab at Yale to build or modify an existing arm which will afford a hardware implementation of the feedback control schemes discussed in this paper. In so doing, we have come up against some fundamental questions which, although invisible from the point of view of the formal analytical techniques presented thus far, go to the heart of what a robot is or should be. Having proposed a control methodology which relies heavily upon feedback, we must create a sufficiently rich sensory environment which delivers enough information sufficiently quickly and sufficiently accurately.

3.1 The Structure of a Position Error Gradient

The jacobian of the kinematic map has been generally recognized as being critical to the implementation of feedback from task space for some time [29] , and increasing attention has been expended in its analytical study [18], [20]. Since this jacobian is a critical component of the error gradients used by our feedback schemes, its structure will be of great importance in any hardware implementation of such controllers as well. We will use insights similar to those of [20], [29] in this section - refer to Appendix B for definitions and computational details. For ease of exposition, we confine our attention to the control of a manipulator which has all revolute joints in the context of a task which calls for the control of end-effector position only. Insights in the ensuing discussion apply to more general circumstances - robots with prismatic joints; tasks requiring orientation control as well.

It is shown in Appendix B that the jacobian of the forward kinematic position transformation, $dr_g(q)$, is given by

$$[J(z_0)r_g, J(z_1)[r_g - r_1], J(z_2)[r_g - r_2], ..., J(z_{n-1})[r_g - r_{n-1}],],$$

where

$$J(a) \triangleq \begin{bmatrix} 0 & -a_3 & a_2 \\ a_3 & 0 & -a_1 \\ -a_2 & a_1 & 0 \end{bmatrix}$$

is a skew symmetric matrix defined by $a \triangleq \begin{bmatrix} a_1 \\ a_2 \\ a_3 \end{bmatrix} \in \mathbf{R}^3$. Recall, now, that the position error gradient of equation (8) is given by

$$[d_q r_g]^{\mathrm{T}} K_1(r_g - r_d) = \begin{bmatrix} [r_g - r_d]^{\mathrm{T}} K_1^{\mathrm{T}} J(z_0) r_g \\ [r_g - r_d]^{\mathrm{T}} K_1^{\mathrm{T}} J(z_1)[r_g - r_1] \\ \vdots \\ [r_g - r_d]^{\mathrm{T}} K_1^{\mathrm{T}} J(z_{n-1})[r_g - r_{n-1}] \end{bmatrix},$$

and this may be interpreted geometrically by various appeals to the following identities from vector algebra. For any $a, b, c \in \mathbf{R}^3$,

$$c^{\mathrm{T}} J(a) b = c^{\mathrm{T}}(a \times b) = |c, a, b| = |a, b, c| = |b, c, a|,$$

where $|a, b, c|$ denotes the determinant of the array whose columns are a, b, c. For instance, if K_1 is the identity matrix, then these identities imply that the ith entry of the position error gradient is given by the determinant

$$|z_i, (r_g - r_i), K_1(r_g - r_d)|.$$

In other words, when its velocity is zero, the ith joint will have a non-zero torque input applied unless its axis of motion is parallel to the plane spanned by the position error vector and the difference vector between its own origin and the base. In the case that $K_1 = I$, this determinant takes an even simpler form, since the reference frame of these measurements is unimportant

$$|z_i, (r_g - r_i), (r_g - r_d)| = |R_j| \, |{}^j z_i, ({}^j r_g - {}^j r_i), ({}^j r_g - {}^j r_d)|,$$

where $|R_j| = 1$. In particular, when $j = i$, we have

$$|{}^i z_i, {}^i r_g - 0, {}^i r_g - {}^i r_d| = |{}^i z_i, {}^i r_g, {}^i r_d|, \tag{9}$$

which is a 2×2 determinant since ${}^i z_i = \begin{bmatrix} 0 \\ 0 \\ 1 \end{bmatrix}$.

3.2 Measure or Compute?

In evaluating possible feedback implementations the performance measure, as always, is some combination of speed and accuracy. Rule of thumb seems to have it that good controllers for linear time invariant sytems should possess time constants of roughly an order of magnitude greater than those of the system being controlled [4]. Paul [19] cites a factor of 15 in the context of robotic systems. If the time constant of a typical dc-servo is 100 ms, then we must achieve a control rate of better than one command every 10 ms. Since the effects of discretization in the context of nonlinear dynamics are not well understood, it seems preferable to attain a rate closer to 1 ms. On the other hand, traditional insight of the control community has been that a closed loop dramatically lowers sensitivity to reference signal errors at the expense of high sensitivity to feedback errors. While this insight may be made fairly precise in the context of linear time invariant systems, no such analysis is available in the present context. Although we are attempting to develop a sensitivity analysis for the schemes presented, the obvious strategy is to achieve a computational accuracy at every step which is better than the resolution of the best joint position sensor on the arm.

The feedback contoller, (8), requires far less information about the underlying plant than inverse dynamics feedforward schemes, e.g. [14], and, in the absence of gravitationally induced torques, there is not even a need for the dynamical parameters of the arm or its load. Accurate estimates of expressions of the form (9), which were shown to derive from the arm kinematics, however, are critical. The somewhat vague term "derive" is used intentionally here since there is a broad range of sensory modalities which will determine the particular form of computation. This range runs all the way from massive computations based upon a priori estimates of the underlying kinematic parameters and state space measurements of $q, \dot{q}$, to a complete reliance upon state space measurements of velocity, and output space measurements of link positions and orientations upon which relatively simple computations might be done using analog electronic components.

Consider the latter extreme first. Equation (9) indicates that two sensors located on every link, one reporting the gripper position, the other reporting the goal position, would suffice to evaluate the jacobian with no computation besides a 2×2 determinant. If the impedance matrix, K_1, is not the identity, then there is a little more computation, but the situation is not essentially different. Unfortunately, it is not clear what sensing technology exists that will deliver unobstructed accurate measurements of n sets of two cartesian positions every millisecond. A large number of frequency separated acoustic sensors may perform sufficiently quickly, but may not deliver the required accuracy particularly if the arm is moving very quickly. High resolution cameras may deliver sufficient accuracy, but will likely cause far too costly delays. Certainly, a robot endowed with cartesian sensors for every link will depart significantly from any anthropomorphically motivated design.

Considering the opposite extreme, if no sensors other than state space positions and velocities, $(q, \dot{q})$ are available - i.e. joint shaft encoders and tachometers - then we must resort to the recursive computation

$$^0r_{i+1}(q_1,...,q_{i+1}) = {}^0r_i(q_1,...,q_i) \cdot {}^ir_{i+1}(q_{i+1}),$$

whose details are given in Appendix B. Essentially, this amounts to computing the forward kinematic transformation for the position of each link with respect to the base frame, then taking differences and determinants. This is an obvious application for parallel computation, and we are beginning to evaluate appropriate architectures for implementation at Yale.

A variety of sensing schemes for delivering partial information regarding cartesian link measurements are currently being evaluated within the laboratory as well. For instance a very accurate picture of task space delivered at a rather low data rate might, nevertheless, be useful in correcting the results of computation. Or very accurate relative position error measurements in the gripper frame alone might be very cheaply delivered by a simple sonar device and used to increase the resolution of the absolute distance, $r_g - r_d$, computed from joint measurements.

4. Conclusion

This paper concerns a stable method of controlling robot arms in the command of tasks which have been defined by geometric descriptions and impedance relations in the task space. The control methodology is entirely feedback based, thus takes into account the intrinsic dynamics of the manipulator without their explicit computation, and obviates the need for inverse kinematic computation. Some insight into a number of difficulties is gained by interpreting these ideas as a scheme to use the robot arm as an analog computer which sequentially solves its own inverse kinematics equations. An initial discussion of instrumentation for these schemes is presented: there is seen to be a tradeoff between sensory and computational complexity, which inverts the traditions of the control community by suggesting a preference for output rather than state feedback.

Although several robotic task methodologies [6], [9], are very simply implemented using such schemes, it is not yet clear how to establish a general framework for translating abstract tasks into the appropriate feedback structure. Very likely, it will prove more effective to resort to some reference signal based encoding for certain task classes, in which cases sensible feedforward pre-compensators will be desirable as well. Since the use and importance of the various sensory modalities available to a robot will depend upon the control methods used in commanding a task, the problem of developing principles for the generation of *dynamically sound* task encoding methodologies should be explicitly addressed in an interdisciplinary fashion by "high level planners" as well as "joint torque level" control theorists.

Acknowledgments

This work is supported in part by the National Science Foundation under grant no. DMC-8505160.

A Some Notation

If $f : \mathbf{R}^n \to \mathbf{R}^M$ has continuous first partial derivatives, denote its $m \times n$ jacobian matrix as df. When we require only a subset of derivatives, e.g. when $x = \begin{bmatrix} x_1 \\ x_2 \end{bmatrix}$, and we desire the jacobian of f with respect to the variables $x_1 \in \mathbf{R}^{n_1}$, as x_2 is held fixed, we may write

$$d_{x_1} f \triangleq df \begin{bmatrix} I_{n_1 \times n_1} \\ 0 \end{bmatrix}.$$

If $A \in \mathbf{R}^{n\times m}$, the "stack" representation of $A \in \mathbf{R}^{nm}$ formed by stacking each column below the previous will be denoted A^{S}. If $C \in \mathbf{R}^{p\times q}$, and A is as above then the Kronecker Product of A and C is

$$A \otimes C \triangleq \begin{bmatrix} a_{11}C & \dots & a_{1m}C \\ a_{21}C & \dots & a_{2m}C \\ & \vdots & \\ a_{n1}C & \dots & a_{nm}C \end{bmatrix} \in \mathbf{R}^{np\times mq}.$$

Finally, if $B \in \mathbf{R}^{m\times p}$, and A and C are as above, then it can be shown that

$$[ABC]^{\mathrm{S}} = (C^{\mathrm{T}} \otimes A)B^{\mathrm{S}}.$$

[3]

B Kinematic Transformations: the Algebra and Calculus of Frames

Define a *frame* to be the ordered pair

$$\mathrm{r} \triangleq (r, R) \in \mathcal{W} \triangleq \mathbf{R}^3 \times SO(3)$$

where

$$SO(3) \triangleq \{R \in \mathbf{R}^{3\times 3} | R^T R = I \ \ and \ \ |R| > 0\}$$

is the set of rotations on $\mathbf{R}^3$ with positive orientation. For the purposes of this paper it will do no harm to confuse $r, R,$ *and* r with their matrix representations in $\mathbf{R}^3, \mathbf{R}^{3\times 3},$ *and* $\mathbf{R}^{3\times 4}$, respectively. Thus, the transpose, and stack conventions of the previous appendix may be applied meaningfully to any frame, r. Moreover, left matrix multiplication of a frame, p by any rotation, $R \in SO(3)$ is well defined, and will be used in computation as required, denoted by square brackets as

$$R[\mathrm{p}] \triangleq [Rp, RP].$$

At times, we shall refer to the columns of $\mathrm{r} = (r, R) = [r, x, y, z]$ as its *position, and orientation*, or *position, and x axis, y axis, z axis*, respectively. The frames form a group under the binary product

$$\mathrm{p} \cdot \mathrm{r} \triangleq (Pr + p, PR)$$

whose identity is $i \triangleq (0, I)$, and where $\mathrm{r}^{-1} = (-R^{\mathrm{T}}r, R^{\mathrm{T}})$.

A *joint transformation* is a map from $\mathbf{R}^1$ into W which relates a coordinate system fixed in link $i-1$ to one fixed in link i through the action of the ith joint. According to the standard conventions, the joint transformation may be written

$${}^{i-1}\mathrm{r}_i = \left(\begin{bmatrix} a_i cos\theta_i \\ a_i sin\theta_i \\ \delta_i \end{bmatrix}, \begin{bmatrix} cos\theta_i & -cos\alpha_i sin\theta_i & -sin\alpha_i sin\theta_i \\ sin\theta_i & cos\alpha_i cos\theta_i & sin\alpha_i cos\theta_i \\ 0 & sin\alpha_i & cos\alpha_i \end{bmatrix} \right)$$

where either θ_i *or* δ_i is the joint variable depending upon whether the joint is revolute or prismatic, respectively, and the other kinematic parameters are defined in the link body, e.g. as in [19]. More generally, a *kinematic transformation* is a map $\mathrm{r}_n : J \to \mathcal{W}$ which is the group product of n joint transformations,

$$\mathrm{r}_g(q) = {}^0\mathrm{r}_1 \cdot {}^1\mathrm{r}_2 \cdot \ldots \cdot {}^n\mathrm{r}_g\,,$$

where n is the number of degrees of freedom of the chain. For any partial product we will use the notation

$${}^i\mathrm{r}_j \triangleq {}^i\mathrm{r}_{i+1} \cdot {}^{i+1}\mathrm{r}_{i+2} \cdot \ldots \cdot {}^{j-1}\mathrm{r}_j\,,$$

which is a representation of the coordinate system j in terms of i. Individual columns will be labeled similarly: e.g. ${}^{i}x_j$ denotes the second column of ${}^{i}r_j$. For $i = 0$, we will drop the preceding superscript.

It is well known that any element, $R \in SO(3)$, may be written in *exponential form*

$$R = e^{J},$$

where J is skew symmetric. As is also well known, there is an isomorphism between $\mathbf{R}^3$ and the set of 3×3 skew symmetric matrices which uniquely identifies a point, $\begin{bmatrix} a_1 \\ a_2 \\ a_3 \end{bmatrix} \in \mathbf{R}^3$, with

$$J(a) \triangleq \begin{bmatrix} 0 & -a_3 & a_2 \\ a_3 & 0 & -a_1 \\ -a_2 & a_1 & 0 \end{bmatrix}$$

whose action on any $b \in \mathbf{R}^3$, $J(a)b = a \times b$, represents the vector cross product of b with a. Note that for any $R \in SO(3)$, we have

$$R(a \times b) = Ra \times Rb$$

which implies

$$R^{\mathrm{T}} J(a) R = J(R^{\mathrm{T}} a).$$

These ideas are presented, for instance, in [1].

After fixing a little more notation, it becomes possible compute the jacobian of kinematic transformations quite readily. From the definition of the joint transformation, above,

$${}^{i-1}r_i = e^{\theta_i J(z_0)}[p]$$

where

$$p \triangleq (x_0 + \delta_i z_0, e^{\alpha_i J(x_0)}),$$

and x_0, y_0, z_0 are matrix representations of the base coordinates with respect to themselves - i.e. unit vectors with the appropriate zero and one entries. It follows that the jacobian of ${}^{i-1}r_i\,(q_i)$ - in this case, the ordinary derivative with respect to q_i - is given by

$$d_{q_i}\ {}^{i-1}r_i = J(z_0)[\,{}^{i-1}r_i\,],$$

if the joint is revolute, and

$$d_{q_i}\,{}^{i-1}r_i = (z_0, 0_{3\times 3})$$

if the joint is prismatic. Note that the jacobian of a joint transformation is a linear map from the tangent space of $\mathbf{R}^1$ at q_i to the tangent space of $\mathcal{W}$ at ${}^{i-1}r_i$. By a slight abuse of notation we may represent an image point of this map in frame notation as a point in $\mathbf{R}^{12}$ for which the group product "$\cdot$" is still defined (although it is not in the group) and write

$$d_{q_i}\,{}^{i-1}r_i = j_i \cdot {}^{i-1}r_i\,,$$

where

$$j_i = \begin{cases} (0, J(z_0)) & \text{i revolute} \\ (z_0, 0_{3\times 3}) & \text{i prismatic} \end{cases}$$

Using this notation, the computation of the jacobian of a single single joint transformation under right and left translation in the group is given by

$$d(p \cdot r \cdot s) = P[j_i \cdot {}^{i-1}r_i \cdot s]$$

where p and s are constant frames.

The jacobian of the kinematic transformation is a linear map from $TJ_q \approx \mathbf{R}^n$ to $T\mathcal{W}_{r_n(q)} \subset \mathbf{R}^{12}$ which may be represented as a $12 \times n$ matrix if we pass to the stack representation of $T\mathcal{W}_{r_n(q)}$. This takes the form

$$dr_g^S = [d_{q_1} r_g^S, d_{q_2} r_g^S, \ldots, d_{q_n} r_g^S],$$

and, since the group product is associative,

$$= [d_{q_1}({}^0r_1 \cdot {}^1r_g)^S, d_{q_2}({}^0r_1 \cdot {}^1r_2 \cdot {}^2r_g)^S, \ldots, d_{q_n}({}^0r_{n-1} \cdot {}^nr_g)^S]$$

$$= [(j_1 \cdot {}^0r_1 \cdot {}^1r_g)^S, (R_1[j_2 \cdot {}^1r_2 \cdot {}^2r_g])^S, \ldots, (R_{n-1}[j_n \cdot {}^{n-1}r_g])^S].$$

To summarize, the general form of the ith column of the jacobian, dr_n^S is given by

$$\begin{bmatrix} R_{i-1} J_i {}^{i-1}r_g + R_{i-1} j_i \\ R_{i-1} J_i {}^{i-1}R_g\, x_0 \\ R_{i-1} J_i {}^{i-1}R_g\, y_0 \\ R_{i-1} J_i {}^{i-1}R_g\, z_0 \end{bmatrix}$$

which evaluates to

$$\begin{bmatrix} R_{i-1} J(z_0)\, {}^{i-1}r_g \\ R_{i-1} J(z_0)\, {}^{i-1}x_g \\ R_{i-1} J(z_0)\, {}^{i-1}y_g \\ R_{i-1} J(z_0)\, {}^{i-1}z_g \end{bmatrix} = \begin{bmatrix} J(z_{i-1})[r_g - r_{i-1}] \\ J(z_{i-1})\, {}^0x_g \\ J(z_{i-1})\, {}^0y_g \\ J(z_{i-1}) z_g \end{bmatrix},$$

when the ith joint is revolute, and

$$\begin{bmatrix} z_{i-1} \\ 0 \\ 0 \\ 0 \end{bmatrix},$$

when the ith joint is prismatic. This computation, is, of course, identical to that obtained by Whitney [29] and Sastry and Paden [20].

References

[1] V. I. Arnold, *Mathematical Methods of Classical Mechanics* Springer-Verlag, NY, 1978

[2] H. Asada, T. Kanade, and I. Takeyama, "Control of a Direct Drive Arm" *ASME J Dyn. Syst.* 105(3):136-142., 1983.

[3] R. Bellman *Introduction to Matrix Analysis* McGraw Hill, NY, 1965

[4] G. F. Franklin and J. D. Powell, *Digital Control of Dynamical Systems* Addison Wesley, Reading MA, 1980.

[5] E. Freund, "Fast Nonlinear Control with Arbitrary Pole-Placement for Industrial Robots and Manipulators" *Int. J. Robotics Res.* 1(1): 65-78, 1982.

[6] N. Hogan, "Impedance Control: An Approach to Manipulation, Part I: Theory" *ASME J. Dyn. Syst. Meas. and Control* Vol 107, pp. 1-7, March 1985.

[7] J. M. Hollerbach, "A Recursive Formulation of Manipulator Dynamics and a Comparative Study of Dynamics Formulation and Complexity", in Brady, et. al. (eds) *Robot Motion*, pp. 73-87, MIT Press, 1982.

[8] S. Jacobsen, J. Wood, D. F. Knutti, and K. B. Biggers, "The Utah/MIT Dextrous Hand: Work in Progress" *Int. J. Rob. Res.* Vol.3, No.4, Winter, 1984

[9] O. Khatib, "Dynamic Control of Manipulators in Operational Space" *Sixth IFTOMM Congress on Theory of Machines and Mechanisms* , New Dehli, 1983, p. 10

[10] D.E. Koditschek, "Natural Motion for Robot Arms" *IEEE Proc.* 23rd *CDC*, Las Vegas, December, 1984, pp. 733-735

[11] D. E. Koditschek, "Natural Control of Robot Arms" Yale Center for Systems Science Technical Report No. 8409, Dec. 1984 (revised, Mar. 1985).

[12] D. E. Koditschek, "Adaptive Strategies for the Control of Natural Motion" *Proc. IEEE 24th CDC*, Fort Lauderdale, Dec. 1985.

[13] R. H. Lathrop, "Parallelism in Manipulator Dynamics", *Int. J. Robotics Res.* 4:2, pp. 80 - 102, summer1985.

[14] J.Y. S. Luh, M. W. Walker, and R. P. Paul, "Resolved Acceleration Control of Mechanical Manipulators" *IEEE Tran. Aut. Contr.* AC-25 pp. 468-474, 1980.

[15] F. Miyazaki and S. Arimoto, "Sensory Feedback Based On the Artificial Potential for Robot Manipulators" *Proc. 9th IFAC* Budapest, Hungary, July, 1984.

[16] K. S. Narendra and Y. H. Lin "Design of Stable Model Reference Adaptive Controllers", in *Applications of Adaptive Control*, Narendra and Monopoli (eds.), Academic Press, 1980.

[17] K. S. Narendra and L. S. Valavani, "Stable Adaptive Observers and Controllers" *Proceedings of the IEEE* vol. 64, no. 8, August, (1976)

[18] D. E. Orin, and W. W. Schrader "Efficient Computation of the Jacobian for Robot Manipulators" *Int. J. Rob. Res.* 3(4), pp. 66-75, Winter, 1984.

[19] R. P. Paul, it Robot Manipulators, Mathematics, Programming, and Control MIT Press, Cambridge, MA, 1981.

[20] B. E. Paden and S. S. Sastry " Geometric Interpretation of Manipulator Singularities" Memo. No. UCB/ERL M84/76, Electronics Research Laboratory, College of Engineering, UC Berkeley, Sept., 1984

[21] J. Reif, "Complexity of the Mover's Problem and Generalizations", Proc. 20th Symposium of the Foundations of Computer Science, 1979.

[22] S. Smale, "The Fundamental Theorem of Algebra and Complexity Theory" *Bull. AMS.* vol. 4, no. 1 pp. 1,36, Jan. (1981)

[23] M. W. Hirsch and S. Smale, "On Algorithms for Solving $f(x) = 0$", *Comm. Pure and Appl. Math.*, vol. XXXII, pp. 281-312 (1979)

[24] J. T. Schwartz and M.Sharir, "On the Piano Mover's Problem. II." NYU Courant Institute, Report No. 41, 1982.

[25] Sir W. Thompson and P. G. Tait, *Treatise on Natural Philosophy*, University of Cambridge Press, Cambridge, 1886.

[26] M. Takegaki, and S. Arimoto, " A New Feedback Method for Dynamic Control of Manipulators" *J. Dyn. Syst.* Vol 102, pp.119-125, June, 1981.

[27] T.J. Tarn, A. K. Bejczy, A. Isidori, Y. Chen, "Nonlinear Feedback in Robot Arm Control" *Proc. 23rd IEEE Conf. on Dec. and Control*, Las Vegas, Dec. 1984, pp. 736 - 751

[28] W. A. Wolovich and H. Elliott, "A Computational Technique for Inverse Kinematics" *Proceedings of the Twenty Third IEEE Conference on Decision and Control* pp. 1359 - 1364 (1984)

[29] D. E. Whitney "The Mathematics of Coordinated Control of Prosthetic Arms and Manipulators" *ASME. J. Dyn. Syst. Contr.*, Vol 122, pp. 303-309, Dec, 1972

AUTHORS' INDEX

SUBJECT INDEX

(Numbers that are underscored denote figures)